WATER AND SANITATION-RELATED DISEASES AND THE CHANGING ENVIRONMENT

WATER AND SANITATION-RELATED DISEASES AND THE CHANGING ENVIRONMENT

Challenges, Interventions, and Preventive Measures

Second Edition

Edited by
JANINE M. H. SELENDY
FOREWORD BY PAUL FARMER
AFTERWORD BY WAFAIE FAWZI

WILEY Blackwell

HORIZON
INTERNATIONAL

This second edition first published 2019

Edition History
Wiley-Blackwell (1e, 2011)

Registered Office(s)
John Wiley & Sons, Inc., 111 River Street, Hoboken, NJ 07030, USA

Editorial Office
The Atrium, Southern Gate, Chichester, West Sussex, PO19 8SQ, UK

For details of our global editorial offices, customer services, and more information about Wiley products visit us at www.wiley.com.

Wiley also publishes its books in a variety of electronic formats and by print-on-demand. Some content that appears in standard print versions of this book may not be available in other formats.

Library of Congress Cataloging-in-Publication Data

Names: Selendy, Janine M. H., editor.
Title: Water and sanitation-related diseases and the changing environment : challenges, interventions, and preventive measures / edited by Janine M.H. Selendy ; foreword by Paul Farmer ; afterword by Wafaie Fawzi.
Other titles: Water and sanitation-related diseases and the environment.
Description: Second edition. | Hoboken, NJ, USA : Wiley-Blackwell, 2019. | Preceded by: Water and sanitation-related diseases and the environment / edited by Janine M.H. Selendy. c2011. | Includes bibliographical references and index. |
Identifiers: LCCN 2018018468 (print) | LCCN 2018018989 (ebook) | ISBN 9781119416203 (Adobe PDF) | ISBN 9781119416180 (ePub) | ISBN 9781119416210 (hardback)
Subjects: | MESH: Water Supply | Disease Outbreaks–prevention & control | Water Pollution–prevention & control | Sanitation | Water Microbiology | Developing Countries
Classification: LCC RA642.W3 (ebook) | LCC RA642.W3 (print) | NLM WA 675 | DDC 363.739/4–dc23
LC record available at https://lccn.loc.gov/2018018468

Cover Image: by Béla P. Selendy
Cover Design: Wiley

Set in 10/12pt Times by SPi Global, Pondicherry, India

10 9 8 7 6 5 4 3 2 1

CONTENTS

FOREWORD

The work of experienced scholars, public-health advocates, and implementers, this new edition of *Water and Sanitation-Related Diseases and the Changing Environment* offers a thorough review of some of the ranking problems of our time. Taken individually, these chapters constitute a critical compendium of ongoing debates among experts and a concise summary of more settled matters. But editor Janine Selendy has also woven these diverse chapters—which include highly focused considerations of specific waterborne illnesses and more broad-ranging matters from climate change to technological innovation—into a powerful and hefty manual to guide collective action going forward.

Much of what appears in these pages is cause for concern, or should be. The structural violence of poverty, climate change, and socially constructed scarcity is evident in the long list of assaults and afflictions detailed in these pages: the preventable deaths from diarrhea of hundreds of thousands of children every year, the lack of access to safely managed water services for over two billion people globally, and the persistence of rudimentary sanitation systems—or none at all, in some areas—heighten risks of waterborne disease and even physical insecurity for over a third of the world's population. But this volume, which will meet the expectations of specialists while instructing (and even entertaining) the lay reader, does more than offer jeremiads about failures. A case study of efforts to eradicate dracunculiasis—an "ancient loathesome disease" that as recently as 1986 infected 3.5 million people annually across a grim belt of Africa south of the Sahel—shows how an end to guinea-worm disease is within our grasp.

Rescuing history and a bit of optimism, this pragmatic volume details numerous examples that serve to redefine what's possible in the struggle for equitable access to clean water and effective sanitation, without which health and productive lives are impossible. Its chapters also take a pragmatic and aspirational view in seeking to identify key social, political, and economic changes essential to efforts to alleviate the harms caused by our collective failure to guarantee the most fundamental human rights. There's little doubt that even the most unpromising settings—impoverished, bereft of clean water and sanitation, plagued by disease and disrupted by catastrophe—can be transformed with the requisite staff, stuff, space, and systems. Keeping these systems working requires ongoing accompaniment and investment, of course.

In the face of unprecedented climate and environmental change, low aspirations on behalf of others accelerate a deadly desertification of the imagination. In 2010, ten months after a devastating and deadly earthquake, ecologically fragile Haiti—one of the world's most water-insecure nations—was struck by a cholera outbreak that rapidly became the world's largest, and remained so for most of the decade. Campaigns to halt it initially echoed control-over-care paradigms commonly imposed in settings of poverty. But cholera spread because Haiti is a public-health desert; cholera killed where Haiti remains a clinical desert. We've seen this before. But paradigms of response—to cholera in Haiti and to other pressing health and human-rights challenges—shift each time health professionals, activists, and policymakers come together to resist the notion that some people are simply too poor to treat, or otherwise just not worth the investment.

In addressing many of these challenges, the need for more data and less dogma is obvious. The first edition of *Water and Sanitation-Related Diseases and the Changing Environment* promised playbooks to guide collective responses to a life-and-death struggle. In assembling and updating it, Selendy has accomplished all of this, and more. It will be an authoritative reference for practitioners and trainees to deliver on the promise of water, sanitation, and health for all.

Paul Farmer, MD, PhD
Kolokotrones University Professor, Harvard University
Chair, Department of Global Health and Social Medicine, Harvard Medical School
Chief, Division of Global Health Equity, Brigham and Women's Hospital
Co-founder, Partners In Health

PREFACE

We are all together in this world, breathing, as it were, as one, for each and every life affects those near to us and in the far reaches of the world.

As is evident throughout much of these two books, while health and living conditions are deplorable and life-threatening for literally billions of children, women and men, it is with the emergence of successful initiatives, largely thanks to collaborations on local, national and international levels, that one can hope for a healthier future. As discussed in these two books, such cooperation has enabled Guinea worm disease to be nearly eradicated worldwide, Onchocerciasis to be eliminated in large regions, and trachoma to be officially validated as eliminated in five countries with a further six countries reporting that they have achieved the elimination targets.

Also, although already seriously complicated by extreme weather conditions and other effects of climate change, collaboration secured safe drinking water, adequate sanitation and hygiene for hundreds of millions of people.

I hope you will read both volumes of our books. You will find references to content in the 36 excellent chapters of the First Edition throughout this edition. Many of the chapters in the Second Edition have been revised and updated, such as with respect to the effects of climate change. Many of the chapters stand as they are in First Edition and are not included in this volume. They include substantial coverage of naturally occurring and anthropogenic pollutants, toxic cyanobacteria, and onchocerciasis, and chapters on sanitation initiatives in India, health in Ghana, cholera in Zimbabwe, small-scale water services in developing countries, ocean pollution from flame retardants, and successful community interventions using kinship structures.

Several authors for both editions were joined by new or additional authors for the Second Edition which was written by authorities from the fields of climate change, epidemiology, environmental health, environmental engineering, demography, global health, medicine, medical anthropology, nutrition, and public health.

New chapters advance our discussions of nutrition and malnutrition, of ecosystems, and of population. They include "Food Systems and Nutrition, In The Context of Climate Change" by Director-General José Graziano da Silva of the UN FAO, "Coping With Water Needs: The Demographic Future," by Guigui Yao and Robert Wyman, and "Ecosystem Health as The Basis for Human Health" by Tom Barker and Jane Fisher. "The Human Right to Sanitation" by Anoop Jain and Jay P. Graham and "Catalyzing Rural Sanitation at Scale: Lessons Learned from The Global Sanitation Fund" by Carolien Van der Voorden and Patrick England add new ideas to discussion of the desperate sanitation situation with 2.4 billion people lacking adequate sanitation. "Antimicrobial Resistance" by Rochelle Rainey advances the discussion of the major problem facing people throughout the world as antibiotics can no longer be relied upon as they had been. And, drinking water is addressed throughout and exclusively in "Toward Universal Access to Basic and Safely Managed Drinking Water: Remaining Challenges and New Opportunities in the Era of Sustainable Development Goals," by Mitsuaki Hirai and Jay Graham.

Complementary material related to both editions are available on the Wiley Companion website for the two volumes https://www.wiley.com/legacy/wileychi/selendy/ with links to sources. This is a continually updated site for which suggested additions are welcome from our readers. Some of the content is also presented in articles on Horizon International's Solutions Site at www.solutions-site.org and on the Global Innovation Exchange, for which Horizon International is a co-creator, which is available at https://www.globalinnovationexchange.org/organizations/horizon-international-yale-university.

As I wrote in the First Edition, the facts we present are often disconcertingly grave, but I believe the discussions in that edition and in the current edition which augment those facts and ideas will generate substantial action to address those concerns and by doing so they will generate hope.

I wish to express my immense gratitude to the 74 authors who not only gave of their time and expertise in writing for these volumes, but also provided me with counsel and inspiration as we together brought these two books to completion. Among them, Professor Robert Wyman, who is our immediate host at Yale. And, I wish to express my gratitude to Yale University for their hosting of Horizon International.

I am grateful to the editors of Wiley for their foresight and belief in our books, their attention to detail during the production stages, and their encouraging comments throughout the process, with a special note of appreciation to Ramprasad Jayakumar, Production Editor, who saw to the myriad details in the final stages of the book's preparation.

All of us who have contributed to these books are thankful to Dr. Paul Farmer, Kolokotrones University Professor of Global Health and Social Medicine, Harvard University and co-founder of Partners in Health, and to Dr. Wafaie Fawzi, Chair, Department of Global Health and Population, Harvard T. H. Chan School of Public Health (HSPH) for expressing their appreciation of our work by giving of their time to write the Foreword and Afterword for the Second Edition. And a special thanks to Dr. Fawzi's colleague Chelsey Canavan, Program Manager for the Nutrition and Global Health Program at HSPH, who co-authored the Afterword.

I wish to recognize the commitment of untold numbers of people who give of their time, energy, and, for many, their lives to improve the lives of others. The direction of my own life was inspired by Dr. P. O. and Shanta Satralker, medical missionaries, with whom I had the privilege of living in Iran when I was 16 years old and who called themselves my Iranian Mom and Dad.

I dedicate this book to those whose love and encouragement are a constant source of support and inspiration: my sons, Philippe and Béla, my grandchildren, Max and Liam, and Nicolas and Linnea, my daughters-in-law, Jennifer and Ulrika, and Béla's wife Helen; and, to Charles R. Dickey, my partner of 20 years, whose loving care and ready words of wisdom and humor are a constant gift, and our dog, Heather, whose good spirits and love are gifts to all who know her.

Janine M. H. Selendy
New Haven, CT, USA
August 2018

CONTRIBUTORS

Jens Aagaard-Hansen, MD, MPH, Diploma TM Specialist in General Medicine. Steno Diabetes Center Copenhagen, Copenhagen, Denmark; MRC Developmental Pathways for Health Research Unit, Department of Paediatrics, School of Clinical Medicine, Faculty of Health Sciences, University of the Witwatersrand, Johannesburg, South Africa

Tom Barker, BSc, PhD Lead Academic, Graduate School of the Environment, Centre for Alternative Technology, Machynlleth, Wales, UK; Department of Geography & Planning, University of Liverpool, Liverpool, UK

Brian G. Blackburn, MD Department of Internal Medicine, Division of Infectious Diseases and Geographic Medicine, Stanford University School of Medicine, Stanford, CA, USA

Marcia C. Castro, PhD Professor of Demography, Department of Global Health and Population, Harvard School of Public Health, Boston, MA, USA

Nikhil Chandavarkar, PhD Founder and CEO, Thirsus Sustainability, Cedar Park, Texas, USA;
Former Chief, Communications and Outreach Branch, Division for Sustainable Development (COB/DSD), Department of Economic and Social Affairs (UNDESA), United Nations, New York, USA;
Former Secretary UN-Water, United Nations, New York, USA

Thomas F. Clasen, JD, MSc, PhD Professor of Environmental Health, Rose Salamone Gangarosa Chair in Sanitation and Safe Water, Rollins School of Public Health, Emory University, Atlanta, Georgia, USA

Joseph A. Cook, MD, MPH, FACP Adjunct Epidemiology Professor, University of North Carolina at Chapel Hill School of Public Health, Chapel Hill, NC, USA; Former Executive Director of International Trachoma Initiative

Patrick England, MSc Knowledge Management and Innovation Support Officer, Technical Support Unit, Water Supply and Sanitation Collaborative Council (WSSCC), Geneva, Switzerland

Paul Farmer, MD, PhD Chair, Department of Global Health and Social Medicine, Kolokotrones University Professor of Global Health and Social Medicine, Harvard University, Cambridge, Massachusetts, USA

Wafaie Fawzi, MBBS, MPH, MS, DrPH Chair, Department of Global Health and Population, Harvard T.H. Chan School of Public Health, Harvard University, Cambridge, Massachusetts, USA

Jane Fisher BSc, M. Res, PhD, PGCE, FHEA, FRGS Lead Academic, Centre for Alternative Technology, Machynlleth, Wales, UK
Learning and Teaching Institute, University of Chester, Chester, UK

Sean Fitzwater, MD, MHS Department of Pediatrics, University of California, Los Angeles, CA, USA

Julio Frenk, MD, PhD President of the University of Miami, Miami, FL, USA;
Former Dean of the Faculty, Harvard School of Public Health, T & G Angelopoulos Professor of Public Health and International Development, Harvard School of Public Health and Harvard Kennedy School, Cambridge, Massachusetts, USA;
Former Minister of Health of Mexico (2000–2006)

Octavio Gómez-Dantés, MD, MPH Researcher, Center for Health Systems Research, National Institute of Public Health, Cuernavaca, Mexico

Jay Graham, PhD, MPH, MBA Program Director, Public Health Institute, Oakland, CA, USA

José Graziano da Silva, PhD Director-General, Food and Agriculture Organization of the United Nations, Rome, Italy

Emma M. Harding-Esch, MPhil, MSc, PhD Associate Professor at London School of Hygiene & Tropical Medicine, University of London, London, UK

Mitsuaki Hirai, DrPH, MPH Evaluation Fellow, Centers for Disease Control and Prevention, Atlanta, Georgia, USA

Donald R. Hopkins, MD, MPH Special Advisor, Guinea Worm Eradication, The Carter Center, Atlanta, Georgia, USA

Anoop Jain, Third Year DrPH Student University of California, Berkeley, CA, USA

Joshua Karliner, International Director, Program and Strategy, Health Care Without Harm, San Francisco, USA

Margaret Kosek, MD Assistant Professor, Johns Hopkins Bloomberg School of Public Health, Baltimore, MD, USA

Barry S. Levy, MD, MPH Adjunct Professor of Public Health at Tufts University School of Medicine, Boston, MA, USA

David C. Mabey, DM, FRCP, FMedSci Professor of Communicable Diseases at London School of Hygiene & Tropical Medicine, University of London, London, UK

Pascal Magnussen, MD, DTM&H Associate Professor, Specialist in Tropical Medicine and Infectious Diseases, Department of Immunology and Microbiology, Centre for Medical Parasitology and Department of Veterinary and Animal Sciences, Section for Parasitology and Aquatic Pathobiology, Faculty of Health and Medical Sciences, University of Copenhagen, Copenhagen, Denmark

M. Danielle McDonald, PhD Associate Professor, Department of Marine Biology and Ecology, Rosenstiel School of Marine and Atmospheric Science, University of Miami, Miami, FL, USA

Anitha Nimmagadda, MD Occupational and Environmental Medicine, University of Illinois Hospital and Health Sciences System, Chicago, IL, USA

Peter Orris, MD, MPH Professor and Chief of Service, Occupational and Environmental Medicine, University of Illinois Hospital and Health Sciences System, Chicago, IL, USA

Rochelle Rainey, PhD Senior Advisor, Environmental Health, USAID Global Health Bureau, Washington, DC, USA

Ernesto Ruiz-Tiben, PhD Director of Dracunculiasis Eradication Program, The Carter Center, Atlanta, Georgia, USA

Mathuram Santosham, MPH, MD Director, Center for American Indian Health; Professor, International Health; Professor, Department of International Health, John Hopkins Bloomberg School of Public Health, Baltimore, MD, USA

Janine M.H. Selendy Co-Chairman, Founder, and Publisher, Horizon International, Yale University, New Haven, Connecticut, USA

Anita Shet, MD, PhD Department of International Health, Johns Hopkins Bloomberg School of Public Health, Baltimore, Maryland, USA

Burton H. Singer, PhD Emerging Pathogens Institute, University of Florida, Gainesville, FL, USA

Anthony W. Solomon, MBBS, DTM&H, PhD, PGCAP, FHEA, FRCP Medical Officer, Neglected Tropical Diseases, Department of Control of Neglected Tropical Diseases, World Health Organization, Geneva, Switzerland

Ivorie Stanley, MD, MPH Clinical Assistant Professor, Occupational and Environmental Medicine, University of Illinois Hospital and Health Sciences System, Chicago, IL, USA

Carolien Van der Voorden, MA Head, Technical Support Unit, Water Supply and Sanitation Collaborative Council (WSSCC), Geneva, Switzerland

Birgitte Jyding Vennervald, MD Department of Veterinary and Animal Sciences, Section for Parasitology and Aquatic Pathobiology, Faculty of Health and Medical Sciences, University of Copenhagen, Copenhagen, Denmark

Mary E. Wilson, MD Clinical Professor of Epidemiology and Biostatistics, School of Medicine, University of California, San Francisco, CA, USA;
Adjunct Professor of Global Health and Population, Harvard T.H. Chan School of Public Health, Boston, MA, USA

Robert Wyman, PhD Professor, Department of Molecular, Cellular and Developmental Biology, Yale University, New Haven, CT, USA

Guigui Yao, PhD Professor, School of Foreign Languages, Jianghan University, Wuhan, Hubei, China; Associate Research Scholar, American Studies, Yale University, New Haven, CT, USA

Alexander T. Yu, MD, MPH Clinical Fellow in Infectious Diseases, Department of Internal Medicine, Division of Infectious Diseases and Geographic Medicine, Stanford University School of Medicine, Stanford, CA, USA

INTRODUCTION

Janine M. H. Selendy and Jens Aagaard-Hansen

This Second Edition of *Water and Sanitation-Related Diseases and the Environment: Challenges, Interventions, and Preventive Measures* is written by authorities from the fields of climate change, epidemiology, environmental health, environmental engineering, global health, medicine, medical anthropology, nutrition, population, and public health. It presents an interdisciplinary picture of the conditions responsible for water and sanitation-related diseases, the pathogens, vectors and their biology, morbidity and mortality resulting from lack of safe water and sanitation, distribution of these diseases, and the conditions that must be met to reduce or eradicate them. Augmenting coverage in the First Edition, this book continues to cover access to and maintenance of clean water, and guidelines for the safe use of wastewater, excreta, and graywater and examples of solutions.

The broad extent, variety, and ramifications of changing environmental factors are covered in many chapters throughout this book. These concerns are a major focus in updated and revised chapters from the First Edition and new chapters introduced in this Second Edition. Many chapters stand as they are in the First Edition to which readers are referred and are not included in this volume.

The chapters on individual diseases cover not only basic information about the disease in question, i.e., causality, pathogenesis, epidemiology, clinical manifestations, treatment, prevention, and control, but also distribution, prevalence, and incidence and interconnected factors such as environmental factors. They address how the diseases are further aggravated by nutritional deficiencies such as anemia and other problems that often exist in the same person such as diarrheal diseases.

The negative ramifications from climate change affecting human health in all corners of our world are being realized in ever greater and growing magnitude. Frequent and intense extreme climatic events are wreaking havoc, causing immediate trauma, death, and destruction with loss of most basic of human needs of water, food, and secure shelter, and huge populations are migrating in search of safer and more secure livelihoods. Extreme flooding and droughts damage agriculture and infrastructure, increased temperatures take their toll on human populations subjected to excessive heat, and life in oceans is depleted because of loss of mangrove and coral reef spawning grounds and increases in harmful algal blooms. Global distribution of disease vectors, influenced by environmental conditions, is rapidly changing. These problems are often exacerbated by ecosystem destruction due to human activities such as construction of dams, poor infrastructure for homes and transport, and irrigation systems.

Since the publication of the First Edition in 2011, environmental changes, population migration, and many conditions addressed by the Millennium Development Goals (MDG) have changed. After 15 years of efforts, while the MDG for drinking water was met in 2010, with 91 % of the global population using an improved drinking water source, there were, however, by 2015, still 2.4 billion people without basic sanitation facilities such as toilets or latrines and 946 million individuals still defecating in the open. A major consequence of this lack of sanitation and adequate hygiene is an estimated 280 000 diarrheal deaths annually. The lack of sanitation and hygiene is also a major factor in "several neglected tropical diseases, including intestinal worms, schistosomiasis, and trachoma."[1]

[1] WHO Fact Sheet, *Sanitation*, Reviewed November 2016. Available at: http://www.un.org/sustainabledevelopment/water-and-sanitation/.

Thus, when, in 2016, the Sustainable Development Goals[2] were established, they were designed to further the MDGs, such as for sanitation and hygiene, and add new objectives. According to the UN, there is sufficient fresh water on the planet to achieve water for all, "But due to bad economics or poor infrastructure, every year millions of people, most of them children, die from diseases associated with inadequate water supply, sanitation and hygiene. Water scarcity, poor water quality and inadequate sanitation negatively impact food security, livelihood choices and educational opportunities for poor families across the world. Drought afflicts some of the world's poorest countries, worsening hunger and malnutrition."[3]

It is essential for the understanding of the diseases and environmental conditions covered in this book to consider the sociocultural and economic context. Human perceptions and practices strongly influence transmission patterns and access to preventive and curative measures in a multitude of ways. Thus, local notions of illness (including perceived etiology, pathogenesis, diagnosis, and preferred treatment) are often not compatible with biomedical lore and they are usually influenced by factors such as gender, age, religion and ethnicity. Social and demographic macro-trends in society such as migration, urbanization, socioeconomic development (or financial crisis), health sector reforms, and political priorities such as privatization and level of control with, for example, environmental development projects play crucial roles in either ameliorating or (usually) aggravating the situation. In some cases, populations may be directly opposed to control interventions that from a biomedical perspective are rational, as shown by examples of cholera[4] and deworming programs addressing soil-transmitted helminths and schistosomiasis[5]. In the case of cholera, environmental factors (presence of *Vibrio* bacteria in lakes and river deltas together with specific fluctuations in rainfall and water level) combined with sociocultural factors (migration and inadequate hygiene) are strong determinants for the outbreaks of epidemics.

Prevention and control are important parts of the discussions. Accordingly, preventive measures and solutions are presented throughout the book and in more detail in a special section at the end of the book. They are also included in articles on Horizon International's Solutions Site at www.solutions-site.org, the Global Innovation Exchange at https://www.globalinnovationexchange.org/organizations/horizon-international-yale-university, and on the websites created by John Wiley & Sons, Inc. for the books in their "Companion Sites."

"...once we can secure access to clean water and to adequate sanitation facilities for all people, irrespective of the difference in their living conditions, a huge battle against all kinds of diseases will be won," said the late Dr Lee when he was Director-General of the World Health Organization.

[2] http://www.un.org/sustainabledevelopment/water-and-sanitation/.

[3] http://www.un.org/sustainabledevelopment/water-and-sanitation/. Goal 6: Ensure access to water and sanitation for all.

[4] Nations, M.K. and Monte, C.M.G. (1996). "I'm not dog, no!": cries of resistance against cholera control campaigns. *Social Science and Medicine*, 1996, 43 (6): 1007–1024.

[5] Parker, M., Allen, T., and Hastings, J. (2008). Resisting control of neglected tropical diseases: dilemmas in the mass treatment of schistosomiasis and soil-transmitted helminths in north-west Uganda. *Journal of Biosocial Sciences*, 40: 161–181.

SECTION I

WATER, SANITATION, AND HYGIENE: MEETING THE NEED

1

TOWARD UNIVERSAL ACCESS TO BASIC AND SAFELY MANAGED DRINKING WATER: REMAINING CHALLENGES AND NEW OPPORTUNITIES IN THE ERA OF SUSTAINABLE DEVELOPMENT GOALS

Mitsuaki Hirai[1] and Jay Graham[2]

[1] *Milken Institute School of Public Health at the George Washington University, DC, USA*
[2] *Public Health Institute, Oakland, CA, USA*

1.1 BACKGROUND

Improvements in water supply, sanitation, and hygiene have greatly advanced the health of industrialized countries [1], where diarrhea, cholera, and typhoid were once the leading causes of childhood illness and death. Access to safe drinking water, adequate sanitation, and good hygiene provides a wide range of explicit and implicit benefits, including increased time, reduced morbidity, and mortality from various diseases, augmented agriculture and commerce, improved school attendance, reduced healthcare costs, and reduced physical burden. The time savings can allow women to engage in non-illness-related tasks, provide more time for childcare and time for socialization and education activities [2]. Further, when water supplies are brought closer to homes, women's savings in energy expenditure can result in a reduction of energy intake. The energy savings may then be transferred to children's intake of food at no extra cost [3]. The implicit benefits of an improved water supply include improved quality of life through an available supply of drinking water and increased potential for communities to engage in other improvements once they have achieved improved access to a safe water supply (Figure 1.1).

Globally, an estimated 2.1 billion people lack access to safely managed water services, defined as "an improved drinking water source that is located on premises, available when needed and free from faecal and priority chemical contamination" [4]. Under the Sustainable Development Goals (SDGs), the proportion of the population using safely managed drinking water services will be promoted and monitored.

The lack of basic access to drinking water results in significant health impacts because of water-related diseases as well as lost productivity. Globally, under-five child mortality from diarrhea – linked to the lack of access to water and sanitation infrastructure and poor hygiene – was estimated at 0.58 million (95% confidence interval, 0.45–0.75), representing an estimated 9.2% of total deaths in 2013 [5].

Improvements to water supply – in terms of quantity, reliability, and quality – are an essential part of a country's overall development, but a number of obstacles may limit success.

Rapid population growth, degradation of the environment, increase of poverty, inequality in the distribution of resources, and misappropriation of funds are some of the factors that have prevented water supply interventions from producing optimal results [6]. Further, numerous studies have shown that resources and time are being spent in water supply interventions that do not take into account beneficiaries' needs, preferences, customs, beliefs, ways of thinking, and socioeconomic and political structures (i.e., the enabling environment).

Water and Sanitation-Related Diseases and the Changing Environment: Challenges, Interventions, and Preventive Measures, Second Edition. Edited by Janine M. H. Selendy.

Companion website: https://www.wiley.com/legacy/wileychi/selendy/

FIGURE 1.1 A woman fetching water in India. *Source*: License information: McKay Savage from London, UK (https://commons.wikimedia.org/wiki/File:India_-_Sights_&_Culture_-_32_-_woman_fetching_water_(2458024353).jpg), "India – Sights & Culture – 32 – woman fetching water (2458024353)," https://creativecommons.org/licenses/by/2.0/legalcode. (*See insert for color representation of this figure*.)

1.2 PAST EFFORTS TO IMPROVE ACCESS TO SAFE WATER

Development interventions began to flourish in the 1970s as disparities became evident in terms of quality of life and access to basic services between wealthy and poor countries. The original motivation for providing water and sanitation to the inhabitants of less developed countries was based upon the consideration of water and sanitation as a cornerstone to public health and as a basic human right [7]. As a human right, those services should be financed by the national government of each country. The governments of economically developing countries, however, did not have the resources needed to provide basic water and sanitation services to their entire population. Thus, developed countries and international organizations provided assistance in the provision of these services [8].

The approach utilized for the design and implementation of most early projects did not typically consider the preferences of beneficiaries as it was perceived that they did not have the knowledge and ability to contribute. Facilities constructed soon fell into disrepair due to the lack of operation and maintenance ensuing from a deficiency in organization, training, and sense of ownership by beneficiaries. Shortly after many water supply and sanitation interventions, communities often found themselves in the same conditions as before the project. The results were not promising and it became evident that community involvement was missing.

TABLE 1.1 Number of people who received improved water supplies (1981–2000).

Water supply category	Avg. number of people who gained access per day (1981–1990)	Avg. number of people who gained access per day (1991–2000)
Urban	100 000	130 000
Rural	270 000	90 000
Total	370 000	220 000

Source: Mara and Feacham [9].

During the International Drinking Water Supply and Sanitation Decade (1981–1990), the international community established a common goal of providing safe water supplies and adequate sanitation services to all the communities around the world. This meant that by 1990 every person worldwide should have their basic water and sanitation needs met. In 1981, it was estimated that 2.4 billion people would need to gain access to improved water supplies – equal to connecting 660 000 people each day for 10 years [9]. Although the goal was far from accomplished, an estimated 370 000 people on average received access to improved water supplies each day (Table 1.1). Following the decade and after two world conferences (New Delhi in 1990 and Dublin in 1992), the international community determined that water and sanitation could no longer be regarded simply as a right. After the Dublin conference, there was a shift to consider water as an economic good because it had an environmental and a productive value. It was made clear that need was no longer a sufficient reason for the provision of water and sanitation to any community [7].

After the World Conference on Water and Sanitation held at the Hague, Netherlands, in March 2000, the international community set a new common goal and published "Vision 21: Water for People." Vision 21 proposed to achieve a world by 2025 where everybody knows the importance of hygiene and education and enjoys safe water and appropriate sanitation services. At the United Nation Summit in September 2000, 189 UN member States adopted the Millennium Declaration, from which the Millennium Development Goals (MDGs) emerged. Target 10 of MDG 7 was to "halve by 2015 the proportion of people without sustainable access to safe drinking water and basic sanitation" (over 1990 estimates) [10]. The MDGs were a significant force in garnering donor support and government commitment to increasing water supply and sanitation.

A very important aspect of Vision 21 and the MDGs, reflecting concerns of the international community, was the recognition of the need for a new approach. The new approach emphasized "buy-in" before the implementation of a water project in any community and a stronger focus on ensuring that improvements made are sustained. Another particular aspect of Vision 21 was the ratification of water and sanitation as basic human rights. After the Water Decade, the international community indicated that water and sanitation could not be seen as basic rights anymore because the beneficiaries of the projects did not value the improvements made and facilities constructed when they were not required to contribute monetarily. In other words, people will not appreciate, continue to utilize, and preserve something that they have not contributed to. Based on previous experiences, the Conference concluded that the lack of a sense of ownership and commitment to project improvements on the part of the beneficiaries was due to the inadequate and often neglected inclusion of beneficiaries' preferences into project design and implementation. Further, beneficiaries of water projects should be responsible for the costs of the operation and maintenance of the system but not for the costs of the water itself, based on the idea that all people on earth have the right to obtain and consume enough water to guarantee their survival.

1.3 TRANSITION FROM THE MILLENNIUM DEVELOPMENT GOALS TO THE SUSTAINABLE DEVELOPMENT GOALS

Under the MDGs, global access to improved water sources increased from 76% of the population in 1990 to 91% in 2015. By reducing the number of people who rely on unimproved water sources from 1.3 billion in 1990 to 663 million in 2015, the MDG target of halving the proportion of the population without access to safe drinking water was met [10]. A major lesson learned from the MDG implementation was that the progress toward and attainment of the MDG target was not equal across regions. While 93% of the population in Southern Asia had gained access to improved water supply by 2015, it was only 68 and 56% in Sub-Saharan Africa and Oceania, respectively (Figure 1.2). Use of unimproved water sources and surface water was also disproportionately higher in rural populations than urban populations (Figure 1.3). Furthermore, the MDG mostly focused on household access to drinking water while people need safe drinking water in all settings.

In September 2015, UN Member States set the 2030 Agenda for Sustainable Development with 17 SDG goals and 169 targets [4]. The SDG Goals 1, 4, and 6 address drinking water, sanitation, and hygiene with respective targets and indicators (Table 1.2) [11]. The SDGs create a number of new opportunities to accelerate global efforts to promote access to water, sanitation, and hygiene (WASH). First, the SDGs address hygiene by monitoring the availability of handwashing facilities and resources while the MDGs only focused on water and sanitation. This addition allows the WASH sector to provide a more holistic picture of access to WASH services at the national, regional, and global

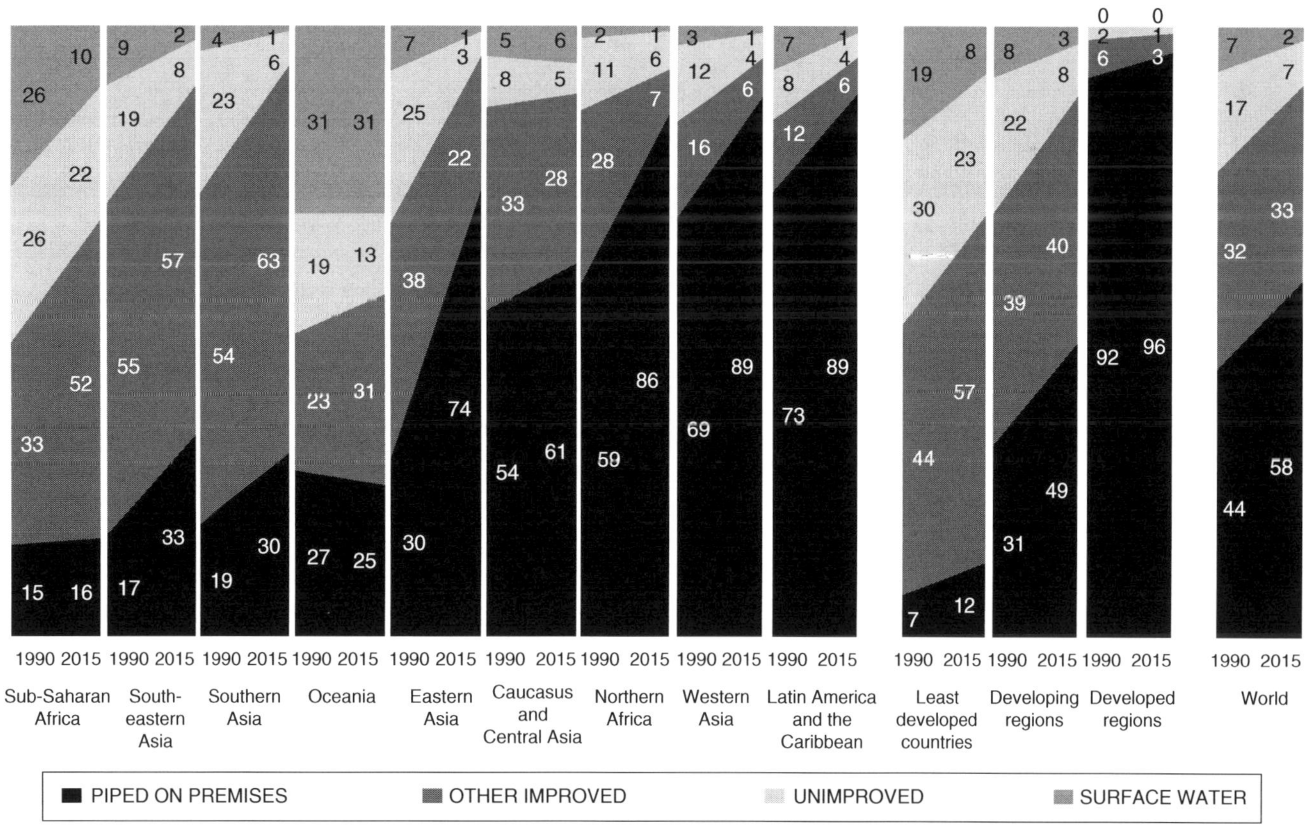

FIGURE 1.2 Regional trends in the percentage of population by drinking water service levels. *Source*: UNICEF/WHO [10].

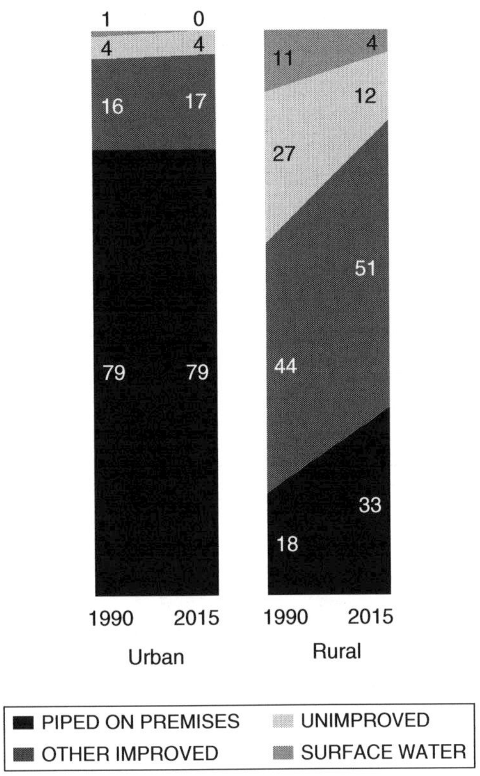

FIGURE 1.3 Trends in the percentage of population by drinking water service levels for urban and rural areas. *Source*: UNICEF/WHO [10].

levels. Second, the SDGs place more emphasis on reducing inequalities in access to WASH services than the MDGs. The MDGs achieved halving the proportion of the global population without access to safe drinking water and basic sanitation, but universal access or equity was not attained. The SDG targets therefore expanded the scope by specifically calling for *universal* and *equitable* access for all. Eliminating spatial and social inequalities is of vital importance to achieve the SDG targets. Third, the SDGs monitor and expand access to WASH services at some institutional settings, including schools and healthcare facilities. Compared to the MDGs, which only addressed household access to water and sanitation services, the SDGs can inform the accessibility, availability, and/or quality of WASH in other key locations.

The Joint Monitoring Programme for Water Supply and Sanitation (JMP) will use service ladders to monitor progress on access to drinking water under the SDGs. The JMP ladder for the MDGs used water piped to premises, other improved sources, unimproved sources, and surface water as the main rungs [10]. The new ladder for the SDGs consists of safely managed, basic, limited, unimproved, and surface water as the service levels [4] (Table 1.3). Drinking water from an improved source can be regarded as safely managed if it is accessible, available, and free from contamination at the same time. More specifically, a drinking water source should be located within the house, yard, or plot (accessibility), drinking water should be sufficiently available in the last

TABLE 1.2 Relevant SDG goals, targets, and indicators to drinking water, sanitation, and hygiene.

Goals	Targets	Indicators
Goal 1: End poverty in all its forms everywhere	Target 1.4: By 2030, ensure that all men and women, in particular the poor and the vulnerable, have equal rights to economic resources, as well as access to basic services, ownership, and control over land and other forms of property, inheritance, natural resources, appropriate new technology and financial services, including microfinance	Indicator 1.4.1: Proportion of the population living in households with access to basic services
Goal 4: Ensure inclusive and equitable quality education and promote lifelong learning opportunities for all	Target 4.a: Build and upgrade education facilities that are child, disability, and gender sensitive and provide safe, non-violent, inclusive and effective learning environments for all	Indicator 4.a.1: Proportion of schools with access to (a) electricity; (b) the Internet for pedagogical purposes; (c) computers for pedagogical purposes; (d) adapted infrastructure and materials for students with disabilities; (e) basic drinking water; (f) single-sex basic sanitation facilities; and (g) basic handwashing facilities (as per the WASH indicator definitions)
Goal 6: Ensure availability and sustainable management of water and sanitation for all	Target 6.1: By 2030, achieve universal and equitable access to safe and affordable drinking water for all SDG Target 6.2: By 2030, achieve access to adequate and equitable sanitation and hygiene for all and end open defecation, paying special attention to the needs of women and girls and those in vulnerable situations	Indicator 6.1.1: Proportion of population using safely managed drinking water services Indicator 6.2.1: Proportion of population using safely managed sanitation services, including a handwashing facility with soap and water

TABLE 1.3 JMP ladder and SDG baseline for drinking water.

Service level	Definitions	2015 global baseline
Safely managed	Drinking water from an improved water source that is located on premises, available when needed and free from fecal and priority chemical contamination	71% (5.2 billion)
Basic	Drinking water from an improved source, provided collection time is not more than 30 minutes for a round trip, including queuing	17% (1.3 billion)
Limited	Drinking water from an improved source for which collection time exceeds 30 minutes for a round trip, including queuing	4% (263 million)
Unimproved	Drinking water from an unprotected dug well or unprotected spring	6% (422 million)
Surface water	Drinking water directly from a river, dam, lake, pond, stream, canal or irrigation canal	2% (159 million)

week or available for at least 12 hours each day (availability), and a drinking water source should be compliant with standards to claim to be contamination free (quality). Access to safely managed drinking water is a key data element for SDG Target 6.1. A basic service includes drinking water from an improved source, and the water collection time should not be more than 30 minutes. Drinking water with a partial fulfillment of safely managed drinking water (e.g., one of three conditions met) will also be included in the basic category. Further, the basic service level will be monitored to assess the attainment of SDG Target 1.4. A limited service means that drinking water comes from an improved water source, but it requires more than 30 minutes for water collection. As with the previous ladder, the new ladder includes unimproved and surface water categories.

As a key update, improved water sources under the MDGs and the SDGs are not the same. Improved sources in the updated ladder include piped water, protected springs, protected dug wells, boreholes or tubewells, rainwater, and packaged or delivered water. Packaged or delivered water was considered an unimproved water source in the previous JMP ladder because data on accessibility, availability, and quality were missing. However, the JMP suggests that packaged or delivered water may be safely managed, so it has been added as an improved source [4]. Another update was the explicit use of water collection time as a determinant of service levels in the updated ladder.

The SDG baseline assessment estimated that 5.2 billion people worldwide have access to safely managed drinking water services. Over 1.3 billion people use basic drinking water services. However, 263 million people spend over 30 minutes to collect drinking water, 422 million people collect drinking water from unimproved sources, and 159 million people drink surface water.

TABLE 1.4 JMP ladders for monitoring access to water in schools and healthcare facilities.

Service level	Schools	Healthcare facilities
Advanced	To be defined at national level	To be defined at national level
Basic (SDG)	Drinking water from an improved source is available at the school	Water from an improved source is available on premises
Limited	There is an improved source, but water was not available at time of survey	Water from an improved source is available off premises or an improved source is on-site, but no water was available
No service	No water source or unimproved source	Unprotected dug well or spring, surface water, or no water source

The JMP ladders have been developed for monitoring access to WASH in schools and healthcare facilities [4]. The service levels consist of advanced, basic, limited, and no service (Table 1.4). For both schools and healthcare facilities, the advanced service level is to be defined by each country, but the definition is not available as of January 2018. The JMP is in the process of establishing the global baseline for WASH in schools and healthcare facilities to be published in the near future.

1.4 IMPACTS OF WATER SUPPLY INTERVENTIONS

A great deal of evidence and literature addresses the impacts associated with improved quality and quantity of water supplies. Between 1980 and 2000, most studies of water quality assessed only the source of water and not the point at which users actually consumed the water (point of use). A review of 67 studies to determine the health impact of water supplies found that the median reduction in diarrheal morbidity from improvements in water availability was 25% and the median reduction based on improvements to water quality at the source, not at the point-of-use, was 16%, with a range of 0–90% [12]. Combinations of water quality at the source and water quantity resulted in a 37% median reduction in diarrheal morbidity. In 1991, the study was updated and covered 144 studies and looked more carefully at studies and the rigor with which they were conducted [13]. In the 1991 analysis, looking only at studies deemed rigorous, improvements in water quantity resulted in a median reduction of 30%, improvements to water quality at the source was 15%, and combinations of water quality at the source and water quantity resulted in a 17% median reduction in diarrheal morbidity. These reviews helped set the agenda for specific interventions that the global community would pursue. There was, however, a growing interest in assessing water quality at the point of use. In 2003, an analysis of 21 controlled field trials dealing with interventions designed to improve the microbiological water quality at the point of use showed a median reduction in endemic diarrheal disease of 42% compared to control groups [14]. Nine studies used chlorine as a method of treating water, five used filtering, four used solar disinfection, and three used a combination of flocculation and disinfection. The results of this study and subsequent studies resulted in donor investments to improve drinking water quality at the point of use and a large number of economically developing countries now have point-of-use products that are being socially marketed.

The impact of water, sanitation, and/or hygiene interventions on diarrhea morbidity among children in low- and middle-income countries has been examined in more recent reviews of studies using experimental and quasi-experimental methods [15, 16]. Waddington et al. analyzed 65 rigorous impact evaluations from 71 distinct interventions for quantitative synthesis and assessed across 130 000 children in 35 developing countries during the past three decades. These studies were evaluated for a range of factors, such as type of intervention, effect size and precision, internal validity, and external validity. The interventions were grouped into five categories: (i) water supply improvements, (ii) water quality, (iii) sanitation, (iv) hygiene, and (v) multiple interventions involving a combination of water and sanitation and/or hygiene. The results challenged the notion that interventions to improve water quality treatment at the point of use are necessarily the most efficacious and sustainable interventions for promoting the reduction of diarrhea. The analysis suggested that while point-of-use water quality interventions appear to be highly effective, and generally more effective than water supply or improving water quality at the source, much of the evidence is from small trials conducted over short periods of time. The review indicated that point-of-use interventions conducted over longer periods of time demonstrated smaller effects as compliance rates fell. The study found that hygiene interventions, particularly the promotion of handwashing with soap, were effective in reducing diarrhea morbidity, even over longer periods of time.

Calculations of the cost-effectiveness of the interventions described above have shown point-of-use and hygiene interventions to be highly efficient for health improvements [17, 18]. In terms of the costs per disability-adjusted life year (DALY) averted, a community connection to improved water supplies results in $94USD/DALY averted, which is less than half that for household water connection, but substantially higher than estimates for point-of-use water quality interventions, estimated at $53USD/DALY averted using chlorination. Estimates from improved hygiene and sanitation suggest that hygiene promotion is the most cost-effective, at $3USD/DALY averted, followed by sanitation promotion, at $11USD/DALY [17].

Water supply interventions have a large number of benefits. For example, improved water supplies enable improved hygiene practices, such as handwashing and better home hygiene, and there are likely considerable spillover effects in terms of environmental health benefits. In Lesotho, use of smaller quantities of water was related to higher rates of *Giardia lamblia* infection [19]. In Taiwan, a reduction of 45% in rates of trachoma was noted, when the water supply was attached to the home compared to a water supply that was 500 or more meters away [20]. Time savings associated with water supply interventions are also significant. In rural Nigeria, the installation of water systems was estimated to reduce collection time from 6 hours to 45 minutes per household per day during the dry season, mainly benefiting adolescent girls and young women [21]. Another study also estimated a time savings of 20 minutes per household per day from a village water supply improvement in China [22]. In the Philippines, water quantity was strongly associated with nutritional status. Children in households that averaged less than 6 l per capita per day were significantly more malnourished than children in households that averaged 6–20 l or more than 20 l per capita per day [23]. A study of Pakistan households showed that increased quantity of water available at the household level was associated with reduced stunting in children [24].

The public health gains stemming from access to increased quantities of water typically occur in steps. The first step relates to overcoming a lack of basic access, where distance, time, and costs involved in water collection result in the use of volumes inadequate to support basic personal hygiene and may be marginally adequate for human consumption. Significant health gains occur largely when water is available at the household level. Other benefits derived from the second step in improving access include increased time available for other purposes. The availability of new or improved supplies does not always translate directly into a significant increase in use. In East Africa, after new water supplies were placed in proximity to households, the amount of water usage did not increase if the original water source was less than 1 km from the home [25].

In practice, the use of water for domestic purposes cannot easily be distinguished from productive use, particularly among very poor communities. When communities design their own water systems, they invariably plan for multiple-use water systems, and this is especially the case if the livelihoods of households depend on livestock [26]. In multiple-use approach interventions, it is critical to: (i) work with the community to assess the range of water needs in collaboration with end users; (ii) examine the water sources available; and (iii) match water supplies to needs based on the quantity, quality, and reliability required for various purposes. There may also be important health and social gains from ensuring adequate quality of service to support small-scale productive use, especially when this involves food production. Access to water used for small-scale productive activity in such areas is therefore important as part of economic growth and may deliver significant indirect health benefits as a result [27].

The water-related indicators in the MDGs and SDGs have tracked people's access to improved water sources. The technologies considered "improved," however, often do not consistently result in high-quality water. There are certain sources of water that the public health community condemns as risky (e.g., unprotected wells) and others they deem safe (e.g., protected wells). Comparing water quality from protected and unprotected supplies across countries, however, has demonstrated that in many cases protected supplies often provide lower water quality than protected wells in other countries. This suggests that certain practices – not certain types of water sources – may be more important for improving water quality [28]. As mentioned above, it is now generally accepted that providing safe water at the source does not imply that water is safe at the point of use. A study by Gundry et al. found that about 40% of water samples from microbiologically safe sources of water were contaminated at the point of consumption. Household water treatment at the point-of-use for most communities is an important intervention, regardless of whether the water comes from an improved source [29, 30].

Boiling is currently one of the most accessible means for water treatment to most populations and has been shown to be effective [31]. However, in the absence of safe storage, boiled water is immediately vulnerable to recontamination, especially when the environment is unhygienic. Further, this mode of treatment can have serious side effects, such as indoor air pollution and depleting environmental resources if biofuels (e.g., wood) are used for boiling.

1.5 RESOURCE REQUIREMENTS

To achieve SDG targets on WASH, adequate and continued financial investment for infrastructure, operation, and maintenance is essential. Hutton and Varughese estimated that \$6.9 billion (range: \$3.6–\$11.6 billion) is required each year from 2015 to 2030 to achieve universal access to basic drinking water services (Figure 1.4). Compared to the provision of basic sanitation and hygiene services, basic drinking water services are less costly than sanitation (\$19.5 billion) and more costly than hygiene (\$2.0 billion). Provision of universal access to safely managed drinking water was estimated to be five times more expensive than that of basic drinking water at \$37.6 billion per year (range: \$29.4–\$45.6 billion). In the initial stage of SDG implementation, capital costs for infrastructure may account for the significant majority of investment while operation and maintenance costs are expected to increase over years [32].

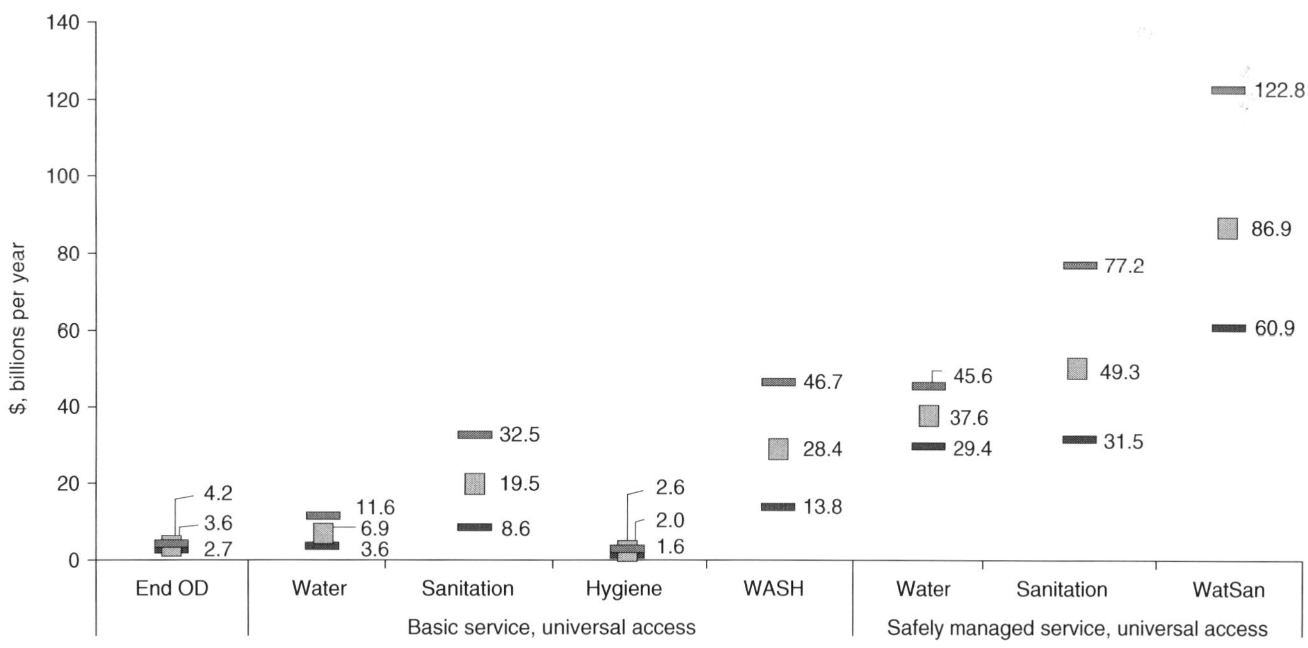

FIGURE 1.4 Estimated annual costs to achieve universal access to basic WASH services and safely managed drinking water and sanitation. Notes: OD = Open Defecation. WASH = Water, Sanitation, and Hygiene. WatSan = Water and Sanitation. *Source*: Hutton and Varughese [32].

1.6 NATURALLY OCCURRING AND ANTHROPOGENIC WATER POLLUTION

While microbiological contamination of water is the main emphasis of this chapter, naturally occurring and anthropogenic sources of chemical pollution can pose serious human health risks. Although no published estimates are available on the global burden of disease resulting from chemically polluted water [33], a number of countries with a growing industry sector have faced water pollution challenges. In addition to anthropogenic pollutants, groundwater commonly contains naturally occurring toxic chemicals, including arsenic and fluoride, which dissolve into the water from soil or rock layers. The most extensive problem of this category is arsenic contamination of groundwater, which has been observed in Argentina, Bangladesh, Chile, China, India, Mexico, Nepal, Taiwan, and parts of Eastern Europe and the United States [34]. Arsenic in Bangladesh's groundwater was first highlighted in 1993 as a result of promoting protected wells in an effort to eliminate diarrheal diseases caused by fecally contaminated surface waters. Millions of shallow wells were drilled into the Ganges delta alluvium in Bangladesh and estimates indicate that an estimated 40 million people were put at risk of arsenic poisoning-related diseases because of high arsenic levels in the groundwater [35]. Fluoride is another naturally occurring pollutant that causes health effects and exposure to high levels in drinking water can detrimentally affect bone development and in some cases can cause crippling skeletal fluorosis. The burden of disease from chemical pollution in specific areas can be large. There are a number of events that have underscored the high levels of disease burden from chemical pollution, including methylmercury poisoning, chronic cadmium poisoning, and diseases of nitrate exposure, as well as lead exposure [33, 36, 37].

1.7 SPATIAL AND SOCIAL INEQUITIES IN ACCESS TO DRINKING WATER

Equity relates closely to the idea of fairness and that all members of a society have equal rights. Perfect equity in intra-country budgets, for example, would be a situation where every citizen is allocated an equal amount of the investment regardless of what part of the country they live in. From this perspective, water supply interventions can be considered equitable if they benefit all parts of society equally. Equal levels of access to clean and safe water would be an equitable outcome [38].

Equity is also concerned with comparing different parts and groups of society. Geography, social or health status, gender, and ethnicity can be used for comparison. Two types of equity – spatial and social – are useful for considering different levels of access to water supply and sanitation [38]. Social equity addresses groups or communities formed by geographic areas of residence, such as urban versus rural or administrative boundaries within a country. Social equity is concerned with groups defined by attributes linked to their identity and traverses spatial boundaries. Particularly vulnerable groups may include women, people living with HIV/AIDS, the elderly, the poor, the disabled, orphans, and widows.

Social and spatial inequities may overlap with each other. A large percentage of urban populations without access to basic water and sanitation services are also poor and a larger proportion of the rural population who spend time collecting water is women. Additionally, equitable investments do not necessarily equate to equitable outcomes and water supply costs may vary according to many factors, such as population density, distance from water sources, and the geology [38]. A number of spatial and social inequities persist and there are many challenges facing efforts to improve equitable access. Population growth is a major barrier to current efforts in the water sector to reduce the number of people living without access to basic or safely managed drinking water. In the last several decades, the population of the world has increased from 3.6 billion in 1970 to roughly 7.3 billion people in 2015. In 1980, the United Nations estimated that 1.8 billion people lacked access to safe water supplies and in 2015, there were still 844 million people without a basic drinking water service [4].

Spatially, more than 9 out of 10 people who drink surface water live in rural areas, with Sub-Saharan Africa and Oceania being the regions most behind in drinking water coverage (Figure 1.5). Just 58% of the population in Sub-Saharan Africa and 52% of the population in Oceania are estimated to use a basic drinking water service. At the country level, the proportion of population with a basic drinking water service was less than 50% in 10 countries, including Angola (41%), Chad (43%), Democratic Republic of the Congo (42%), Eritrea (19%), Ethiopia (39%), Mozambique (47%), Niger (46%), Papua New Guinea (37%), Somalia (40%), and Uganda (39%) (Figure 1.5). Within many countries, disparities in access to basic drinking water can be revealed by comparing urban and rural residence, wealth levels, and subnational regions (Figures 1.6 and 1.7) [4].

Even when a piped supply becomes accessible, typically in urban areas, unreliable water availability can be an issue. Less than 10% of people in many South Asian cities receive a 24-hour piped water supply. While governments and international institutions have provided financial and technical assistance with the goal of making water more accessible to the poor, many municipal pipelines reach wealthiest clients first. Thus, a significant number of urban populations without utility connections must rely on alternatives, such as service from Small-Scale Water Providers (SSWPs). Currently SSWPs are most prevalent in Southeast Asia

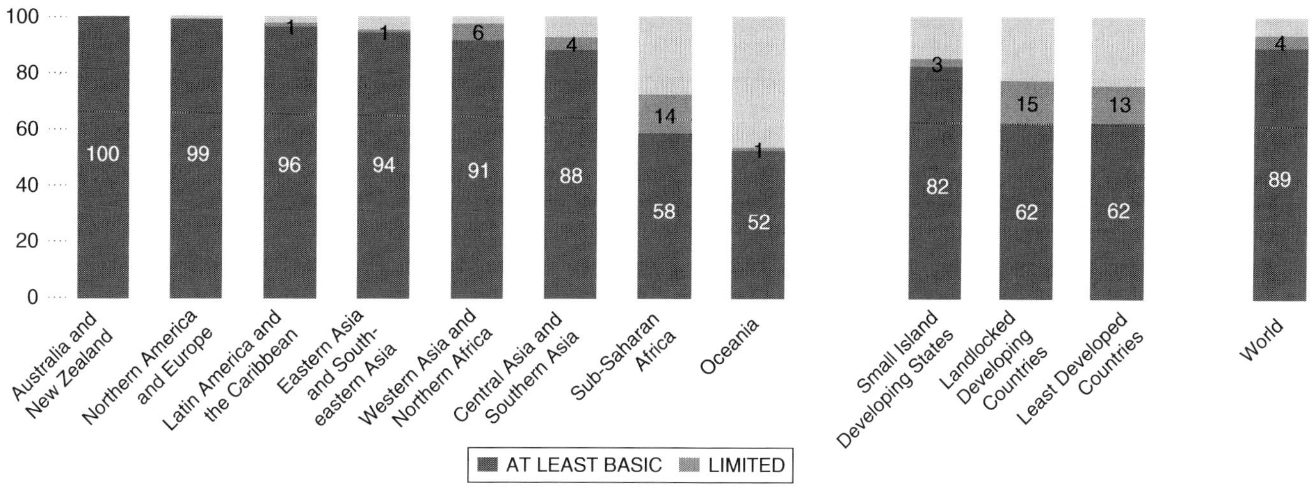

FIGURE 1.5 Regions represented by the percentage of population using at least basic drinking water services in 2015. *Source*: WHO/UNICEF [4].

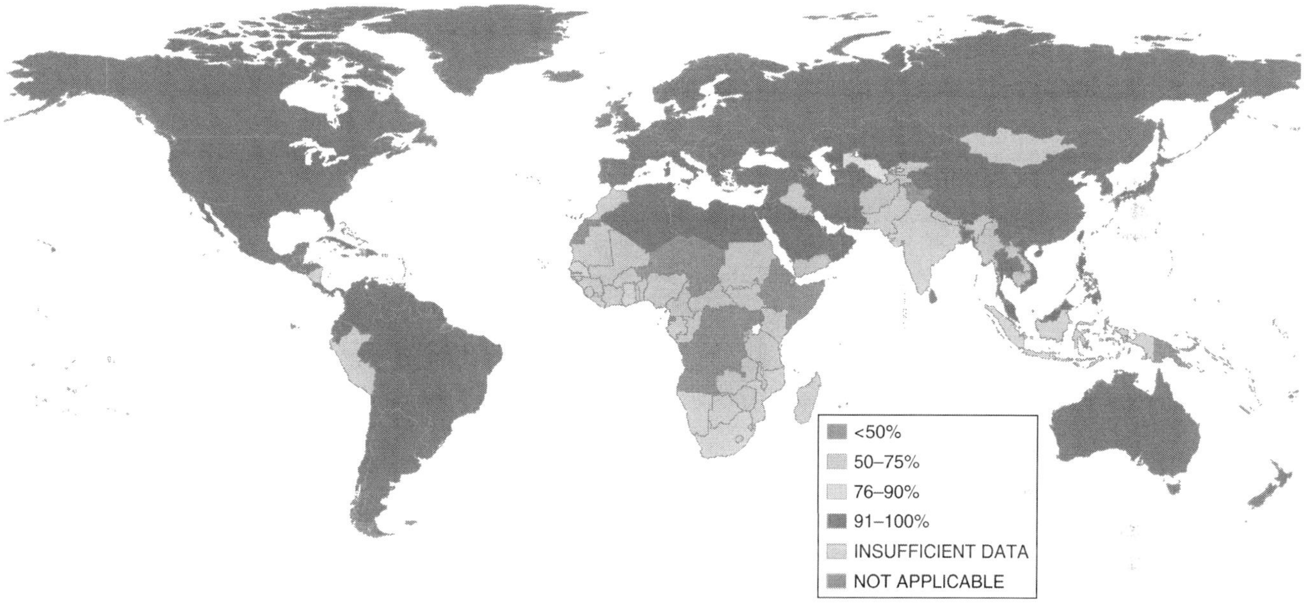

FIGURE 1.6 Countries represented by the percentage of population using at least basic drinking water services in 2015. *Source*: WHO/UNICEF [4]. (*See insert for color representation of this figure.*)

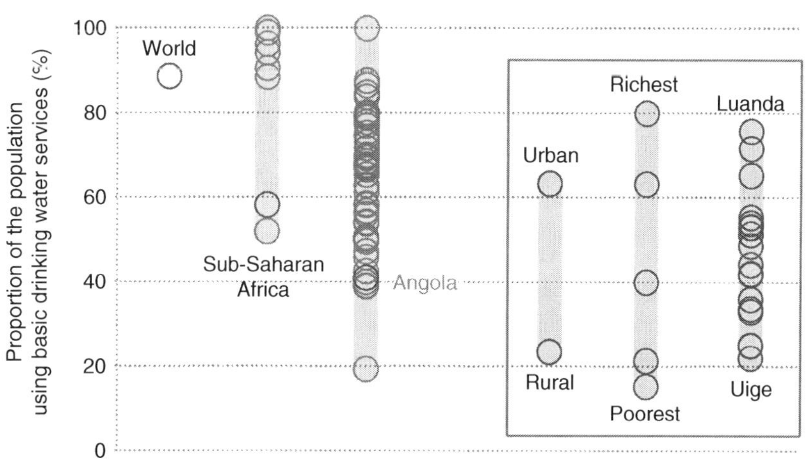

FIGURE 1.7 A summary of the percentage of population using at least basic drinking water services in 2015 by region, country, urban and rural residence, wealth levels, and subnational region. *Source*: WHO/UNICEF [4].

where a quarter of households in Cebu (Philippines), Ho Chi Minn City, Jakarta, and Manila may use these services [39].

In urban areas of the developing world, governments have favored large water utilities. Unfortunately, existing tariffs and management structures have caused these systems to fail to provide piped water coverage to entire populations. Connection fees are frequently too high or total available water is insufficient to support an urban area. Many utilities choose not to equip poor neighborhoods because of the high percentage of unpaid bills, fraudulent consumption, low levels of individual consumption, and because network maintenance costs are high. Additionally, people who occupy land illegally may also be excluded from public services. In cases where water companies are allowed or mandated to serve poor households, water is not always affordable or payment schedules may not be feasible. Thus, many people are forced to illegally draw their water from "spaghetti networks" that connect to the border of a municipal grid system or to purchase expensive, and commonly contaminated, water from SSWPs.

Of further importance are the inequalities surrounding the cost of water for the urban poor. While SSWPs offer a more flexible payment schedule, their water is usually more costly and consumes a large portion of household expenses. In some cities, the poor pay huge premiums to water vendors over the standard water price of those hooked up to municipal systems: 60 times more in Jakarta, Indonesia; 83 times more in Karachi, Pakistan; and 100 times more in both Port-au-Prince, Haiti and Nouakchott, Mauritania.

Socially, the poor, women, and children suffer disproportionately. In most developing countries, the provision of water and sanitation are women's responsibility [40]. Often, rural women from poor households must walk long distances to provide their families with water for drinking, cooking, and domestic and personal hygiene. Interventions to increase access often diminish the time that women spend gathering water and have provided participants with opportunities to learn new skills and spend more time cultivating crops in the time they had previously used for water collection. These classes of changes can have positive impacts on the local

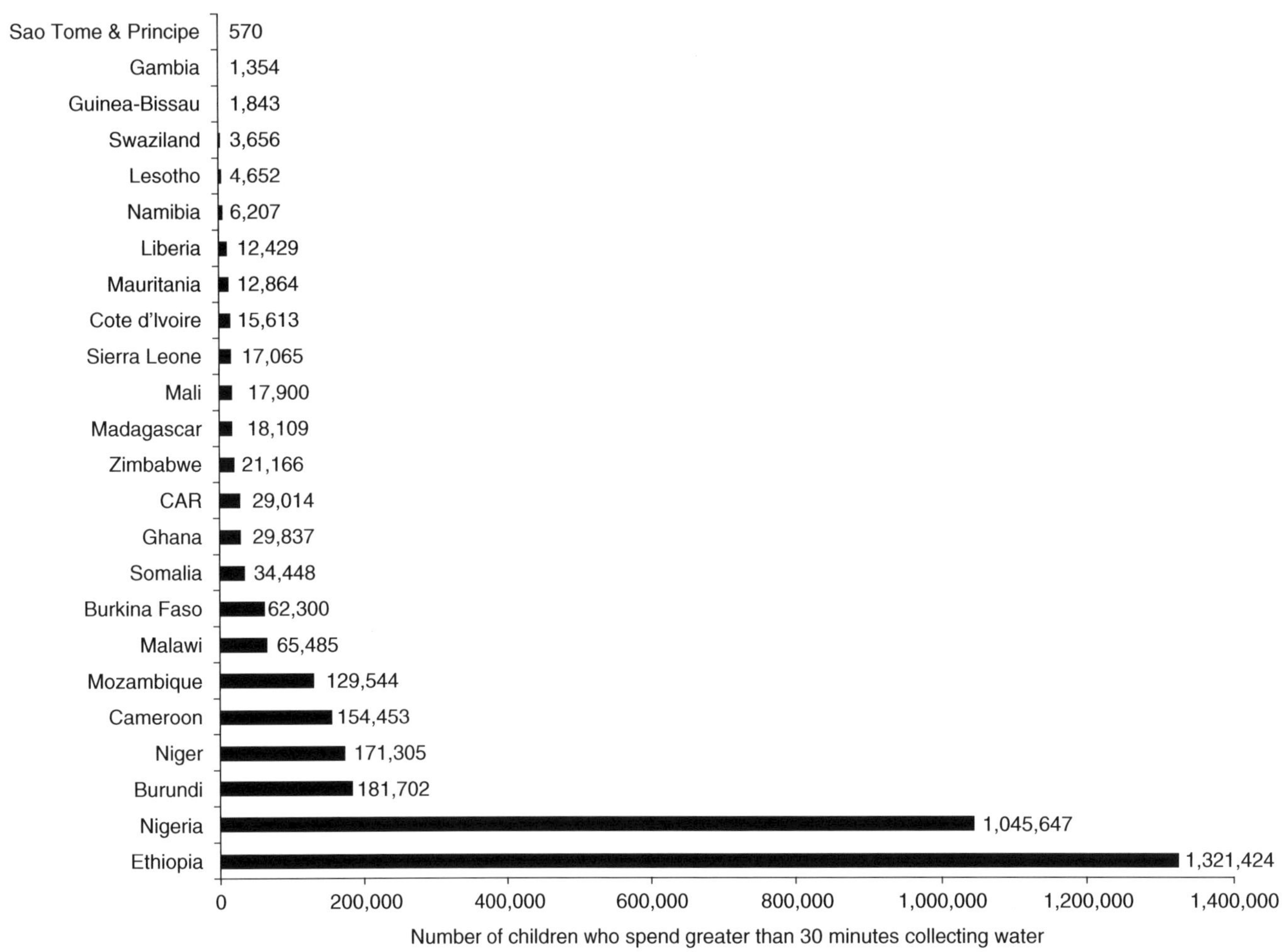

FIGURE 1.8 Number of children who spend more than 30 minutes for household drinking water in 24 countries in Sub-Saharan Africa. *Source*: Graham et al. [42].

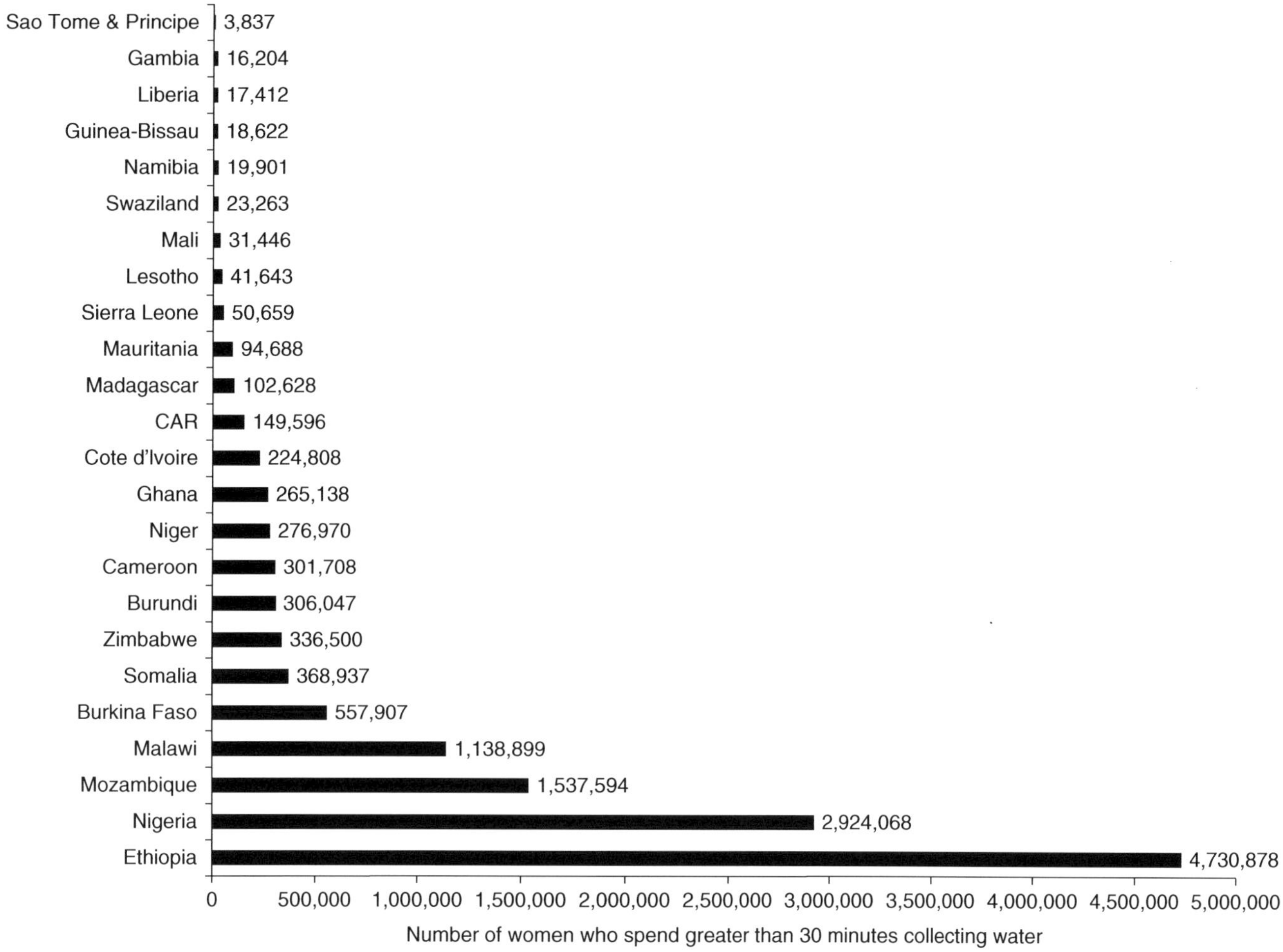

FIGURE 1.9 Number of adult women who spend more than 30 minutes for household drinking water in 24 countries in Sub-Saharan Africa. *Source*: Graham et al. [42].

economy, especially when income-earning involves tasks such as laundry work and other types of activities that use water. By decreasing the time for water collection, new opportunities enable women to effectively contribute to the communities' economic growth.

In an analysis, of more than 40 developing countries, women collected water for almost two-thirds of homes, versus a quarter of households where men collected water. In 12% of homes, children were responsible for collecting water and girls under 15 years of age were twice as likely to collect water as boys of the same age category [40, 41]. A recent study of 24 countries in Sub-Saharan Africa also revealed that 3.36 million children and 13.54 million women were responsible for water collection, which required more than 30 minutes per trip (Figures 1.8 and 1.9) [42]. Burundi, Cameroon, Ethiopia, Mozambique, Niger, and Nigeria included more than 100 000 households where children were responsible for water collection and spent more than 30 minutes for this task.

1.8 SUSTAINABILITY

Sustainability of water supplies is especially difficult in rural areas because of the lack of support through monitoring systems, training, human resource back-up support and availability of spare parts and services. Throughout rural Sub-Saharan Africa, numerous water systems have been developed every year, such as boreholes equipped with motorized or hand/foot pumps. These systems often fall into disrepair shortly after installation. Over 50 000 water supply systems were once estimated to be non-functional across Africa – representing an investment of nearly $300 million USD. This problem resulted from one reason, lack of operations and maintenance. Operations and maintenance, however, is multifaceted.

Many of the negative results in past interventions were linked to: (i) lack of community participation; (ii) utilization of inappropriate technologies; (iii) lack of a sense of ownership on the part of the beneficiaries; (iv) failure to provide the institutional support required for the project; and (v) dissatisfaction of the community with project outcomes [43]. In order to design a more effective and responsive approach for the provision of water and sanitation, development organizations and donor agencies are utilizing a series of participatory methodologies and techniques that focus on getting intended users actively involved in all stages of the project cycle. Fundamentally, community participation increases the probability of success and the sustainability of the projects implemented.

Participatory approaches evolved from disciplines such as anthropology and sociology, which have tried to fill in the existing gap between technology (hardware) and operations and maintenance (software). These approaches were based on the flaws identified and the lessons learned while implementing the supply-driven approach for the provision of safe water and sanitation services. The underlying principle was, and continues to be, the involvement of all stakeholders, especially the main users of the system, in all the phases of water and sanitation programs or projects, with the intention of improving their sustainability and probability of success. The primary objective was to be more responsive to the needs and preferences of users and more appropriate to given local conditions and the environment. Another important characteristic of these participatory methodologies was the significant change in the role that users of the system played during the design, implementation, construction, operation, and maintenance of the systems. Participatory methodologies were developed to facilitate the process of empowerment and capacity building of the communities benefiting from development interventions [44].

Community participation can bring about numerous benefits to development interventions but must be weighed against the time and costs related to their implementation. For effective participation and commitment on the part of the community, availability of financial and human resources at the beginning of the process is vital for success [45]. At the same time, no single approach for community participation can work in all situations. The approaches utilized in the water sector have to be flexible enough to incorporate site-specific information about environmental, social, and cultural factors as well as stakeholders' needs and priorities into the design and implementation of water and sanitation projects.

One of the most commonly used models for developing rural water interventions involves village-level coordination and the development of a system for cost recovery for operations and maintenance. Typically, a community bank account is opened and a community member is appointed to collect the fees. The selection of the technology and who has the skills to operate and maintain it are also part of the operations and maintenance system in place. Other models have been developed and experimented with and include public and private sector arrangements that aim to provide support to community systems following construction.

The community management model has brought many benefits, but it has not always resulted in a sustainable water

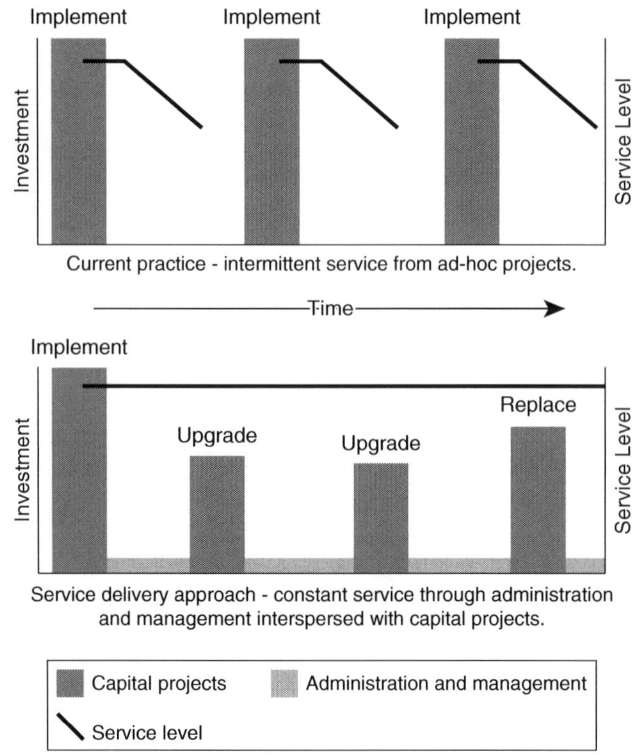

FIGURE 1.10 A model of rural water supply interventions (top half) versus the service delivery approach (bottom half) that provides for constant service through ongoing support interspersed with capital projects.

supply at scale. It became evident that communities often cannot manage the variety of tasks that arise after the construction of water systems, including repairs, accounting, conflict resolution, legal issues, and system replacement. A relatively recent model, the Service Delivery Approach, was developed for improving rural water services and aims to better incorporate enabling environment factors to increase sustainability and scale. The approach considers the whole life cycle of service, from design, day-to-day operations and maintenance, to eventual replacement. For millions of rural people, the top half of Figure 1.10 represents a standard water supply intervention. Following construction of a new system, users have access to an improved source, but due to lack of follow-up support, the system quickly deteriorates until it is non-functional. In the Service Delivery Approach, once the water supply access is improved, it will be maintained through a proper understanding of the full life-cycle costs and institutional support needs [46].

1.9 FINAL REMARKS

Drinking water, adequate sanitation, and good hygiene are essential requirements for the well-being of all humanity. Global efforts under the Millennium Development Goals reduced the number of population who relied on unimproved water sources from 1.3 billion people in 1990 to 663 million people in 2015 [10]. The Sustainable Development Goals will further expand access to drinking water, sanitation services, and hygiene facilities by aiming for universal and equitable access in household and institutional settings, including schools and healthcare facilities. A great deal of financial investment is necessary to ensure accessible, available, and safe drinking water everywhere. Spatial and social inequities also persist and need to be addressed through context-specific approaches with adequate stakeholder participation. Significant progress toward universal access to basic and safely managed drinking water can be made through collective actions at the individual, household, communal, social, international, and global levels.

REFERENCES

1. Preston, S.H. and Van de Walle, E. (1978). Urban French mortality in the nineteenth century. *Population Studies* 32: 275–297.
2. Esrey, S.A. (1996). Water, waste, and well-being: a multicountry study. *American Journal of Epidemiology* 143: 608–623.
3. Diaz, S., Herreros, C., Aravena, R. et al. (1995). Breast-feeding duration and growth of fully breast-fed infants in a poor urban Chilean population. *The American Journal of Clinical Nutrition* 62 (2): 371–376.
4. WHO/UNICEF. Progress on drinking water, sanitation and hygiene: 2017 update and SDG baselines, 2017. Available from: https://www.unicef.org/publications/index_96611.html.
5. Liu, L., Oza, S., Hogan, D. et al. (2015). Global, regional, and national causes of child mortality in 2000–13, with projections to inform post-2015 priorities: an updated systematic analysis. *Lancet* 285: 430–440.
6. Briscoe, J. (1993). When the cup is half full: improving water and sanitation services in the developing world. *Environment* 35 (4): 6–37.
7. Black, M. (1998). *Learning What Works: A 20 Year Retrospective View on International Water and Sanitation Cooperation: 1978–1998*. Washington, DC: UNDP-World Bank Water and Sanitation Program.
8. Yacoob, M. (1990). Community self-financing of water supply and sanitation: what are the promises and pitfalls? *Health Policy and Planning* 5 (4): 358–366.
9. Mara, D. and Feacham, R. Taps and toilets for all: two decades already, and now a quarter century more, 2001. Available from: https://www.ircwash.org/resources/taps-and-toilets-all-two-decades-already-and-now-quarter-century-more.
10. UNICEF/WHO. Progress on sanitation and drinking water: 2015 update and MDG assessment, 2015. Available from: http://www.unicef.org/publications/index_82419.html.
11. United Nations. SDG indicators. 2017. Available from: https://unstats.un.org/sdgs/indicators/database.
12. Esrey, S.A., Feachem, R.G., and Hughes, J.M. (1985). Interventions for the control of diarrhoeal diseases among young children: improving water supplies and excreta disposal facilities. *Bulletin of the World Health Organization* 63 (4): 757–772.
13. Esrey, S.A., Potash, J.B., Roberts, L., and Shiff, C. (1991). Effects of improved water supply and sanitation on ascariasis, diarrhoea, dracunculiasis, hookworm infection, schistosomiasis, and trachoma. *Bulletin of the World Health Organization* 69: 609–621.
14. Clasen, T., Roberts, I., Rabie, T., and Cairncross, S. (2006). Interventions to improve water quality for preventing infectious diarrhea. *Cochrane Database of Systematic Reviews* 3: CD004794.
15. Waddington, H., Snilstveit, B., Fewtrell, L., and White, H. (2009). *Water, Sanitation and Hygiene Interventions to Combat Childhood Diarrhea in Developing Countries*. International Initiative for Impact Evaluation Available from: http://www.3ieimpact.org/media/filer_public/2012/05/07/17.pdf.
16. Darvesh, N., Das, J.K., Vaivada, T. et al. (2017). Water, sanitation and hygiene interventions for acute childhood diarrhea: a systematic review to provide estimates for the lives saved tool. *BioMed Central Public Health* 17 (Suppl. 4): 776.
17. Cairncross, S. and Valdmanis, V. (2006). Water supply, sanitation, and hygiene promotion. In: *Disease Control Priorities in Developing Countries*, 2e (ed. D.T. Jamison, J.G. Breman, A.R. Measham, et al.). New York: Oxford University Press.
18. Clasen, T., Cairncross, S., Haller, L. et al. (2007). Cost-effectiveness of water quality interventions for preventing diarrhoeal disease in developing countries. *Journal of Water and Health* 5 (4): 599–608.
19. Esrey, S.A., Collett, J., Miliotis, M.D. et al. (1989). The risk of infection from *Giardia lamblia* due to drinking water supply, use of water, and latrines among preschool children in rural Lesotho. *International Journal of Epidemiology* 18 (1): 248–253.
20. Assaad, F.A., Maxwell-Lyons, F., and Sundaresan, T. (1969). Use of local variations in trachoma endemicity in depicting interplay between socio-economic conditions and disease. *Bulletin of the World Health Organization* 41 (2): 181.
21. Blum, D., Emeh, R., Huttly, S. et al. (1990). The Imo State (Nigeria) drinking water supply and sanitation project, 1. Description of the project, evaluation methods, and impact on intervening variables. *Transactions of the Royal Society of Tropical Medicine and Hygiene* 84 (2): 309–315.
22. Wang, Z.S., Shepard, D.S., Zhu, Y.C. et al. (1989). Reduction of enteric infectious disease in rural China by providing deep-well tap water. *Bulletin of the World Health Organization* 67 (2): 171–180.
23. Magnani, R., Tourkin, S., and Hartz, M. (1984). *Evaluation of the Provincial Water Project in the Philippines*. International Statistical Programs Center, Bureau of the Census, US Department of Commerce.
24. van der Hoek, W., Feenstra, S.G., and Konradsen, F. (2002). Availability of irrigation water for domestic use in Pakistan: its impact on prevalence of diarrhoea and nutritional status of children. *Journal of Health, Population and Nutrition* March 1, 20 (1): 77–84.
25. White, G.F., Bradley, D.J., White, A.U., and Ahmed, T. (1972). *Drawers of Water*. Chicago: University of Chicago Press.

26. Adank, M., Jeths, M., Belete, B., Chaka, S., Lema, Z., Tamiru, D., et al. The costs and benefits of multiple uses of water: the case of Gorogutu woreda of East Hararghe zone, Oromiya Regional States, eastern Ethiopia. RiPPLE Working Paper 7, 2008.
27. World Health Organization. (2003). *Domestic Water Quantity, Service, Level and Health*. Available from: http://www.who.int/water_sanitation_health/publications/wsh0302/en/.
28. Sutton, S. (2008). The risks of a technology-based MDG indicator for rural water supply. 33rd WEDC International Conference, Accra, Ghana.
29. Clasen, T.F. (2011). Household water treatment and safe storage. In: *Water and Sanitation-Related Diseases and the Environment: Challenges, Interventions, and Preventive Measures*, 1e (ed. J.M.H. Selendy). Hoboken, NJ: Wiley Blackwell.
30. Clasen, T.F. Household water treatment and safe storage. In: *Water and Sanitation-Related Diseases and the Environment: Challenges, Interventions, and Preventive Measures*, 2e (ed. J.M.H. Selendy). Wiley-Blackwell, 2011.
31. Rosa, G., Miller, L., and Clasen, T. (2010). Microbiological effectiveness of disinfecting water by boiling in rural Guatemala. *The American Journal of Tropical Medicine and Hygiene* 8 (3): 473–477.
32. Hutton, G. and Varughese, M. (2016). *The Costs of Meeting the 2030 Sustainable Development Goal Targets on Drinking Water, Sanitation, and Hygiene*. Washington, DC: World Bank.
33. Kjellstrom, T., Lodh, M., McMichael, T. et al. (2006). Air and water pollution: burden and strategies for control. In: *Disease Control Priorities in Developing Countries*, 2e (ed. D.T. Jamison, J.G. Breman, A.R. Measham, et al.). New York: Oxford University Press.
34. WHO (2001). *Environmental Health Criteria 224: Arsenic and Arsenic Compounds*. Geneva: WHO.
35. BGS, DPHE (2001). Arsenic contamination of groundwater in Bangladesh. In: *British Geological Survey Technical Report WC/00/19* (ed. D.G. Kinniburgh and P.A. Smedley). Keyworth: British Geological Survey.
36. Backer, L.C., Kish, J.K., Solo-Gabriele, H.M., and Fleming, L.E. (2011). Naturally occurring water pollutants. In: *Water and Sanitation-Related Diseases and the Environment: Challenges, Interventions, and Preventive Measures*, 1e (ed. J.M.H. Selendy). Hoboken, NJ: Wiley- Blackwell.
37. Shannon, K.L., Lawrence, R.S., and McDonald, M.D. (2011). Anthropogenic sources of water pollution: Parts 1 and 2. In: *Water and Sanitation-Related Diseases and the Environment: Challenges, Interventions, and Preventive Measures*, 1e (ed. J.M.H. Selendy). Hoboken, NJ: Wiley Blackwell.
38. Taylor, B. (2008). *Water: More for some....or some for more?* WaterAid.
39. McIntosh, A.C. (2003). *Asian Water Supplies: Reaching the Urban Poor*. Asian Development Bank. London: IWA Publishing.
40. WHO/UNICEF. Progress on sanitation and drinking-water: 2010 update, 2010. Available from: http://www.who.int/water_sanitation_health/publications/9789241563956/en/.
41. MICS and DHS surveys from 45 developing countries, 2005–2008.
42. Graham, J.P., Hirai, M., and Kim, S.-S. (2016). An analysis of water collection labor among women and children in 24 sub-Saharan African countries. *PLoS One* 11 (6): e0155981.
43. Elmendorf, M.L. and Isely, R.B. (1983). Public and private roles of women in water supply and sanitation programs. *Human Organisation* 42 (3): 195–204.
44. Simpson-Herbert, M., Sawyer, R., and Clarke, L. The PHAST Initiative – Participatory Hygiene and Sanitation Transformation, A New Approach to Working with Communities. WHO Geneva and UNDP – World Bank Water and Sanitation Program, 1997: pp. 1–22.
45. White, A. Community Participation in Water and Sanitation: Concepts, Strategies and Methods. Technical Paper Series No. 17. Hague, The Netherlands: IRC International Water and Sanitation Center.
46. Lockwood, H., Smits, S., Schouten, T., and Moriarty, P. Providing sustainable water services at scale, 2010. Available from: https://www.ircwash.org/sites/default/files/Lockwood-2010-Providing.pdf.

2

THE HUMAN RIGHT TO SANITATION

ANOOP JAIN[1] AND JAY GRAHAM[2]

[1] *University of California, Berkeley, CA, USA*
[2] *Public Health Institute, Oakland, CA, USA*

2.1 INTRODUCTION

The United Nations took responsibility for an outbreak of cholera in Haiti, which began in October 2010, after it was revealed that cholera-infected troops based at a UN camp – set up after the devastating Haitian earthquake in January 2010 – were using outhouses that were draining sewage in to a nearby river. This river was the source of drinking water for local communities in the surrounding area. By March 2017, the outbreak had infected over 700 000 Haitians and killed nearly 10 000 people in Haiti, Mexico, the Dominican Republic, and Cuba combined [1, 2].

The virulence of this event might have been exacerbated by the fact that large segments of Haiti's population were already living without access to improved sanitation services. The Joint Monitoring Programme reported that only 17% of Haitians had access to at least basic sanitation in 2000 [3]. The remaining 83% were using some form of unimproved sanitation, including open defecation [2]. While a significant percentage of Haitians gained access to improved sanitation between 2000 and 2010 [3], enough people continued living without this access, thus worsening the propagation of deadly fecal contamination throughout Haiti and beyond.

The Haitian example underscores the fact that equitable and universal access to sanitation is at the bedrock of public health, while also typifying the many ways in which the right to sanitation is violated around the world. This chapter will first examine why access to sanitation is critical to the full enjoyment of life and all human rights. Next, this chapter will describe the various types of sanitation, global trends in sanitation access, and what this means in terms of the human right to sanitation. Finally, this chapter will also critically examine current efforts from around the world that aim to improve access to sanitation and how they might be incongruent with a rights-based approach to sanitation. Thus this chapter will help paint a broad picture of where the development community is as it moves forward in trying to achieve Sustainable Development Goal 6, availability and sustainable management of water and sanitation for all by 2030.

2.2 THE BENEFITS OF SANITATION

This section will first discuss the dangers associated with untreated human excreta, a common consequence of unimproved sanitation. This section will then briefly discuss studies that have tried to measure the impact of sanitation interventions, and why these studies exemplify the pitfalls of a singular focus on health impact. Finally, this section will discuss the other, less often discussed, benefits of this vital infrastructure, which further underscore why sanitation is a human right.

2.2.1 The Dangers of Untreated Human Excreta

Readers of the *British Medical Journal* recently voted sanitation the single most important medical milestone since 1840 [4]. This makes sense when one considers the fact that 1 g of feces can contain 10^6 viral pathogens, 10^{6-8} bacterial pathogens, 10^4 protozoan cysts, and $10–10^4$ helminth eggs [5].

This contamination causes diarrheal disease, which remains the second leading cause of death in children under

Water and Sanitation-Related Diseases and the Changing Environment: Challenges, Interventions, and Preventive Measures,
Second Edition. Edited by Janine M. H. Selendy.

Companion website: https://www.wiley.com/legacy/wileychi/selendy/

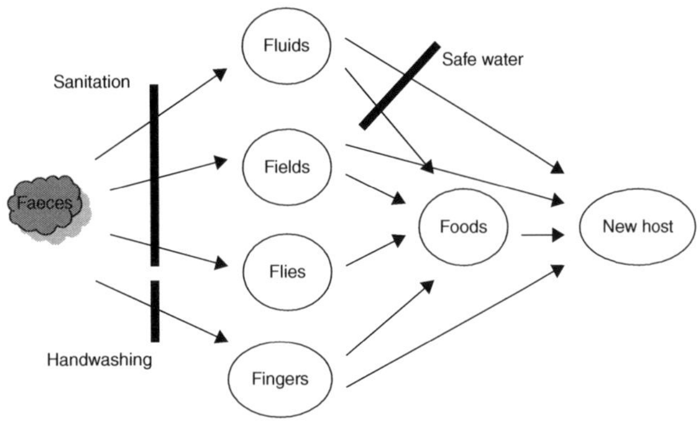

FIGURE 2.1 The diagram describes the various ways in which fecal contamination can spread through the environment. *Source*: Mara et al. [5].

five years of age globally, and nearly one in every five child deaths, around 1.5 million a year, is due to diarrhea [5]. Additionally, nearly one billion people worldwide are infested with intestinal worms and as a result suffer nutritional deficiencies and poor growth [6]. Feces – primarily human, but in some cases animal – is at the core of these diseases, especially when left untreated in the environment (Figure 2.1).

Furthermore, soil-transmitted helminths are a fundamental cause of disease associated with health and nutrition problems beyond gastrointestinal tract disturbances. Pre-school age children account for 10–20% of the one billion people worldwide who are infected with soil-transmitted helminths, which include: *Ascaris lumbricoides* (roundworm), *Trichuris trichiura* (whipworm), and *Ancylostoma duodenale/Necator americanus* (hookworms) [6]. Of the 200–400 million preschool age children infected with roundworm, whipworm, and hookworm worldwide, a large proportion of these children are found in the East Asia region. Helminths consume nutrients from children they infect, thus stifling their cognitive and physical development. For example, studies in India reveal that a 10% increase in open defecation leads to a 0.7% increase in stunting and severe stunting [7], both of which are proxy indicators for long-term social and economic outcomes [8].

De-worming school children through drug treatment is a curative approach for expelling the heavy worm load, but is often only a temporary intervention for reducing worm infection and re-infection is frequent. Improved sanitation, handwashing, and hygiene are also needed control measures to prevent infection and re-infection. More recently, there is research suggesting that a subclinical disorder of the small intestine, tropical enteropathy, which is linked to exposures to large quantities of fecal bacteria, may be a significant cause of child malnutrition globally, and that sanitation and hygiene may be more important to malnutrition, and thus child mortality, than previously thought [6].

2.2.2 The Health Impact of Sanitation Interventions

Health impacts are typically considered the most important effects associated with sanitation. As such, a number of high-quality studies that have used experimental and quasi-experimental designs have attempted to measure the health impact associated with sanitation interventions. The motivation for these studies makes sense given the dangers associated with untreated human excreta, as noted above.

Some of these studies, however, have failed to find statistically significant improvements in health outcomes associated with the given sanitation intervention. For example, Patil et al. conducted a randomized, controlled study of the Indian Government's Total Sanitation Campaign (TSC) in Madhya Pradesh (a central state in India). This study found that TSC increased toilet coverage by 19% in treatment villages when compared to control villages, which led to a decrease in open defecation by 10% among adults [9]. However, TSC did not result in consistent improvements in child health, most notably incidence of diarrhea and helminth infections, and child growth [9].

Similarly, Clasen et al. conducted a cluster-randomized trial to assess the impact of TSC in Odisha, another state in India. The study found that the intervention dramatically improved latrine coverage from 9% of households to 63% in treatment villages, compared to an increase from 8% of households to 12% in control villages [10]. However, the difference in under a five to seven day prevalence of diarrheal disease between treatment and control villages was not statistically significant [10].

It would be a mistake to use these results to undermine the importance of sanitation as a method for improving health and overall quality of life. This is because there are limitations to these high-quality and well-designed studies. For example, Patil et al. point to the short duration of their study as one possible explanation as to why health improvements were not observed. In other words, there might have remained a significant bacterial and viral load in the environment despite increased latrine coverage and use. The short duration of the study did not outlast the fecal contamination that persisted in the surrounding environment (as protozoan cysts and helminth eggs can survive in the environment for prolonged durations), even in treatment villages. This might have contributed to continued morbidity [9].

Using the aforementioned study, results to undermine the importance of sanitation is further problematic because there are other benefits – perhaps equally important, in the eyes of sanitation users – associated with sanitation. These benefits were not captured by these studies. The remainder of this section will focus on these other, less-often, discussed benefits of sanitation.

2.2.3 The Social and Economic Impact of Sanitation Interventions

It is important to note that access to sanitation has benefits beyond improved health. Documenting these benefits bolsters the argument for the human right to sanitation, especially when the health impacts are not always discernable.

One such benefit is economic. For example, the Water and Sanitation Programme (WSP) estimates that open defecation costs India nearly $54 billion annually, of which over $10 billion is due to lost productivity (often from time spent looking for a safe place to defecate) [11]. Thus increasing sanitation coverage, not only at home but also in market places and places of employment, can help people earn more money.

Increasing access to sanitation also promotes feelings of safety and security. This is particularly true for women and girls who defecate in the open. They are often subjected to threats of sexual violence, or verbal abuse, as they search for a safe place to relieve themselves. Thus increasing sanitation coverage bolsters a sense of safety, while helping people preserve their sense of dignity, both of which are essential to the full enjoyment of life.

Safe and hygienic sanitation facilities in schools can help with teacher and student retention. Again, this is especially important for young girls who are reaching puberty. The absence of school toilets means that girls often skip school while menstruating, stifling their ability to succeed in school.

Improved access to sanitation can also help communities reclaim public spaces. Again, this is particularly true for those communities that primarily defecate in the open in public fields. Access to sanitation, and a subsequent decrease in open defecation, means that public spaces are no longer covered in fecal contamination. In many cases, this enables communities to reclaim public spaces, which are often used by children to play sports, for building community gardens, or for organizing community-wide events. This is yet another example of how sanitation can impact the full enjoyment of life.

Thus, while the motivation to improve access to, and use of, sanitation has primarily been seen as a method to improve health outcomes, the "other" benefits of sanitation cannot be understated. This is especially true when one considers the fact that measuring the health impact of sanitation is often difficult to quantify and that community members might derive a different set of benefits from sanitation beyond just improved health.

2.3 SANITATION ACCESS AND THE SANITATION LADDER

The above section highlights the various reasons why sanitation is essential for the full enjoyment of life and human rights. This section will discuss global trends in access to sanitation, the types of sanitation that are available to people, and why this matters from a human rights perspective.

2.3.1 Global Trends in Access to Sanitation

In 1980, the United Nations estimated that 3.9 billion people lacked access to adequate sanitation [3]. Significant gains toward reducing that number have been made since then. Over 2.1 billion people gained access between 1990 and 2015 [3]. This is impressive considering that the world's population increased by 50% between 1981 and 2000 – an increase by nearly 2.4 billion people [3].

Despite this success, the Millennium Development Goal (MDG) target to halve the proportion of people without access to improved sanitation by 2015 was missed [3]. As a result, 2.4 billion people continue living without access to even basic sanitation [3] (Figure 2.2).

This segment represents nearly a third of the world's population, of which 600 million people use limited sanitation services and nearly 900 million people still resort to open defecation [3] (Figure 2.3).

The figures above highlight the fact that inadequate access to improved forms of sanitation is most common in Sub-Saharan Africa, Oceania, and Central/Southern Asia. However, even within these regions there exists heterogeneity of outcomes when stratifying by urban versus rural and by wealth quintiles (Figure 2.4).

The figure above, for example, highlights differences in access to basic sanitation infrastructure in Panama. Richer households in urban areas are more likely to have access to at least basic sanitation in Panama.

Similarly, there are significant inequality rates of open defecation. Again, these are rooted in the urban versus rural and wealth divides. The figure below highlights how this is the case in Bangladesh, where poor households in rural areas are more likely to defecate in the open (Figure 2.5).

Thus, while inadequate access to sanitation remains very much a global challenge, it is important to note that it disproportionately impacts poor households in rural areas. Efforts need to be made to preferentially treat these areas to actuate the human right to sanitation and to meet global targets for improved sanitation.

2.3.2 The Sanitation Ladder

The sanitation ladder is a common framework for evaluating sanitation progress. As the name implies, the ladder provides steps that go from the lower rungs of sanitation, such as open defecation, to more sophisticated and effective sanitation (Figure 2.2). It is increasingly recognized that there is value in motivating people to climb up the sanitation ladder, even if the resulting sanitation facility is not ideal (i.e., not meeting the standards of the WHO/UNICEF Joint Monitoring Programme (JMP) for water supply and sanitation).

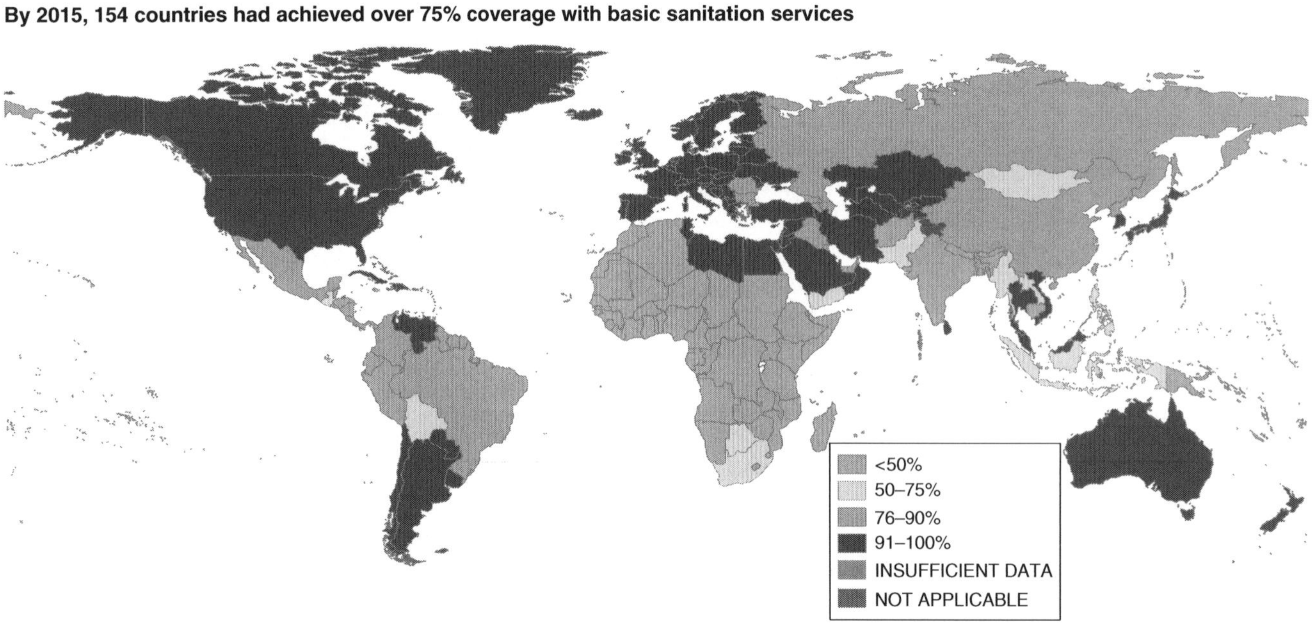

FIGURE 2.2 Global sanitation coverage as of 2015. *Source*: World Health Organization [3]. (*See insert for color representation of this figure.*)

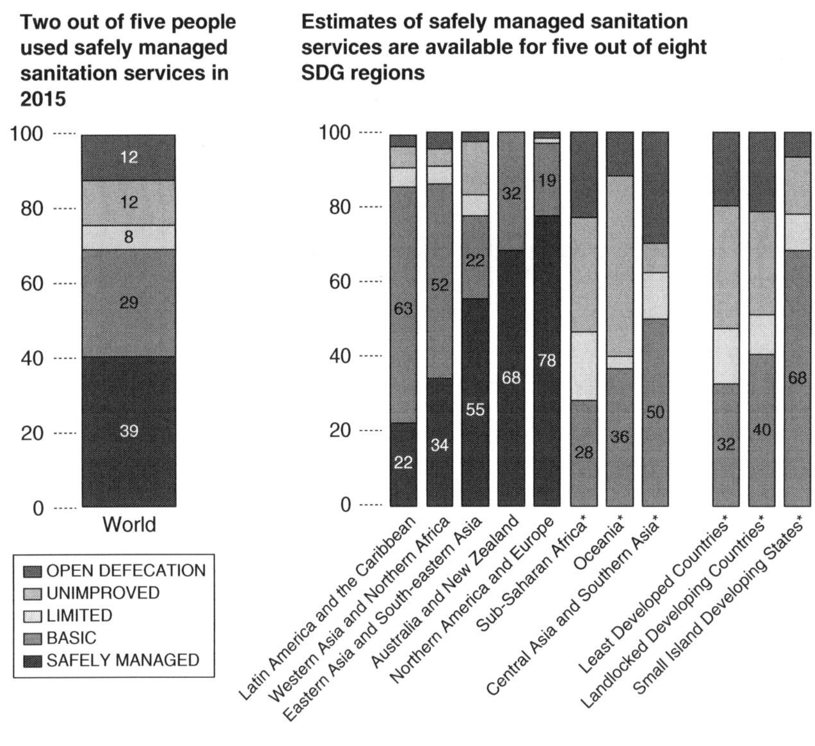

FIGURE 2.3 Global access to safely managed sanitation. *Source*: World Health Organization [3]. (*See insert for color representation of this figure.*)

The remainder of this section will be dedicated to defining these various rungs and why this matters in the context of the human right to sanitation.

2.3.2.1 ***Open Defecation*** Open defecation is the lowest rung of the sanitation ladder and is defined as the practice of defecating in fields, forests, open bodies of water, and other open spaces [3]. Globally, 892 million people still defecate in the open [3]. Open defecation is most common in Sub-Saharan Africa, Oceania, and Central/Southern Asia. India alone is home to nearly 500 million people who defecate in the open.

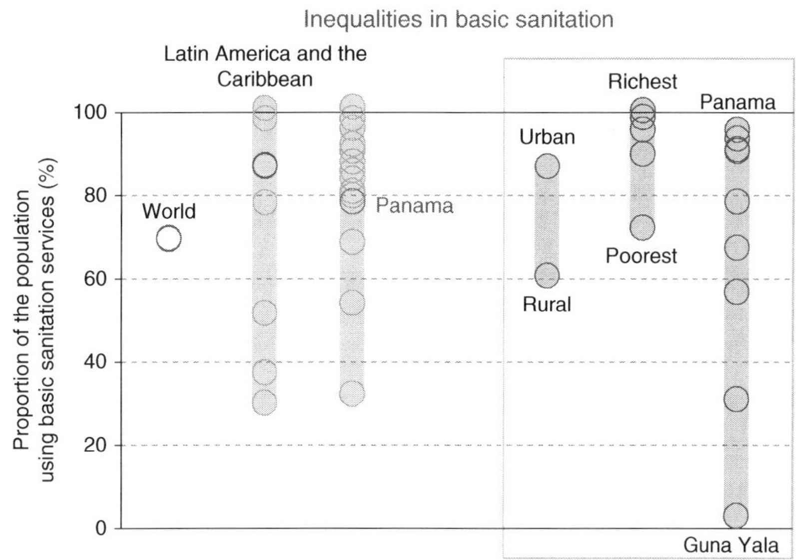

FIGURE 2.4 Inequalities in basic sanitation (box is Panama). *Source*: World Health Organization [3].

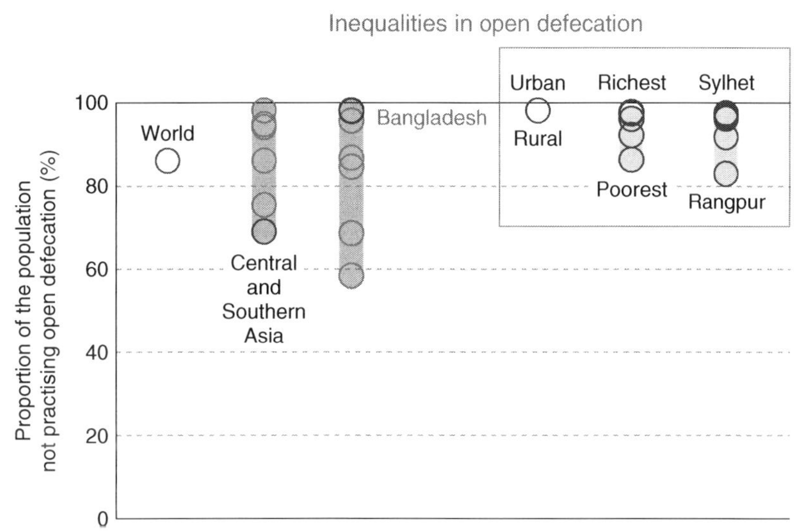

FIGURE 2.5 Inequalities in open defecation (box is Bangladesh). *Source*: World Health Organization [3].

2.3.2.2 Unimproved Sanitation Unimproved sanitation is defined as pit latrines that do not have a slab or platform, hanging latrines, or bucket latrines. The JMP estimates that over 850 million people around the world still use some form of unimproved sanitation [3]. While unimproved sanitation might provide basic shelter that helps ensure privacy and safety, it does ensure the hygienic disposal of waste. Thus fecal contamination more easily spreads through the environment, which can lead to increased rates of morbidity and mortality.

2.3.2.3 Improved and Limited Sanitation Improved sanitation facilities are those that hygienically separate human waste from human contact. Access to limited sanitation, then, is the use of improved sanitation that is shared between two or more households. This distinction is made because there is an underlying assumption that there is a general lack of maintenance when sanitation facilities have to be shared. This might lead to unhygienic conditions or it might cause people to revert to defecating in the open (Figure 2.6).

2.3.2.4 Basic Sanitation The JMP defines basic sanitation as the use of improved sanitation that is not shared by more than one family. However, this type of sanitation might not have the provision for the safe management of human excreta. The JMP estimates that 29% of the world's population uses basic sanitation.

SERVICE LEVEL	DEFINITION
SAFELY MANAGED	Use of improved facilities that are not shared with other households and where excreta are safely disposed of in situ or transported and treated offsite
BASIC	Use of improved facilities that are not shared with other households
LIMITED	Use of improved facilities shared between two or more households
UNIMPROVED	Use of pit latrines without a slab or platform, hanging latrines or bucket latrines
OPEN DEFECATION	Disposal of human faeces in fields, forests, bushes, open bodies of water, beaches or other open spaces, or with solid waste

Note: improved facilities include flush/pour flush to piped sewer systems, septic tanks or pit latrines; ventilated improved pit latrines, composting toilets or pit latrines with slabs.

FIGURE 2.6 The sanitation ladder. *Source*: World Health Organization [3].

2.3.2.5 Safely Managed Safely managed sanitation is the highest rung on the sanitation ladder and constitutes any sanitation infrastructure that is not only improved but also safely manages human excreta either *in situ* or is in a transported/treated offsite. The JMP estimates that 39% of the world's population has access to safely managed sanitation services.

2.3.3 Why this Matters

As noted earlier, there is value as people, households, and communities move up the sanitation ladder. For example, using shared sanitation, albeit not improved, is generally recognized as better than defecating in the open.

This notion bolsters the argument that sanitation should be considered a broad human right, not just infrastructure that improves health. Thus, while there is evidence that suggests that shared sanitation, for example, might not be the best type of sanitation to improve health, this infrastructure does certainly help provide other benefits to communities, such as safety, privacy, preserved dignity, and the improved cleanliness of public spaces. Therefore, this infrastructure, even if not classified as "improved" by the JMP, might still enable the full enjoyment of life in a variety of important ways.

2.4 CURRENT EFFORTS TO IMPROVE SANITATION OUTCOMES

In the past, the provision of sanitation was chiefly a supply-driven approach. New approaches, however, are generally no- or low-subsidy (financial and hardware) programs. Implementers justify this approach in several ways, including:

- Improvements in sanitation coverage typically stop once subsidy budgets run out.
- Subsidies lead to inappropriate facility designs that are often too expensive.
- Subsidies are often not captured by the poor who need sanitation.
- Subsidies can potentially destroy a developing sanitation market by creating perverse incentives.
- Households often do not use and maintain latrines that are heavily subsidized.

When subsidies are used effectively, it is usually to offer low interest loans or to remove certain administrative fees.

This section will discuss two of these current efforts that take this approach and will also critically examine these programs from a human rights perspective.

2.4.1 Community Led Total Sanitation

Community-Led Total Sanitation (CLTS) is a community-wide participatory approach to improve sanitation based on stimulating a collective sense of disgust and shame. The approach applies simple, visual methods to make obvious the spread of fecal–oral contamination, highlighting the fact that community members are eating each others' feces [12]. The approach has many of the following characteristics:

- Allows community members to analyze their own sanitation situation, using transect walks, mapping of open defecation sites, and community-wide fecal load calculations (more commonly known as "shit" calculations).

- Raises the perception of risks associated with open defecation.
- Generates a collective sense of disgust and shame, which is used to mobilize entire communities to end open defecation.
- Hardware subsidies are usually not provided and it generally promotes use of local sanitation options based on affordability and durability.

The approach, developed in 1999 in Bangladesh, has been transferred and adapted to nearly two dozen countries in Sub-Saharan Africa, beginning around 2005 [13]. Community-wide approaches, such as CLTS and its variants, are often referred to as *Total Sanitation* approaches because they aim to achieve universal use of toilets and end open defecation in the communities targeted. The Total Sanitation approach is rapidly gaining acceptance in rural parts of Sub-Saharan Africa and a number of organizations and governments are using the approach to jumpstart sanitation activities [14].

2.4.1.1 Examining CLTS from a Human Rights Perspective CLTS has achieved a certain measure of success since its design in 1999. Nevertheless, it has also been criticized for not always accounting for basic human rights. For example:

1. Bartram et al. note that some communities have used the shaming technique in sometimes violent ways. For instance, community members in one Bangladeshi village resorted to throwing stones at women as a way to discourage them from defecating in the open. In other extreme cases, government officials have locked people out of their own homes until they sign an agreement saying that they will build a household toilet and no longer defecate in the open. Both of these examples underscore the ways in which the basic tenets of CLTS egregiously violate basic human rights [14].
2. Cavill et al. note that CLTS often excludes already vulnerable populations from community-wide sanitation decisions. CLTS facilitators are not always trained to include those with disabilities, women, or girls as they lead communities through CLTS activities. The failure to capture these voices, opinions, insights, and perspectives can lead to inappropriately designed sanitation interventions and can reinforce inequalities in sanitation access [15].
3. Spears et al. note that the success of CLTS often hinges on community homogeneity. In other words, in order for CLTS to achieve its community-wide sanitation goals, it requires entire communities to work together toward these goals. However, this is not always possible, especially in places such as rural India, where the caste system, which remains deeply entrenched, functions as a mechanism that severely fragments communities. Again, CLTS does not necessarily account for these community-level power imbalances and thus might reinforce disparities in sanitation access [15].

2.4.2 Swachh Bharat Abhiyan (Clean India Mission)

India is home to the world's largest share of people who defecate in the open. The Joint Monitoring Programme estimates that over 500 million Indians defecate in the open every single day [3], leading to the spread of fecal contamination.

India has long tried to bring an end to its open defecation crisis. In fact, Mahatma Gandhi was once quoted as saying, "sanitation is more important than political independence" [4]. As such, the Indian government has engaged in various efforts to end this risky health behavior. For example, the Central Rural Sanitation Programme (CRSP), which was founded in 1986, worked with state governments to provide subsidies for individual household toilet construction [5]. This program evolved into the Total Sanitation Campaign (TSC) in 1999, which in 2014 was renamed Swachh Bharat Abhiyan (Clean India Mission). These programs aim to end open defecation by focusing primarily on helping families overcome financial barriers to toilet construction. More specifically, this program provides households with a USD 188 incentive to construct a toilet. This amount is only meant to cover a portion of the cost to build a toilet and households are supposed to cover the remaining part. Furthermore, this financial assistance is disbursed to families in the form of a reimbursement, after toilet construction. Government data suggest that over 40 million toilets have been built throughout India since Swachh Bharat Abhiyan was launched in late 2014.

2.4.2.1 Examining Swachh Bharat Abhiyan from a Human Rights Perspective Like CLTS, the Government of India's Swachh Bharat Abhiyan also fails to account for basic human rights considerations in its design. For example:

1. While Swachh Bharat Abhiyan does provide some financial assistance to households for toilet construction, this funding comes in the form of a reimbursement. In states such as Bihar, where the average annual household income is just $450 per year in the rural areas, households are often unable to afford the upfront cost of toilet construction and are therefore unable to take advantage of the government's program. Thus, Swachh Bharat Abhiyan perhaps does not help all households actuate their right to sanitation.
2. Swachh Bharat Abhiyan also fails to address a variety of other factors that might influence a family's ability to own a toilet. For example, rural households in Bihar own on average just 330 ft^2 of land for their

dwelling [16]. This might not be enough space for a family to build a toilet. Swachh Bharat Abhiyan also provides no assistance to households for waste management. Households are thus asked to shoulder this enormous responsibility, which ultimately might deter families from building a toilet. By not addressing some of these factors, Swachh Bharat Abhiyan again might not help some households actuate their right to sanitation.

2.5 CONCLUSION

Access to sanitation is a basic human right. Not only does it prevent morbidity and mortality, it enables a fuller enjoyment of life by helping to ensure safety, privacy, improved productivity, and cleaner communities. Unfortunately, 2.4 billion people continue living without access to this vital infrastructure. While safely managed sanitation, as defined by the JMP, might be the gold standard, there is utility in helping people move up the sanitation ladder to ensure that at least some of their rights and needs are met. Moving forward, sanitation interventions need to be designed in such a way that they actually help people actuate their right to sanitation. This will help lead the development community to achieve its Sustainable Development Goals and will lead to a more just future.

REFERENCES

1. UN General Assembly. The human right to water and sanitation: resolution/adopted by the General Assembly, 3 August 2010, A/RES/64/292. Available at: http://www.refworld.org/docid/4cc926b02.html. Accessed 2 December 2017.
2. Editorial board (12 August 2015). UN must step up, apologize, and help drive cholera from Haiti. *The Boston Globe Newspaper*. Retrieved 2 December 2017.
3. World Health Organization (2017). Progress on drinking water, sanitation, and hygiene: 2017 update and SDG baselines. S.1.: World Health Organization. Available from: https://www.unicef.org/publications/files/Progress_on_Drinking_Water_Sanitation_and_Hygiene_2017.pdf.
4. Ferriman, A. (2007). BMJ readers choose the "sanitary revolution" as greatest medical advance since 1840. *British Medical Journal* 334: 111.
5. Mara, D., Lane, J., Scott, B., and Trouba, D. (2010). Sanitation and health. *PLoS Medicine* 7 (11): e1000363. doi: 10.1371/journal.pmed.1000363.
6. Albonico, M., Allen, H., Chitsulo, L. et al. (2008). Controlling soil-transmitted helminthiasis in pre-school-age children through preventive chemotherapy. *PLoS Neglected Tropical Diseases* 2 (3): e126.
7. Spears, D., Ghosh, A., and Cumming, O. (2013). Open defecation and childhood stunting in India: an ecological analysis of new data from 112 districts. *PLoS One* 8 (9): e73784. doi: 10.1371/journal.pone.0073784.
8. Spears, D. and Lamba, S. (2013). *Effects of early-life exposure to sanitation on childhood cognitive skills: evidence from India's Total Sanitation Campaign*. The World Bank Retrieved from: http://elibrary.worldbank.org/doi/book/10.1596/1813-9450-6659.
9. Patil, S.R., Arnold, B.F., Salvatore, A. et al. (2013). *A randomized, controlled study of a rural sanitation behavior change program in Madhya Pradesh, India*. The World Bank. doi: 10.1596/1813-9450-6702.
10. Clasen, T., Boisson, S., Routray, P. et al. (2014). Effectiveness of a rural sanitation programme on diarrhoea, soil-transmitted helminth infection, and child malnutrition in Odisha, India: a cluster-randomised trial. *The Lancet Global Health* 2 (11): e645–e653. doi: 10.1016/S2214-109X(14)70307-9.
11. Water and Sanitation Program (2011). *Economic Impacts of Inadequate Sanitation in India*. New Delhi: Write Media.
12. Bongartz, P. (2009). *'AfricaSan – One Year on from Workshop on CLTS in Africa*. Institute of Development Studies and Plan.
13. Peal, A., Evans, B., Van der Voorden, C. et al. (2009). *Introduction to Hygiene and Sanitation Software: A Selection of Approaches*. Geneva, Switzerland: Water Supply and Sanitation Collaborative Council.
14. Bartram, J., Charles, K., Evans, B. et al. (2012). Commentary on community-led total sanitation and human rights: should the right to community-wide health be won at the cost of individual rights? *Journal of Water and Health* 10 (4): 499. doi: 10.2166/wh.2012.205.
15. Bongartz, P., Vernon, N., and Fox, J. (eds.) (2016). *Sustainable Sanitation for All: Experiences, Challenges, and Innovations*. Practical Action Publishing. doi: 10.3362/9781780449272.
16. 69th Round National Sample Survey (2012). Key Indicators of Drinking Water, Sanitation, Hygiene, and Housing Conditions in India. Ministry of Statistics and Prgramme Implementation, Government of India.

3

COPING WITH WATER NEEDS: THE DEMOGRAPHIC FUTURE

GUIGUI YAO[1,2] AND ROBERT WYMAN[3]

[1] *American Studies, Yale University, New Haven, CT, USA*
[2] *Jianghan University, Wuhan, Hubei, China*
[3] *Department of Biology, Yale University, New Haven, CT, USA*

3.1 INTRODUCTION

Lack of access to water already "claims more lives through disease than any war claims through guns" [1, p. 9]. Globally, at least 1.8 billion people use a drinking water source contaminated with feces [2, p. 1] and by 2025, half the world's population will be living in water stressed areas [2, p. 1]. We have reached this situation because both population and per capita water use have increased. However, population has increased at more than double the rate of per capita water use. In the last century, population has quadrupled while per capita water use has increased by a factor of 1.75. As a result, total water use has grown by a factor of seven [5, p. 47].

The World Health Organization (WHO) lists population growth and demographic change as two of the five major challenges for water supply systems [2, p. 2]. A billion more people will be added to the globe's population in the next dozen years [3, p. 2]. The water and sanitation needs of so many more people in such a short time period may overwhelm all the water security amelioration efforts that can be achieved in a similar span of time.

Opinions on the role of population in water scarcity are contentious. Even within the same report, one can see contending opinions. For instance, the UN's 2006 Human Development Report [1], repeatedly stresses the difficulties encountered due to population growth, but then "debunk[s] the myth that the water crisis is the result of scarcity, arguing that poverty, power, and inequality are at the heart of the problem" [1, back cover]. This contention is discussed in Chapter 4 in this book, "Water, Food, and the Environment" [5, p. 47].

The use of water to provide food overwhelms all other uses. Agriculture and animal husbandry accounts for two-thirds of total extractions [6, Figure 7] and 90–92% of consumptive use [6, 7, Figure 7]; the remainder is consumed by personal use (drinking, washing, waste disposal) and industrial use. Water scarcity for agriculture results in poor food yields, which in turn leads to malnutrition (a key risk factor for disease) and to poverty (also a key risk factor for disease) for those who rely economically on the production of food and for those who have little money to purchase food. The relationship between human population and the water supply for food is the subject of Chapter 4, "Water, Food, and the Environment" [5].

3.2 PERSONAL WATER NEEDS AND POPULATION

Compared to agriculture, the personal needs (drinking, cleaning, waste disposal) for water are small. This allows room for much progress in supplying clean water and sanitation to people previously without it [4]. However, to achieve this progress, "the challenges [that had to be overcome] were huge," including addressing the interrelated factors of "poverty, instability and rapid population growth" [4, p. 2]. The 2015 UNICEF/WHO 2015 Progress Report [4] describes "population growth" as a major inhibiting factor

Water and Sanitation-Related Diseases and the Changing Environment: Challenges, Interventions, and Preventive Measures, Second Edition. Edited by Janine M. H. Selendy.

Companion website: https://www.wiley.com/legacy/wileychi/selendy/

11 times (see especially the sidebar, p. 17, "Keeping up with Population Growth," in that document).

3.2.1 Water for Drinking

From 1990 to 2015, 2.6 billion people have gained access to an improved drinking water source[1] [4, p. 7], outdistancing the global population increase in that time period of about 2 billion (Figure 3.1). Already, in 1990, 76% of the world's population had access to an improved drinking water source; by 2015 91% had that access [4, p. 2 and 6]. Nevertheless, "663 million people still lack improved drinking water sources." These are mostly in rural areas and in the least developed countries [4, p. 4].

3.2.2 Sanitation

In meeting sanitation needs, the progress during that time period has been less impressive. "The world has fallen short on the Millennium Development Goal for sanitation, leaving 2.4 billion without access to improved sanitation facilities" (Forward in [4]). While "2.1 billion people have gained access to an improved sanitation facility,"[2] this just keeps up with the simultaneous population growth of about 2 billion.

3.2.3 Open Defecation

One-eighth (13%) of the world's population [4, Figure 4 and Table, p. 25], almost a billion people [4, p. 16], have little choice but to resort to open defecation [8].

Most of these people live in regions where rapid population growth outpaces the increase in facilities. Two-thirds of people practicing open defecation live in Southern Asia; with its rapid population growth, "the number of people practicing open defecation has declined only moderately, from 771 million to 610 million. During the same period the number of people practicing open defecation has actually increased in sub-Saharan Africa" [4, pp. 16–17]. In India, as late as their 2011 census, half of the population practiced open defecation; this triggered large-scale campaigns to improve the situation [4, Box 8, p. 39].

"In Eastern Asia, the population has increased by a fifth (20 per cent) since 1990, whereas in sub-Saharan Africa it has almost doubled (94 per cent), and in Oceania it has increased by 68 per cent. In the latter two regions, the increase in access to sanitation has not kept up with population growth, with only 36 per cent of the additional population gaining access in each region" [4, p. 17].

The shortfall occurs especially in "urban areas [which] were struggling to keep pace with population growth" [4, p. 37]. "Population growth in urban areas has been a key feature of population dynamics in all regions, with the most pronounced changes taking place in four regions where the urban population has more than doubled: sub-Saharan Africa (an increase in urban population of 169 per cent), Eastern Asia (136 per cent), South-eastern Asia (115 per cent) and Western Asia (109 per cent). In Eastern Asia, gains in access to drinking water and sanitation not only kept up with, but far exceeded, population growth. By contrast, sub-Saharan Africa registered a decline in water or sanitation coverage in urban areas in 14 out of 46 countries" [4, Figure 25, p. 17].

Among the billion people that have to defecate openly are a half billion women who lack even a suitable place for managing menstruation, let alone basic sanitation and clean water facilities for menstrual hygiene management [4, p. 45].

3.3 THE TIME SCALE OF FORECASTS

Water security specialists need to have some idea of future demand and this depends critically on the size of the population served. WHO argues that "by 2025, half of the world's population will be living in water-stressed areas" [2, p. 1]. A statement like this requires knowledge of both the water situation in 2025 and the population magnitude at that time. 2025 is only nine years from the 2016 date of that forecast and so it is quite likely to be correct on both counts. With respect to the future population, the young people who will do the bulk of childbearing in the next 30 or so years have already been born. Hence population forecasts for about a generation, out to almost 2050, will possibly be approximately correct. It is fairly certain that more than 2 billion people will be added in this time frame [3]. However, forecasts at longer time scales, which are frequently necessary for foreseeing water demand and planning remediation projects, are more difficult to construct and more uncertain in their predictions. Hence, this chapter discusses the recent progression of human population and then some of the difficulties, disagreements, and divergent possibilities for the future trajectory of population [9–13]. Refer to the chapter "Water and Health: The Demographic Background" in the first edition of this book [14] for basic information for understanding the present and future state of global population and its relationship to water scarcity. The current chapter, after updating global population statistics, mostly discusses issues not covered in that edition.

[1] Unimproved *drinking water* is defined as surface sources and unprotected dug well, spring, small tank, tanker truck and bottled water. An improved drinking water source is one that, by the nature of its construction, adequately protects the source from outside contamination, particularly fecal matter [4, Figure A1-1, p. 50].

[2] Unimproved *sanitation facilities* are those that do not ensure hygienic separation of human excreta from human contact. It includes open defecation, latrines, and any that are shared. An improved sanitation facility is one that hygienically separates human excreta from human contact. Sanitation facilities shared with other households are not considered to be improved [4, Figure A1-1, p. 50].

3.4 THE CURRENT SITUATION AND THE NEAR FUTURE TO 2050

The world's population has been growing rather steadily for the past 70 years and is projected to continue on that trajectory (Figure 3.1). Looked at more closely (Figure 3.2), the number of people added each year grew from 1950 to 1990 but has now eased slightly off that peak. (The dip just before 1960, shown in Figure 3.2, is the result of the famine that occurred during China's Great Leap Forward.) The United Nations data indicates a more rapid growth than shown here. Their 2015 World Population Prospects (WPP) reports a global population of 7349 million [15, Table 1] while the 2017 WPP reports 7550 million [3, Table 1], a growth of 201 million in two years. Elsewhere [3, p. 2], it reports a growth of 83 million a year. All the data shows global population to be increasing more rapidly than it did in the 1960s and 1970s, during the peak of the "population explosion" scare [16].

While the overall world population continues to grow, the number of children born to each woman has been falling (Figure 3.3). However, it is still above the replacement level, so each birth cohort (people born in the same year) is larger than the previous one. As older women age out of their reproductive years, they are replaced by a larger cohort of young women just reaching their reproductive ages. Consequently, each year, there are more women of childbearing age than in the previous year. Thus, even if average fertility falls, the number of children born may rise. It is the balance between the rate of fertility fall and the rise in number of fertile women that determines the future birthrate. In Nigeria now, there are three times as many women coming into reproductive ages as are leaving them. In 1980, China also was facing a tripling of the childbearing population; that was part of the reason for China's One-Child Policy.

The birth and the death rates have both been falling, yet they are nowhere near balance. There are now 2.5 births in the world for every death (20/1000 births versus 8/1000 deaths) [17]. Even in China, after 35 years of the One-Child

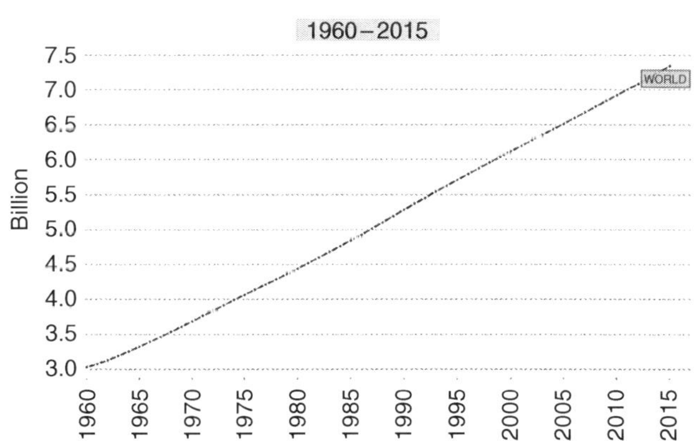

FIGURE 3.1 World population: 1960–2015. World population has increased at an extraordinarily constant rate from 1950 until now. See text for important developments that might cause population growth to increase even faster than shown here. *Source*: World Bank: data available from worldbank.org/indicator/SP.POP.TOTL. See also US Census Bureau, International Data Base, August 2017 update, for a century long series also showing a constant rate if increase and projection of this increase until 2050. Available from: https://www.census.gov/programs-surveys/international-programs.html.

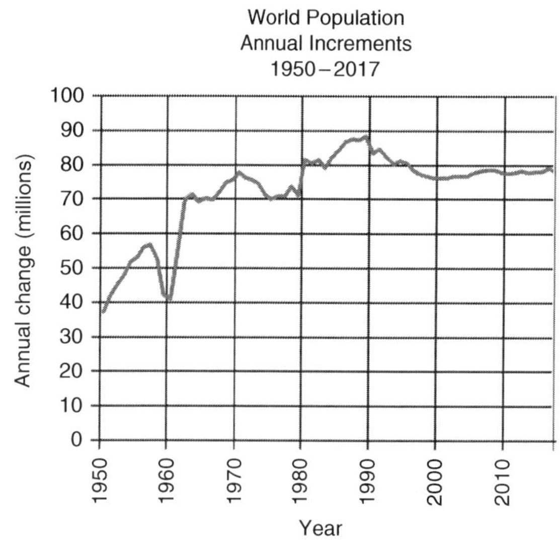

FIGURE 3.2 World population: Annual increments 1950–2017. The annual growth of the global population is now even greater than it was during the "Population Bomb" scare of the late 1960s and 1970s, yet public concern is much less now than it was then. UN data show population growing at 100 million a year during the last two years and the Population Reference Bureau shows an increase of 118 million in the last year. Both indicate much faster growth than shown here. *Source*: US Census Bureau, International Data Base, August 2017. Update available from: https://www.census.gov/programs-surveys/international-programs.html.

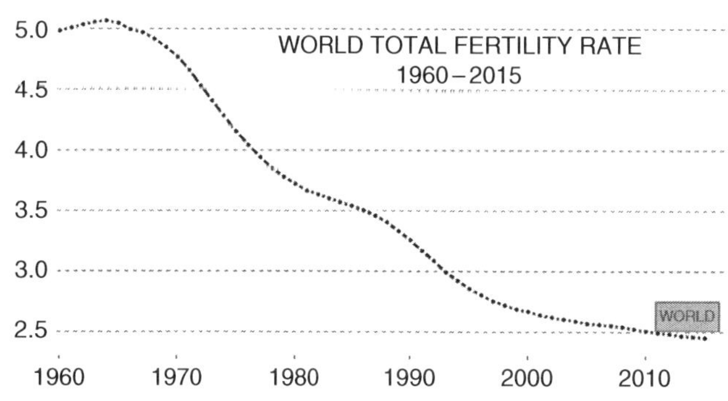

FIGURE 3.3 World total fertility rate 1960–2014 (TFR for lifetime births per woman). At 2.5 births per woman, population grows by 25% per generation due to just this factor. Since population has been growing in the past, there are now more women than in the previous generation; this raises the population growth rate above 25%. *Source*: World Bank. Available from: http://data.worldbank.org/indicator/SP.DYN.TFRT.IN.

Policy, there are almost twice as many births as deaths (in the most recent data available, 2014, there were 12.37 births and 7.16 deaths per 1000 population) [18]. Hence, the Earth is a long way from population stabilization.

3.5 THE FUTURE SITUATION TO 2100

The view of the future that is most often referred to is the United Nations (UN) biennial series of World Population Prospects. The UN calculates possible future population levels under a variety of scenarios. In the 2010 Revision, for instance, the UN provided eight projection variants [10]. At one point, The International Institute for Applied Systems Analysis (IIASA) presented 27 different scenarios [19]. Five of the UN projections are shown in Figure 3.4.

The constant fertility variant (top line in Figure 3.4) assumes that fertility remains at its current level. Each of the other scenarios assume a continuing decrease in fertility by various amounts (lower lines in Figure 3.4). There is no consideration of the possibility that fertility may rise.

The "medium variant" assumes that countries will experience the same rate of fertility fall as other, now more advanced, countries experienced when they were at that socioeconomic level; in short, fertility will keep falling in the same way it has in the recent past. This expectation is based on the 150-year history of falling fertility rates. Rates fell first in the West from 1870 to 1930 and then, after WWII the rates began to fall in the developing world, with an acceleration due to the development of the Birth Control Pill and other modern contraceptive technologies and the expansion of international aid and organizations promoting family planning [20].

This assumption of constantly falling fertility may not hold for the future. For instance, fertility has been declining in Africa over the past decade, but only at one-quarter of the rate at which it declined in Asia and Latin America in the 1970s, when these regions were at a comparable stage of the fertility transition. Indeed, in some African countries, the decline seems to have stalled or even reversed [11].

To accommodate uncertainty about future fertility levels, the UN projections also provide "high" and "low" variants, showing the results if global fertility turns out to be 0.5 child more or less than in the medium variant. Another variant assumes that in every country fertility immediately goes to replacement level. Note the large range of possible outcomes, covering a factor of 4.5, from 6 to 27 billion possible people in 2100.

Projections also vary between successive biennial rounds of estimation. For example, the projections made in 2010 are higher than those made two years earlier in 2008; just two

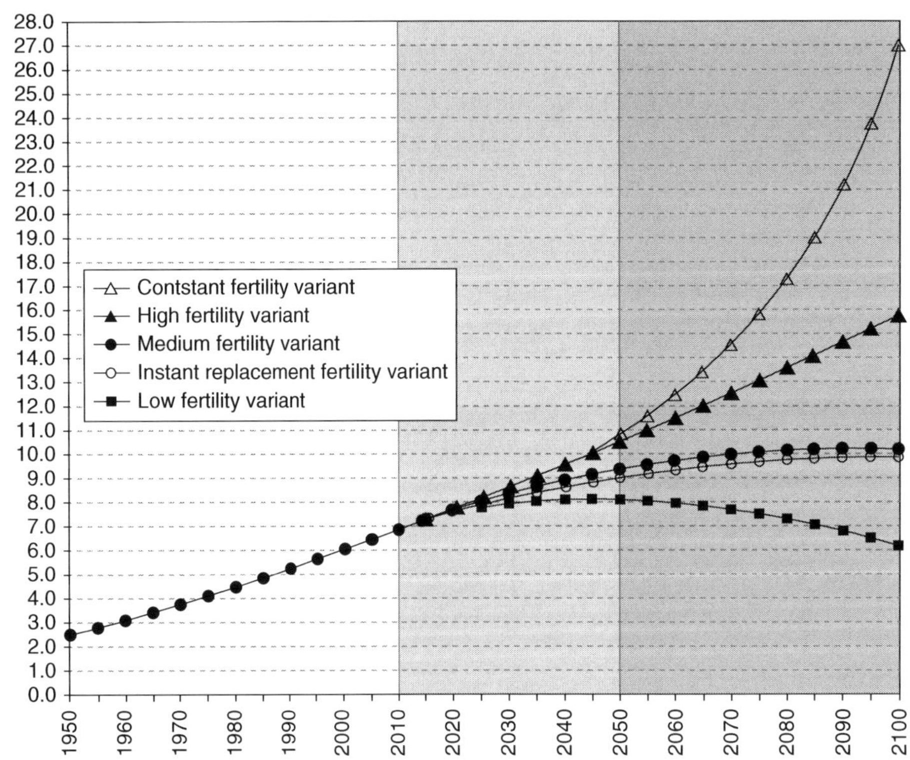

FIGURE 3.4 United nations projections of future world population. Historical estimates and projections of possible futures under different assumptions. Scale: billions. The 2012 revision showed slightly higher levels of projected population. Starting in 2015, the UN changed its method of projection to a probabilistic one and no longer presented a "Constant Fertility" projection. *Source*: United Nations, Department of Economic and Social Affairs, Population Division World Population Prospects: The 2010 Revision. Updated 15 April 2011.

years later, the 2012 Revision apologized for underestimations in the 2010 series and raised the "medium variant" projection for 2100 by almost three-quarters of a billion [21] and raised the "constant fertility" projection by over a billion people. Forecasting water needs becomes very difficult when the future level of population is so uncertain.

One point of agreement in all the projections is that, in order for world population to level off, fertility must continue to decrease and must decrease by 20%. All the projections that reach a leveling out of population (let alone a decline) assume a rapid decrease in fertility. Whether that will actually happen is unknown.

Which of these very divergent scenarios should one consider most likely? In the current authors' opinion, the most conservative assumption is that fertility continues as it is now (constant fertility variant: top line in Figure 3.4). Current families have made their choices based on current conditions and that is what sets the current Total Fertility Rate (TFR). Unless we can predict future socioeconomic changes and understand what the demographic result of these changes would be, we have no reason to expect that fertility will be either higher or lower than now. The assumption that fertility will remain at its current unprecedented low, much lower than ever before in human history, may be an optimistic assumption. Millions of years of biological evolution and thousands of years of cultural evolution (e.g., "be fruitful and multiply") have pushed humans to high reproduction rates. Attainment of the current low depends on many factors that may not be sustainable. In a later section, we consider factors that may mitigate against a further fall in fertility.

Under this assumption of constant fertility, the 2010 UN World Population Prospects indicated a quadrupling of population, from 7 to 27 billion (raised to 28 billion in the 2012 Revision) by the end of this century. Such a quadrupling of the global population is a dire possibility for the prospect of avoiding and/or remediating all kinds of human and environmental misery.

Although the UN does not explicitly state this, most users assume that the "medium variant" is the most likely. The assumptions underlying the medium variant lead to some very optimistic projections. Figure 3.5 is from the UN's 2012 medium variant projection of TFR for Nigeria [11]. The vertical dotted line is the date of the projection (2012). The actual fertility rate, graphed to the left of the dotted line, shows that Nigeria had a high TFR of about 6.4 until about 1970, then experienced a short-lived increase before falling back and possibly stabilizing at 6.0, just below the levels of the 1950s and 1960s. In the projection, the solid line to the right of the dotted line, the stabilization continues for a few years but then a dramatic decline takes over and Nigeria's TFR falls to the replacement level by 2100. Is such an abrupt change in trajectory likely? Only the future will tell.

All projections start with what is believed to be the current population. If the current population estimate is inaccurate,

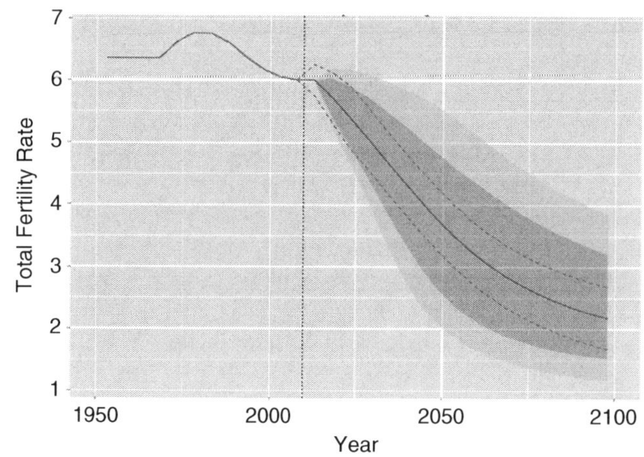

FIGURE 3.5 Total fertility rate projection for Nigeria. The UN's 2012 World Population Prospects includes this projection for Nigeria. Even with this startling drop in fertility, there is still a 90% probability that Nigeria's population in 2100 will exceed 532 million, a more than three-fold increase. Without the fertility decrease shown, population would quintuple. *Source*: Gerland et al. [11].

then all projections into the future may amplify this original error. Although projections like those shown in Figures 3.5 to 3.7 include uncertainties for the future, none show error bars for the current estimate. The effect of any such inaccuracies on the projections are unknown.

The Total Fertility Rate estimates what completed fertility will be for women who are currently in their reproductive years. When calculated for any particular date, the TFR assumes that currently younger women will continue to have children at the same rate that currently older cohorts have them. However, women may change their reproductive behavior as time goes on. The actual number of children born in a lifespan, which can only be known after a woman has completed her reproductive years, may not match the number that was predicted when they were younger. At the present time, women are increasingly delaying childbearing until older ages. In fact, some recent research estimates that that the TFR underestimates the actual experience of populations by some 20% [22, p. 32]. If this bears out, the population may be growing much more rapidly than currently presumed.

Satellite imagery is now being used to produce what may be the most detailed population maps ever produced for developing countries. The first of these, for Nigeria, shows that many urban areas are more populated than the census data indicates and other areas that are both under- and overestimated [23].

3.6 THE FUTURE MAY SURPRISE US

For the foreseeable future, all regions of the world are going to be much more crowded than they are now. Except for that, the future is unknowable. For many years, demographers have expected a slowdown of population growth. A large reason for this optimism is the spread of education, especially for women [13]. However, history does not always unfold as we expect. There is, as yet, no such slowdown (Figure 3.1). There are many events that may put a continuation of the decline in fertility in doubt. The current chaos in the Middle East was unexpected, as were the continuing difficulties in much of Sub-Saharan Africa. Also, not foreseen were the continuing world financial crises caused by mortgage securities fraud in the US, the rise of conservative religiosity in much of the world (including the US), and the elimination of US foreign aid for most family planning organizations. As the numbers of people in the richer, low fertility cultures decline, the proportion of world population born into the poorer, high fertility cultures rises. This pushes the global fertility rate upward. These factors provide reasons to doubt a continuing rapid fertility decline.

Much of the past decline in global fertility is due to its decline in the massive Chinese population, but that will now slow or end. Since China's fertility is already at a very low level, there is little room for further decline. In addition, the government has now relaxed its One Child Policy, allowing all couples to have two children. From November 2013, couples, only one of whom was an only child, were allowed to have a second child. Even though more than 11 million couples were eligible to have a second child the number of births in 2015 declined to 16.5 million, which was about 320 000 less than the previous year.

From January 1 of 2016, all Chinese couples are allowed to have two children. Leading Chinese demographers write, "The lukewarm response of couples [to the 2013] relaxation confirmed that very low fertility in China is more the result of choice than of policy restrictions." Hence, "This end of China's one-child policy is unlikely to increase births in China by a significant number in the years to come" [24]. However, it is likely to slow the rate of decline in Chinese fertility. In fact, in 2016, 18.4 million babies were born in China, an increase of about 8% from 2015 and the highest number since 2000. However, 2016 was the Year of the Monkey. Many Chinese believe "that people born in the year of the monkey are 'smart and confident'" [25]. Most commentators attribute the upsurge in births more to the Year of the Monkey than to elimination of the One Child Policy. Many of those births may also have been catch-up births by families that have long wanted more children; once this catch-up demand is satisfied, the number may drop again. The fertility rate that had stayed between 1.5 and 1.6 through the years 2000–2015 rose to 1.7 children/woman [26]. In any case, 1.7 is still far below the 2.1 needed to keep China's population from eventually falling.

A possibly more important factor may be that the "missing girls" in China have been found. The sex ratio of *reported* births has been strongly biased in favor of males. The normal sex ratio at birth is 105–106. In China, in1990, it amounted to 112 males for every 100 females. In 2000 and 2010 it was 117 and 118. The gap of up to 13% indicates the proportion of females that were presumed "missing," largely via sex selective abortion. Estimates of the total number of missing girls ran as high as 30 million. However, an analysis of the 2010 census, just reported, showed that about 75% of the girls that appeared to be missing in the 1990 census have appeared in either the 2000 or 2010 census [27]. It appears that local officials colluded in allowing the non-reporting of many female births. These girls were subsequently registered, but only when necessary. Village children can attend the local primary school without having a registration. However, when they go to middle school at about 12 years old they must have an ID card (hukou); some get registered then and, if born in the 1980s, appeared in the 2000 census. Others wait until marriage when over 22 (minimum legal age) to be registered and, if born in the 1980s, do not appear until the 2010 census. It can take 20 or more years before "missing" girls finally get registered. That explains the long delay in figuring this out.

The significance of this finding, for this discussion, is that China has had a higher birthrate than reported and has millions more young women than previously estimated. Each of these is a potential child bearer. The birthrate will almost certainly be higher than current projections.

There are a number of other countries where we might expect an increase in the birth rate. Iran, whose Minister of Health received the UN's 1999 prize for developing one of the best family planning programs in the world [28], has reversed course and eliminated its family planning program [29, 30]. Into the 1980s, Iran's TFR was about 6.5 and the population grew at over 3% per year. Beginning in the late 1980s, the Health Ministry introduced a comprehensive and effective family planning program. By 2005–2010, the TFR had dropped to below 2, but in 2012, Iran reversed direction and eliminated its family planning program. Ayatollah Ali Khamenei wants Iran's population doubled. Parliament is expected to approve bills outlawing voluntary sterilization and restricting access to contraceptives [31].

Bangladesh has one of the world's highest rates of child marriage, but it had been gradually dropping under steady pressure from the government. According to UNICEF, as of 2000, 65% of girls were married before the age of 18 and 38% were married before 15. Now those rates have dropped to 52 and 18%. However, in 2017, Bangladesh's Parliament softened its landmark law against underage marriage, the Child Marriage Restraint Act, which has been in existence since 1929. This could roll back the country's decades-long campaign to curtail teenage pregnancy. A provision in the new law allows girls under the age of 18 to marry in some circumstances. The change was met with praise from Islamist groups, which said that it fell more in line with traditional religious practices [32].

Even more worrisome is the situation in Sub-Saharan Africa. "Optimism that a rapid fertility transition was imminent, a common view among scholars in the 1990s, was dashed by survey evidence that steadily accumulated through the 2000s. The future course of fertility in sub-Saharan Africa remains highly uncertain" [33].

Among other problems, Sub-Saharan Africa has some of the highest rates of child marriage in the world. While marriage of girls aged 15–18 has declined over time, more than one-third of girls continue to marry before 18 in well over half the countries. However, the proportion of girls who marry before their 15th birthday has remained unchanged for 20 years throughout much of the region and has increased in a few countries [34, Abstract].

Among water-scarce regions, the Middle East is the most populous and it is in turmoil. Population is rising in the "Arab World" (22 countries, World Bank regional designation, 2016 population 406 million) at about 2% per year [35, 36]. The total fertility rate of the Arab World, which had been falling rapidly, has now stagnated at about 3.4 children/woman [37]. If continued, both statistics will result in almost a 75% increase in the reproductive age population in a generation. "The number of people in the region living in countries at high risk of conflict is projected to grow from about 250 million in 2010 to over 305 million in 2020; the number is likely to double over 2010–2050" [38, Figure 8.6, p. 176].

One-third of the current Arab World population is in the reproductive ages of 15–29 years. Another one-third is below 15 years [38, pp. 5, 7, 22]. The youth population is growing fast, but unemployment, poverty, and marginalization are growing even faster. The economy cannot accommodate the current youth population, much less a rapidly increasing one. Unemployment among youth in Arab countries is the highest in the world, 29% in 2013 versus 13% worldwide [38, p. 31]. Almost half of young Arab women looking for jobs fail to find them (against a global average of 16%) [39]. The region needs to create more than 60 million new jobs in the next decade to absorb the large number of workforce entrants and stabilize youth unemployment [38, p. 32].

The global drop in fertility has been largely caused by a rising age at marriage and the attendant drop in the possible number of children born during the marriage. This is not necessarily a sign of progress, or educational conditions, or of a changed attitude about family and childbearing. Rather than being caused by improved economic conditions, it is often the opposite – the inability to support a family, high unemployment, the inability to provide a dowry, and the high cost of housing [38, p. 33]. This rise in the marriage age, without an obvious change in cultural attitudes, is most pronounced in the Arab world where conditions are worsening and the economy, in places, is collapsing. Almost half the men aged 25–29 in the region are still unmarried, the highest proportion among developing regions and up from 37% a generation ago [38, p. 33]. The same is true for women; in Lebanon, for instance, the average age of a woman's first marriage climbed from 21 in 1970 to 32 in 2008 [38, p. 33].

This imbalance between population growth and the ability of the economy to accommodate it contributes to the instability in the region. "Never before has the region had such a large share of youth; youth of ages 15–29 make up around 30 percent of the population, or some 105 million people. Rapid population growth has placed massive pressures on societies and the entire infrastructure of Arab States. It is youth who often translate broader social problems into an explosive and radicalizing mixture" [38, p. 22]. Although home to only 5% of the world's population, in 2014 the Arab world accounted for 45% of the world's terrorism, 68% of its battle-related deaths, 47% of its internally displaced, and 58% of its refugees [38, Figure 8.5, p. 176].

Egypt, the most populous country in the region (population, 95 million) may be instructive. From a TFR of 6.6 in 1960, fertility fell to 3.1 in 2006 [40]. There was not a single year in which the TFR increased. However, since 2006, the TFR has risen a half a child to 3.6 and has been maintained at that level from 2011 to 2015. A half child rise in fertility can have an enormous impact if it is widespread and maintained. Notice, from Figures 3.4 and 3.6, the billions of people difference between the "medium" variant and the "high" (+0.5 child) variant.

The severe disruption in the Middle East prevents the gathering of fertility data in many countries. Afghanistan's first and only census occurred in 1979 [41], so any current estimates have enormous uncertainties. UNESCO estimates that its TFR is 5 and that 44% of its population is 14 and younger [42]. If this estimate is even approximately correct, this large pre-fertile population will lead to a large increase in the number of reproductive age women and a comparable increase in the birthrate.

Of special interest is the effect of war and displacement on fertility. There is one study of Iraq from 1997 to 2010 that showed a rise in early marriage among less-educated women, with adolescent fertility rising over 30%, while the TFR remained stable at about 4.5 children/woman [43]. Iraq's fertility had been dropping rapidly from 7.4 children/woman in 1969 to 4.6 in 2004, but during the US invasion (2003–2011), it stalled at that level [44].

Syrian refugees living in camps in Jordan have a crude birth rate of 42 births per 1000 people. That is near double the birthrate in either Syria or Jordan [45, 46] and a higher rate than even Nigeria, which has one of the fastest growing populations in the world. Before the 2011 start of the Syrian Civil War, 13% of all marriages were to girls under the age of 18 in both Syria and Jordan. By 2014, among Syrian refugees in Jordan, the rate more than doubled to 32% [47–49]. Given the large number of currently displaced people in the world, these numbers portend not only a rapid increase of people in the camps but a significant contribution to global population growth.

The rise of political and religious conservatism in the US has led to support of pro-natal policies, which act to increase population. Under the presidency of Donald Trump, the US has returned to the Mexico City Policy [50], which, under the umbrella of forbidding US funding of abortion, actually blocks the vast majority of US bilateral global health assistance, of which approximately $600 million has been used for family planning [51]. In the US, there are currently about one million abortions each year [52]. If the US makes abortion illegal, possibly another million births will occur. If the US defunds Planned Parenthood, some fraction of its 2.5 million patients [53], who depend on Planned Parenthood for contraceptives and family planning advice, may become pregnant and bear a child. If the US expels immigrants, back to locales in countries where a high birthrate is socially sanctioned and where contraception may be less available, their birthrate may rise.

The current authors have no more of a crystal ball than anyone else and do not want to be doomsayers. However, we believe it is important to be aware of the myriad situations that may not allow the current optimistic projections to eventuate.

3.7 PROBABILISITIC PROJECTIONS

Since the 1990s, there have been an increasing number of attempts to place error bars on population projections. Recent efforts use probabilistic methods to produce a spread of possible future populations with percent likelihoods (confidence intervals) that the population will be within various population ranges. Figure 3.6 shows the UN's 2017 projection for world population. (Figures 3.5 to 3.7 are also probabilistic projections.) The range of possible populations diverge over time within each forecast and also diverge between the projections of different research groups. A major point of contention is whether population will level off within this century. The UN statistical group believes that is an unlikely scenario [11] while the group at IIASA argues that it is a near certainty (88% probability) [54]. None of the current readers may be alive at the end of the century to know which will be correct.

In the traditional high and low projections, *every* country has one-half child more or one-half child less. All the deviations from the median add in the same direction, producing projections that are much higher or much lower than the median. The new probabilistic projections realize that some countries will exceed their median value while others will fall below. The deviations from the median will somewhat cancel each other out. Hence, the probabilistic projections present a more restricted range of likely outcomes.

This consideration does not apply to the "constant fertility" projection, which assumes that each country will maintain its current fertility and there is no assumption that all countries will deviate in one direction or the other. In the future, fertility likely will vary from its present levels, but as

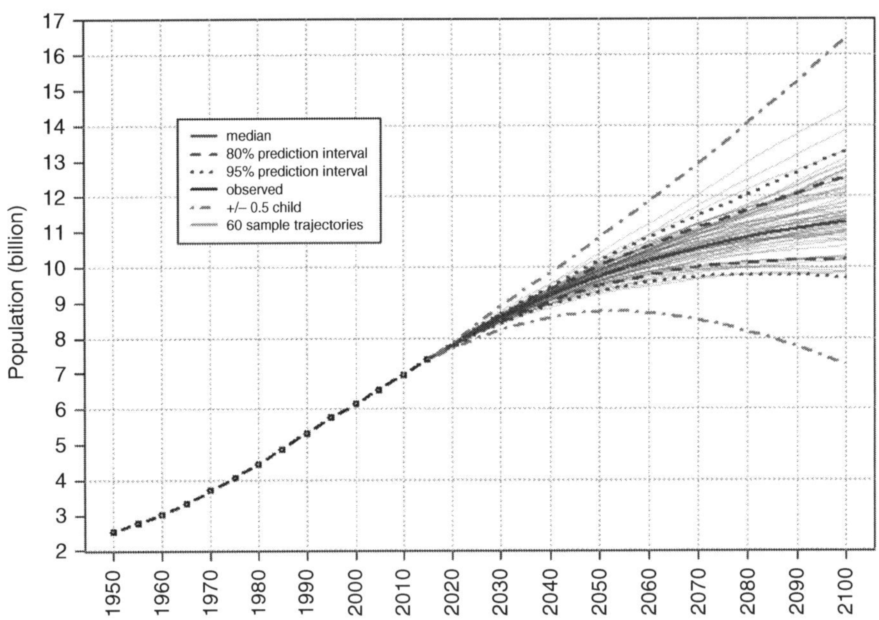

FIGURE 3.6 Total world population projection, UN 2017. In this "probabilistic" projection, the many light lines are iterations of a computer simulation of population growth with different probabilities assigned to events for each iteration. The median projection (half the iterations resulted in a higher population projection and half a lower projection) is the solid line in the center; 80% of projections lay between the dashed lines surrounding the median and 95% of the projections lay within the dotted lines. The outermost lines are the results from the UN's traditional high and low variant projections. *Source*: United Nations, Department of Economic and Social Affairs, Population Division World Population Prospects: the 2015 revision. Available from: https://esa.un.org/undp/wpp/Graphs/Probabilistic/POP/TOT.

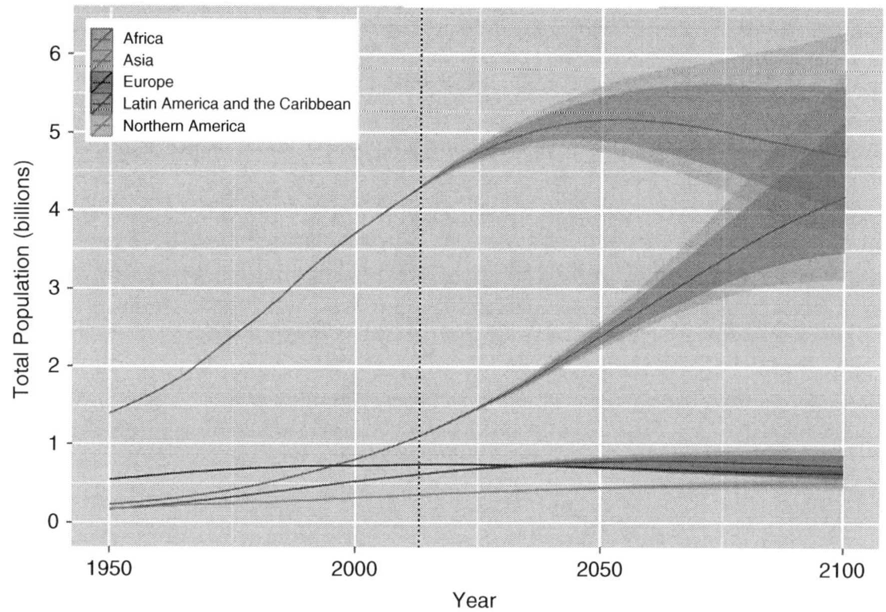

FIGURE 3.7 UN projections of the future population of the five continents. More than half of the anticipated growth in global population between now and 2050 is expected to occur in Africa [3]. Europe is the only region expected to have a smaller population in 2050 than in 2017. After 2050, essentially all the growth is expected to occur in Africa; the other continents will have a stable, or even in Asia, a declining population. By the end of the century, Africa will have as many people as Asia. *Source*: Gerland et al. [11]. See also Figure 3 in [3], which depicts an updated projection from the UN's 2017 World Population Prospects.

long as there is no overall increase or decrease, the projection should be approximately correct.

3.8 HIGH FERTILITY–LOW FERTILITY

Developing countries often have difficulties managing water scarcity and/or water quality. Currently, there are about 4.5 times as many people living in less developed countries than there are living in more developed countries. This ratio will become even more extreme, since essentially all population growth is in the less developed countries while developed countries face a decline. We are currently in a period of heavy migration; migration from poor and strife-torn regions to more developed and more stable regions provides a significant degree of equalizing population growth. Because of the heavy migration currently taking place in many parts of the world, many regions of the world are going to be much more crowded than they are now.

Some cultures have adopted contraception resulting in a lower fertility rate, even one below replacement level. Other cultures that have only hesitantly adopted contraception, are experiencing fertility rates that remain far above replacement. As time goes on, low fertility cultures become a smaller proportion of the world's population while high fertility cultures increase their proportion. This acts to maintain a high global fertility rate.

These shifting population proportions are shown strikingly in recent projections (Figure 3.7), which indicate that by 2100 the population projections for Africa will overlap with those for Asia. This is the case even with the assumption of rapid fertility declines generally in Africa (similar to those for Nigeria in Figure 3.5). In 1950 Africa's population was only about a half billion while Asia's was three times that size. By 2012 the gap grew to a factor of almost four (1.1 billion versus 4.3 billion). However, with the impending stabilization of the East Asian population while Africa's population growth accelerates, the gap narrows and given the large uncertainties for both areas, the projections largely overlap by 2100. In comparison, the populations of Europe and the Americas shrink dramatically as a percentage of the total population.

At the country level, an important example of the shifting population proportions is the China/India comparison. China is now a low fertility country. Its TFR has been falling since 1970 and has been below replacement level since 1990 [55]. It has since fallen to approximately 1.6 children per woman [17]. India, although having had a falling fertility rate since 1960, still has an above replacement fertility level of 2.3 [17, 56]. As a result, China is expected to lose about 34 million people from 2016 to 2050 (maybe less as this estimate is from before the "missing girls were found), but India is expected, in that same time period, to gain 379 million people [17]. India will take over as the world's most populous country. Hence, even though both countries have had falling fertility for about half a century, their sum population will still add almost 350 million to the globe's population between now and 2050.

Many countries have very young populations; over 25% of the world's population is less than 15 years old and, in the least developed countries, 41% are under 15. In these countries, with many more young people coming into their reproductive ages than older people who are leaving their reproductive ages, the population will continue to explode. In the 2016 to 2050 timeframe, Nigeria's population is expected to increase by 211 million, Pakistan by 141 million, and Indonesia by 101 million. Africa's population is expected to double [17].

As a result of the current age structure, with a predominance of young people in developing countries, the Population Reference Bureau expects the combined population of the world's 48 least developed countries to double by 2050 and global population to increase by 33% by 2050 [17].

3.9 ECONOMIC CONSIDERATIONS

For the sake of people's health and the planet's environment, the water supply should be improved and sanitation facilities increased. This, of course, requires money. Some support can be provided by the international community, but the main burden will always be borne by the individual countries themselves. Unfortunately, the countries most in need are often the poorest countries – and also the ones with the most rapid population growth. The correlation between high population growth rates and poor economic performance (poverty) is not coincidental.

It used to be accepted that "development is the best contraceptive," meaning that, if a country develops economically, then its birthrate would fall automatically as a result [57]. Recent empirical evidence shows that the reverse is true [58–60]. Contrary to the theory that economic development initiates fertility decline, major fertility drops have *preceded* economic booms in all the "economic tigers," countries that have had spectacular economic growth [61]. Examples include China, Korea, Taiwan, Indonesia, Thailand, Malaysia, Japan, Mexico, and Ireland. The more rapid the fertility decline, the more rapid the subsequent economic rise. Countries like India, which have experienced a slower fertility drop, have experienced a slower economic rise [62]. Each country is a special case with different contributing factors (e.g., economic liberalization) and different triggers for the fertility decline. For example, in Mexico and Ireland, the trigger was legalization of contraception, in Japan it was legalization of abortion, and in Korea and Taiwan it was a successful national family planning program.

The major economic effect of fertility reduction (called the Demographic Dividend) works via an improvement in the dependency ratio: the ratio of working age people to dependents. In poor countries with birth rates of six or seven children per couple, and few or no jobs available for women, the dependency ratio can be very high. Since life expectancy is low in poor countries, there are not many old-age dependents. Most dependents are children. With many dependents, most income goes to support the family and savings are minimal. Savings are the source of funds for investment, so, without savings, there is no capital to invest. The economy goes nowhere.

Fertility drops occur rapidly. For example, a halving of fertility took just a decade in China and occurred even more rapidly in Japan. The large cohorts of children born before the drop remain dependents until they grow up; the younger cohorts behind them are much smaller. In the course of 10–20 or so years, the large cohort of children enters working ages. As they enter the workforce, the dependency ratio improves and this is the period when the economy takes off. For East and Southeast Asia, it is estimated that the fertility decline accounted for 44% of the region's economic boom [63]. A fertility decline seems essential for the economic take-offs, but does not act by itself. For example, in China, the economic liberalization of Deng Xiaoping (the paramount leader from 1978 until 1989) provided jobs for the increased number of workers. In other countries a variety of factors act in synergy with the demography: for example, improved governance, foreign investment and technology transfer.[3] The implication for policy is that if a country wants to have sufficient financial resources to solve the host of development problems that they face, of which water is only one, a moderation of rapid population growth is a prerequisite.

3.10 UNMET NEED FOR CONTRACEPTION

Until less than a century ago, contraception was banned almost everywhere on moral, legal, and religious grounds. Technological advances in contraception and the spread of its use since WWII has been nothing less than spectacular. Now, about 60% of the world's women practice contraception [64]. However, in some parts of the world, couples are not availing themselves of modern methods to achieve their desired family size.

Even as far back as the 1970s and 1980s, people in all the less developed countries were bearing more children than

[3] The above description of the mechanism of economic development has gained a large degree of acceptance. However, it requires much more discussion than can be devoted to it here. Some have argued that other East Asian populations, not subject to the One Child Policy, also had rapid fertility drops and economic "miracles." However, these were small countries with very large American and European cultural influences and investments: South Korea and Taiwan from the US, Hong Kong and Singapore from the British. Yong Cai found China's rapid fertility drop in the 1970s was due to Wan, Xi, Shao (the population policy at the time advocating Later Marriage, Longer birth interval, and Fewer total children), but its later and slower drop under the One Child Policy could be explained by socioeconomic factors [55].

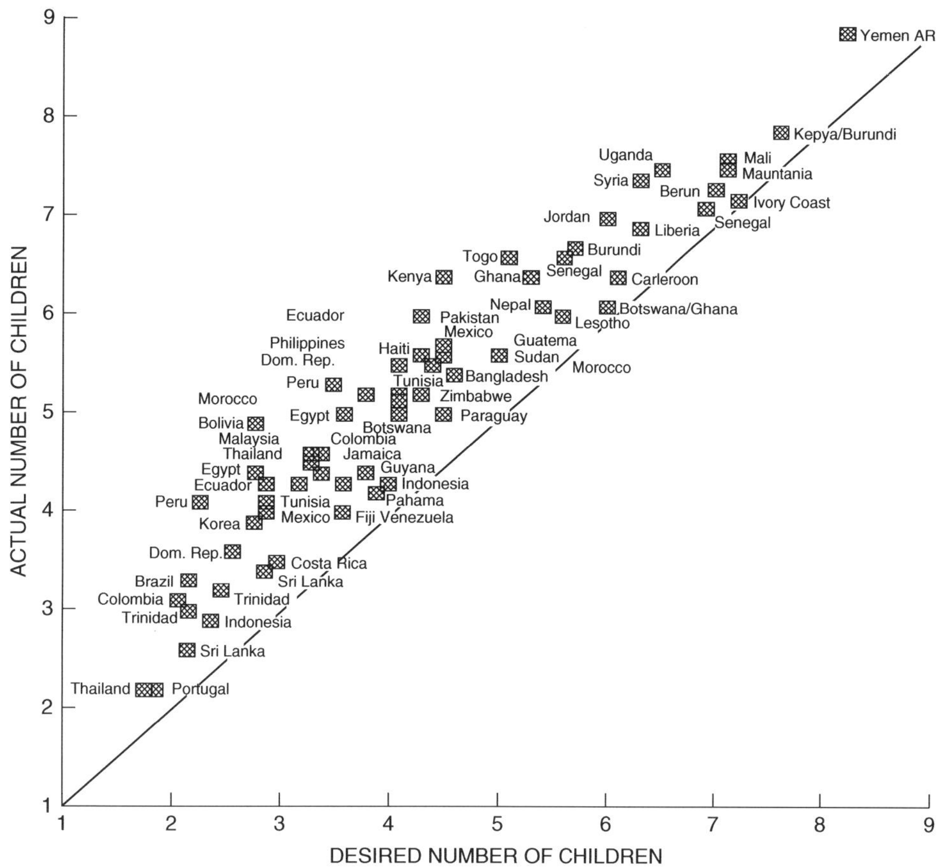

FIGURE 3.8 Actual fertility above desired fertility. The Total Fertility Rate (actual) compared to the Desired Total Fertility Rate for developing countries. The diagonal line (added by RW) shows where points would lie if actual and desired fertility matched. All countries are above the line; actual fertility is higher than desired fertility. *Source*: Data from the World Fertility Surveys and from the Demographic and Health Surveys, 1974–1991. Adapted from [58].

they wanted (Figure 3.8) [65]. Now, with the extraordinary spread of contraception, fertility has come down (Figure 3.3). The number of children that families want has also come down. Nevertheless, many women are still having unplanned pregnancies and bearing children that they do not want.

Because of its size, the most populous country in Africa, Nigeria, is an important case. As is usual in Sub-Saharan Africa, Nigerian women want large families; their ideal number of children averages to 6.5 [66, p. 85]. This is even larger than its current TFR of 5.5 children [63, p. 64]. However, within these generally high fertility preferences, there are still many women who are bearing more children than they want. The high levels of abortion, which is illegal and therefore extremely dangerous in Nigeria, attests to a large number of unwanted pregnancies.

There are several ways of measuring this excess fertility. The most commonly used number is "unmet need": the number of fertile women who want to delay childbearing or who do not want any more children but are not using contraception. At present, only about one-third of the potential demand for family planning in Nigeria is being met by using modern methods [63, p. 106]. The total fertility rate is 15% higher than it would be if unwanted births were avoided. About 40% of Nigerian women undergoing abortion experience complications serious enough to require medical treatment. Nevertheless, Nigerian women have about one abortion for every five and a half live births [67]. This indicates a strong desire to control their own fertility. Hence, even in a culture with high fertility desires, there is room for the increasing adoption of contraception. Because fertility desires are strongly influenced by local norms, any reduction in fertility tends to spread as couples see the social acceptability and desirability of having fewer children [68, 69].

3.11 IMPLICATIONS FOR POLICY

As with all environmental problems, there are a variety of strategies for amelioration. These different strategies do not necessarily compete with each other, but complement each other. For environmentalists, an important consideration is which "fix" is inexpensive, technologically feasible, and politically acceptable. Amelioration of population growth may satisfy these requirements better than other alternatives.

The rapid and voluntary worldwide fall in fertility indicates that this is a change wanted by the world's citizens, not one that is perceived as entailing great sacrifice (for the partial exception of China, see [70] and references therein). The high levels of desire to delay or stop childbearing and the high levels of abortion also show how people around the world want to be able to control their fertility.

The cost of contraception is minimal and more than pays for itself via its boost to economic development. Contraceptive technology, while not perfect, has proven acceptable to the very demanding consumers of industrialized countries and, increasingly, to those of developing regions.

In short, reducing population growth is a win–win proposition from many angles. Environmentalists would do well to consider efforts to assist fertility reduction as perhaps the most practicable, effective, and welcome means of ameliorating environmental problems.

REFERENCES

1. United Nations Development Programme, Human Development Report (2006). *Beyond Scarcity – Power, Poverty and the Global Water Crisis*. New York: UNDP.
2. WHO Drinking-Water Fact Sheet, Reviewed November 2016. Available from: http://www.who.int/mediacentre/factsheets/fs391/en. Retrieved 26 November 2016.
3. United Nations Department of Economic and Social Affairs, Population Division (2017). World Population Prospects. Key findings and advance tables. 2017 revision.
4. WHO/UNICEF (2015) 25 Years Progress on Sanitation and Drinking Water, 2015 Update and MDG Assessment. Data and Country Files available at WHO/UNICEF (2015b).
5. Wyman, R.J. and Yao, G. (2019). Water, Food, and the environment, Chapter 4 in this volume.
6. OECD (2017). Global Forum for Food and Agriculture, Berlin.
7. Hoekstra, A.Y. and Mekonnen, M.M. (2012). The water footprint of humanity. *Proceedings of the National Academy of Sciences, USA* 109: 3232–3237.
8. Sengupta, S. (2002). In Bombay, public indignity is poverty's partner. *New York Times*, February 9, 2002.
9. Wyman, R.J. (2003). The projection problem. *Population and Environment* 24: 329–337.
10. UN, World Population Prospects, 2010, 2012, 2015 and 2017. Revisions.
11. Gerland, P., Raftery, A.E., Ševčíková, H. et al. (2014). World population stabilization unlikely this century. *Science* 10 October 2014 346 (6206): 234–237. doi: 10.1126/science.1257469.
12. Lutz, W., Butz, W., and Samir, K.C. (2014). Population growth: peak probability. *Science* 346: 561. doi: 10.1126/science.346.6209.561-a. Available at: http://www.sciencemag.org/content/346/6209/561.1.short.
13. Lutz, W., Butz, W.P., and Samir, K.C. (eds.) (2014). *World Population and Human Capital in the 21st Century*. Oxford: Oxford University Press.
14. Wyman, R. (2011). Water and health: the demographic background. In: *Water and Sanitation-Related Diseases and the Environment* (ed. J.M.H. Selendy). Wiley-Blackwell.
15. United Nations Department of Economic and Social Affairs, Population Division (2015). World Population Prospects. Key findings and advance tables. 2015 revision.
16. Ehrlich, P. (1968). *The Population Bomb*. New York: Ballantine Books.
17. PRB (2016). *World Population Data Sheet*. Washington, DC: Population Reference Bureau Text available from: http://www.prb.org/Publications/Datasheets/2016/2016-world-population-data-sheet.aspx.
18. China Statistical Yearbook (2015). Available from: http://www.stats.gov.cn/tjsj/ndsj/2015/indexeh.htm.
19. Lutz, W. (1994). *The Future Population of the World. What Can We Assume Today?* London: Earthscan Publications and Revised 1996 Edition.
20. Coale, A. and Watkins, S. (1986). *The Decline of Fertility in Europe*. Princeton University Press.
21. Tabulated in Lutz et al. (2014), op.cit. Ref. [13], pp. 528–530.
22. Myrskla, M., Goldstein, J.R., and Cheng, Y.-h.A. (2013). New cohort fertility forecasts for the developed world: rises, falls, and reversals. *Population and Development Review* 39: 31–56.
23. Tollefson, J. (2017). Satellite images reveal gaps in global population data. *Nature* 545: 141–142.
24. Feng, W., Baochang, G., and Cai, Y. (2016). The end of China's one-child policy. *Studies in Family Planning* 47 (1): 83–86.
25. Xinhua (3 March 2016). China Focus: hospitals under pressure amid new year baby boom. Available from: http://news.xinhuanet.com/english/2016-03/03/c_135152826.htm.
26. National Health and Family Planning Commission of the PRC (2017). Fertility rate edges closer to two children. Available from: http://en.nhfpc.gov.cn/2017-03/11/c_71341.htm. Accessed 14 May 2017.
27. (a) Cai, Y. and Laverly, W. (2003). China's missing girls: numerical estimates and effects on population growth. *China Review* 3 (2): 13–29. (b) Shi, Y. and Kennedy, J.J. (2016). Delayed registration and identifying the "missing girls" in China. *The China Quarterly* 228 (December 2016: 1018–1038.
28. United Nations (1999). Iran and Vietnam win 1999 United Nations population award. Available from: http://www.un.org/press/en/1999/19990405.pop721.html. Retrieved 29 November 2016.
29. Roudi, F. (2012). *Iran is Reversing Its Population Policy*. Woodrow Wilson International Center.
30. Yeganeh, T. (2012) Iran aims for baby boom, but are Iranians in the mood? *Reuters World News*, 14 November 2012. Available from: http://www.reuters.com/article/us-iran-population-idUSBRE8AD0UZ20121114.
31. Dehghan, S.K. (2015). Iran aims to ban vasectomies and cut access to contraceptives to boost births. *The Guardian*, London.

10 March 2015. Available from: https://www.theguardian.com/world/2015/mar/11/iran-ban-voluntary-sterilisation-contraceptive-access-block-boost-population.

32. Sattar, M. and Barry, E. (2017). Bangladesh weakens long-standing law against underage marriage. *The New York Times*, February 27, 2017.
33. Casterline, J.J. and Bongaarts, J. (eds) (2017). Fertility transition in sub-Saharan Africa. *Population and Development Review*, Supplement to vol. 43.
34. Koski, A., Clark, S., and Nandi, A. (2017). Has child marriage declined in sub-Saharan Africa? An analysis of trends in 31 countries. *Population and Development Review* 43: 7–29.
35. World Bank (2017). Available from: https://data.worldbank.org/indicator/SP.POP.TOTL?locations=1A. Retrieved November 11, 2017.
36. World Bank (2016). Available from: http://data.worldbank.org/indicator/SP.POP.GROW?locations=1A. Retrieved November 16, 2016.
37. World Bank (2016). Available from: http://data.worldbank.org/indicator/SP.DYN.TFRT.IN?locations=1A. Retrieved November 16, 2016.
38. UNDP (2016). *Arab Human Development Report: Youth and the Prospects for Human Development in Changing Reality*. New York: United Nations Development Programme.
39. *The Economist*, Another Arab awakening is looming, warns a UN report, 29 November 2016.
40. World Bank (2017). Available from: http://data.worldbank.org/indicator/SP.DYN.TFRT.IN?locations=EG&order=wbapi_data_value_2013+wbapi_data_value&sort=asc. Somewhat different estimates at US Census Bureau, International Data Base. Available from: https://www.census.gov/population/international/data/idb/informationGateway.php.
41. Ministry of Information and Culture (1979). *First Afghan Population Census*. Kabul: Afghanian Publicity Bureau.
42. UNESCO Institute for Statistics (2018). *Country Profiles*, Afghanistan. Available from: http://uis.unesco.org/en/country/af?theme=culture.
43. Cetorelli, V. (2014). The effect on fertility of the 2003–2011 war in Iraq. *Population and Development Review* 40: 581–604.
44. World Bank (2016). Available from: http://data.world bank.org/indicator/SP.DYN.TFRT.IN?locations=1A-IQ. Retrieved November 16, 2016.
45. National Public Radio, Patrick Adams (2016) A baby boom in a refugee camp is a mixed blessing. Available from: https://www.npr.org/sections/goatsandsoda/2016/04/16/474213390/a-baby-boom-in-a-refugee-camp-is-a-mixed-blessing.
46. Alabaster, O. (2016) Birth rate soars in Jordan refugee camp as husbands discourage wives from contraception. *The Independent* (UK), 3 December 2016. Available from: http://www.independent.co.uk/news/world/middle-east/birth-rate-soars-in-jordan-refugee-camp-as-husbands-discourage-wives-from-using-contraception-a6928241.html.
47. UNICEF (2014). A study on early marriage in Jordan, 2014. Jordan Country Office, Amman. Available from: https://www.unicef.org/jordan/UNICEFJordan_EarlyMarriageStudy2014-E_COPY_.pdf.
48. UNICEF, Middle East and North Africa (2014). Concern over child marriage among vulnerable girls in Jordan. Available from: https://www.unicef.org/mena/media_9469.html.
49. Save the Children (2014). Too young to wed. The growing problem of child marriage among Syrian girls in Jordan. London. Available from: https://www.savethechildren.org.uk/content/dam/global/reports/education-and-child-protection/too-young-to-wed.pdf.
50. (a) Trump, D. (2017). Presidential Memorandum regarding the Mexico City Policy. The White House, Office of the Press Secretary. Available from: https://www.whitehouse.gov/the-press-office/2017/01/23/presidential-memorandum-regarding-mexico-city-policy. (b) Congressional Research Service (CRS) Report for Congress. International Family Planning: The "Mexico City" Policy. April 2, 2001.
51. Kaiser Family Foundation (2017). The Mexico City Policy: An Explainer. Available from: https://www.kff.org/global-health-policy/fact-sheet/mexico-city-policy-explainer/.
52. Guttmacher Institute, Induced abortion in the United States. September 2016 Fact Sheet.
53. Planned Parenthood Federation of America. 2014–15 Annual Report.
54. Lutz, W., Sanderson, W., and Scherbov, S. (2008). The coming acceleration of global population aging. *Nature* 451: 716–719. The IIASA projection may be too low; the actual global population now is already at the upper limit of their previous (2007) projection (click total population in http://webarchive.iiasa.ac.at/Research/POP/proj07/).
55. Cai, Y. (2010). China's below-replacement fertility: government policy or socioeconomic development? *Population and Development Review* 36: 419–440.
56. World Bank India (2018). Fertility Rate, Total (Births per Woman), India. Available at: http://data.worldbank.org/indicator/SP.DYN.TFRT.IN?locations=IN . Accessed 27 April 2018.
57. See critical discussion in Teitelbaum, M. S. (1974). Relevance of demographic transition theory for developing countries. *Science* 188: 420–425.
58. (a) Bloom, D.E. and Williamson, J.G. (1998). Demographic transitions and economic miracles in emerging Asia. *World Bank Economic Review* 12: 419–455.(b) D., B. (2011). 7 billion and counting. *Science* 333: 562–568.
59. David, B. (2011). 7 billion and counting. *Science* 333: 562–568.
60. Lee, R. and Mason, A. (2006). What is the demographic dividend? *International Monetary Fund: Finance and Development* 43 (3): 1–5.
61. Bloom, D.E. and Canning, D. (2004). Global demographic change: dimensions and economic significance 2004. National Bureau of Economic Research Working Paper 10817. Available from: http://www.nber.org/papers/w10817.
62. Aiyar, S. and Mody, A. (2011). The demographic dividend: evidence from the Indian states. 2011 International Monetary Fund Working Paper 11/38. Available from: https://www.imf.org/external/pubs/ft/wp/2011/wp1138.pdf.

63. Lee, R. and Mason, A. (2006). What is the demographic dividend? *International Monetary Fund: Finance and Development* 43 (3): 1–5.

64. United Nations (2017). Model-based estimates and projections of family planning indicators 2017. Available from: http://www.un.org/en/development/desa/population/publications/pdf/family/Figure_Model-based_estimates_Regions_Run20170313.pdf.

65. Pritchett, L.H. (1994). Desired fertility and the impact of population policies. *Population and Development Review* 20: 1–55.

66. National Population Commission Federal Republic of Nigeria (2013). Nigeria, demographic and Health Survey, pp. 1–565. Abuja, Nigeria.

67. Guttmacher Institute (2015). Abortion in Nigeria. Available from: https://www.guttmacher.org/fact-sheet/abortion-nigeria.

68. Coale, A.J. and Watkins, S.C. (eds.) (1986). *The Decline of Fertility in Europe*, 484. Princeton University Press.

69. Amin, S., Diamond, I., and Steele, F. (1997). Contraception and religiosity in Bangladesh, Chapter 12. In: *The Continuing Demographic Transition* (ed. G.W. Jones et al.), 268–289. Oxford University Press.

70. Nie, Y. and Wyman, R. (2005). The one-child policy in Shanghai: acceptance and internalization. *Population and Development Review* 31: 313–336.

4

WATER, FOOD, AND THE ENVIRONMENT

ROBERT WYMAN[1] AND GUIGUI YAO[2,3]

[1] *Department of Biology, Yale University, New Haven, CT, USA*
[2] *Jianghan University, Wuhan, Hubei, China*
[3] *American Studies, Yale University, New Haven, CT, USA*

> Violent protests broke out in the southern Indian state of Karnataka after the Indian Supreme Court ordered the state to release water from a local river to the neighboring state of Tamil Nadu. The river is the main source of water for the states' thirsty rice crops and the dispute has turned the two states into bitter foes. To ensure that the shutdown on the city was complete gangs of thousands of muscular menacing men on motorbikes patrolled the streets screaming that Karnataka was great and anyone who stood in its way would be vanquished [1, 2].

4.1 WHERE WILL THE WATER COME FROM?

The scarcity of fresh water has become a major impediment to economic development [3], personal health, and now, increasingly, a cause of social conflict and disruption [4, p. 948]. In addition to disputes within nations, as above, "inter-state and regional conflicts are also emerging due to water scarcity and poor management structures [5, 6]. It is noteworthy that 158 of the world's 263 transboundary water basins lack any type of cooperative management framework" [7, p. 11].

Of the many desperately important environmental problems that the world faces, the scarcity of fresh water for agriculture may have the largest repercussions and be the most immediately impending. Already, by 1999, almost one-fifth of the world's population lived in areas where water is physically scarce [8]. Now, WHO says that "by 2025, half of the world's population will be living in stressed areas" [9].

The water supply is the most critical factor in achieving world food security. This security is absent for 3.5 billion people in 40 countries whose lives are under threat because of increasing water scarcity [10, 11]. 67% of the world's total water abstractions and 90% of the world's consumption of fresh water [12, pp. 3232 and 3234; 13, Figure 7] are already devoted to agriculture and livestock. As will be made clear in this chapter, fresh water available for agriculture is being depleted at an enormous rate. It is not at all certain that water use, even for current needs, is sustainable.

UNESCO lists population growth as the first of several factors that influence the world's growing demand for water [7, pp. 2 and 10]. "Over the past century, the development of water resources has been largely driven by the demands of expanding populations for food, fiber, and energy. Strong income growth and rising living standards of a growing middle class have led to sharp increases in water use, which can be unsustainable" [7, p. 10]. Rapid, unchecked population growth and rising quality of life, along with increasing demand for food and energy, all contribute to far greater levels of stress on limited groundwater resources [4, p. 947].

As population increases at a billion people every dozen years (see Yao and Wyman, Chapter 3 in this volume), the scale of agriculture will have to expand. By 2050 agriculture will need to produce 60% more food globally and 100% more in developing countries [7, p. 48]. Further intensification will be required to improve the quantity and quality of the diet of the one-ninth of the population (795 million people) who are currently undernourished [14].

Water and Sanitation-Related Diseases and the Changing Environment: Challenges, Interventions, and Preventive Measures,
Second Edition. Edited by Janine M. H. Selendy.

Companion website: https://www.wiley.com/legacy/wileychi/selendy/

Where will the water for the necessary increase in the food supply come from?

In large swaths of the world, rainfall is insufficient and unreliable. In many regions, spring rains initiate the growing season, but a large part of the growth occurs during the dry season. Much of Asia receives almost 90% of its annual rainfall in less than 100 hours [15, p. 135]. The season of most rapid growth is fed from either river water, which in many places originates in the summer melting of high altitude glaciers, or from groundwater, mined from below the surface. Without these sources, the crops fail.

4.2 THE DEPLETION OF STORED FRESH WATER

The major storage compartments for fresh water are glaciers and groundwater (the general term for all subsurface water). Both are being rapidly depleted. The rate of use of groundwater for irrigation now greatly exceeds the rate of recharge from rain and surface water, resulting in large drawdowns of groundwater. Global warming is melting glaciers at an unprecedented rate and the glaciers are retreating rapidly under the onslaught. Both temporarily increase water availability, but in the long run, availability diminishes.

In a sustainable hydrologic system, the natural irregularity of precipitation is smoothed by the use of water stored in glaciers and groundwater and, in much smaller amounts, in lakes or behind dams. Sustainability in this system means that the storage areas are used only as a buffer – drawn down during periods of drought and replenished in wet periods. Over the long run, the volume stored does not change. The current hydrologic regime is unsustainable; there is a monotonic drawdown of the stored water.

Groundwater is depleted by irrigation. Increases in the water needed to satisfy the increasing demand for food have been met largely by increasing use of groundwater. The increasing demand for food is driven partly by increasing affluence and therefore increased per capita consumption of food, but it is mostly driven by population growth (see below).

Glaciers are depleted by melting, a global warming effect, and by the deposition of black carbon (soot), discussed below. Global warming is most commonly attributed to the CO_2 produced in affluent countries by the high consumption of manufactured goods and by the high use of fossil fuels for transportation.

How do these two drivers compare? What are the magnitudes of water drawdown caused by population pressure and that caused by affluent consumption?

Data are now available that can separately measure the depletion of water by these two distinct routes. Data presented below suggest that the effect of food needs and agriculture, and hence of population pressure, may be significantly greater than that attributable to the global warming caused by industrial production and transport.

4.2.1 Groundwater

"Worldwide, 2.5 billion people depend solely on groundwater resources to satisfy their basic daily water needs" [7, p. 13; 16, p. 13] and for approximately two billion people it is the primary source of fresh water [17, p. 2]. "Groundwater reportedly provides drinking water for at least 50% of the global population" [7, p. 13].

Hundreds of millions of farmers rely on groundwater to sustain their livelihoods and contribute to the food security of so many others. The fraction of farmland that is irrigated has grown rapidly. The earth's irrigated area has increased at 2% per year since 1960, almost doubling between 1961/63 and 1997/99 [18]. In recent years groundwater has accounted for 43% of all water used for irrigation [7, p. 13] and 44% of the world's food has been supplied by irrigated land [13, slide 13; 19]. For countries that have rainfall-limited regions, the proportion can be much larger. For instance, groundwater accounts for half of South Asia's irrigation and supports two-thirds of grain crops produced in China [19, p. 1]. A large part of progress against hunger was the result of the green revolution, which was notably dependent on a massive increase in irrigation [20]. However, the increase in irrigation has exceeded the ability of aquifers to recharge and thus has led to depletion. In China 39% comes from groundwater depletion, 34% in India, and 14% in the US [13, slide 16].

Expansion of irrigation using water pumped from nonrenewing or incompletely renewing underground stores temporarily increases the amount of water available, but eventually the subsurface reservoirs will be drawn down, reducing the water available and increasing the amount of energy that must be expended to raise any remaining water to the surface.

A successful strategy to support this increase in irrigation has been the impoundment of water in artificial reservoirs behind dams. Between 1993 and 2008, the increase in reservoir storage was roughly equal to the lower estimates of groundwater loss [21, Table 1]. Storage thus replaced depletion. However, dam-building reached a peak in the late 1970s with the construction of 800 major dams a year and has fallen sharply in the past 30 years down to 25–50 per year recently, partly because of environmental concerns [22]. As the demand for food and irrigation increases, water impoundment is falling behind [23, p. 10; 24, p. 5]. The difference between irrigation water demand and human-engineered water storage has been filled in by a 10-fold increase since 1961 in groundwater extraction [25]. Extraction now far exceeds recharge rates and groundwater stores are being drawn down; current groundwater use for irrigation is unsustainable [23].

4.2.2 Glaciers

Glaciers accrete water from precipitation and lose it through melting and calving (collapse into the ocean of the terminal edge of a glacier). Global warming (and black carbon deposition, see below) increases melting, but warmer air is also moister and thus allows more precipitation in some places. The balance for each glacier is different. However, the data are now clear that, summed over all glaciers, global warming is causing a massive retreat of the ice.

While the increased melting may temporarily increase the amount of water running into rivers and thus available for irrigation, in the longer term it will reduce the amount of glacial runoff. As meltback proceeds, the decreased glacier surface area exposed to spring and summer sunlight, and the colder air temperatures at the higher altitude of the remaining glacier, will reduce the rate of melting and the amount of water available [26].

The contribution of glacier melt to river flow varies in different seasons and in different river basins [27]. Snow from the previous winter is the first to melt. Later, glaciers take over as the principle source, supplying water in the hottest season. In snow-dominated regions (like the mountain sources of the Indus [28], see below), global warming will shift the peak melting to earlier in the season and away from the time of maximum need, increasing the dependence on groundwater irrigation [29].

4.2.3 Case Study: The Indo-Gangetic Plain

The Indo-Gangetic plain is home to 800 million people, more than 10% of the world's population [30, 31]. The included countries, extending from Afghanistan to Bangladesh, add approximately 26 million people annually to the world's population and account for almost one-third of global population growth [32]. The northwest Indian region alone encompasses 114 million people [33]. The Indo-Gangetic plain is one of the regions most vulnerable to water scarcity.

Even though many Indians migrate from villages to the cities, India's agricultural population continues to grow at about 1% per year. Net sown area, however, has remained more or less constant [34]. Over the decade 1990–2000, the net sown area per person in the agriculture-dependent population has decreased by 10% (from 0.29 ha per person to 0.26 ha per person). Subsistence on ever-decreasing plot sizes has been made possible by the expansion of groundwater irrigation. A 1% increase in rural population density was associated with a 5% increase in the proportion of groundwater-irrigated area in total sown area; rural population density alone significantly accounts for the expansion of groundwater irrigation [35].

The gross irrigated area of India has quadrupled since 1950 [36]. Use of groundwater has grown even faster – by more than sixfold – until it now supplies two-thirds of irrigated land [37, 38]. In India, 91% of the groundwater extracted is used for irrigation [36].

According to satellite measurements, the Gangetic plain is losing 54 billion metric tons (gigatons or Gt) of groundwater per year [31]. (A metric ton is 10% larger than a US, or short, ton; where 1 Gt is the mass of 1 km^3 of water). 54 Gt is much more than the total storage capacity (39.5 Gt) of the 203 main reservoirs plus the largest lake in the Indian grain belt [33]. The region encompassing the most heavily glaciated mountains of the Himalayas is losing only 3 Gt per year and the neighboring high altitude region to the north (including Tibet and the East part of central Asia) is gaining slightly.

4.2.4 Case Study: The Indus Basin of Pakistan

A counter example to the general phenomenon of melting glaciers is the Karakoram range, north of the western Himalayas (Figure 4.1). The Karakorams are the most heavily glaciated non-polar region in the world [39]. K-2, the second highest mountain on earth is located there. The glaciers of the Karakorams, although quite inaccessible and therefore only poorly studied [40, p. 42], are generally thought to be stable or even increasing in mass [41, p. 601]. A possible reason is that, as the overall climate warms and increases melting, two other changes have the opposite effect: (i) the wet monsoon winds increasingly invade the mountain chains supplying the Indus and push out the dry westerly winds and (ii) monsoonal clouds hovering over the region that is normally hot and dry in the summer have a cooling effect. As a result of the opposite warming and wetting effects, "There has been a widespread belief that water resources in the region are stable, at least for now. Research by several groups even suggested that climate change might provide some relief in the short or medium term, thanks to faster melting of the glaciers that supply the river, and increase precipitation" [41, p. 601].

The Karakorams are the main source of the Indus River, which flows the whole length of Pakistan, north to south. The Indus supports the lives of 300 million people [41]. The Indus is fed by a huge basin of over a million square km, over 20 000 sq. km of which are glaciated [27, Table 1, p. 4; 28, p. 3]. The region contributes approximately 135 Gt to the annual flow of the Indus River of which 24 Gt is glacier runoff [28, pp. 1, 3]. The mountain headwaters of the Indus River contribute about 60% of the mean annual flow of the river and, importantly for agriculture, approximately 80% of this volume entering the river system during the summer months of June–September [28, p. 5]. The peak flow is in July, which is the end of the snowmelt season. For the rest of the summer, the runoff peak is maintained at a slightly lower volume by glacier melt through August [28, p. 6].

However, even in the Indus basin, with a possibly stable water supply, the growth of population is making the agricultural situation precarious. As a small example, consider the northern Pakistan administrative territory of Gilgit Baltistan where the Indus originates. The rapidly growing population makes the small landholdings (<0.075 ha = <1/5 acre) insufficient for household food security [42]. As in most of Pakistan, aridity and insufficient irrigation limits the

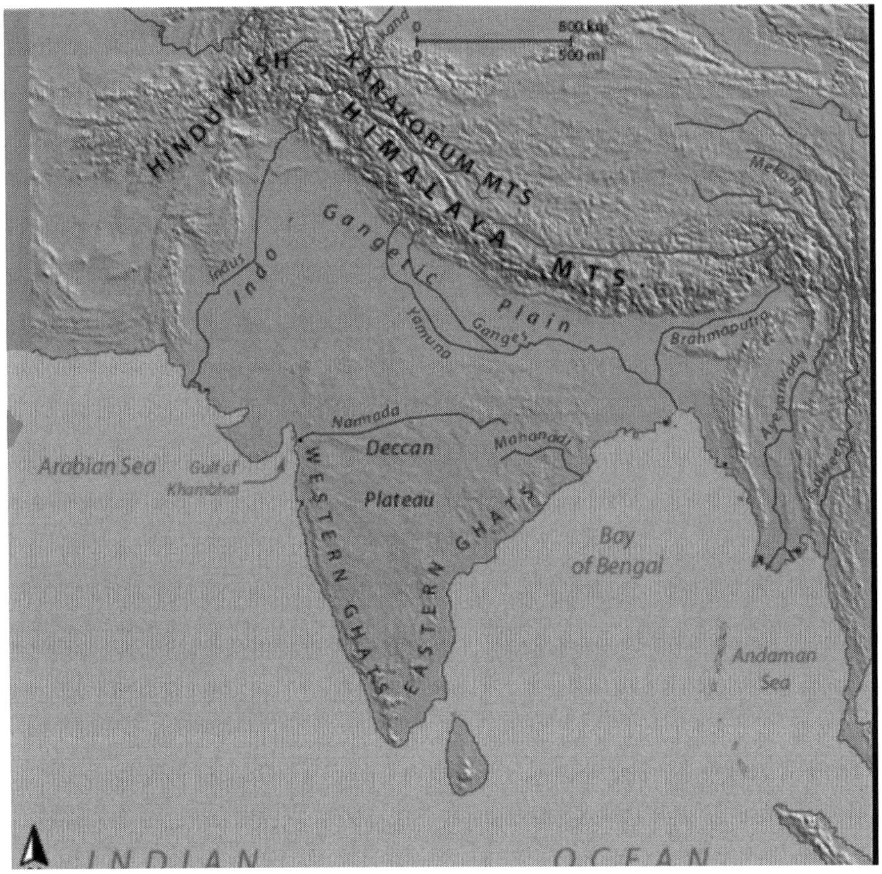

Physically, South Asia consists of the Deccan Plateau, the Indo-Gangetic Plain and the Himalaya, Karakorum, and Hindu Kush mountain ranges. These data are distributed by the EROS Data Center Distributed Active Archive Center (EDC-DAAC), located at the U.S. Geological Survey's EROS Data Center in Sioux Falls, SD.

FIGURE 4.1 The Central Asian glacial origins of the major rivers of South and East Asia. From the west, the rivers, and the countries they mostly supply, are the Indus (Pakistan), the Yamuna (India), the Ganges (India and Bangladesh), the Brahmaputra (Bangladesh), Irrawaddy and Salween (Myanmar), Mekong (China, Laos, Thailand, Vietnam), and (in the northeast and unlabeled on the map) Yangtze and Yellow (China) Rivers. Available from: http://wc4-india.weebly.com/landforms.html.

amount of arable land. However, there is not enough water left in the Indus to increase the amount of arable land.

For ten months of the year, the Indus dries up before it reaches the sea [41, p. 600]. According to the World Bank's World Development Indicators, the renewable freshwater flows/year in Pakistan have increased slightly from 52 GT in 2003 to 55 GT in 2014 [43]. Total withdrawals, however, are 183.5 GT, 3⅓ times the amount available from renewable sources (up from 3 times in 2003). The difference is made by groundwater withdrawals. However, because of population growth, the per capita renewable freshwater resources have diminished by one-seventh from 350 to 297 m^3. Actual water use in Pakistan is about 991 [43, 2017] to 1326 m^3/year [12, Figure 3]. Again, the difference is made by non-renewable groundwater extraction. Barring major technical advances, there is little room for savings, as the actual use is very close to the minimum requirement for personal use and to grow enough food for a person, estimated very approximately at 1150 m^3/year [44, Table 8, p. 87].

To estimate the social importance of seasonally delayed glacier melt it has to be set in relation to the potentially affected population. In an index that combines both factors, the Indus basin comes in second only to the disappearing Aral Sea. This "Population Impact Index" for Pakistan is twice as high as that for the Ganges at third and 5 times as high as the Po in Italy at fourth. No other river basin comes close [27, Table 1].

4.2.5 Estimates of Global Depletion

India and Pakistan are not unique. Twenty-one of the world's 37 largest aquifers are severely overexploited in locations from China and India to France and the United States [17, p. 5225; see also 45]. Eight of these basins were not being naturally replenished at all, and another five were only slightly replenished [46]. The north-central Middle East, including portions of the fertile crescent (the Tigris and Euphrates river basins), lost 91.3 Gt of groundwater from January 2003 to December 2009 [47]. Focusing on a smaller, but intensely irrigated, region, the Central Valley of California (one of the world's most productive agricultural regions, encompassing 52 000 km^2) lost 100 Gt of groundwater from 1962 to 2014, 80 Gt of that in the last 30 years [48, Figure 2]. For comparison, the total capacity of the largest reservoir in the US, Lake Mead, holds 32 Gt [49]. Over large areas, water tables are falling. Declines in water tables range from centimeters to meters per year [33]. Water levels in parts of the Central Valley have dropped more than 120 m [50]. The breadbasket regions of the world's most populous countries are in crisis situations (for China see [51] and [52]). Because of overdrafts for agriculture, some of the world's largest rivers (Colorado, Rio Grande, Yellow, Indus, Ganges, Nile, Amu Darya, and Murray) are now seasonally depleted so that they do not reach the sea [53].

As groundwater extractions, and the concomitant depletion, continually increase, the total drawdown grows. The cumulative global groundwater depletion from 1900 to 2008 is estimated at ~4500 Gt [54, pp. 1 and 4]. A recent estimate for depletion of the 9 major aquifers in arid and semiarid mid latitudes is 77.4 Gt/year [4, Table 1]. The rate of depletion has increased markedly since about 1950, more than tripling from 1960 to 2000 [23, p. 1]. The most recent global estimate is 204 Gt/year. for the year 2000 with an expectation to rise to 358 Gt/year by 2050 [24]; different estimates are summarized in [55].

Estimates of water loss for all glaciers and ice caps (not including the agriculturally irrelevant Antarctic and Greenland ice sheets [2] are of approximately the same magnitude and range as those for groundwater. A consortium review, written by a number of prominent authorities [56] (and used by the IPCC [57], Table 4.4), concludes that the global loss of water from glaciers amounts to 215 ± 26 Gt per year.

It is coincidental, but very striking, that the estimated rate of mass loss of groundwater (204 Gt /year) is so similar to the loss from glacier recession (215 Gt/year). Glacier melt and groundwater loss are depleting by about equal amounts from the Earth's major fresh water stores.

Both losses are accelerating. All estimates of recent global groundwater loss find drastic increases over previous time periods. Groundwater extraction is increasing at 1–2% a year [3, p. 19] and the rate at which glaciers are losing mass appears to have accelerated over roughly the last decade [40, Figure 2.2, p. 9; 58, p. 2].

The rate of depletion of these critical water sources is staggering. To get an idea of these magnitudes, Himalayan glaciers provide about 180 Gt of water each year for the Indus and other rivers of Pakistan [28], but the total storage capacity of the 203 main reservoirs and the largest lake in the Indian grain belt is only 39.5 Gt [59]. The total volume of water in aquifers is not measurable [60] so it is not possible to estimate the number of years that aquifer water will last in any region. Nevertheless, we know the loss is large compared to other storage compartments.

The measured mass loss is the net difference between discharge and refill of the aquifers or between accretion and ablation of glaciers. The global losses of glacier and groundwater are net losses. Some groundwater, extracted for irrigation, is replaced as water slowly drips into underground spaces from rainwater and other surface sources. The net loss is the excess of discharge over recharge. Glaciers grow from the accumulation and compression of snowfall. They are depleted by melting; some fraction of the meltwater is trapped and refreezes, while the remainder runs off in surface streams and, often, in under-glacier streams. The net loss is the difference between accumulation on the one hand and runoff and calving on the other. The water "lost" from the glaciers and aquifers does not just disappear. After its use for irrigation, some water seeps back into the aquifers or rivers. The remainder evaporates and returns as rain; 97% of the "lost" water ends up in the oceans, adding to sea-level rise [16, 24, 61].

4.2.6 The Effects of Food and Agriculture

Since almost all groundwater is used for farmland irrigation, its loss can be attributed to the food needs of vast populations. Glacier melting is often attributed to global warming via greenhouse gas emissions. The situation, however, is actually more complicated.

Food and agriculture have been estimated to contribute between 19 and 29% of global anthropogenic greenhouse gas emissions [62]. Of this, 80–86% originates on farms and in livestock husbandry. The remainder is largely from food transport and fertilizer manufacture.

Agriculture also produces vast quantities of "black carbon," the soot produced when burning fossil fuels or biomass. A large amount is produced in the annual fires set by farmers to clear the land and provide the ashes as fertilizer for the next crop. In 2016, the Parliamentary Standing Committee on Science and Technology in India announced that the number of forest fires had risen by 55% [63]. Hundreds of thousands of farmers in Punjab and Haryana, just two of India's 36 States and Union Territories, burned about 32 million tons of straw [63, 64] (Figures 4.2 and 4.3). In Indonesia, fires consumed 10,000 square miles of forests and sickened hundreds of thousands of people as the land was cleared for palm oil plantations [65] (Figure 4.4). NASA detected hundreds of fires burning simultaneously in central Africa. "The location, widespread nature, and number of these fires

FIGURE 4.2 A farmer burned a harvested wheat field last month on the outskirts of Jalandhar, India. *Source*: Credit Shammi Mehra/Agence France-Presse – Getty Images. Available from: http://www.nytimes.com/2016/11/03/world/asia/farmers-unchecked-crop-burning-fuels-indias-air-pollution.html?_r=0. (*See insert for color representation of this figure.*)

FIGURE 4.3 Agricultural fire in India. *Source*: From NASA – The Future of Fires. Available from: https://www.youtube.com/watch?v=_lSO02pkQZU&index=9&list=PLiuUQ9asub3TaqZy6_07xXDHK_uufgt0k At time marker 1 : 22 of 2 : 05 total running time.

suggest that they were deliberately set to manage land" [66]. Similarly burning areas stretch from west to east in a line across Africa (Figure 4.5). In western Africa, the agricultural burning season continues for almost half the year, usually running from December, after the year's primary crops are harvested, through April or May, when the next growing season begins. Similar widespread fires burning simultaneously have been documented to occur throughout Peru, Ecuador, Eastern Brazil, across the whole of Madagascar as well as eastern China on into eastern parts of Russia. Smaller agricultural fires occur in the Nile delta [67a]. In addition to the burning of previously cropped land, 150,000 sq kms of tropical forests, the total area of Bangladesh, were destroyed by burning in each of 2016 and 2017, to create more farm and pasture land. These are the largest numbers on record and about a 50% increase from the 2001–2015 average [67b].

Black carbon is produced in even larger quantities by the widespread use of biomass in cooking fires. WHO estimates that indoor air pollution was linked to 4.3 million deaths in 2012 among the estimated 2.9 billion people living in homes using wood, coal, or dung as their primary cooking fuel [68]. Each household that uses biomass cooking releases 1–2 tons of CO_2 per year [69].

Black carbon is the dominant absorber of visible solar radiation in the atmosphere [70], but it is a particulate solid and thus is not considered a greenhouse gas. Globally, 60% of black carbon emitted is from biofuel burning in developing countries and the widespread burning of crop residues [71].

At least in South Asia, black carbon appears to be a specific cause of Himalayan glacier melt [72]. The Indo-Gangetic plain is the source of more than half of the 8 million tons per year of black carbon released globally [73, 74].

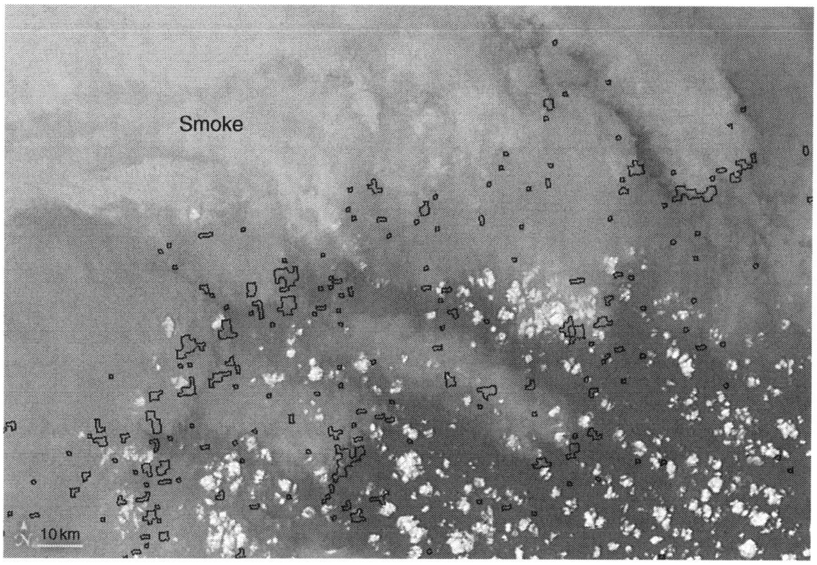

FIGURE 4.4 Heavy smoke blankets Borneo. In September and October farmers engage in frequent burning of rainforest to clear the way for crops or grazing animals. In southern Borneo, the intent is often to make room for new plantings of oil palm and acacia pulp. *Source*: Image taken October 19, 2015. Available from: https://earthobservatory.nasa.gov/IOTD/view.php?id=86847&eocn=image&eoci=related_image.

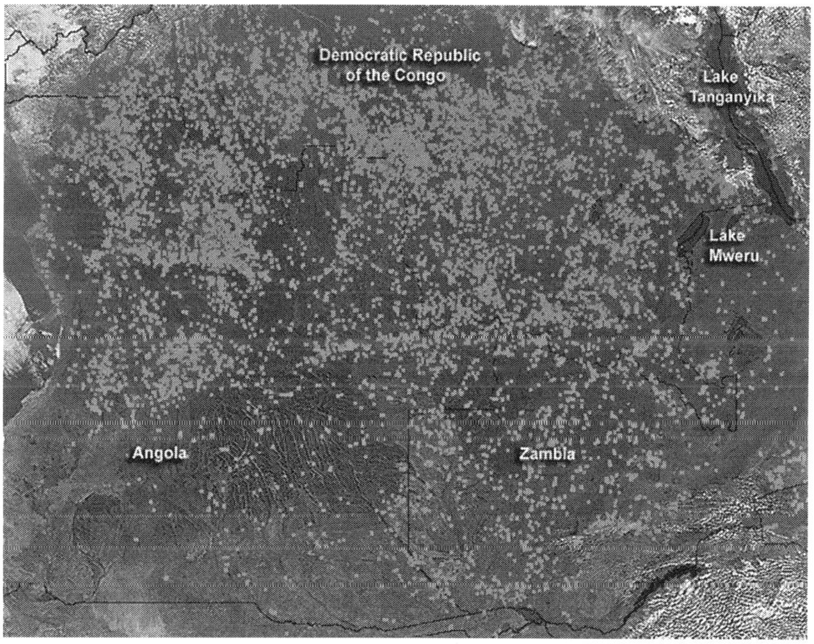

FIGURE 4.5 Widespread agricultural burning throughout central Africa. Fires burning in the southern region of the Democratic Republic of the Congo, Zambia, and Angola. Actively burning areas are outlined in red. *Source*: Photo from the NASA's Aqua satellite June 3, 2016. NASA, NOAA, and the US Department of Defense. Available from: https://www.nasa.gov/sites/default/files/styles/image_card_4x3_ratio/public/thumbnails/image/centralafricasnpp66_0.jp. (*See insert for color representation of this figure.*)

The air, heated by black carbon, flows over the Himalayas and causes an estimated 60% of the warming that occurs in that region [74–76]. In addition, deposition of black carbon on Himalayan slopes increases the melting rate of a typical glacier by up to one-third [73]. It has been estimated that the warming effect of black carbon is as important as greenhouse gases in the observed retreat of over two-thirds of the Himalayan glaciers [76].

More detail on the contributions of the major agriculturally produced greenhouse gases and black carbon to glacier melting may be found in "The Effects of Population on the Depletion of Fresh Water" [55].

The global estimates of water released by glacier melting and from groundwater irrigation are extraordinarily similar (see above). Groundwater is overwhelmingly used for agriculture. Glacier melting, however, is only partly driven by greenhouse gas emissions from industrial production and transport. There is a large contribution from food chain greenhouse gases and from food chain black carbon. In an accounting that distinguishes water loss due to agriculture and cooking from that due to fossil fuel burning [summarized in reference 55, Figure 3], water loss due to the vast human population's need to eat dominates the loss resulting from industrial production and transport.

4.2.7 Food, Population, and Affluence

Population size is the main determinant of food consumption. Grains provide more than half the calories in the world's diet. Globally, income elasticity for grains is small across all income levels [14, 77, 78]: richer people do not consume much more grain than poorer people. Total consumption is fairly closely proportional to population and not very sensitive to affluence.

Meat is more sensitive to affluence; it has a higher elasticity that depends on income level [78]. Meat is currently about 6.5% of the global diet (by weight) and is expected to rise to 7% by 2030 [79, 80]. The FAO [81] warns that "there is less of a meat revolution than commonly asserted, mainly because of lack of development and income growth in many countries [and] cultural and religious factors. … The perception of revolutionary change in the meat sector reflects the extraordinary performance of world production and consumption of poultry meat." Poultry is also the most efficient converter of grain to meat. For chicken, when optimally produced, the feeding efficiency, kg grain fed/kg edible meat, is as low as 1.5 to 2.0 [80].

The current world population, through its demand for food and fiber, is probably a larger contributor to the drawdown of freshwater stocks than the greenhouse gas emissions and global warming attributable to industrial production and transport. Absent major positive changes in the contexts discussed above, as population increases steadily (see Figure 3.1 in Chapter 3) and as per capita consumption of both food and manufactured goods increases, we can expect further acceleration of both glacier melting and aquifer drawdown.

4.3 IS POPULATION A PROBLEM?

As with most environmental problems, water scarcity has a variety of causes. In addition to the fundamental need of humans for food, water scarcity is exacerbated by a number of economic and political factors. For example, in India the government subsidizes water for rice farming in an arid area as does Kazakhstan for cotton growing [82]. The southwest US has similar misallocation issues. Political factors make these misallocations notoriously difficult to fix.

The concern in this chapter for total water availability, and for the effect of the large and growing human population on demand for that water, contradicts many of the mainstream statements on water scarcity. As an example, this chapter runs directly opposite to the thrust of the UN's Human Development Report 2006, "Beyond scarcity: power, poverty and the global water crisis" [15]. That report argues (p. 2):

> Some commentators trace the global challenge in water to a problem of scarcity. The spirit of Thomas Malthus, who in the 19th century disconcerted political leaders by predicting a future of food shortages, increasingly pervades international debates on water. With population rising and demands on the world's water expanding, so the argument runs, the future points to a "gloomy arithmetic" of shortage.
>
> We reject this starting point. The availability of water is a concern for some countries. But the scarcity at the heart of the global water crisis is rooted in power, poverty and inequality, not in physical availability.

Then (p. 3):

> There is more than enough water in the world for domestic purposes, for agriculture and for industry."

In a further discussion (p. 13), it states:

> In the early 21st century debates on water increasingly reflect a Malthusian diagnosis of the problem. Dire warnings have been posted pointing to the "gloomy arithmetic" of rising population and declining water availability. Is the world running out of water?
>
> Not in any meaningful sense. … Viewed at a global level, there is more than enough water to go around and meet all of humanity's needs.

It is possible that this attitude is appropriate for the small fraction of total water that is for personal use – drinking, cooking, sanitation. A minor redistribution can increase this small fraction to a sufficient level. Indeed, extremely encouraging progress has been made [83].

However, the Human Development Report (HDR) continues (p. vi):

> Water for livelihoods poses a different set of challenges. The world is not running out of water, but many millions of its most vulnerable people live in areas subject to mounting water stress. Some 1.4 billion people live in river basins in which water use exceeds recharge rates. The symptoms of

> overuse are disturbingly clear: rivers are drying up, groundwater tables are falling and water-based ecosystems are being rapidly degraded. Put bluntly, the world is running down one of its most precious natural resources and running up an unsustainable ecological debt that will be inherited by future generations.

Although in attacking scarcity concerns, the HDR emphasizes the global abundance of water, they also realize that the global abundance of water is not the relevant issue:

> Unfortunately, the international average is a largely irrelevant number (p. 135).
>
> It does not help water-stressed countries in the Middle East that Brazil and Canada have more water than they could ever use. Nor does it help people in drought-prone areas of northeast Brazil that average water availability in the country is among the highest in the world (HDR, p. 14). ... Water stressed regions in China and India are not relieved by Iceland's [extraordinary] water availability. ... Because water ... is not readily [transportable] in bulk quantities ... What matters is local availability (HDR, p. 135).

Since local availability is a severe problem for the billions of people living in water-scarce regions, as the HD report emphasizes in many places, it is not clear what to make of the statements (p. 2) that "the heart of the global water crisis is ... not in physical availability," or that availability is not a problem "in any meaningful sense" (p. 13).

Not only does the report deny a role for physical scarcity while providing many statistics showing a dire crisis, it also denies a role for population growth in some places while admitting it elsewhere:

> Future water management in agriculture faces pressure. ... On the demand side industrialization, urbanization and changing diets will increase demand for food and the water used in its production (p. 175).

The absence of any mention of population growth is striking. However, on the next page:

> Population will increase by some 80 million people a year over the next three decades .... Maintaining food supplies will require large productivity gains ...." (p. 176).

In arguing against "trends [that] have lent a superficial credence to Malthusian concerns" the report states that "Growing water demand outstrips population growth" (p. 137):

> Water use has been growing much faster than population for at least a century – and that trend is continuing. Over the past hundred years population quadrupled, while water use grew by a factor of seven (p. 137).

This comparison implies a serious misunderstanding of the quantities. Water use is the use per person times the number of people. The change in population times the change in water consumption per person gives the change in total water use.

If the population has grown by a factor of 4, then per capita water consumption must have grown by a factor of 1¾ for total consumption to have multiplied by 7.

$$\frac{\text{Water consumption}}{\text{Person}} \times \text{Number of people} = \text{Total water use}$$

$$4 \times 1\frac{3}{4} = 7$$

Thus, population has grown more than twice as fast (4× versus 1¾×) as has per capita water consumption. How does this provide only "superficial credence to Malthusian concerns?" From these numbers, population growth is THE major factor in increasing demand for water.

Perhaps the denial of the centrality of population growth to developmental problems is fading. The 2015 UNICEF/WHO Progress Report on Sanitation and Drinking Water [83] describes "population growth" 11 times as a major factor inhibiting the provision of adequate water for drinking and sanitation (see especially sidebar, p. 17, "Keeping up with Population Growth" in that document).

4.4 POLICY CONSIDERATIONS: IS POPULATION A SOLUTION?

Human demand on the Earth's resources are the product of a large (and rapidly increasing) population times a high (and rapidly increasing) per capita consumption. World fertility (children/woman) has fallen to lows never before seen (see Figure 3.3 in Chapter 3). Yet there are still 2½ births for every death [32]. If fertility stays at its current unprecedented low, population will rise from 7 billion this year to 27 billion by 2100 (see UN Constant Fertility Variant in Figure 3.4 in Chapter 3). Essentially all the increase will occur among poor families who already suffer dire stress. Only by assuming further rapid drops in fertility do population projections come down to the 6–16 billion people that is often cited as the likely 2100 global population (other variants are shown in Figure 3.4 in Chapter 3). Now that world attention and effort has been diverted from "the population explosion" to AIDS and global climate change will such fertility declines actually occur?

This question should reignite the long-standing discussion among environmentalists over how much concern and effort to devote to the size and rate of increase of human population (mostly among less affluent people) and how much to devote to the level and rate of increase (mostly among more affluent people) of per capita consumption. In practice, environmental scholars and activist organizations devote very little of their activity to human population increase. Although this controversy can generate vehement arguments, there is, in fact, little data available allowing comparison between the two phenomena. China, for instance, is the world's largest emitter of CO_2, producing 1½ [84] to 2 [85, p. 5 and Figure 2.2, p. 14] times the US output. However, China's emissions involve a complex interplay of a very large population, rapid industrial development, and a large component of export to affluent consumers elsewhere. The factors cannot be disentangled. As China, India, and other developing countries approach Western levels of consumption, their huge populations will produce vastly more CO_2. A recent upwards revision in China's official emission statistics [86, 87] increased China's stated CO_2 emissions by almost a billion tons a year, more than the whole German economy emits annually from fossil fuels [88].

However, this chapter provides one of the very few cases where an apples-to-apples comparison can be made of the effects of population and of industrialized consumption. The depletion of the same substance, fresh water, occurs by two distinct routes. Groundwater is depleted by irrigation for agriculture. Glacier melting is caused both by global warming and through biomass burning driven by food demand. The quantitative contributions of each can be separately estimated. This chapter has compared the water loss that is driven by fossil fuel burning for industrialized consumption and transport with the water loss driven by food requirements of vast populations. The data show that food demands, and thus population size, predominates.

4.4.1 The Population/Consumption Issue

Population is clearly a major factor in the Earth's environmental problems. The depletion of fresh water was chosen for discussion because it is so critical and because it is one of the rare areas where the contributions of population and industrial consumption can be reasonably disentangled. It is not unique; most environmental problems include a major contribution from high population levels.

For the last 50 years, global population growth has remained in the band between 70 and 87 million added annually; the increment in 2017 was ~78 million (US Census Bureau, 2018). Even though global population has continued to grow at a higher rate than it did during the "population bomb scare" of the 1960s [89], the phenomenon has fallen off the radar of environmentalists. An editorial in *Nature Climate Change*, discussing the UN population projections (shown in Figure 3.4 in Chapter 3), used words like "scary," "severe," "especially worrisome," yet noted that "the topic of population growth can be a bit passé in sustainability circles" [90].

This neglect of the demand factors arising from the scale of human population is perhaps unjustified. The environmental footprint of a billion more people every dozen or so years swamps all environmental efforts. On the other hand, efforts to reduce fertility have been extraordinarily effective and well accepted around the world: fertility has dropped to half of what it was 50 years ago [91]; for a more complete discussion, see the last two sections of Chapter 3 in this volume. Fertility decline has been one of the few bright spots in an otherwise dismal history of humanitarian and environmental efforts. Although rarely on the agenda of water policy experts, family planning may provide the most cost and resource efficient route to moderating the demand for ultimately limited water resources.

REFERENCES

1. Najar, N. (2016). Violence erupts in Southern India over order to share water. *New York Times* 12 September 2016.
2. Manu, J. (2012). 2 States split bitterly by one river. *New York Times*, October 10, 2012.
3. United Nations (2016). World Water Development Report 2016: Water and Jobs. Paris, France: UNESCO.
4. Famiglietti, J.S. (2014). The global groundwater crisis. *Nature Climate Change* 4: 945–948.
5. Food and Agriculture Organization of the United Nations (FAO), International Fund for Agricultural Development, UNICEF, World Food Programme, World Health Organization (2017). The State of Food Security and Nutrition in the World. See especially Part 2 on Conflict.
6. *The Economist* (2017). Water tortured. The romance between Bangladesh and India is star-crossed, 14 April 2017.
7. United Nations. World Water Development Report 2015 (2015). Water for a Sustainable World. Paris, France: UNESCO.
8. World Health Organization (2009). 10 Facts about water scarcity. Available from: www.who.int/features/factfiles/water/en/.
9. World Health Organization. Drinking Water Fact Sheet, 2016. Available from: http://www.who.int/mediacentre/factsheets/fs391/en/; accessed 15 March 2017.
10. United Nations. World Water Development Report 2014 (2014). Water and Energy, vol. 1. Paris, France: UNESCO.
11. Atwater, R., Anderson, L., and Derry, C. (2016). Agricultural risk management of a peri-urban water recycling scheme to meet mixed land-use needs. *Agricultural Water Management* 176: 266–269.
12. Hoekstra, A.Y. and Mekonnen, M.M. (2012). The water footprint of humanity. *Proceedings of the National Academy of Sciences of the United States of America* 109: 3232–3237.
13. OECD. Global Forum for Food and Agriculture. (2017). Agriculture and Groundwater Feeding Billions from the Ground Up. Berlin: International Food Policy Research Institute (IFPRI). Available from: https://www.slideshare.net/ifpri/agriculture-and-groundwater-feeding-billions-from-the-ground-up.
14. World Food Program, United Nations (2017). 2017 Hunger Statistics. Available from: https://www.wfp.org/hunger/stats. Accessed 3 January 2017.
15. United Nations Development Programme. Human Development Report (2006). Beyond Scarcity – Power, Poverty and the Global Water Crisis. New York: UNDP.

16. OECD (2015). Drying Wells, Rising Stakes: Towards Sustainable Agricultural Groundwater Use. Paris: Organization for Economic Cooperation and Development. doi: 10.1787/9789264238701-en.
17. Richey, A.S., Thomas, B.F., Lo, M.H. et al. (2015). Quantifying renewable groundwater stress with GRACE. *Water Resources Research* 14 July doi: 10.1002/2015WR017349.
18. IPCC, Bates, B., Kundzewicz, Z., Wu, S., and Palutikof, J. (2008). Climate change and water, Technical Paper of the Intergovernmental Panel on Climate Change.
19. OECD (2016). Tackling the challenges of agricultural groundwater use. Retrieved 4 January 2017. Available from http://www.oecd.org/tad/sustainable-agriculture/Challenges%20of%20groundwater%20use.pdf.
20. Bhaduri, A., Amarasinghe, U., and Shah, T. (2009). Groundwater expansion in Indian agriculture: past trends and future opportunities. In: Strategic Analyses of the National River Linking Project (NRLP) of India (ed. U. Amarasinghe, T. Shah and R.P.S. Malik). Colombo, Sri Lanka: International Water Management Institute.
21. Church, J.A., White, N.A., Konikow, L.F. et al. (2011). Revisiting the Earth's sea-level and energy budgets from 1961 to 2008. *Geophysical Research Letters* 38 (L18601): 1–8.
22. Chao, B.F., Wu, Y.H., and Li, Y.S. (2008). Impact of artificial reservoir water impoundment on global sea level. *Science* 320: 212–214.
23. Wada, Y., van Beek, L.P.H., and Bierkens, M.F.P. (2012). Nonsustainable groundwater sustaining irrigation: a global assessment. *Water Resources Research* 48: W00L06. doi: 10.1029/2011WR010562.
24. Wada, Y., van Beek, L.P.H., Weiland, F.C.S. et al. (2012). Past and future contribution of global groundwater depletion to sea-level rise. *Geophysical Research Letters* 39: 1–6.
25. Davidson, D.J. and Andrews, J. (2013). Not all about consumption. *Science* 320: 212–214.
26. Jansson, P., Hock, R., and Schneider, T. (2003). The concept of glacier storage: a review. *Journal of Hydrology* 282: 116–129.
27. Kaser, G., Grosshauser, M., and Marzcion, B. (2010). Contribution potential of glaciers to water availability in different climate regimes. *Proceedings of the National Academy of Sciences of the United States of America* 107: 20223–20227.
28. Yu, W. and Yang, Y.-C. (2013). Hydrology and Glaciers in the Upper Indus Basin of Pakistan. Herndon: US World Bank Publications.
29. Barnett, T.P., Adam, J.C., and Lettenmeier, D.P. (2005). Potential impacts of a warming climate on water availability in snow-dominated regions. *Nature* 438: 303–309.
30. Bolch, T., Kulkarni, A., Kaab, A. et al. (2012). The state and fate of Himalayan glaciers. *Science* 336: 310–314.
31. Tiwari, V.M., Wahr, J., and Swenson, S. (2009). Dwindling groundwater resources in northern India, from satellite gravity observations. *Geophysical Research Letters* 36: 1–5.
32. PRB (2016). World Population Data Sheet. Washington, DC: Population Reference Bureau.
33. Rodell, M., Velicogna, I., and Famiglietti, J.S. (2009). Satellite-based estimates of groundwater depletion in India. *Nature* 460: 999–1002.
34. Bhaduri, A., Amarasinghe, U., and Shah, T. 2008). Future of foodgrain production in India, International Water Management Institute. Available from: http://nrlp.iwmi.org/PDocs/DReports/Phase_01/08-Future%20of%20Food%20grain%20Production-Anik%20Bhaduri.doc.
35. Bhaduri, A., Amarasinghe, U., and Shah, T. (2009). Groundwater expansion in Indian agriculture: past trends and future opportunities. In: Strategic Analyses of the National River Linking Project (NRLP) of India (ed. U. Amarasinghe, T. Shah and R.P.S. Malik), 181–196. Colombo, Sri Lanka: International Water Management Institute.
36. Ministry of Agriculture, India (2013). Agricultural statistics at a glance 2012, Table 13.2, Directorate of Economics and Statistics. Available from: eands.dacnet.nic.in/latest_2006.htm.
37. Amarasinghe, U.A., Shah, T., and Malik, R.P.S. (2009). India's water futures: drivers of change, scenarios and issues. In: Strategic Analyses of the National River Linking Project (NRLP) of India (ed. U.A. Amarasinghe, T. Shah and R.P.S. Malik), 3–24. Sri Lanka, Colombo: International Water Management Institute.
38. Gandhi, V.P. and Namboodiri, N.V. (2009). Groundwater irrigation in India: gains, costs and risks, Working Paper 2009-03-08, Indian Institute of Management, Ahmedabad.
39. See, for instance, Karakoram in Wikipedia. Available from: https://en.wikipedia.org/wiki/Karakoram.
40. Global Glacier Change Bulletin No. 1 World Glacier Monitoring Service, 2015
41. Qiu, J. (2016). Stressed Indus River threatens Pakistan's water supplies. *Nature* 534: 600–601. doi: 10.1038/534600a.
42. Pakistan Council of Research in Water Resources, International Centre for Integrated Mountain Development (Nepal), World Wildlife Fund (2017). An innovative approach to agricultural water management in the upper Indus basin: the water, energy, food nexus at the local level.
43. World Bank (2005 and 2017). World Development Indicators. Washington, DC: World Bank.
44. Gleick, P.H. (1996). Basic water requirements for human activities meeting basic needs. *Water International* 21: 83–92.
45. Gleeson, T., Wada, Y., Bierkens, M.F.P., and van Beek, L.P.H. (2012). Water balance of global aquifers revealed by groundwater footprint. *Nature* 488: 197–200. doi: 10.1038/nature11295.
46. Staff of Science Magazine. (2015). Running out of groundwater. *Science* 348–1405.
47. Voss, K.A. et al. (2013). Groundwater depletion from the middle east from GRACE with implications for trans-boundary water management in the Tigris–Euphrates–Western Iran region. *Water Resources Research* 49: 1–11.
48. Faunt, C.C., Sneed, M., Traum, J. et al. (2015). Water availability and land subsidence in the Central Valley, California, USA. *Hydrogeology Journal* doi: 10.1007/s10040-015-1339-x.

49. Famiglietti, J.S., Lo, M., Ho, S.L. et al. (2011). Satellites measure recent rates of groundwater depletion in California's Central Valley. *Geophysical Research Letters* 38: 1–4.
50. Hiscock, K. (2005). Hydrogeology Principles and Practice. Blackwell Publishing.
51. Li, J. (2011). Water shortages loom as northern China's aquifers are sucked dry. *Science* 328: 1462–1463.
52. Yu, C. (2011). China's water crisis needs more than words. *Nature* 470: 307.
53. Postel, S. (2010). Water: adapting to a new normal. In: The Post Carbon Reader: Managing the 21st Century's Sustainability Crises (ed. R. Heinberg and D. Lerch). Healdsburg, CA: Watershed Media.
54. Konikow, L.F. (2011). Contribution of global groundwater depletion since 1900 to sea-level rise. *Geophysical Research Letters* 38 (L17401): 1–5.
55. Wyman, R.J. (2013). The effects of population on the depletion of fresh water. *Population and Development Review* 39: 687–704.
56. Gardner, A.S. et al. (2013). A reconciled estimate of glacier contributions to sea level rise: 2003–2009. *Science* 340: 852–857.
57. Intergovernmental Panel on Climate Change. 5th Assessment Report. Working Group I: The Physical Science Basis.
58. US Environmental Protection Agency (2016). Climate Change Indicators: Glaciers.
59. Rodell, M., Velicogna, I., and Famiglietti, J.S. (2009). Satellite-based estimates of groundwater depletion in India. *Nature* 460: 999–1002.
60. Richey, A.S., Thomas, B.F., Lo, M.-L. et al. (2015). Uncertainty in global groundwater storage estimates in a total groundwater stress framework. *Water Resources Research* 51: 5198–5216. doi: 10.1002/2015WR017351.
61. Church, J.A., White, N.A., Konikow, L.F. et al. (2011). Revisiting the Earth's sea-level and energy budgets from 1961 to 2008. *Geophysical Research Letters* 38 (L18601): 1–8.
62. Vermeulen, S.J., Campbell, B.M., and Ingram, J.S.I. (2012). Climate change and food systems. *Annual Review of Environment and Resources* 37: 195–222.
63. NASA (2016). Blowing smoke across the northern plains. Available from: https://www.nasa.gov/image-feature/goddard/2016/forest-fire-numbers-rise-in-india-in-2016.
64. Anand, G. (2016). Farmers' unchecked crop burning fuels India's air pollution. *New York Times*, November 2, 2016.
65. NASA (2015). Agricultural Fires Across Central Africa. Available from: https://www.nasa.gov/image-feature/goddard/agricultural-fires-across-central-africa.
66. Cochrane, J. (2016). Indonesia's orangutans suffer as fires rage and businesses grow. *New York Times*, April 5, 2016.
67. (a) https://www.nasa.gov/mission_pages/fires/main/world/index.html.
67. (b) New York Times (2018). https://www.nytimes.com/reuters/2018/06/27/world/americas/27reuters-environment-forests.html; Global Forest Watch (2017). https://blog.globalforestwatch.org/data/global-tree-cover-loss-rose-51-percent-in-2016; Global Forest Watch (2018). https://blog.globalforestwatch.org/data/2017-was-the-second-worst-year-on-record-for-tropical-tree-cover-loss.
68. http://www.who.int/mediacentre/news/releases/2014/air-pollution/en/.
69. NASA (2016) World Fire Image Gallery. Available from: https://www.nasa.gov/mission_pages/fires/main/world/index.html.
70. Kerr, R.A. (2009). From burning dung to global warming and back again. *Science* 326: 362–363.
71. Weaver, A.J. (2011). Climate change. Toward the second commitment period of the Kyoto Protocol. *Science* 332: 795–796.
72. Menon, S., Koch, D., Beig, G. et al. (2010). Black carbon aerosols and the third polar ice cap. *Atmospheric Chemistry and Physics* 10: 4559–4571.
73. Qiu, J. (2013). Pollutants capture the high ground in the Himalayas. *Science* 339: 1030–1031.
74. Ramanathan, V., Li, R., Ramana, M.V. et al. (2007). Atmospheric brown clouds: hemispherical and regional variations in long-range transport, absorption, and radiative forcing. *Journal of Geophysical Research* 112: 1–26.
75. Meehl, G.A., Arblaster, J.M., and Collins, W.D. (2008). Effects of black carbon aerosols on the Indian monsoon. *Journal of Climate* 21: 2869–2882.
76. Ramanathan, V. and Carmichael, G.H. (2008). Global and regional climate changes due to black carbon. *Nature Geoscience* 1: 221–227.
77. Amarsinghe, U.A. and Singh, O.P. (2009). Changing consumption patterns of India: Implications on future food demand. In: Strategic Analyses of the National River Linking Project (NRLP) of India (U.A. Amarasinghe, T. Shah and R.P.S. Malik), 131–146. Colombo, Sri Lanka: International Water Management Institute.
78. Baffes, J., van der Mensbrugghe, D., Burns, A., and Osorio, I. (2009). The macroeconomic environment for agriculture and rural areas to develop over the next 50 years. Table: Income elasticity for grains is small even at low income levels. Expert Meeting on How to Feed the World in 2050, Rome, FAO.
79. Alexandratos, N. and Bruisma, J.(2012). World Agriculture Towards 2030/50: The 2012 Revision. Food and Agriculture Organization.
80. Smil, V. (2002). Eating meat: evolution, patterns and consequences. *Population and Development Review* 28: 599–639.
81. Food and Agriculture Organization (2012). World Agriculture Towards 2030/50: The 2012 Revision.
82. Micklin, P. (2007)). The Aral Sea disaster. *Annual Review of Earth and Planetary Sciences* 35: 47–72.
83. UNICEF, WHO (2015). Progress on Sanitation and Drinking Water – 2015 update and MDG Assessment.
84. Liu, Z. (2016). China's Carbon Emissions Report. Kennedy School, Belfer Center, Harvard Available from: http://www.belfercenter.org/sites/default/files/files/publication/China%20Carbon%20Emissions%202016%20final%20web.pdf.

85. PBL Netherlands Environmental Assessment Agency and European Commission Joint Research Center (2016). Trends in Global CO_2 Emissions 2016 Report.

86. National Bureau of Statistics of China. NBS (2015). China Statistical Yearbook 2015 (in Chinese). China Energy Statistical Yearbook 2014. Energy Statistics Office. Available from: http://tongji.cnki.net/overseas/EngN.

87. IEA (2016, 2017). World Energy Statistics and Balances online data service 2016 and 2017 editions. Internet: https://www.iea.org/statistics/relateddatabases/worldenergystatisticsandbalances.

88. Buckley, C. (2015). China burns much more coal than reported, complicating climate talks. *New York Times*, November 3, 2015.

89. Ehrlich, P.R. (1968). The Population Bomb. New York: Ballantine Books.

90. Nature Climate Change (2011). A scary statistic. 1: 321.

91. UN 2015). United Nations. World Population Prospects, 2015 Revision

5

WATER AND ARMED CONFLICT

BARRY S. LEVY
Tufts University School of Medicine, Boston, MA, USA

> The challenge of securing safe and plentiful water for all is one of the most daunting challenges faced by the world today ….Too often, where we need water, we find guns instead. Population growth will make the problem worse. So will climate change. As the global economy grows, so will its thirst. Many more conflicts lie just over the horizon.
>
> • Ban Ki-moon
> Former Secretary General of the United Nations at the World Economic Forum, Davos, Switzerland, 2008.

The global challenges to ensure availability and accessibility of water are daunting. This chapter explores these challenges in the context of preventing armed conflict over water and protecting access to water as a basic human right.

5.1 AVAILABILITY AND ACCESSIBILITY OF WATER

Although water covers two-thirds of the Earth's surface, less than 0.01% of water is accessible to people. Less than 3% of water worldwide is freshwater, two thirds of which is contained in glaciers and the polar ice caps. Of the remaining one-third of freshwater, about 70% is used for agricultural purposes.

The sharing of freshwater resources between countries is very common. About three-fifths of water flowing in all rivers is shared by two or more countries – in 263 river basins in 145 countries [1]. Two-fifths of the world's population lives in these river basins [1].

The World Bank estimates that, on average, people require between 100 and 200 L of water daily to meet their basic needs (for drinking, cooking, personal hygiene, and sanitation), amounting to about 36–72 m^3 of water per person annually. If one includes other needs for water, such as agriculture, industry, and energy production, the total annual average requirement of water per person is 1000 m^3 [2].

Water scarcity (annual water supplies below 1000 m^3 per person) affects at least 700 million people. Another 1.6 billion people face economic water shortages, in which countries lack the necessary infrastructure to utilize water from rivers and aquifers.

During 1990, six countries in Africa and five countries in the Middle East – all of which are in arid or semi-arid regions – had less than 1000 m^3 of freshwater available per person [2] and each of these 11 countries are projected to have substantially less water per person during 2025 than during 1990 due to major population increases (Table 5.1).

* This chapter represents an update of a similar chapter in the first edition of this book, which was co-authored by Barry S. Levy and Victor W. Sidel. Dr Sidel, who was a champion for health, peace, and social justice, passed away in early 2018.

Water and Sanitation-Related Diseases and the Changing Environment: Challenges, Interventions, and Preventive Measures,
Second Edition. Edited by Janine M. H. Selendy.

Companion website: https://www.wiley.com/legacy/wileychi/selendy/

TABLE 5.1 Annual per-capita water availability in selected countries of North Africa, the Middle East, and Southwest Asia per-capita water availability (cubic meters per person per year).

Country	1990	2025
Rwanda	880	350
Algeria	750	380
Burundi	660	280
Kenya	590	190
Tunisia	530	330
Israel	470	310
Jordan	260	80
Yemen	240	80
United Arab Emirates	190	110
Libya	160	60
Saudi Arabia	160	50

Source: Klare [4].

TABLE 5.2 Estimated annual population growth rates, 2017, selected countries sharing a major water source.

Major water source	Country	Annual growth rate
Nile	Egypt	2.5%
	Ethiopia	2.9%
Jordan	Jordan	2.1%
	Israel	1.5%
Tigris-Euphrates	Iraq	2.6%
	Syria	7.9%
Indus	Pakistan	1.4%
	India	1.2%

Source: US Central Intelligence Agency. The World Factbook. Available from: https://www.cia.gov/library/publications/the-world-factbook/rankorder/2002rank.html. Accessed September 27, 2017.

In addition to its direct impact on the health and well-being of individuals, an inadequate supply of water may lead to shortages in food and energy and may limit industrial development [3]. These problems will become more acute as the world population grows from approximately 7.5 billion people now to approximately 9.7 billion by 2050. The population growth rate is even higher in countries that share a major source of freshwater with one or more other countries (Table 5.2) [4]. For example, the Nile River Basin is shared by 11 countries: Sudan, Ethiopia, Egypt, Uganda, Tanzania, Kenya, Congo, Rwanda, Burundi, Eritrea, and the Central African Republic, all of which have high population growth rates.

5.2 ACCESS TO WATER AS A BASIC HUMAN RIGHT

For many years, the human right to water did not receive adequate attention, although it was specifically recognized in the Convention on the Elimination of Discrimination Against Women (1979), the Convention on the Rights of the Child (1989), and the African Charter on the Rights and Welfare of the Child (1990). In 1992, Stephen McCaffrey analyzed the human rights framework of the United Nations and related international law and concluded that there is a right to an amount of water to sustain life. In addition, he concluded that each nation has the "due diligence obligation" to safeguard this right [5].

In 1999, Peter Gleick built on this analysis and concluded that international law, international agreements, and evidence from the practice of nations broadly and strongly support the human right to a basic water requirement [6]. He has also stated the benefits of explicitly acknowledging a human right to water:

1. To encourage the international community and individual governments to renew their efforts to meet the basic water needs of their populations.
2. By acknowledging such a right, pressures to translate that right into specific national and international legal obligations and responsibilities are more likely.
3. To spotlight the deplorable state of water management in many parts of the world.
4. To help focus attention on the need to more widely address international watershed disputes and to help resolve conflicts over the use of shared water by identifying minimum water requirements and allocations for all basin parties.
5. Explicitly acknowledging a human right to water can help set specific priorities for water policy: Meeting a basic water requirement for all humans to satisfy this right should take precedence over other water allocations and investment decisions [7].

5.3 CONFLICTS OVER WATER

The number of conflicts over water has been rapidly increasing, with a remarkable sharp increase from 2010 through 2017, which may partially be an artifact of increased reporting [8] (Table 5.3). Most of these conflicts have occurred within, rather than between, countries and have often been related to subnational disputes, terrorism, and/or local violence concerning equitable access to water [8]. Although most water-related conflicts overall have arisen in areas in which violence is widespread, few have become violent.

In recent years, conflicts over water have included the following:

- Conflicts occurred in northeastern Brazil after severe drought reduced water availability; news agencies reported that one person a day was killed in "water wars" that involved local people fighting over scarce supplies.
- Violence occurred over access to a water source in Indonesia, with rival mobs from two villages attacking

TABLE 5.3 Water conflicts, 1900–2017, by year begun.

Time period	Number of conflicts begun	Avg. number per year
1900–1959	32	0.5
1960–1989	52	1.7
1990–2009	144	7.2
2010–2017	258	32.3

Source: Adapted from Gleick [8].

each other with sharp weapons, guns, and explosives, causing several deaths and injuries.
- Farmers in Egypt held more than 200 tourists hostage to protest against inadequate irrigation water.
- Violence occurred between farmers and pastoralists in a region in southeastern Tanzania that was affected by drought.
- Fights occurred between animal herders and farmers with competing needs in Burkina Faso, Ghana, and Ivory Coast.
- Thousands of farmers breeched security and stormed the area of Hirakud Dam in India to protest allocation of water to industry.
- Tamil Tiger rebels cut the water supply to government-held villages in Sri Lanka and government forces then launched attacks on the Mavil Aru Reservoir.
- Clashes occurred in Kenya and Ethiopia over water as well as livestock and grazing land in which at least 40 people died [8, 9].

Throughout much of the Middle East, access to oil has been the major resource concern for many years. However, there has also been a conflict over water rights in the Jordan River Basin for many years, and now access to water is a major resource concern throughout the region. Climate change is warming the Fertile Crescent, where the Tigris and Euphrates rivers are located, adversely affecting Iraq, Syria, Lebanon, and southeast Turkey.

In Asia, water problems are becoming serious. If current trends continue, by 2030, several low- and middle-income countries in Asia will probably face an unprecedented crisis in water quality. Between now and 2025, some cities with more than 10 million residents, such as Karachi, Mumbai, Manila, Shanghai, and Jakarta, will face severe freshwater shortages [10].

Climate change was likely a contributing factor to the start of the civil war in Syria, which has killed more than 500 000 people and caused more than 5 million people to flee to other countries. Between 2006 and 2010, a severe drought affected Syria, causing close to 1 million farmers and their family members to abandon their farms and migrate to cities, which contributed to political and socioeconomic instability [11].

Water scarcity is infrequently the sole cause of armed conflict over water. The major underlying reasons for conflicts over water include:

- Low rainfall, inadequate water supply, and dependency on one major water source.
- High population growth and rapid urbanization.
- Modernization and industrialization of societies.
- History of armed combat, and poor relations between countries and among groups within countries [4].

In addition, issues concerning transnational pollution, resource scarcity, and inequitable access to resources and their use are often interwoven with these underlying reasons [9].

The immediately precipitating causes of conflicts over water are often sociopolitical tensions, including border disputes; disputes concerning dams, reservoirs, and other large-scale projects; and disputes concerning environmental and resource issues. According to Peter Gleick, the relationship between water and conflict can be placed in the following four categories:

1. *Water as a military tool (state actors).* Where water resources, or water systems themselves, are used by a nation or state as a weapon during a military action.
2. *Water as a military target (state actors).* Where water resources or water systems are targets of military actions by nations or states.
3. *Terrorism (non-state actors).* Where water resources or water systems are the targets or tools of violence or coercion by non-state actors.
4. *Development disputes (state and non-state actors).* Where water resources or water systems are a major source of contention and dispute in the context of economic and social development [12].

5.4 HEALTH CONSEQUENCES OF VIOLENT CONFLICT

Violent conflict, in general, has numerous serious health consequences to individuals and to the health of the public:

1. Deaths, injuries, and illnesses and long-term physical and mental impairment.
2. Destruction of the health-supporting infrastructure of society, including medical care and public health services, safe food and water supply, sanitation and waste disposal systems, transportation, communication, and energy supply.
3. Forced migration, creating refugees to other countries and internally displaced persons within countries.

4. Violation of human rights.
5. Diversion of human and financial resources away from health and other human services.
6. Promotion of further violence [13].

5.5 PREVENTING ARMED CONFLICTS OVER WATER

One set of approaches to preventing conflicts over water are measures to increase the availability of water. These include:

1. Reducing the use of water, such as reducing wasteful uses and increasing efficiency of water use.
2. Increasing availability of clean water, such as by reducing industrial pollution and sewage contamination of water.
3. Establishing and maintaining new groundwater wells.
4. Designing and implementing improved methods of desalinization.

Another set of approaches to addressing these problems is resolving water conflicts before they "boil over" -- before they become violent or develop more serious consequences. These measures include:

- Laws and regulations, at the local, state (or province), national, or international level.
- Proactive cooperation among nations or among states or provinces within nations.
- Mediation and arbitration.

By resolving conflicts over water, countries and groups within countries can help to build sustainable peace. Water plays an important role in improving and maintaining health, food security, and economic prosperity. Therefore, sound water management practices help to support social, environmental, and economic stability and help prevent violent conflict over water. Sustainable water management is a prerequisite for establishing and maintaining peace.

There have been more than 3800 unilateral, bilateral, or multilateral declarations or conventions concerning water, including at least 286 treaties, at least 61 of which refer to more than 200 international river basins [14]. Most of these multilateral declarations or conventions focus on the use, development, protection, and conservation of water resources. In addition, there are many examples of cooperation and peacebuilding that have helped prevent or resolve conflicts over water.

There were 1831 water-related events between nations in the Middle East between 1948 and 1999 [9]. Only 37 incidents led to acute conflict [15]. Two-thirds resulted in cooperation among these nations and most of the rest did not escalate beyond verbal arguments [15].

One example of such cooperation has been the Good Water Neighbors project, established in 2001, which has used transboundary water resources to build peace. It has brought together Israeli, Jordanian, and Palestinian communities to protect shared water resources and has significantly improved the local water sector and helped to build peace at the local level. The project identifies cross-border communities and uses their mutual dependence on shared water resources as a basis for developing dialogue and cooperation on sustainable water management [16].

There have been numerous examples of cooperation elsewhere, including the following:

- The Nile Basin Initiative, now among 10 countries that share the Nile River, has, since its inception in 1999, helped to develop the Nile in a cooperative manner, share substantial socioeconomic benefits, and promote regional peace and security [17].
- The Binational Autonomous Authority of Lake Titicaca was created in 1996 by Bolivia and Peru, which share Lake Titicaca. These countries have recognized how crucial it is to work together on management of water resources [18].
- The freshwater basin of the Aral Sea, which is shared by Kazakhstan, Uzbekistan, and four other countries. The surface of the Aral Sea had shrunk to less than one-half of its original size by diversion of water, which drained two rivers feeding it and devastated the environment. With the completion in 2005 of the Kok-Aral Dam, a World Bank project, the Aral Sea has now begun to fill again [19].

5.6 ROLE OF WOMEN

The role of women is especially important in promoting and maintaining cooperation among and within countries concerning peaceful access to water. Women produce about 70% of food in most low- and middle-income countries. They are major stakeholders in all development issues related to water. Yet, up until now, they have often remained on the periphery of planning and management decisions concerning water resources [20].

5.7 "A SOFT PATH"

During the twentieth century, water policies in low- and middle-income countries often focused on the construction of major infrastructure projects, such as dams, pipelines, and centralized water treatment plants. These projects yielded major benefits to many people; however, they also created major socioeconomic and ecological problems. As Peter Gleick has described, a transition is in progress to "a soft path," which complements centralized infrastructure projects with "lower cost community-scale systems, decentralized and open decision-making, water markets and equitable pricing, application of efficient technology, and environmental protection" [21].

5.8 CONCLUSION

Despite the great challenges to peace that are posed by current and imminent conflicts over water, there is reason to be optimistic that these dangers can be transformed into opportunities. As the United Nations has stated: "Despite widespread perceptions that water basins shared by countries tend to engender hostility rather than collaborative solutions, water is an often untapped resource of fruitful cooperation" [14].

REFERENCES

1. Wolf, A.T., Natharius, J.A., Danielson, J.J. et al. (1999). International river basins of the world. *Int. J. Water Resour. Dev.* 15: 387–427.
2. Gleick, P.H. (1993). Water and conflict. *Int. Security* 18 (1): 90,100–90,101.
3. Solomon, S. (2010). *Water: The Epic Struggle for Wealth, Power and Civilization*, 371. New York: HarperCollins.
4. Klare, M.T. (2001). *Resource Wars: The New Landscape of Global Conflict*. New York: Henry Holt and Company.
5. McCaffrey, S.C. (1992). A human right to water: domestic and international implications. *Georgetown Int. Environ. Law Res.* 5: 1–24.
6. Gleick, P.H. (1999). A human right to water. *Water Policy* 1: 497–503.
7. Gleick, P.H. (2000). *The World's Water, 2000–2001: The Biennial Report on Freshwater Resources*. Washington, DC: Island Press.
8. Gleick, P.H. (2018). The water conflict chronology. Available from: http://worldwater.org/water-conflict/. Accessed June 25, 2018.
9. Gleick, P.H. and Heberger, M. (2014). Water conflict chronology. In: *The World's Water: The Biennial Report on Freshwater Resources*, vol. 8 (ed. P.H. Gleick), 173–208. Washington, DC: Island Press.
10. Tacio, H.D. (2017). *Asia's Looming Water Crisis*. Gaia Discovery. Available form: http://www.gaiadiscovery.com/latest-planet/asias-looming-water-crisis.html. Accessed September 27, 2017.
11. NPR staff (2017). How could a drought spark a civil war? Available from: http://www.npr.org/2013/09/08/220438728/how-could-a-drought-spark-a-civil-war. Accessed September 27, 2017.
12. Gleick, P.H. (1998). *The World's Water: The Biennial Report on Freshwater Resources, 1998–1999*, 107. Washington, DC: Island Press.
13. Levy, B.S. and Sidel, V.W. (eds.) (2008). *War and Public Health*, 2e. New York: Oxford University Press.
14. United Nations (2017). From water wars to bridges of cooperation: exploring the peace-building potential of a shared resource. 10 stories the world should hear more about. Available from: http://www.un.org/events/tenstories/06/story.asp?storyID=2900. Accessed September 27, 2017.
15. Postel, S. and Wolf, A.T. (2009). Dehydrating conflict, *Foreign Policy*, November 18, 2009. Available from: http://foreignpolicy.com/2009/11/18/dehydrating-conflict/. Accessed September 27, 2017.
16. Kramer, A. (2008). *Regional Water Cooperation and Peacebuilding in the Middle East*. Berlin: Adelphi Research.
17. Nile Basin Initiative. Available from: http://www.nilebasin.org. Accessed September 27, 2017.
18. Delli Priscoli, J. and Wolf, A.T. (2009). *Managing and Transforming Water Conflicts*, 211–214. Cambridge, UK: Cambridge University Press.
19. The World Bank. Saving a Corner of the Aral Sea. Available from: http://www.worldbank.org/en/results/2005/09/01/saving-a-corner-of-the-aral-sea. Accessed September 27, 2017.
20. Gleick, P.H. (2006). *The World's Water: The Biennial Report on Freshwater Resources, 2006–2007*, 126. Washington, DC: Island Press.
21. Gleick, P.H. (2003). Global freshwater resources: soft-path solutions for the 21st century. *Science* 302: 1524–1528.

6

ADDITIONAL MEASURES TO PREVENT, AMELIORATE, AND REDUCE WATER POLLUTION AND RELATED WATER DISEASES: GLOBAL WATER GOVERNANCE

NIKHIL CHANDAVARKAR

Thersus Sustainability, Cedar Park, Texas, USA and
Communications and Outreach Branch, Division for Sustainable Development, Department of Economic and Social Affairs, United Nations, New York, USA

Ensuring clean drinking water with improved sanitation is perhaps the single most important means to prevent disease across the world. Efforts must be undertaken at the national, regional, and international levels in both water supply and sanitation to ensure health and well-being for all.

Water is essential for life. Unfortunately, some 661 million people around the world – 1 in 10 – lack access to clean drinking water and some 2.5 billion people – 1 in 3 – do not have access to basic toilets [1]. Consequently, water-borne diseases are spreading worldwide and know no borders.

Ensuring health through clean water and improved sanitation is a global challenge and requires a global solution, through effective global water governance. Accordingly, ensuring clean water forms an integral part of the water and sanitation agenda of the most representative global governance institution, the United Nations. The UN system's water and sanitation policy and operational work in support of member states is coordinated by UN-Water, an interagency mechanism that brings together more than 30 UN entities that deal with water and sanitation [2]. Among the many accomplishments of UN-Water are annual assessments of the water and sanitation situation in the world, several editions of the World Water Development Report, which provide a strategic overview to policy-makers and policy-shapers on global water challenges, and a series of policy briefs for decision-makers on such subjects as water and climate change, gender, and water and transboundary water cooperation.

Water governance refers to political, social, economic, and administrative systems that regulate the development and management of water resources and provision of water services. How water resources and services are governed depends on the socioeconomic, political, and historical situations of individual countries [3].

Most nations are beset by comparable water challenges, such as addressing water scarcities from climate change, rapid urbanization, and population growth, and ensuring effective water infrastructure for all, including pipe systems, pumping stations, filtering plants, and sewage systems. Additionally, countries must address political and social issues of improving service coverage, effectiveness, and affordability, raising quality standards, ensuring transparency and accountability of water operations, and resolving international water conflicts. Nations need to meet public health needs as well as ecosystem needs of water.

The massive capital costs and economies of scale that characterize water resources management make it a natural monopoly, which needs government regulation no matter what type of governance is put in place. All forms of water governance need to meet two basic practical and ethical goals: improving the water supply and defending the public interest, recognizing that water is a public good essential for life.

Possible governance structures range from a fully public option, to public–private partnerships, to a completely private option. Water governance arrangements vary greatly across

Water and Sanitation-Related Diseases and the Changing Environment: Challenges, Interventions, and Preventive Measures,
Second Edition. Edited by Janine M. H. Selendy.

Companion website: https://www.wiley.com/legacy/wileychi/selendy/

the world. However, globally, in both developing and developed countries, some 85% of drinking water is publicly provided.

Fully public governance of water is usually the responsibility of national or municipal authorities. Public officials make decisions and funding is provided from public coffers, loans, or user charges. Public authorities oversee water operations, set quality and performance standards, and promote public information and participation.

Public–private water partnerships imply privatizing some of the assets or operations of a public water system [4]. Such partnerships involve service contracts, management contracts, leases, and concessions. Additionally, ownership of water systems can be split among private and public shareholders in a corporate utility. Majority ownership, however, is usually maintained within the public sector, while private ownership is often legally restricted.

At the other extreme, private governance, which is highly exceptional in today's world, implies that the government transfers the water operations to the private sector. As such, infrastructure, capital investment, commercial risk, and operations and management become the responsibility of the private provider.

Fully private businesses and entrepreneurs tend to occur in areas where public water service is poor or nonexistent. These private businesses range from large multinational water companies to small-scale private water vendors at local level.

Good national governance of water resources, however, is only part of the story. Climate change is making the whole world aware of the interdependence of water resources across boundaries. Ensuring safe water supplies for current and future generations will depend on effective supranational governance of water.

Supranational governance of water and sanitation involves non-binding, hortatory resolutions of international governing bodies, particularly the United Nations General Assembly, which brings together 193 countries, as well as binding instruments under international law, such as the Protocol on Water and Health under the Convention on the Protection and Use of Transboundary Watercourses and International Lakes, which require ratifying countries to enact legislation in line with the provisions of the convention once they enter into force [5]. The 2030 Agenda and its sustainable development goal (SDG) 6, approved by the United Nations General Assembly in 2015 [6], now provides an overarching intergovernmental framework for the global governance of water and sanitation issues. SDG 6 seeks to "ensure availability and sustainable management of water and sanitation for all" by the year 2030 and is composed of six substantive targets and two means of implementation targets [7].

The Convention on the Protection and Use of Transboundary Watercourses and International Lakes, adopted by the UN General Assembly in May 1997, ratified by 36 countries and the European Union, seeks to promote local, national, and regional measures to protect and ensure the ecologically sustainable use of transboundary surface waters and groundwater, preventing transboundary impacts, protecting ecosystems, including from hazardous substances, and preserving ecosystems for future generations, as well as promoting public participation in decision-making.

The Convention is intended to strengthen national measures for the protection and ecologically sound management of transboundary surface waters and groundwater. The Convention obliges Parties to prevent, control, and reduce water pollution from point and non-point sources. The Convention also includes provisions for monitoring, research and development, consultations, warning and alarm systems, mutual assistance, institutional arrangements, and the exchange and protection of information, as well as public access to information [8].

Moreover, several bilateral or multilateral agreements between European countries are based on the principles and provisions of this Convention. A first example was the Danube River Protection Convention in 1994, which develops the Convention's provisions in a more specific subregional context. Other examples are the agreements on the rivers Bug, Meuse, Rhine, and Scheldt, on Lake Peipsi, as well as on Kazakh–Russian and Russian–Ukrainian transboundary waters. Recent examples include the 1999 Rhine Convention and the European Union Water Framework Directive.

The Protocol on Water and Health under the above water convention tackles the issue of ensuring access to safe water and adequate sanitation. The Protocol goes beyond the millennium development goal of halving the percentage of those without access to water and sanitation and calls for access to affordable and potable drinking water for all people across the world.

Water is now a human right. On 28 July 2010, by its resolution 64/292, the United Nations General Assembly formally recognized the human right to water and sanitation and acknowledged that clean drinking water and sanitation are essential to the realization of all human rights [9]. A UNFCCC Synthesis Report [10] analyzing national climate plans and commitments of 119 countries identified water as their top climate change adaptation priority, followed by agriculture, health, and ecosystems.

Ultimately ensuring human health through clean water and adequate sanitation will involve both national and supranational governance, whether at the regional or global levels. International legislative outcomes such as resolutions and conventions on water and sanitation, ranging from the non-binding to the legally binding, will be essential to ensure that water and sanitation as a public good and as a human right are provided worldwide to all humanity.

REFERENCES

1. UN-Water Joint Monitoring Programme Report (2010). Available from: http://www.unwater.org/activities_JMP2010.html.
2. UN-Water website, http://www.unwater.org/flashindex.html.
3. UNDP, Water Governance for Poverty Eradication. Available from: http://www.undp.org/water/pdfs/241456_UNDP_Guide_Pages.pdf.
4. WaterWiki, Public Private Partnerships in Water Sector. Available from: http://waterwiki.net/index.php/Public-Private_Partnerships_in_the_Water_Sector.
5. UN-Water, Institutional Capacity Development in Transboundary Water Management. Available from: http://www.unwater.org/downloads/05_Institutional_Capacity_Development_in_Transboundary_Water_Management.pdf.
6. United Nations, Transforming Our World. Available from: https://sustainabledevelopment.un.org/post2015/transformingourworld.
7. United Nations, Sustainable Development Knowledge Platform, SDG 6. Available from: https://sustainabledevelopment.un.org/sdg6.
8. UNECE, The Convention on the Protection and Use of Transboundary Watercourses and International Lakes. Available from: http://www.unece.org/env/water/text/text.htm.
9. United Nations, Human Right to Water and Sanitation, General Assembly Resolution A/64/292. Available from: http://www.un.org/es/comun/docs/?symbol=A/RES/64/292&lang=E.
10. World Water Day website, http://www.worldwaterday.org/.

SECTION II

WATER AND SANITATION-RELATED DISEASES

7

INFECTIOUS DIARRHEA

Sean Fitzwater[1], Anita Shet[2], Mathuram Santosham[2], and Margaret Kosek[2]

[1] Department of Pediatrics, University of California, Los Angeles, Los Angeles, CA, USA

[2] Department of International Health, Johns Hopkins Bloomberg School of Public Health, Baltimore, MD, USA

7.1 INTRODUCTION

Diarrhea is one of the leading causes of morbidity and death globally. Children and young infants are at highest risk. The majority of diarrheal illnesses are caused by viral, bacterial, and parasitic pathogens, which disrupt the gut's normal secretory and absorptive functions. Most infectious diarrheal episodes are acute in nature and can lead to rapid loss of bodily fluids, dehydration, and death if proper supportive care is not provided promptly. Repeated bouts of acute diarrhea and persistent diarrhea can exasperate malnutrition and lead to poor growth, impaired immune function, and increased susceptibility to diarrhea as well as other infections. Fortunately, most infectious diarrheal illnesses can be successfully managed using oral rehydration therapy and, in severe cases, with intravenous therapy. In cases of dysentery and cholera, antimicrobial therapy is indicated. Diarrhea can be prevented to a large extent by disrupting transmission of pathogens through vaccination against leading pathogens, nutritional interventions, and sanitary interventions including improving personal hygiene, water quality, disposal of waste, and appropriate preparation of food. Together, improved case management and prevention of diarrhea has led to a dramatic reduction in diarrheal deaths over the last three decades, most notably in children. In the early 1980s approximately 5 million pediatric diarrheal deaths were estimated to occur; by 2015 there were less than 600 000 deaths a year from diarrhea in children.

7.2 EPIDEMIOLOGY

Infectious diarrhea is a leading cause of disease and death worldwide. Each year there are approximately 2.4 billion episodes of diarrhea resulting in 1.3 million deaths [1]. Globally, infectious diarrhea is responsible for 2.3% of all deaths. Most of this morbidity and mortality occurs among young children (Table 7.1). In 2015 the average child under the age of five had a mean of 1.4 episodes of diarrhea each year, which led to an estimated 498 000 deaths, making diarrhea the fourth most common cause of under-five deaths [1].

Multiple factors are associated with increased risk of diarrheal diseases, including lack of access to clean water, inadequate disposal of human waste, overcrowding, limited quality control in food handling and processing, and malnutrition [2]. As a result, the burden of diarrheal illnesses varies widely globally (Table 7.2 and Figure 7.1). The highest incidence of diarrheal disease occurs in low- and middle-income countries that lack basic sanitary preventive measures. Additionally, deaths due to diarrhea occur almost exclusively in low- and middle-income countries due to lack of access to appropriate medical care (Figure 7.2). The WHO estimates that in 2015, over 98% of deaths due to diarrhea occurred in low or low-middle-income countries [1]. In total, the 10 countries with the highest number of deaths due to diarrhea (India, Nigeria, Indonesia, Pakistan, Ethiopia, Democratic Republic of the Congo, Kenya, Niger, Tanzania, and Chad) accounted for 71% of all diarrheal deaths globally [1].

Water and Sanitation-Related Diseases and the Changing Environment: Challenges, Interventions, and Preventive Measures, Second Edition. Edited by Janine M. H. Selendy.

Companion website: https://www.wiley.com/legacy/wileychi/selendy/

TABLE 7.1 Morbidity and mortality associated with diarrheal disease stratified by age group.

Age group (years)	Deaths	Deaths (per 100 000)	Cases (millions)	Cases (per person)
0–4	498 641	74.7	955	1.43
5–14	51 611	4.2	394	0.32
15–49	171 438	4.6	679	0.18
50–69	223 527	18.1	213	0.17
>70	365 762	93.8	145	0.37

Source: Number of deaths adapted from The Global Burden of Disease: Update 2015; WHO [1].

TABLE 7.2 Morbidity and mortality associated with diarrheal disease stratified by income group.

Income group	Deaths (total)	Deaths (per 100 000)	Cases (millions)	Cases (per person)
High	26 481	2.2	21	0.02
Upper middle	57 123	2.2	364	0.14
Lower middle	890 597	30.4	1454	0.50
Lower	336 779	55.4	546	0.90
Global	1 310 980	18.0	2386	0.33

(Number of deaths and cases adapted from The Global Burden of Disease: Update 2015; WHO [1]. Income stratification is based on World Bank classification, 2015).

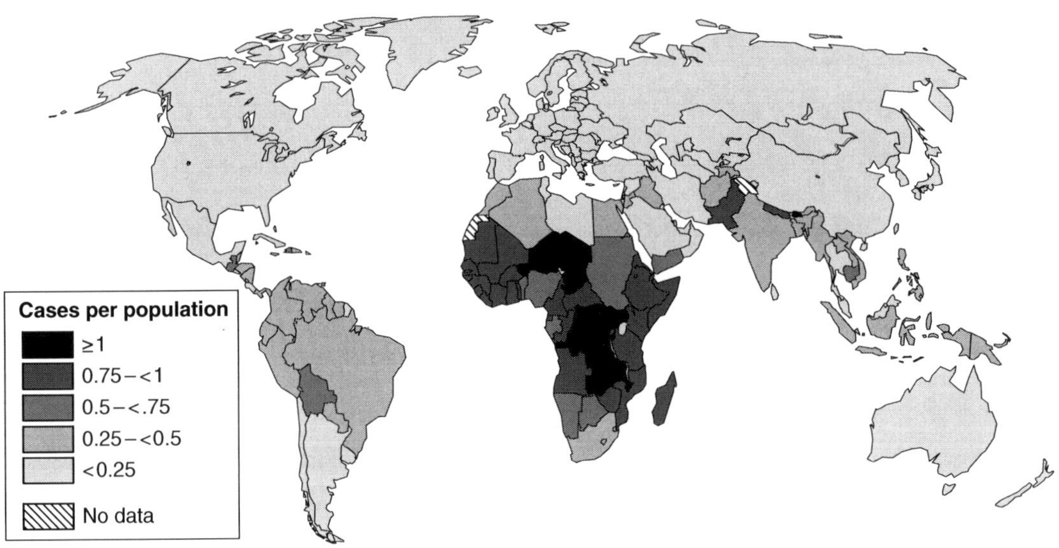

FIGURE 7.1 Incidence on diarrhea per population. *Source*: Adapted from The Global Burden of Disease: 2015 Update; WHO [1]. *(See insert for color representation of this figure.)*

By comparison, less than 27 000 deaths due to diarrhea occurred in all high-income countries combined. Although diarrhea rarely causes death in high-income countries, it still is a significant health problem and accounts for considerable utilization of healthcare services. In the US, between 1997 and 2000, there were 150 000 childrenhospitalized each year with diarrhea, accounting for 13% of all hospital admissions for children under five years old [3].

The burden of diarrheal disease varies within countries, with the highest rates of disease and deaths occurring in impoverished and neglected populations [4, 5]. In high-income countries, minority populations tend to have higher rates of diarrhea than the general population due largely to poorer socioeconomic conditions, although most severe disease in all social strata are concentrated in children and the elderly. In the United States, Native American and Alaskan Native infants have higher rates of diarrhea-associated outpatient visits than the general population [6]. In all countries, natural disasters and other humanitarian crises put populations at increased risk of diarrheal disease by breaking down the regular sanitation infrastructure, displacing populations, and facilitating the spread of pathogens while concurrently interrupting access to optimal case management [7]. In Bangladesh, the floods of 1988, 1998, and 2004 resulted in a doubling of diarrheal cases seen at the International Center for Diarrheal Disease Research, from an average of 249

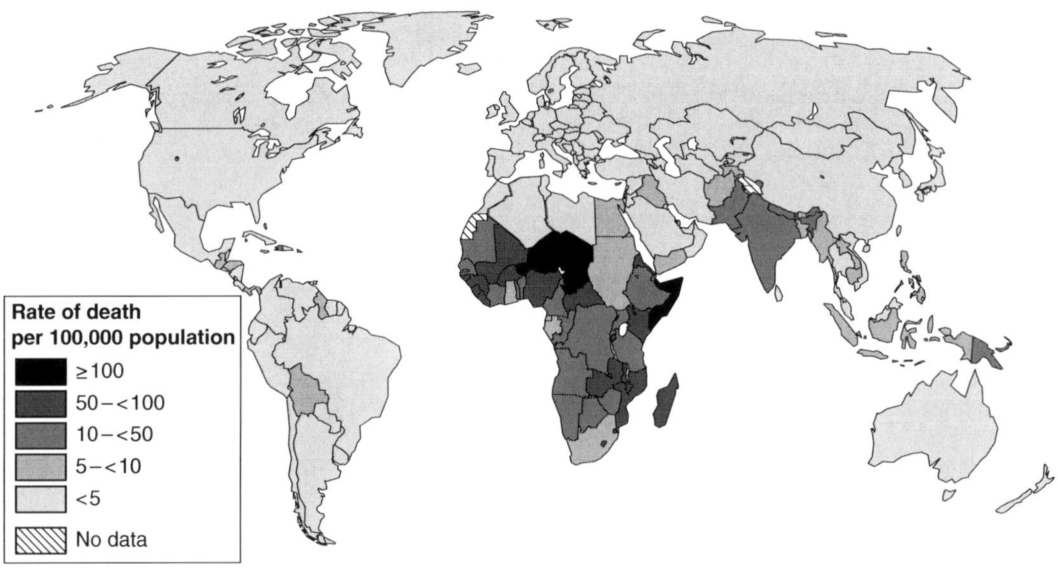

FIGURE 7.2 Rate of death due to diarrhea per 100 000 population. *Source*: Adapted from The Global Burden of Disease: 2015 Update; WHO [1]. Population estimates from World Bank database 2015. (*See insert for color representation of this figure.*)

cases per day during non-flood times to 542 cases per day during flooding [8]. Well-planned responses to crises have been demonstrated to limit the mortality caused by diarrhea. A study of refugee camps in Tanzania from 2005 to 2007 found increased rates of diarrheal illness compared to the general population, but a decreased risk of death. The rate of death in refugee camps varied yearly from 0.05 to 0.07 deaths per 1000 diarrheal cases, while the national average was over 12 times higher (0.88 per 1000) [9]. The effects of climate change on diarrhea are unclear and based on limited data [10]. However, any increase in natural disasters or instability in local climate are likely to exasperate diarrhea, at least transiently, in already vulnerable populations [7, 8, 11].

7.3 DEFINITION AND PHYSIOLOGY

Diarrhea is defined by the WHO as having at least three loose or watery stools per day [12] and is often accompanied by an increased sense of urgency, abdominal discomfort, and alterations in stool characteristics [13]. These symptoms are caused by an alteration in the gastrointestinal tract's ability to absorb nutrients, augmented secretion of fluids, or both. Infectious agents accomplish this through a combination of inflammatory and non-inflammatory processes [14].

Non-inflammatory processes leading to diarrhea are non-invasive, and are caused by enterotoxin-producing organisms or by organisms that disrupt the normal absorptive or secretory functions of the intestinal lining without causing significant inflammation or mucosal damage [14]. A variety of pathogens produce enterotoxins, substances that change the permeability of the cells lining the intestinal tract by altering the function of specific cellular pathways [15–18]. The specific target of an enterotoxin is pathogen-dependent, but the result is an alteration to the function of cell membrane bound transport proteins and ion channels. This leads to a dysregulated outflow of water into the lumen of the intestine, resulting in diarrhea [14]. Other organisms act through less specific processes that lead to damage to the intestinal mucosa. Pathogen adherence and replication can lead to an increased rate of apoptosis in enterocytes with minimal inflammation [14, 19, 20]. This damage leads to a disruption in the epithelial barrier and changes the pathology of the intestinal mucosa, resulting in a reduction in its ability to selectively absorb and secrete water and nutrients.

Inflammatory diarrhea is a result of pathogen-produced cytotoxins or active invasion of the intestinal tissue by pathogens [14]. Cytotoxin-producing pathogens release toxins that damage epithelial cells or disrupt the junctions between cells [18, 21–23]. Some organisms directly invade the intestinal mucosa and, often with the aid of toxins, induce an acute inflammatory reaction [23–26]. In both cases this leads to ulcerations in the intestinal mucosa. This results in damage to the mucosal surface, disruption of absorption and secretion, and the presence of blood, leucocytes, and/or mucus in the stool. Furthermore, ulcerations can act as a portal for other intestinal organisms to enter the body, increasing the likelihood of sepsis.

7.4 MANIFESTATIONS OF INFECTIOUS DIARRHEA

7.4.1 Acute Watery Diarrhea

The most common manifestation of infectious diarrhea is acute watery diarrhea, which accounts for approximately 85% of all diarrheal cases globally [27]. Acute watery diarrhea is categorized by the WHO as episodes of loose stool with a duration less than 14 days in duration and is characterized by discharges of watery feces with or without the presence of mucous [28]. The main clinical complication associated with acute diarrhea is dehydration, which is often more severe in children than adults [29]. Most uncomplicated cases of acute watery diarrhea will resolve with appropriate rehydration therapy and do not require specific medicinal treatments or laboratory diagnostics [13, 30]. Since there is considerable loss of nutrients during diarrheal episodes, it is critical to include continued age-appropriate feeding as part of the treatment of childhood diarrhea [31]. Although the case fatality rate is relatively low, <1% even in low income settings, the large number of cases results in the majority of deaths due to diarrhea being caused by acute watery diarrhea [32].

7.4.2 Persistent Diarrhea

Although most cases of infectious diarrhea resolve within two weeks, 3–19% of acute cases continue beyond 14 days, becoming persistent diarrhea [28, 30]. Persistent diarrhea is characterized by chronic or relapsing infections with an infectious agent [33]. Progression from acute to persistent diarrhea is most strongly associated with the nutritional status of the host (such as low weight for age in children), where severe acute diarrheal cases, shigellosis, and dysentery are more likely to become persistent [30, 34, 35]. As with acute diarrhea, persistent diarrheal patients are at risk of dehydration if not provided rehydration therapy [31]. Additionally, persons with persistent diarrhea are at an increased risk of malnutrition because of chronic malabsorption and are at an increased risk of developing secondary infections spreading beyond the gut [31]. The outcome and duration of persistent diarrhea can be improved if the causative pathogen is identified and treated [33]. In developing countries, specific diagnostics and treatment are generally not available, resulting in relatively high risk of death. For example, up to 12% case fatality rates have been reported for non-dysenteric persistent diarrhea in rural India [32].

7.4.3 Bloody Diarrhea

Bloody diarrhea, also known as dysentery, is typically accompanied by abdominal pain, mucous in the stool, and often systemic toxicity [27, 36]. Blood in the stool can accompany chronic or acute infectious diarrhea; approximately 5–10% of all diarrheal cases are dysenteric or become dysenteric [27, 32]. Bloody diarrhea is typically caused by bacterial or parasitic pathogens, although parasitic dysentery is relatively rare in developed countries [36, 37]. Dysenteric patients are at risk of dehydration, but much less so than cases with acute or persistent watery diarrhea. However, since bloody stool is indicative of a break in the gastrointestinal tract, dysenteric patients are at an increased risk of developing sepsis, especially in young infants and severely malnourished children [31]. Additionally, damage to the gastrointestinal mucosa disrupts nutrient absorption more than non-complicated diarrhea, which puts dysenteric patients at a higher risk of developing malnutrition [31]. Treatment with effective antibiotics in addition to rehydration is indicated as it shortens illness, decreases mortality, and limits disease transmission [31]. Empiric antimicrobial therapy should be directed at the most common bacterial causes of dysentery, *Shigella* and *Campylobacter*, as rapid and accurate diagnostic tests are generally not available in contexts where they are most needed. Therapy must be guided by regional antimicrobial resistance data. With proper treatment, the likelihood of death from dysentery is low, although increasing antibiotic resistance to both *Shigella* and *Campylobacter* make options for treatment progressively more limited [38, 39].

7.5 ETIOLOGY OF INFECTIOUS DIARRHEA

The etiology of infectious diarrhea varies globally and is caused by a heterogeneous mix of viral, bacterial, and parasitic pathogens (Table 7.3). The understanding of the etiology of diarrhea is an evolving field, as more sensitive diagnostics and testing strategies based on nucleic acid testing has greatly expanded our ability to detect pathogens. Determining the exact etiology of diarrheal illnesses is further complicated as classically identified enteropathogens may be present in stool in the absence of diarrhea and symptomatic diarrheal disease likely a complex interplay between multiple potential pathogens and host factors. Using modern sensitive diagnostics targeted at a broad array of enteropathogens, putative agents are encountered in most stools evaluated, even in the absence of symptoms, and finding multiple pathogens is common [4, 37, 58, 59]. A recent multicenter study found 65% of non-diarrheal stools to have detectable pathogens, compared to 77% of diarrheal stools [58]. Two or more pathogens were identified in 41% of diarrheal and 29% of non-diarrheal stools [58].

The attribution of a single episode of diarrhea is therefore challenging in many populations living in countries with high endemicity of diarrheal pathogens. Several methods have been developed to attribute burden to specific pathogens to guide public health interventions. Classically the ratio of prevalence of diarrhea compared to non-diarrheal stools was used to estimate pathogenicity of organisms, sometimes called a pathogenicity index [60]. More recently population attributable fractions are employed [61], which combine the effect of the relative ratio of pathogens in diarrhea to non-diarrheal stools with the prevalence of the etiology under evaluation in diarrheal stools. Using these

TABLE 7.3 Etiology of infectious diarrhea.

Viruses	Syndrome	Global burden
Rotavirus	Acute diarrhea Often severe	Prior to vaccination: 29–47% of diarrhea hospitalizations [40] 114 million episodes of gastroenteritis, 24 million outpatient visits, 2.4 million hospitalizations each year [41]. 475 000–580 000 deaths annually [42]
Norovirus	Acute diarrhea	7–22% of childhood acute gastroenteritis [43]; 68–80% of out breaks of gastroenteritis in developed countries [44]; ~684 million cases of diarrhea and 212 000 deaths annually [45]
Astrovirus	Mild acute diarrhea	2–9% of childhood acute diarrhea [43]
Adenovirus	Acute, persistent, and bloody diarrhea	2–6% of childhood acute gastroenteritis [43]; 7–22% of moderate to severe diarrhea [46]
Measles	Disposes for severe acute and persistent diarrhea	Complication in 12–63% of all measles cases [47–51] Implicated in 2–5% of diarrheal cases in unvaccinated children [51]
Bacteria		
Escherichia coli	Acute watery, persistent, bloody diarrhea	30–40% of diarrhea cases [37]. 23–30% of persistent diarrhea [28]
Enterotoxigenic	Acute diarrhea	15–30% of diarrhea illness [52]; ~240 million cases of diarrhea and ~74 000 deaths annually [45]
Enteropathogenic	Acute and persistent diarrhea	4–8% of diarrhea illness [52]. <30% of persistent diarrhea [28, 53]; ~81 million cases of diarrhea and ~122 000 deaths annually [45]
Enteroaggregative	Acute, persistent, and bloody diarrhea	<30% of persistent diarrhea [28, 53]
Enterohemorrhagic	Bloody diarrhea, hemorrhagic colitis	~2.8 million cases of diarrhea annually [54]
Enteroinvasive	Bloody diarrhea	<4% of diarrheal cases [55].
Shigella	Acute watery diarrhea, severe bloody diarrhea	5–15% of diarrheal cases [46, 56]; 10% diarrheal deaths [56], 30–64% of dysentery [46, 56]; ~190 million cases of diarrhea and 66 000 deaths annually [45].
Salmonella, non-typhi	Acute watery diarrhea, rarely bloody diarrhea	0–15% of diarrheal cases [52] 1.4 million cases in US [38]
Campylobacter	Acute diarrhea, bloody diarrhea	2–14% of all diarrhea disease [46, 52]; ~166 million cases of diarrhea and 36 000 deaths annually [45]
Vibrio cholerae	Severe acute watery	~2.9 million cases of diarrhea; ~95 000 deaths annually [57]
Aeromonas	Dysentery	Significant cause of moderate and severe diarrhea and dysentery in Southwest Asia [58]
Parasites		
Cryptosporidium	Acute diarrhea, persistent diarrhea, dysentery	1–17% of moderate to severe diarrhea [46] 16% persistent diarrhea [30]; ~64 million cases of diarrhea and ~27 000 deaths annually [45]
Giardia lamblia	Mild acute diarrhea, persistent diarrhea	<10% persistent diarrhea [28]; ~183 million cases of diarrhea annually [45]
Entamoeba hystolytica	Severe dysentery,	~104 million cases and ~5000 deaths annually [45]

methods, two recent large studies inform the etiologic burden estimates for enteropathogens in children living in impoverished settings. The first, The Global Enteric Multicenter study (GEMS), was a health-center-based case control study of moderate to severe diarrhea in children under the age of five [46]. This was followed by the Etiology, Risk Factors, Interactions of Enteric Infection, and Malnutrition and the Consequences for Child Health and Development (MAL-ED) study, a community-based longitudinal cohort of children under the age of two years of age, which assessed the etiology for all causes of diarrhea, regardless of severity [58]. These large multisite studies provide the most complete and unbiased data available to inform enteropathogen-specific disease estimates in populations with a high burden of disease.

7.5.1 Viral Pathogens

The majority of viral gastrointestinal infections are caused by rotavirus, norovirus, astroviruses, and adenovirus [62]. Numerous other viral pathogens are known to cause diarrhea, including toroviruses, coronaviruses, picornaviruses, and pestiviruses, but are of limited (or unknown) epidemiological importance [43, 44, 62]. The etiology of viral diarrhea has been best characterized in developed counties, where they represent the majority of acute diarrheal cases and are responsible for most food-borne outbreaks of gastroenteritis [63]. The role of specific viral pathogens in developing countries was, until recently, limited due largely to the availability of diagnostic testing [64]. Recent large multinational studies have greatly improved our understanding of the role of all enteric pathogens in developing nations [58, 65]. The symptoms of viral gastroenteritis are similar regardless of the specific pathogen, manifesting with acute non-inflammatory watery diarrhea, fever, and vomiting, although the severity of symptoms varies by virus [66, 67]. Symptoms typically resolve within a week of disease onset; specific diagnostic tests and treatment are not necessary, although supportive care using rehydration therapy is critical [68]. Endemic viral diarrhea occurs mainly in young children, peaking during cooler months, although seasonality is less pronounced in tropical settings [59, 69, 70]. The fecal oral route is thought to be the main transmission route in developing countries, but there is an increasing consensus that viral pathogens are also spread by person to person transmission or aerosolization of viral particles [2, 71].

7.5.1.1 **Rotavirus *[Image 7.1]*** Rotaviruses are a group of spherical non-enveloped viruses characterized by a double-stranded, segmented, RNA genome [15, 72]. Rotavirus infection is the most common cause of diarrhea in all nations without routine rotavirus vaccination. Prior to widespread use of the vaccines there were an estimated 114 million episodes of gastroenteritis, 24 million outpatient visits, and 2.4 million hospitalizations each year [41]. Rotavirus disease occurs almost exclusively in children and infects nearly every child by the age of five years [73]. The rate of rotavirus illness is similar in both developed and developing countries; rotavirus is responsible for 29–45% of all hospitalizations due to diarrhea, with higher proportions seen in countries with a higher average income [40]. In total, there where over 500 000 deaths attributed to rotavirus in 2004, resulting in 5% of all deaths in children <5 years of age [42, 74]. More recent estimates since the wide introduction of rotavirus vaccine indicate that the number of rotavirus associated deaths among children under five years of age approximated 197 000 in 2011 [75].

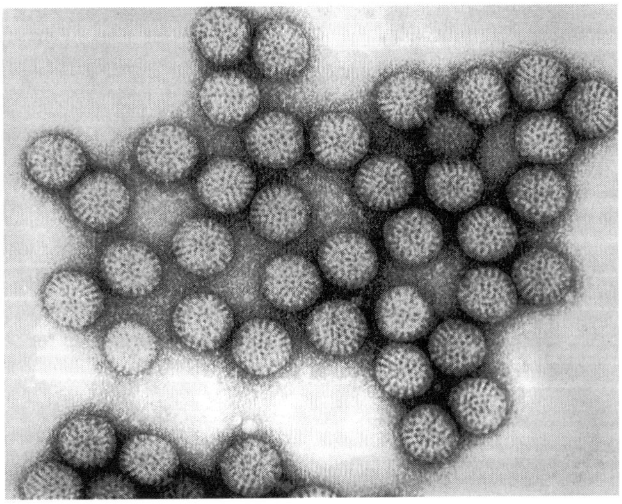

IMAGE 7.1 Colorized transmission electron micrograph of rotavirus virions. (Content provider: CDC/Dr Erskine Palmer. *Source*: CDC, Public Health Image Library, Image ID # 178.)

The main transmission route in developing countries is thought to be fecal oral, as viral particles are shed profusely in feces of patients [15]. However, the prevalence of rotavirus in developed countries with excellent sanitation suggests that non-fecal routes play a large role in transmission [2, 76]. Unlike other pathogens, improvements in water and sanitation do not reduce the risk of rotavirus diarrhea. In countries where there have been significant improvements in water and sanitation as well as economic development, the proportion of diarrheas caused by rotavirus have also increased.

The viral capsid, composed of a glycoprotein (G-type antigen) coat with protease-sensitive protein (P-type antigen) surface spikes, is the major antigenic target in humans. Over forty G and P antigen combinations have been observed in humans, although globally, six G/P combinations account for 57–100% of rotavirus disease, depending on region [77]. This diversity results in decreased cross-protection from other rotavirus serotypes after infection with one serotype [15]. Rotavirus causes disease by disrupting gastrointestinal function using a variety of strategies including reducing expression of digestive enzymes, altering tight junctions between cells, and stimulating the secretion of chlorine ions into the intestinal lumen using a potent enterotoxin [15]. After a brief incubation period of one to three days, the disruption to the gastrointestinal tract leads to the non-specific symptoms of viral diarrheal infection [15, 73]. No pathogen specific treatment exists, but most cases resolve in two to seven days after onset of symptoms. Severity varies widely from case to case, but compared to other causes of viral diarrhea, rotavirus is associated with an increased severity of gastroenteritis symptoms [66, 67]. However, repeat infection with rotavirus provides increasing protection against disease regardless of serotype, with two natural infections providing protection against severe disease and three infections protecting against any rotavirus diarrhea [78].

7.5.1.2 **Norovirus and Other Caliciviruses *[Image 7.2]*** The viral family Caliciviridae is a diverse group of single-stranded RNA viruses [44]. Norovirus is the most well-known and prevalent member of the group, although sapovirus also causes disease [64]. Norovirus is the second most common cause of gastroenteritis worldwide and is

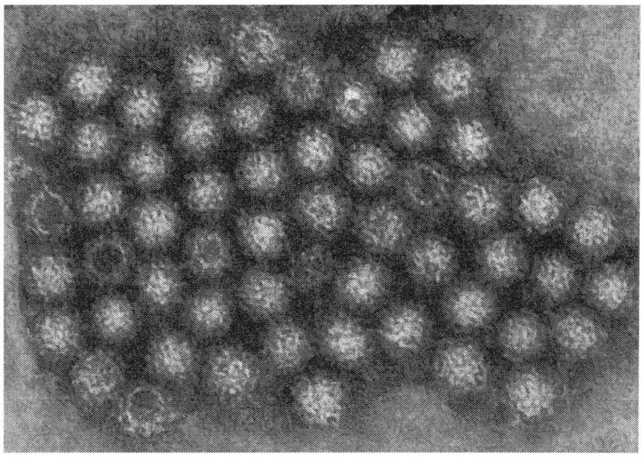

IMAGE 7.2 Colorized transmission electron micrograph of norovirus virions. (Content provider: CDC/Charles D. Humphrey. *Source*: CDC, Public Health Image Library, Image ID # 10708.)

replacing rotavirus as the most common cause of gastroenteritis in countries with routine rotavirus vaccination [79, 80]. It is estimated that norovirus causes 684 million cases of gastroenteritis and 212 000 deaths annually worldwide [45, 80]. Norovirus infection in children in developing countries is extremely prevalent. In the MAL-ED multisite longitudinal cohort conducted in eight countries, 89% of children had a documented norovirus infection before the age of 24 months, while norovirus was present in 23% of all diarrheal stools [58]. Similarly, a multiyear study in China of winter diarrhea demonstrated norovirus to be present in 24% of children who tested negative for rotavirus, while a longitudinal community cohort study in Peru found norovirus in 21% of children with diarrhea [81, 82]. The burden of norovirus and other caliciviruses is best described in developed countries in the context of outbreaks, where it is reported to be responsible for 68–80% of gastroenteritis outbreaks, predominately due to contaminated food [44].

The genera of Caliciviridae are composites of many antigenically distinct subtypes [64]. Norovirus alone has six described genogroups and 26 known genotypes [83]. Additionally, there is constant viral evolution among these groups driven by viral evasion of human immune processes, resulting in novel viruses emerging every two to four years, which replace previous dominant viral strains [83]. The best-studied norovirus genotype, GII.4, is the predominant cause of pediatric infections and healthcare-associated outbreaks amongst the elderly. Protective immunity against norovirus is not well understood; it appears that immunity against natural infection is short-lived, lasting between six months to five years, and is genotype-specific, with little or no protection conferred across genogroups [84]. The clinical presentations of calicivirus infections are indistinguishable from those of other viral causes of acute gastroenteritis. The

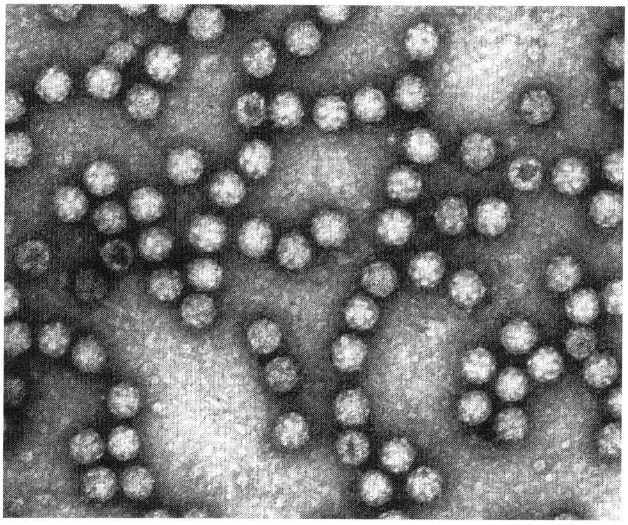

IMAGE 7.3 Transmission electron micrograph of *Astrovirus virions*. (Content provider: Graham Colm. *Source*: http://en.wikipedia.org/wiki/File: Astrovirus_4.jpg.)

classic clinical presentation of norovirus is vomiting with diarrhea, when compared to rotavirus diarrhea [69, 81]. The routes of virus transmission include fecal-oral, food-borne or water-borne modes, and can also occur from person-to-person. Efforts are ongoing to develop specific treatments and vaccines for norovirus [85, 86].

7.5.1.3 **Astrovirus** ***[Image 7.3]*** Astroviruses are a ubiquitous group of small non-enveloped single-stranded RNA viruses closely related to caliciviruses [87]. Astroviruses are the third most common cause of viral diarrhea in children, accounting for 2–16% of children hospitalized with diarrhea [67]. Additionally, astroviruses are an important cause of nosocomial outbreaks among children and immunocompromised patients [87]. Clinical manifestations are generally mild symptoms compared to rotavirus or norovirus, and of shorter duration. As a result, astrovirus infections are more common in diarrhea identified in the community compared to hospitalized patients with diarrhea [44, 67]. Most people are exposed to astroviruses in childhood. Greater than 90% of children develop antibodies against astrovirus serotype 1 in the US, Japan, and the Netherlands by the age of nine years [67]. However, different serotypes of astrovirus result in different patterns of infection with specific serotypes predominating in infants while others are more common in older children [88]. Astroviruses have a more variable seasonality than other viral causes of gastroenteritis: cases can peak in the winter or early summer depending on geography [87].

7.5.1.4 **Adenovirus** ***[Image 7.4]*** Adenovirus is a group of lytic non-enveloped DNA viruses that cause a wide array of diseases [62, 89]. Although many serotypes of adenovirus

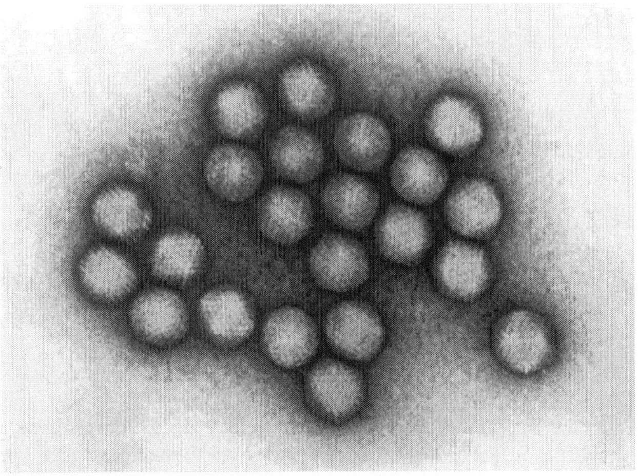

IMAGE 7.4 Colorized transmission electron micrograph of *Adenovirus virions*. (Content provider: CDC/Dr G. William Gary, Jr. *Source*: CDC, Public Health Image Library, Image ID # 10010.)

can cause diarrhea, most acute gastroenteritis infections are caused by serotypes Ad40 and Ad41, collectively known as "enteric adenoviruses" [44]. In developed countries, adenoviruses account for 5–10% of pediatric hospitalizations due to diarrhea, with the incidence peaking in the second year of life [43, 90], and can be severe in immunosuppressed individuals [44]. Adenovirus is highly prevalent in developing countries and is attributed to 7–22% of moderate to severe diarrhea across multiple countries [46] and is associated with dysentery [58]. Compared to other causes of viral gastroenteritis, adenovirus has a notably long incubation period, 8–10 days, and long duration, which lasts from 5 to 12 days [90]. Up to one-third of cases persist past 14 days [44]. Additionally, adenovirus is unique in that is shows little seasonal variation when compared with other viruses [44].

7.5.1.5 **Measles *[Image 7.5]*** Measles virus is an enveloped, single-stranded RNA virus that is spread by respiratory secretion and is well known for its characteristic rash [91]. Although generally considered a mild disease in developed countries, severe complications are common in the developing world. Diarrhea is one of the most common complications of measles infections, reported in 12–63% of all measles cases identified by community surveillance [47–51]. In unvaccinated populations, measles has been estimated to be a factor in 2–5% of all diarrhea cases [92]. The measles virus itself is not generally considered the causative agent of diarrhea. Rather, infection with the measles virus produces long-lasting immunosuppression, which predisposes children to secondary bacterial and parasitic infections that manifest as diarrhea [93–95]. This immunosuppression can last for weeks and, as a result, the majority of measles associated diarrheal complications occur after the measles rash itself has cleared [51, 93, 94]. Additionally, measles associated diarrhea is known to persist longer and has increased severity compared to non-measles associated diarrhea, increasing the likelihood of poor outcomes and making diarrhea a leading cause of death in measles patients [51, 94, 95].

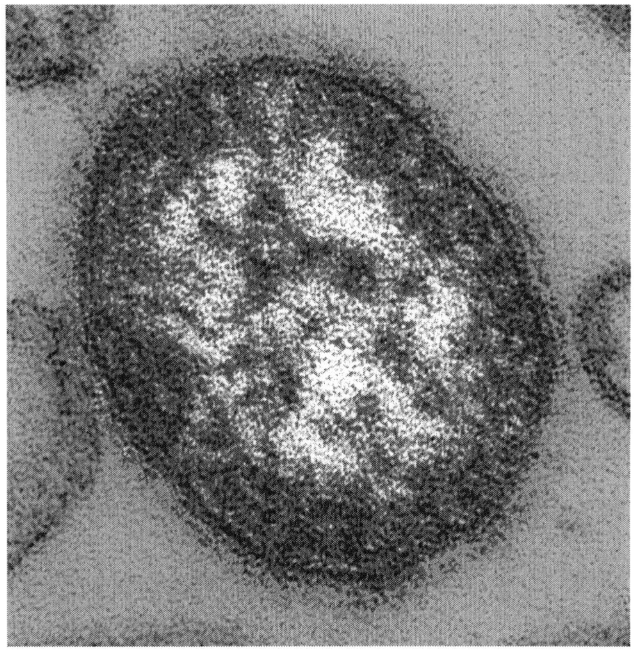

IMAGE 7.5 Colorized thin-section transmission electron micrograph transmission electron micrograph of a single measles virion. (Content provider: CDC/Cynthia S Goldsmith; William Bellini, PhD. *Source*: CDC, Public Health Image Library, Image ID # 10707.)

7.5.2 Bacterial Pathogens

Globally, most disease and death caused by bacterial diarrhea are the result of *Escherichia coli*, *Shigella*, *Vibrio cholerae*, *Salmonella*, and *Campylobacter* infections. Bacteria, much more than viruses, have a strong association with poor sanitary conditions and are exclusively transmitted by the fecal-oral route. Many bacterial pathogens have zoonotic reservoirs [37, 38]. As a result, the burden of bacterial diarrhea is located almost exclusively in developing countries, where they have been found to be responsible for about half of all cases of acute diarrhea [96]. Some bacteria, notably *Clostridium difficile*, cause significant disease in patients with pre-existing medical conditions, which has been documented predominately in developed countries [38].

Endemic bacterial diseases tend to predominate in warmer months, although outbreaks can occur in any season [37]. Bacteria are a relatively minor cause of diarrhea in developed countries, being responsible for only 2–10% of diarrheal cases, and the majority of these cases in the US are food-related illnesses [70]. Additionally, bacterial gastroenteritis is more common in patients returning from travel to developing countries, causing an estimated 80% of traveler diarrhea [38]. When gastroenteritis develops, most bacterial diarrhea is indistinguishable from viral diarrhea, causing acute watery diarrhea, fever, abdominal pain, and vomiting [70]. However, inflammatory and persistent diarrhea is common with bacterial diarrhea, which can progress to overt dysentery and sepsis [28, 70]. Leukocytes in the stool can be a useful indicator of bacterial and invasive diarrhea, and etiology can be confirmed using stool cultures, alone or in combination with polymerase chain reaction tests (PCR). However, etiological diagnosis is unnecessary in uncomplicated bacterial diarrhea, which can be treated with rehydration therapy [70]. Antibiotic treatment is only indicated in cases complicated with dysentery or sepsis or the development of chronic symptoms. Stool culture or enteric bacterial PCR panels for establishing a microbiological diagnosis in complicated cases is warranted. Stool culture is encouraged for developing regional empiric treatment guidelines as antibiotic resistance is of growing concern across all species [37]. However, it is not necessary to perform diagnostic tests for bacteria in uncomplicated acute watery diarrhea.

7.5.2.1 **Escherichia coli *[Image 7.6]*** Members of the species *Escherichia coli* (*E. coli*) group of microbes, the majority of which are non-pathogenic, are Gram-negative bacterial flora of the healthy human gastrointestinal tract [97]. However, there are five major classes of enterovirulent *E. coli*: enterotoxigenic *E. coli* (ETEC), enterohemorrhagic *E. coli* (EHEC), enteropathogenic *E. coli* (EPEC), enteroinvasive *E. coli* (EIEC), and enteroaggregative *E. coli* (EAEC) [70], all with distinct virulence factors that cause diarrhea. Collectively, these pathogens are the most common bacterial cause of diarrhea in developing countries and are implicated in 30–40% of cases of diarrhea [37]. In addition they are the most common pathogen associated with persistent diarrhea, causing 23–30% of persistent diarrhea cases [28]. The clear attribution of the burden of disease caused by these organisms is evolving as multiplex PCR assays make diagnosis in research settings more practical and affordable than previously used probe assays or cell culture techniques.

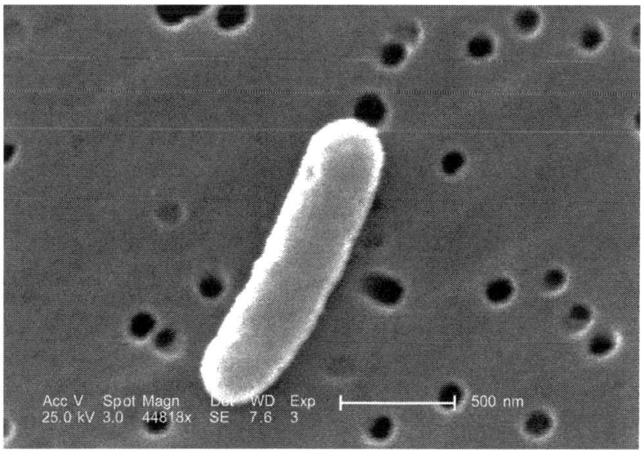

IMAGE 7.6 Colorized scanning electron micrograph a single enterotoxigenic *Escherichia coli* bacterium. (Content provider: CDC/Janice Haney Carr. *Source*: CDC, Public Health Image Library, Image ID # 10577.)

ETEC, EPEC, and EAEC predominately cause watery diarrhea in children younger than two years old [37]. ETEC is the most common cause of bacterial gastroenteritis and is alone responsible for 15–30% of diarrhea in developing countries [52, 97]. ETEC is estimated to cause 240 million cases of diarrhea and 74 000 deaths annually [45]. ETEC causes disease by producing several enterotoxins, including a cholera-like toxin, but invasive and persistent ETEC disease is rare [38]. In developed countries ETEC is the main cause of traveler diarrhea [70]. EPEC is the second leading cause of *E. coli* acute watery diarrhea and is responsible for 4–8% of cases in developing countries [52], accounting for an estimated 81 million cases of diarrhea and 122 000 deaths annually [45]. Disease is caused by tight attachment to the intestinal lining, causing loss of microvilli and resulting in malabsorption [38]. EPEC is commonly seen in younger children, particularly in children <2 years of age [98]. EAEC is a heterogeneous group of *E. coli* subtypes that express many possible virulence factors including enterotoxins and cytotoxins. Together, EPEC and EAEC account for the majority of persistent *E. coli* diarrheas [53].

EHEC (also known as Shiga toxin producing *E. coli*) is estimated to cause 2.8 million cases of diarrhea annually [54]. EHEC is naturally found in the gastrointestinal tract of healthy cattle living in intense feed lots and in the US most outbreaks are the result of contamination of food during processing [70, 99]. Disease with EHEC begins as acute non-bloody gastroenteritis characterized by cramping, vomiting, and diarrhea, which can become dysenteric and, in up to 8% of infections, progresses to hemorrhagic colitis [59, 70]. Pathogenic factors in EHEC include proteins for attachment to the intestinal epithelial cells that are like those expressed in EPEC. EHEC expresses Shiga-like toxins 1 and 2, which cause injury to the colonic microvasculature leading to bloody diarrhea and injury to renal endothelial cells, resulting in hemolytic uremic syndrome [100]. The strains most commonly associated with hemolytic uremic syndrome are *E. coli* O157:H7, which is most common in the US, and non-O157 EHEC serotypes, such as *E. coli* O104:H4 that is reported in parts of Europe [101, 102]. Antibiotic use with EHEC infections is controversial, as there is evidence indicating an association between receipt of antibiotics and development of hemolytic uremic syndrome [103].

EIEC is a little studied pathogen with a clinical course similar to shigellosis [38]. EIEC and *Shigella* appear to share the same virulence genes [104]. Thus, rapid molecular testing that targets virulence factors cannot distinguish between EIEC and *Shigella* species. However, compared to *Shigella*,

EIEC has a lower expression of these genes and causes less severe disease [104]. It is considered of minor relevance compared to other classes of *E. coli*; there are isolated instances where EIEC appears to be more common, although its uniform distribution across all ages in such instances suggest protracted outbreaks rather than endemic transmission [55].

7.5.2.2 **Shigella** ***[Image 7.7]*** *Shigella* is a genus of nonmotile, encapsulated, Gram-negative bacteria of the family Enterobacteriaceae [56].

The majority of disease is caused by two species: *Shigella sonnei* and *Shigella flexneri* [59]. *Shigella dysenteriae* is less common, but remains an important pathogen due to its propensity for high mortality outbreaks in developing countries, which affect all ages. *Shigella* is implicated in 5–15% of acute watery diarrheal cases, with symptoms ranging from mild to severe [46, 56]. *Shigella* is estimated to cause 190 million cases of diarrhea and 66 000 deaths annually [45], predominantly occurring as endemic disease among toddlers and older children in developing countries [58, 59]. *Shigella* is most well-known for causing dysentery. In the MAL-ED study *Shigella* spp., or EIEC expressing *Shigella* genes, were attributed as the cause of 64% of all cases of dysentery [46]. Mainly due to dysentery and the resulting bacteremia, *Shigella* infections are associated with 10% of all diarrheal deaths globally [56]. *Shigella* have a complex set of mechanisms to invade intestinal cells, multiply intracellulary, and spread between cells. This is achieved with the help of several toxins that have enterotoxic, cytotoxic, and neurotoxic effects [56]. The combination of toxins and invasion results in a highly inflammatory disease that begins with acute diarrhea and often develops into dysentery and, occasionally, seizures [56]. Bacterial shedding lasts for a few weeks, but chronic carriage is rare [70, 96]. Malnutrition increases the persistence and severity of dysentery [37]. Epidemic disease is common in settings of extreme poverty, natural disasters, and wars, and it affects all ages [37]. *Shigella* disease is less common in developed countries and occurs predominately in recent travelers to developing countries and food-borne outbreaks [70]. In the US alone, *Shigella* species are responsible for over 400 000 cases per year [63]. Antibiotic treatment is indicated in dysenteric cases, but multidrug resistance in *Shigella* is a global problem making knowledge of antibiotic susceptibility key for treatment [37].

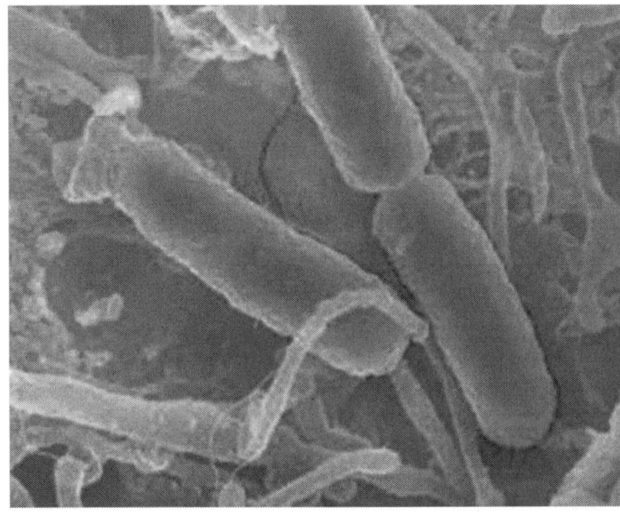

IMAGE 7.7 Colorized scanning electron micrograph a *Shigella* sp. bacteria. (*Source*: Copyright Institut Pasteur.)

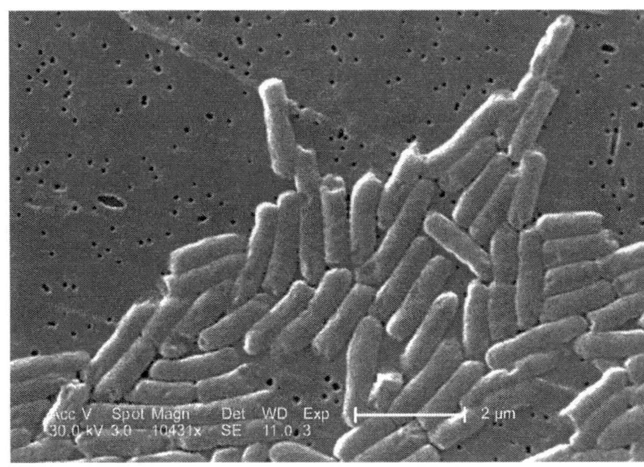

IMAGE 7.8 Colorized scanning electron micrograph of a *Salmonella* sp. colony. (Content provider: CDC/Janice Haney Carr. *Source*: CDC, Public Health Image Library, Image ID # 10896.)

7.5.2.3 **Salmonella** ***[Image 7.8]*** Salmonella is a large genus of bacteria, all of which are pathogenic to humans [59]. The most important disease associated with *Salmonella* is typhoid fever, which is caused by *Salmonella enterica enterica*, serovar Typhi and rarely manifests with diarrheal symptoms [105]. However, non-typhi strains of *Salmonella* are common residents of the gastrointestinal tracts of many wild and domesticated animals and cause gastroenteritis in humans similar to other bacterial pathogens. In developed countries, non-typhi *Salmonella* is a common cause of gastroenteritis outbreaks, resulting in 1.2 million cases of gastroenteritis each year, mostly caused by contaminated poultry and unpasteurized dairy products [38, 70]. Non-typhi *Salmonella* is less well characterized in developing countries, but is associated with endemic diarrhea in children, sepsis in neonates, and sporadic outbreaks in all ages [37, 70]. Dysentery is uncommon. Prolonged asymptomatic carriage in children is common. In developing countries, up to 45% infected children under five can transmit *Salmonella* for up to 12 weeks after the primary infection [68, 70]. Treatment of complicated *Salmonella* infections has been hindered in recent years due to a large increase in antimicrobial resistance caused by the practice of feeding antibiotics to poultry [106].

7.5.2.4 **Campylobacter** *[Image 7.9] Campylobacter* are a motile genus of Gram-negative bacteria [70]. Two species of *Campylobacter*, *Campylobacter jejuni*, and *Campylobacter coli*, are most commonly implicated in human disease. There is growing evidence that other species of *Campylobacter* are important in the developing world, where they are a common cause of diarrhea [25, 58, 107]. *Campylobacter* is naturally found in the gastrointestinal track of animals, with poultry being the main source of bacteria for human infections [70]. *Campylobacter* is highly prevalent in developing countries. *C. jejuni* and *coli* were found to be responsible for 6–12% of moderate and severe diarrhea in children less than one year old across multiple developing countries [46]. Globally there is an estimated 166 million cases of diarrhea and 36 000 deaths annually [45]. Most cases self-resolve in about one week and only require supportive care, although 20% persist beyond a week or later relapse [70] and *Campylobacter* infections are associated with dysentery [58]. In developed countries *Campylobacter* disease is often unreported: in Europe the incidence of campylobacteriosis is thought to be 8–30 times the reported numbers [25]. An estimated 1.4–2.4 million cases per year occur in the US, mostly associated with food contamination and undercooked poultry [38]. Of note, *Campylobacter* infections can cause autoimmune sequelae in patients, particularly Guillain-Barre syndrome, which occurs in 1 in 1000 *Campylobacter* patients after resolution of infection [25].

7.5.2.5 **Vibrio cholerae** *[Image 7.10]* Cholera is caused by ingesting *Vibrio cholerae,* a Gram-negative bacterium that naturally reside in brackish rivers and coastal waters. All epidemic cholera disease is caused by the O1 serogroup of *V. cholerae* [108]. In the past multiple serotypes of O1 have caused pandemic disease, but the current global pandemic is caused by the *El Tor* serotype, which spread out of Southeast Asia in the early 1960s. However, in 1993 a new serotype, O139, emerged in Bangladesh and has spread throughout Asia. Cholera is characterized by voluminous diarrhea, which can lead to rapid dehydration, hypokalemia, acidosis, and shock within as little as 4–12 hours. It is believed to cause an estimated 2.9 million cases of diarrhea and 95 000 deaths annually, most of which occur in cholera endemic countries [57].

Cholera is most famous for its role in epidemic disease, which, throughout history, has occurred in settings of extreme poverty, natural disasters, and wars, where case fatality may exceed 20% [37, 108]. However, cholera is minimally inflammatory, is never invasive, and is rarely fatal if appropriate rehydration therapy is promptly instituted. Unfortunately, care is often limited in endemic areas and case fatality rates exceeding 20% are not uncommon. Currently, South Asia and Africa are the only areas endemic for cholera; outside of these regions the disease is extremely rare [37].

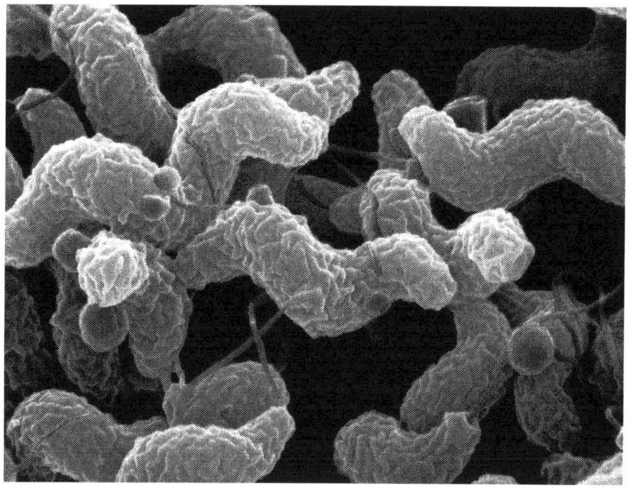

IMAGE 7.9 Colorized scanning electron micrograph of *Campylobacter jejuni* bacteria. (*Source*: Agricultural Research Service (ARS) is the US Department of Agriculture's Chief Scientific Research Agency.)

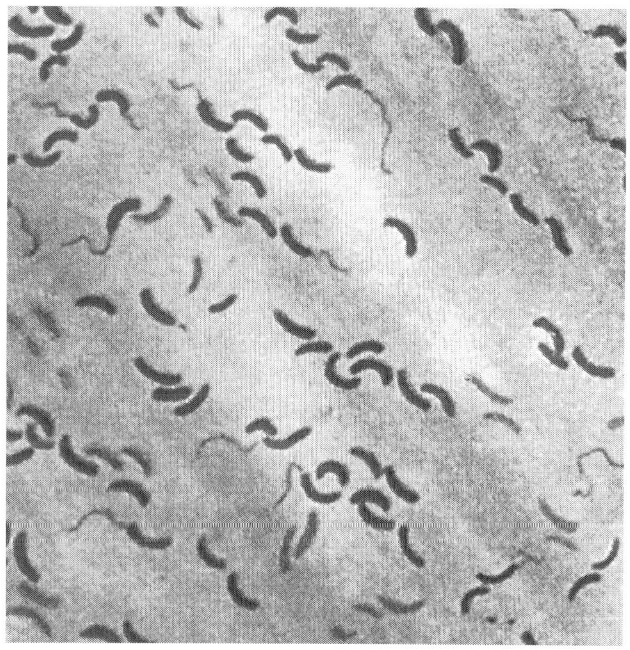

IMAGE 7.10 Gram-stained light micrograph of *Vibrio cholerae*. (Content provider: CDC. *Source*: CDC, Public Health Image Library, Image ID # 5324.)

Antibiotics should also be used as part of a comprehensive case management strategy as it decreases the amount of fluid purged, shortens illness duration, limits the use of resources per patient, and limits disease transmission.

7.5.2.6 **Aeromonas** *[Image 7.11] Aeromonas* is a genus of facultative anaerobic Gram-negative bacilli that naturally occur in water, soil, and food [109]. Multiple Aeromonads

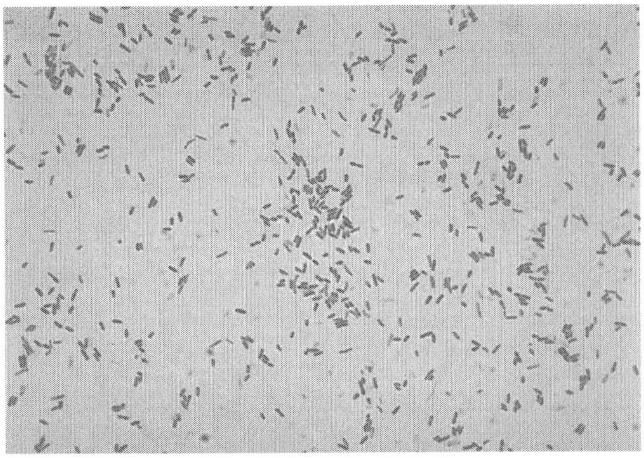

IMAGE 7.11 Gram-stained light micrograph of *Aeromonas hydrophila*. (Content provider: CDC. *Source*: CDC, Public Health Image Library, Image ID # 1255.)

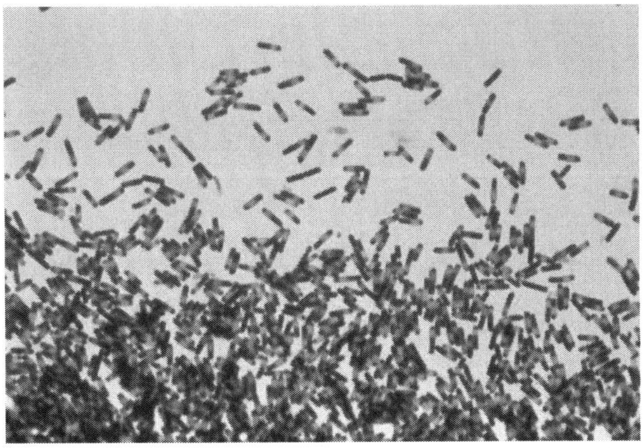

IMAGE 7.12 Gram-stained light micrograph of *Clostridium difficile*. (Content provider: CDC. *Source*: CDC, Public Health Image Library, Image ID # 3876.) *(See insert for color representation of this figure.)*

have been suggested as a cause of acute, chronic, and dysenteric diarrhea. Their commonality in asymptomatic persons and in the environment has limited our understanding of its role in these syndromes. However, a recent multicenter study found *Aeromonas* to be present in 22% of children with moderate and severe diarrhea in Pakistan and Bangladesh [110]. After controlling for other infections, it was still found to be highly associated with dysentery and disease in co-occurrence with *Shigella* [110]. Other countries in the study had limited disease attributed to *Aeromonas*, suggesting that they are important pathogens in certain regions of the world but not others [65].

7.5.2.7 **Clostridium difficile** ***[Image 7.12]*** *Clostridium difficile* is a common cause of healthcare-associated diarrhea, resulting in an estimated 450 000 cases each year in the US [111]. The organism is a spore-forming Gram-positive anaerobic bacillus. Pathogenic strains produce two major toxins, toxin A and B, that are mediators of disease [112]. Exposure to toxigenic *C. difficile* alone does not cause disease. Between 0 and 15% of healthy adults are asymptomatically colonized with *C. difficile*, over half of whom will have a toxigenic strain [113]. Symptomatic *C. difficile* only develops when the healthy microbial flora is disrupted, often by antibiotics and gastric acid-suppressing agents, allowing toxigenic *C. difficile* to replicate in an uncontrolled manner [114].

Clinical manifestations of disease consist of mild or moderate diarrhea, rarely with blood or mucous, occasionally accompanied by fever and cramping abdominal pain [112]. Complications may include sepsis, severe dehydration, renal failure, and death. Standard treatment includes metronidazole or oral vancomycin. Newer agents such as fidaxomicin and non-antimicrobial therapies such as probiotics and fecal transplants are being studied with promising results [115].

7.5.3 Parasitic Pathogens

Until recently parasites were thought to cause a relatively low burden of infectious diarrhea when compared to viruses and bacteria. This notion has been overturned by recent studies which show that parasites are increasingly recognized as important causes of acute diarrhea [58, 65] as well as an important cause of persistent diarrhea [70]. Numerous types of helminths, which are covered in detail in another chapter in this book, and protozoa have been identified as causes of diarrhea, but *Giardia lamblia*, *Cryptosporidium*, and *Entamoeba hystolytica* are the most common. All are transmitted through the fecal-oral route. Etiology is confirmed through visual detection of eggs or cysts in the stool; however, sensitivity is low even with multiple stool samples [33]. Improved tests including PCR and enzyme-linked immunosorbent assay are available, but are not commonly used in clinical situations in impoverished areas.

7.5.3.1 **Cryptosporidium** ***[Images 7.13 and 7.14]*** The genus *Cryptosporidium* is composed of numerous species of protozoa that find hosts in more than 40 mammalian species, although 97% of *Cryptosporidium* infections in humans are the result of one species: *Cryptosporidium parvum* [16, 116]. *Cryptosporidium* is estimated to cause 64 million cases of diarrhea and 27 000 deaths annually [45]. Most symptomatic cases are self-resolving and clinically similar to other enteric pathogens, with the exception of an increase in mild respiratory symptoms: cough occurs in 20–33% of children [116]. *Cryptosporidium* may cause up to 16% of persistent diarrhea in children [30]. Bloody stools are common, with 14% of acute cases and 31% of persistent cases showing signs of blood in their stool [16]. HIV-positive patients are at an increased risk of severe and persistent disease. Asymptomatic carriage is common in developing countries,

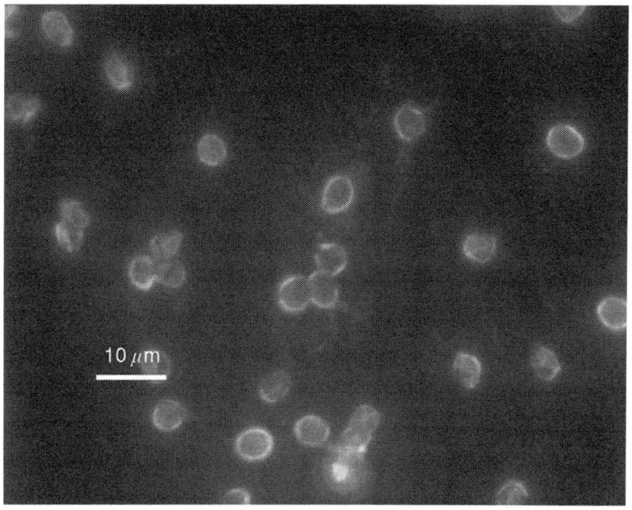

IMAGE 7.13 Immunofluorescence light micrograph of purified *Cryptosporidium parvum* oocysts. (Photo credit: H.D.A Lindquist, US EPA. *Source*: http://www.epa.gov/nerlcwww/cpt_seq1.html.)

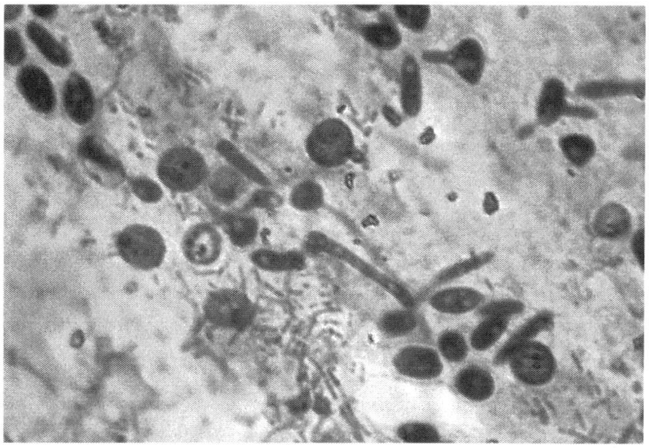

IMAGE 7.14 Light micrograph of *Cryptosporidium* sp. oocysts in direct fecal smear, acid-fast stained red. (Content provider: CDC/*J. Infect. Dis.*, 1983, May, 147 (5): 824–828. *Source*: CDC, Public Health Image Library, Image ID # 5242.)

and is present in 4–5% of healthy children in Southeast Asia and Latin America [30].

In developing countries *Cryptosporidium* is a major cause of endemic diarrhea, but the prevalence varies widely between countries, with attributable fractions of moderate to severe diarrhea ranging from 1 to 17% in children younger than 12 months old [46]. The GEMS study, which used highly specific testing, found *Cryptosporidium* to have the second highest attributable-incidence of all diarrheal pathogens in children under one year old, second only to rotavirus, suggesting that *Cryptosporidium* likely plays a larger role in moderate and severe childhood diarrhea than previously thought [65]. Similarly, in the MAL-ED it was the

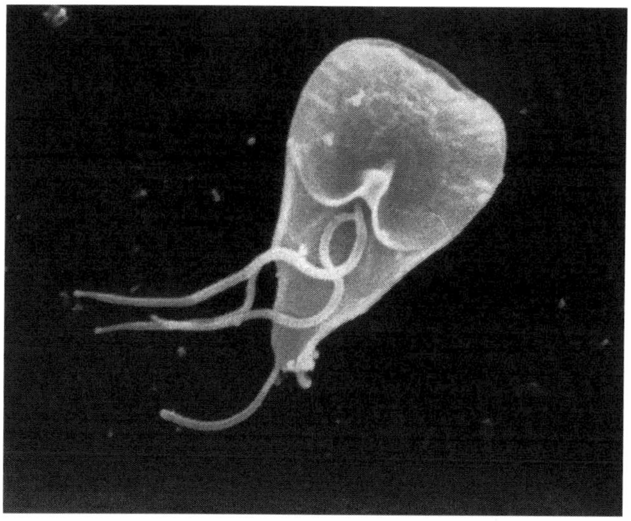

IMAGE 7.15 Scanning electron micrograph of a *Giardia lamblia* trophozoite. (Content provider: CDC/Janice Carr. Source: CDC, Public Health Image Library, Image ID # 8698.)

enteropathogen with the fifth greatest disease burden in community diarrheal cases [58]. Zoonotic reservoirs, particularly cattle, are a major source of infection in developed countries, but human feces are a more important vector of transmission in developing countries. In developed countries, drinking water contaminated with *Cryptosporidium* spores have been the cause of several major outbreaks; one, which occurred in Milwaukee, resulted in over 400 000 cases of cryptosporidiosis [16]. High prevalence of anti-*Cryptosporidium* antibodies in the US, 25–35% of children and adults, suggests exposure may be more common in developed countries as well, but low carriage of the protozoan in the general population suggests most *Cryptosporidium* is eliminated from the gastrointestinal tract without apparent disease [116].

7.5.3.2 **Giardia lamblia** ***[Images 7.15 and 7.16]*** *Giardia lamblia* is a non-invasive flagellated protozoa whose transmission occurs through ingestion of giardial cysts [28]. Humans are thought to be the principle reservoirs, although *Giardia* can be detected in animals and zoonosis has been implicated in some outbreaks [19, 70]. The prevalence of *Giardia* in the intestinal tracts of children is low in developed countries, but can reach 20–30% in children in developing countries [59]. However, most persons infected with *Giardia* do not show symptoms of disease [28]. In symptomatic cases, an incubation period of 1–2 weeks precedes symptoms that include acute diarrhea characterized by foamy, foul smelling stools, flatulence, abdominal distention, and anorexia [33, 70, 117]. Most cases resolve spontaneously, although a subset becomes persistent. *Giardia* is the second most common cause of persistent

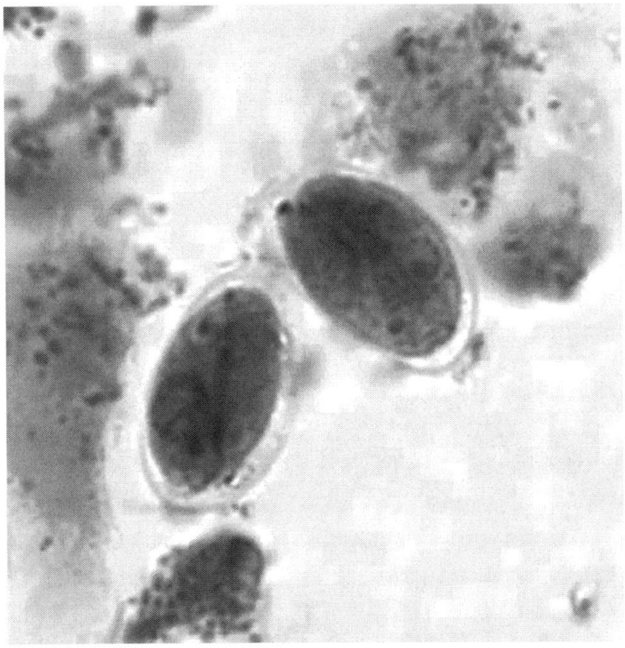

IMAGE 7.16 Light micrograph of *Giardia lamblia* stained with trichrome. (Content provider: CDC/Division of Parasitic Diseases. *Source*: www.dpd.cdc.gov/dpdx/HTML/ImageLibrary/Giardiasis_il.htm.)

parasitic infections, although it accounts for less than 10% of all persistent diarrhea cases in developing countries [28] and is estimated to cause 183 million cases of diarrhea annually [45]. Persistent diarrhea and anorexia put children at an increased risk of malnutrition [59].

7.5.3.3 ***Entamoeba histolytica [Images 7.17 and 7.18]*** *Entamoeba histolytica* is a species of amoeba that infects the human gut, subsisting on intestinal bacteria and food particles. Transmission occurs through ingestion of cysts from fecal contamination of raw or undercooked food. Greater than 90% of infections are asymptomatic, although *E. histolytica* has the ability to invade the intestinal lining, causing highly inflammatory disease [118]. Globally, *E. histolytica* is estimated to cause 104 million cases of diarrhea and 5000 deaths annually [45]. Invasion may occur days or years after infection, manifesting as amoebic colitis, causing bloody diarrhea and abdominal pain [118]. Rarely, the disease will progress, resulting in necrotizing colitis and disseminated disease, which can lead to amoebic liver abscesses.

7.6 COMPLICATIONS AND MANAGEMENT

Although infectious diarrhea is one of the most common causes of deaths globally, most cases can be easily treated with simple, inexpensive rehydration, appropriate feeding, and nutritional supplements. Key treatment strategies are summarized in Table 7.4.

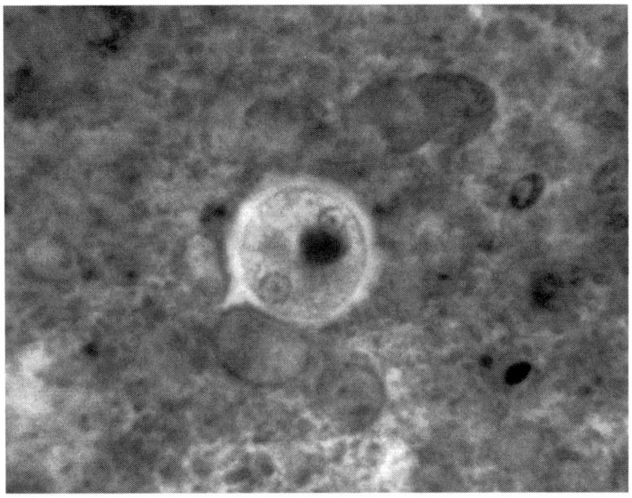

IMAGE 7.17 Light micrograph of *Entamoeba histolytica* cyst stained with chlorazol black. (Content provider: CDC/Dr George Healy. *Source*: CDC, Public Health Image Library, Image ID # 1474.)

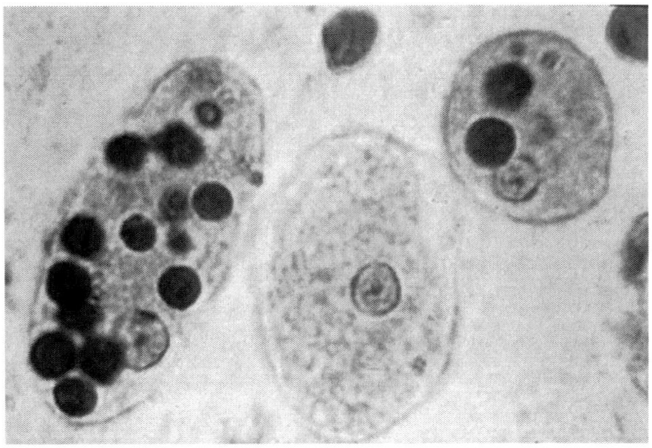

IMAGE 7.18 Light micrograph of *Entamoeba histolytica* trophozoites with ingested erythrocytes stained with trichrome. (Content provider: CDC/Division of Parasitic Diseases. *Source*: www.dpd.cdc.gov/dpdx/HTML/ImageLibrary/A-F/Amebiasis/body_Amebiasis_il6.htm.)

7.6.1 Dehydration

Approximately 90% of deaths caused by infectious diarrhea are the result of loss of bodily fluids [119]. Mild and moderate dehydration in children is characterized by thirst, restlessness, irritability, decreased skin turgor, sunken eyes, and, in infants, sunken fontanelle [31, 138]. Signs of severe dehydration include altered consciousness, lack of urine output, quickened weak pulse, low blood pressure, and peripheral cyanosis [139]. As dehydration is the primary cause of mortality from diarrhea, the main strategy for treatment is the replacement of fluids along with appropriate feeding and nutritional supplements.

TABLE 7.4 Treatment and prevention strategies for acute diarrhea aimed at developing countries.

Treatment	
Original ORS	Effective in treating 90% of diarrhea cases [119]
Reduced Osmolarity ORS	Same efficacy original ORS [120]
	20% lower stool output [120]
	30% reduction in vomiting [120]
	40% reduction in the need for unscheduled IV treatment [120]
Zinc	12 hours mean reduction in duration of acute diarrhea episodes [121]
	27% lower relative risk of diarrhea continuing to day 7 [121]
	51% decrease in overall mortality [122]
Probiotics	Duration of illness reduced by 25 hours [123]
Ondansetron	50% reduction in failing oral rehydration therapy [124]
	47% reduction in hospitalization [124]
Prevention	
Zinc	34% decreased prevalence of diarrhea in the 2–3 months following treatment [125]
Vitamin A	32% reduction in diarrhea associated mortality [126]
Breastfeeding	2–3-fold reduction in diarrhea incidence [127–130]
Handwashing promotion	31% decrease in risk of diarrheal illness [131]
Installation of latrines	30% reduction in diarrheal illness [132]
Water treatment	11% reduction in diarrheal illness when treating the water source [132]
	39% reduction when treating at the household level [132]
Rotavirus vaccine	98–100% efficacy against severe rotavirus gastroenteritis [133, 134]
	74–85% protection against all rotavirus gastroenteritis [133, 134]
	45% reduction in hospitalizations due to acute gastroenteritis across all ages [66]
Measles vaccine	18% lower incidence of diarrhea [135]
Cholera vaccine	85–86% effective against cholera [136, 137]

7.6.1.1 ***Intravenous Fluid Therapy*** Thomas Latta was the first to document the use of intravenous (IV) fluids in 1830 to manage a young woman with severe dehydration secondary to cholera. Latta subsequently treated several other patients with severe diarrhea; however, most of his patients died either because of a lack of replacement of ongoing fluid loss or from nosocomial infections [140]. As a result, IV fluids were not commonly used for the next nearly 100 years and diarrheal mortality of patients with severe dehydration continued to be close to 100% [141]. In the early 1900s, an IV solution containing sodium and bicarbonate was developed, which resulted in a dramatic reduction in mortality among hospitalized individuals with diarrhea. However, the case fatality rate from diarrhea even in the best hospitals was still approximately 20%. In the mid 1940s, Daniel Darrow described the importance of potassium loss with diarrhea, and the addition of potassium into IV fluids resulted in the virtual elimination of diarrheal deaths among hospitalized patients with diarrhea in the US [142].

However, with the advent of oral rehydration solution (ORS), IV fluids are now recommended only in cases of severe dehydration in which a patient may be obtunded and unable to drink. In such cases, IV or intraosseous fluids should be used in the initial management if appropriate equipment is available. ORS should then be instituted as soon as the patient can drink fluids. In addition to delivering the appropriate quantity of ORS to replace the calculated fluid deficit, ongoing fluid losses should be replaced. Table 7.5 summarizes the recommended quantities of fluid for different levels of dehydration [138].

7.6.1.2 ***Oral Rehydration Solution*** Harold Harrison first demonstrated the successful use of an oral rehydration solution (ORS) containing 62 mmol/l of sodium and 3.3% glucose for the management of diarrhea in Baltimore during the mid-1940s [144]. A commercial company marketed a modified solution with increased carbohydrate content to improve the taste. However, due to inappropriate reconstitution and the hyperosmolarity of the solution, this formula caused many cases of hypernatremia throughout the country [145]. Thus, ORS was not accepted by healthcare providers as standard therapy for diarrhea for decades.

In the mid-1960s ORS received renewed attention as physiologic studies demonstrated that glucose was essential to facilitate absorption of water from the gut. Subsequently, in the summer of 1971, ORS received major attention for successfully saving the lives of thousands of cholera patients during the Bangladeshi refugee crisis. ORS was shown to be safe and effective even when given by untrained family members; the case fatality rate of diarrhea in the ORS intervention areas was 3%, compared with 20–30% in areas that used only intravenous fluids [146]. In 1978, the World Health Organization (WHO) began recommending a standard formulation of ORS containing 111 mmol/l of glucose,

TABLE 7.5 Summary of treatment based on degree of dehydration.

Degree of dehydration	Rehydration therapy	Replacement of losses	Nutrition
Minimal or no dehydration	Not applicable	<10 kg body weight: 60–120 ml ORS for each diarrheal stool or vomiting episode >10 kg body weight: 120–140 ml ORS for each diarrheal stool or vomiting episode	Continue breastfeeding, or resume age-appropriate normal diet after initial hydration, including adequate caloric intake for maintenance
Mild to moderate dehydration	ORS, 50–100 ml/kg body weight over 3–4 hours	Same Monitor stool output and alter maintenance appropriately	Same
Severe dehydration	Lactated Ringer's or 0.9% saline in 20 ml/kg body weight intravenous boluses until perfusion and mental status improve Then administer 100 ml/kg body weight ORS over 4 hours or 5% dextrose 0.45% saline intravenously at twice the stool output rate	Same Monitor stool output and alter maintenance appropriately If unable to drink, administer through nasogastric tube or administer 5% dextrose 0.45% saline with 20 mEq/l potassium chloride intravenously.	Same

Source: Adapted from King et al. [143].

90 mmol/l of sodium, 80 mmol/l of chloride, 20 mmol/l of potassium, and 20 mmol/l of citrate as the principal strategy for diarrhea management. Multiple studies demonstrated that ORS could effectively manage over 90% of diarrhea episodes, and ORS was adopted and used by numerous national governments and international relief agencies [119, 144, 147–152]. The extensive use of this solution worldwide is credited with saving millions of lives in the past three decades.

The original WHO-ORS elicited concerns that hypernatremia could occur in cases of non-cholera diarrhea, where salt loss is lower. This led to the development of ORS formulations with lower osmolarity, which contained 75–90 mmol/l glucose and 60–75 mEq/l of sodium [153]. In the early 1990s, laboratory studies showed that hypotonic reduced osmolarity (245 mmol/l) ORS resulted in increased intestinal water absorption when compared with standard WHO-ORS. Several efficacy studies comparing low osmolarity and standard ORS demonstrated a further reduction in stool output and a reduction in diarrhea duration [154–157]. A meta-analysis of trials of low osmolarity ORS for acute diarrhea in children found that there was a 20% reduction in stool output, a 30% reduction in vomiting, and close to a 40% reduction in the need for unscheduled IV treatment in patients receiving the low osmolarity ORS compared to those receiving the original ORS [120]. Although data are more limited, a similar analysis of studies conducted in people with cholera concluded that there was no difference in the safety and efficacy of low osmolarity ORS and original ORS in those patients [158]. The new low osmolarity ORS has been shown to be more acceptable among children [159]. In May 2002, based on the recommendations of an expert panel, the WHO changed its recommendation to the use of low osmolarity ORS containing 75 mmol/l of glucose, 75 mmol/l of sodium, 65 mmol/l of chloride, 20 mmol/l of potassium, and 10 mmol/l of citrate for the treatment of children and adults with acute diarrhea. Although ORS can be prepared from home ingredients, errors in mixing can occur, which may result in serious electrolyte imbalances. Therefore, commercially available ORS packets should be recommended whenever possible.

7.6.2 Malnutrition

Malnutrition and diarrhea have long been known to form a vicious cycle. Malnutrition increases the susceptibility of children to pathogens that cause diarrhea and other disease, and diarrhea decreases the gut's ability to absorb nutrients, leading to malnutrition. The malnutrition/diarrhea cycle is so strong that 61% of children who die of diarrhea have malnutrition as an underlying risk factor [160]. Diarrhea in malnourished children is on average of longer duration. A case control study in India found that malnourished children have a 3.25 increased odds of developing persistent diarrhea compared to non-malnourished controls [161]. Additionally, the case fatality rate of diarrhea increases progressively as malnutrition worsens, with a 24-fold increase in mortality in severely malnourished children compared to normally nourished children [32]. Beyond increased rates of mortality, the malnutrition caused by diarrhea leads to significant physical and mental growth shortfalls, which negatively affect children throughout their lives [162]. The number of diarrheal episodes in infancy has been shown in a Brazilian cohort study to negatively impact cognitive test scores four to seven years later after controlling for a variety of socioeconomic indicators [163]. A similar study by the same group found a strong association with an increased number of diarrhea episodes in the first two years of life and decreased school performance, as well as height and weight for age six to nine years later [164]. Thus, feeding and other nutritional interventions are a critical component of treatment to ameliorate the short and long-term consequences of diarrhea. Nutrition considerations are covered in greater detail in Chapter 11 of this book.

7.6.2.1 Continued Feeding Until the 1970s the medical community endorsed the belief that feeding should be withheld during diarrhea to "rest the gut." In the 1980s, a series of studies demonstrated that continued feeding during a diarrheal episode is safe and improves outcome [165–170]. Based on these studies, the WHO incorporated early refeeding into their recommendations for diarrhea treatment [171]. Breast-fed infants should continue breast-feeding and formula-fed infants should continue their usual formula. Commercial carbonated beverages, sugary drinks, and undiluted fruit juices should be avoided because the high carbohydrate content results in increased osmolarity of intestinal contents, which can aggravate diarrhea.

7.6.2.2 Zinc Supplementation Zinc is a micronutrient that plays key roles in effective immune response and healing. In developing countries zinc supplementation as a treatment for diarrhea, when given in addition to ORS, has been shown to be highly effective at decreasing the duration and severity of diarrheal episodes [172, 173]. Studies have shown that zinc decreased the duration of diarrhea by 12 hours (95% CI: 2–23), decreased use of unnecessary antibiotics, and reduced medical visits for acute diarrhea [174–176]. Furthermore, a 10–14-day course of zinc supplementation decreased the prevalence of diarrhea in the two to three months following treatment (OR: 0.66; 95%CI: 0.52–0.83) [125]. Introduction of zinc along with ORS led to a decrease in hospitalizations for diarrhea as well as other illnesses in two large-scale studies [122, 174]. A study in Bangladesh reported a 24% reduction in the diarrhea hospitalization rate (95% CI: 0.59–0.96) and resulted in an overall decrease in all-cause mortality (risk ratio of 0.49 [95% CI: 0.25, 0.94]) [122]. An Indian study showed a reduction in all-cause diarrhea (OR 0.56; 95% CI: 0.41, 0.75) and all-cause hospitalizations (OR 0.41; 95% CI: 0.29–0.57) [174] with zinc supplements. Zinc therapy is particularly important in the case management of persistent diarrhea and intervention trials have shown that it lowers treatment failure and death in cases of persistent diarrhea by 42% (95% CI: 10–63%) [177]. Based on proven safety and efficacy of zinc supplementation as an adjunct in diarrhea treatment, the WHO recommends 10–14 days of oral zinc at 20 mg per day for children six months to five years of age and 10 mg per day for children <6 months of age for acute diarrhea episodes [121]. There is currently no clear evidence of the benefits of zinc supplementation in older children and adults with acute diarrhea [178].

7.6.3 Antimicrobial Therapy

Antimicrobial therapy should be used only in specific cases of infectious diarrhea with complications, such as dysentery, severe persistent diarrhea, sepsis, or in patients at increased risk of developing severe complications, such as immunocompromised patients and those with cholera, although health and medical organizations in developed countries often recommend the use of antibiotics for traveler diarrhea. Due to widespread prevalence and variation in antimicrobial resistance, it is critical for therapeutic decisions to be based on regional antibiotic susceptibility patterns of likely pathogens or specific pathogens if etiology is confirmed. Guidebooks, such as the *Academy of Pediatrics' Red Book®*, are available in many regions to provide regularly updated guidelines for treating specific pathogens. Table 7.6 summarizes the therapies indicated for otherwise healthy individuals.

7.6.4 Non-antimicrobial Drug Therapy

7.6.4.1 Ondansetron Nausea and vomiting significantly increase the dehydration in diarrheal illnesses by reducing a patient's ability to take in liquids. Ondansetron, an antiemetic, has been increasingly used to decrease vomiting and increase oral intake among children with diarrhea. A meta-analysis demonstrated that single doses of ondansetron in

TABLE 7.6 Empiric antimicrobial recommendations for infectious diarrhea caused by specific pathogens.

Organism	Antimicrobial regimen
Campylobacter[a]	Azithromycin (3 days); Erythromycin (5 days)
Cryptosporidium parvum[a]	Nitazoxanide (3 days)
Entamoeba histolytica[a]	Metronidazole or tinidazole. followed by Iodoquinol or paromomycin
Escherichia coli (non EHEC)[a]	Azithromycin (3 days) or ciprofloxacin (adults, 3 days)
Giardia lamblia[a]	Metronidazole (5 days) or nitazoxamide (>1 year old, 3 days) or paromomycin (pregnant women) or tinidazole (>3 years old, single dose)
Shigella[a]	Azithromycin (3 days) or ceftriaxone (2–5 days) or ciprofloxacin (adults, 3 days)
Vibrio cholerae[a]	Azithromycin (1 dose) or ciprofloxacin (adults, 3 days) or doxycycline (>8 years old, one dose) or erythromycin (3 days); or tetracycline (3 days).

[a] Adapted from American Academy of Pediatrics Red Book, 2015 [179].

children increased the intake of ORS by a mean of 43 ml after 1 hour (95% CI: 15.5–70.5 ml), reduced the relative risk (RR) of failure of oral rehydration by half (RR 0.5; 95% CI: 0.37–0.69) and reduced the risk of hospitalization (RR 0.53; 95% CI: 0.29–0.97) [124]. There are no clear benefits to therapy beyond one dose. There is some evidence that ondansetron may transiently and mildly increase the frequency of stools when given during a diarrheal episode, but this is more than made up for by the increase in oral intake [180]. A single dose of ondansetron as part of rehydration efforts is a reasonable measure given limited side effects and proven benefits [124, 180]. Nevertheless, the most important aspect of the treatment of diarrhea is the prompt implementation of rehydration therapy.

7.6.4.2 ***Probiotics*** Probiotics, or microorganisms that are ingested to benefit the taker, are a promising treatment for acute diarrhea. Multiple randomized control trials have examined the use of probiotics in acute diarrhea and have found that they shorten the duration of diarrhea, reduce the number of diarrheal episodes, and reduce the duration of intravenous hydration [123]. A Cochrane review found that probiotics reduce the duration of illness by 24.8 hours (95% CI: 15.9–33.6 hours), as well as reduce stool frequency on the second day of illness by 0.8 stools (95% CI: 0.45–1.14) [181]. There were no clear negative effects of the treatment. The American Academy of Pediatrics states that probiotics is likely to be useful as an adjunct therapy in addition to rehydration in the treatment of acute viral diarrhea in otherwise healthy children, but note that research is needed to guide the use of specific probiotic formulations and to determine the safety of probiotics in specific patient groups, such as chronically ill or immunocompromised children [123].

7.6.4.3 ***Other Non-antimicrobial Drug Therapy*** Several over-the-counter and prescription antimotility (such as loperamide), antisecretory (such as racecadotril), and toxin-binding agents (such as cholestyramine) are available for symptomatic relief of diarrhea [182–184]. Because of the limited evidence and uncertain side-effect profiles, most experts do not recommend the use of these agents, particularly for pediatric patients [185, 186]. In fact, many experts believe that the use of antimotility agents could be harmful, especially in children. For example, antimotility agents have been implicated in prolonging fever in *Shigellosis* and in gastrointestinal complications or death when given initially prior to antibiotic treatment for *C. difficile* colitis [187, 188]. In general, the use of these agents should be discouraged due to lack of efficacy in most cases, except loperamide, which may provide symptom relief in chronic diarrhea, or in addition to antibiotic therapy in traveler diarrhea.

7.7 PREVENTION

A large proportion of morbidity and mortality due to infectious diarrhea can be prevented using well-characterized interventions including vaccines, nutritional supplementation, breastfeeding, sanitation, and hygiene. Key treatment prevention strategies are summarized in Table 7.4.

7.7.1 Vaccines

7.7.1.1 ***Rotavirus Vaccines*** The first rotavirus vaccine that was licensed was Rotashield® in 1998 was withdrawn soon after introduction because of the increased risk of intussusception after initial use of the vaccine [189]. Subsequently two rotavirus vaccines have been licensed based on large efficacy trials that included 65 000–70 000 children in each of the trials. Rotateq™ (manufacture by Merck) is an oral human–bovine pentavalent reassortant vaccine that was shown in one large clinical trial to prevent 74% of cases of rotavirus gastroenteritis and 98% of severe rotavirus gastroenteritis in a child's first rotavirus season [133]. Rotarix® (GlaxoSmithKline) is an oral live attenuated monovalent vaccine that was shown to prevent 85% of cases of rotavirus hospitalizations and 100% of severe rotavirus gastroenteritis [134]. Furthermore, Rotavac® (Bharat Biotech), developed from a natural neonatal monovalent human–bovine reassorted rotavirus strain [190], has recently been introduced in several states in India. The efficacy of Rotavac was shown to be similar to the efficacy of Rotarix and Rotateq in low-income countries [189, 190]. Preliminary evaluations of Rotateq and Rotarix indicate that vaccine immunogenicity and efficacy do not differ by breastfeeding status [191–193], and there is no reduction in immune response to either vaccine with concomitant administration with oral polio vaccine [194, 195]. Post-marketing surveillance for both vaccines show a small increase in the rates of intussusception in vaccinated children, between 1 and 6 cases per 100 000 children vaccinated, but the risk is small compared to the decrease in disease and deaths prevented by vaccination [196]. The WHO strategic advisory group of experts now recommends the inclusion of rotavirus vaccination into all national immunization programs [197].

Over the last decade, the world has seen an extensive global rollout of rotavirus vaccines in high- and low-income countries (Figure 7.3), dramatically altering the epidemiology of rotavirus infections and reducing diarrheal deaths in children globally. The US has seen a 42–74% reduction, depending on the region, in the number of positive rotavirus tests at sentinel hospitals compared to before rotavirus vaccine was introduced [199]. More importantly, rotavirus vaccines have been shown to be efficacious when used in developing countries. In the first few years after introduction of rotavirus vaccines, South Africa saw a 34% reduction in

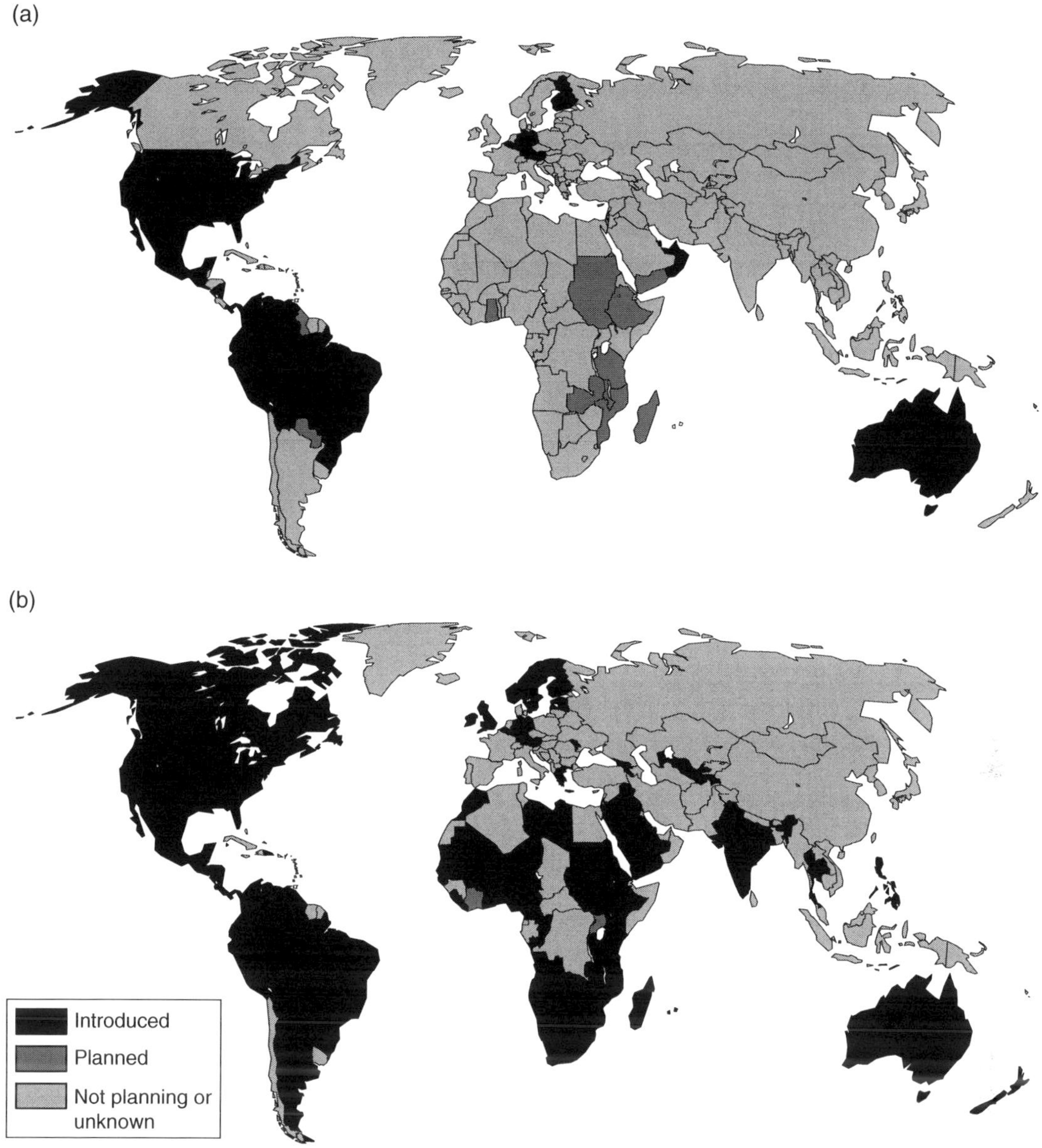

FIGURE 7.3 Global introduction of Pediatric Rotavirus Vaccines. (a) National Introduction of Rotavirus Vaccine, December 2009 [189]. (b) National Introduction of Rotavirus Vaccine, September 2016 [198].

diarrheal hospitalizations for children less than five years old [200], while Ghana saw a 44% (95% CI: 28–64%) reduction [201]. Since introduction in Nicaragua there has been a 23% reduction in acute gastroenteritis in vaccinated children compared to non-vaccinated children aged 0–11 months [202], with similar results seen in Brazil after introduction [66]. A study in Mexico showed a 35% reduction (95% CI: 29, 39) in the diarrhea-related mortality rate among children less than five years of age following rotavirus vaccine introduction [203], while a 33% reduction in all hospitalized diarrheal deaths in children under five years of age was seen in Zambia after the first full year of vaccination [204]. Serotype surveillance has yet to demonstrate disease replacement with new or previously rare serotypes, although surveillance is ongoing [205].

7.7.1.2 ***Cholera Vaccines*** There are two currently available licensed oral cholera vaccines. The first contains killed whole-cell *V. cholerae* O1 and a recombinant cholera toxin β-subunit (Dukoral®, Crucell), given in two doses to persons six years and over or in three doses to children two to six years of age. Studies in Peru and Bangladesh showed vaccine efficacies of 86% (95% CI: 37, 97) and 85% (95% CI: 62, 94), respectively, in the first six months [136, 137].

The second is a bivalent oral vaccine containing killed whole-cell *V. cholera* O1 and 0139 (manufactured in two countries by Vabiotech (Viet Nam, mORC-Vax™) and Shantha Biotechnics Ltd (India, Shanchol™) [206]. This vaccine has been shown to be safe and immunogenic in trials in Vietnam and India [207, 208]. Poor availability of the vaccine continues to limit its impact as does the need for two doses separated by two to four weeks. Strategies to effectively deploy it using the ring vaccination approach and evaluating the use of the vaccine in outbreaks are needed to maximize the impact of this vaccine for the improved control of cholera [209].

7.7.1.3 ***Measles Vaccine*** No randomized control trials have been carried out to determine the effect of measles vaccination on diarrhea, but an observational review of Demographic and Health Surveys found that the incidence of diarrhea was 18% lower in children who received the measles vaccine compared to unvaccinated counterparts [135]. A case control study in Bangladesh found measles vaccination to be protective, reducing deaths due to diarrhea by 57% in measles vaccine recipients aged 10–60 months [210].

In recent years, measles vaccination promotion programs have successfully increased global measles vaccination dramatically; from 20% in 1980 to 80% in 2007, which has resulted in a decrease of measles deaths from 7 to 8 million per year prior to vaccination to 73 000 in 2015 [1, 211]. However, most of the remaining deaths occur in a few countries in Southeast Asia and Africa where immunization rates remain low – the same areas that also have the highest number of diarrhea deaths [212]. In these countries, measles will remain a significant factor in diarrheal disease until vaccination coverage improves.

7.7.2 Nutritional Interventions

7.7.2.1 ***Zinc Supplementation*** Prophylactic zinc supplementation has been shown to reduce the relative risk of diarrhea by 27% in studies that enrolled children with a mean age of at least 12 months (RR: 0.73; 95% CI: 0.61, 0.87) [213]. The additional benefits of zinc on reduction in pneumonia incidence and improvement in growth in children add to the body of evidence supporting the importance of dietary zinc for overall child health [213]. Additional operational research is needed to develop successful delivery strategies for routine zinc supplementation as zinc needs to be delivered daily, or at least several times weekly, to be effective.

7.7.2.2 ***Vitamin A*** Vitamin A supplementation has been shown to result in a 32% (95% CI: 20, 43%) reduction in diarrhea-specific mortality and a 23% (RR: 0.77; 95% CI: 0.68, 0.88) pooled overall reduction in the mortality rate of children 6–59 months of age [126]. A meta-analysis showed no effect of vitamin A on diarrhea incidence (relative risk 1.00; 95% CI: 0.94, 1.07) [214]. The effect of vitamin A on diarrhea duration or severity is less clear [215–218]; there appears to be an effect on diarrhea severity, explaining the effect on mortality in spite of the lack of effect on diarrhea incidence [219]. Because of the recognized overall benefits of vitamin A, the WHO recommends vitamin A supplementation every four to six months in children <5 years of age who live in areas of inadequate vitamin A intake.

7.7.2.3 ***Breastfeeding*** Breastfeeding is considered one of the most basic and effective interventions in the prevention of infectious diseases in children. One meta-analysis showed that in children <6 months of age, any breastfeeding decreased diarrheal disease by over sixfold (OR: 6.1; 95% CI: 4.1, 9.0) [220]. Lack of breastfeeding in the first two months of life has also been shown to increase the risk of diarrhea mortality by over 20-fold [221]. Based on the overall health and nutrition benefits, the WHO recommends exclusive breastfeeding up to six months of age and breastfeeding along with complementary feeds through two years of age or beyond.

7.7.3 Water, Sanitation, and Hygiene

The WHO estimates that 88% of diarrheal diseases are attributable to unsafe water, sanitation, and hygiene, including food contamination [222]. Improvement in water quality through water treatment was shown to reduce diarrheal illness by 11% when the water source was treated (RR: 0.89; 95% CI: 0.42–1.90) and by 39% when treated at the household level (RR 0.61; 95% CI: 0.46, 0.81) [132]. Studies have demonstrated benefits of improved personal and home-level hygiene in reducing diarrheal disease episodes. A meta-analysis showed hygiene interventions can reduce diarrheal disease incidence by a pooled relative risk of 0.55 (95% CI: 0.40, 0.75) [132, 223, 224]. With the promotion of handwashing specifically, the risk of diarrheal illness was reduced by 31% (IRR: 0.69; 95% CI: 0.55, 0.87) [131]. However, antimicrobial soap does not impact diarrheal disease more than plain soap. A randomized trial of handwashing promotion of 36 squatter settlements in Karachi comparing regular and antimicrobial soap to controls found an overall 53% lower incidence of diarrhea in settlements that received handwashing interventions, but no difference between settlements using antimicrobial soap or regular soap [225]. Although efficacy studies of handwashing have been impressive, effectiveness studies have shown considerably less benefits than those seen in short-term research settings.

7.7.4 Global Control Efforts

In 1978 WHO launched its Program for the Control of Diarrheal Disease, which emphasized home and clinic diarrheal illness management using fluids available at home,

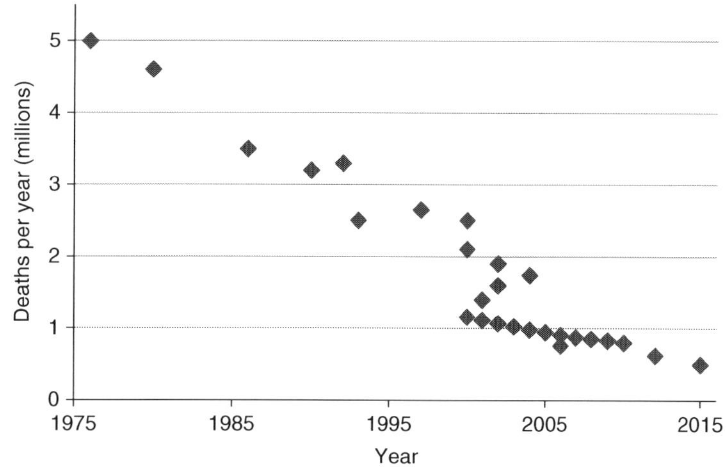

FIGURE 7.4 Mean estimates from studies of global burden of deaths due to diarrhea in children. *Source*: Adapted from Bryce et al. [160], Liu et al. [227], The Global Burden of Disease: 2004, 2012, and 2015 Updates; WHO [1, 228], and Podewils et al. [59].

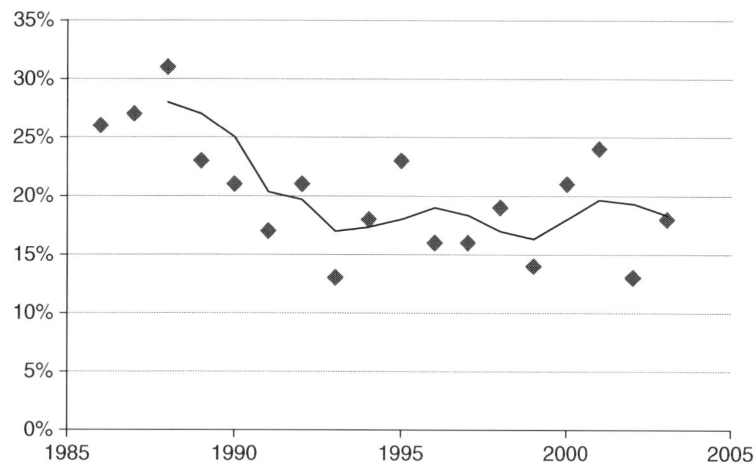

FIGURE 7.5 Proportion of children in Demographic Health Surveys reporting an episode of diarrhea in the previous twoweeks. Line represents three year moving average of proportion of children reporting diarrhea. *Source:* Adapted from Forsberg et al. [229].

oral rehydration therapy, and continued feeding, and, to a lesser extent, on improving personal hygiene, breast feeding, drinking water, sanitation, outbreak control, and measles vaccination [92, 119, 226]. Continued feeding/early re-feeding was added to the WHO's recommendations in the 1980s as its importance became clear. The global effort to control diarrheal disease met with remarkable success between the 1980s and the 2010s, decreasing the number of diarrheal deaths in children from five million deaths per year in 1976 to 620 thousand in 2012 (Figure 7.4), and a modest decrease in diarrheal incidence (Figure 7.5) [59, 228].

Ongoing challenges in diarrhea management control have limited achievements in preventing diarrheal illness and related mortality. In most countries, although caretakers and healthcare providers have knowledge of ORS, use rates remain poor in most countries (lower than 30%) [12]. In addition, among those that use ORS, it is often prepared inappropriately or given below the recommended quantities [230–232]. In India, which has had an active campaign focusing on raising awareness of ORS for decades, a 2007 survey showed that over 73% of Indian women were familiar with ORS. However, only 26% of children with diarrhea were given ORS [233]. Similarly, Ellis and colleagues showed that in southern Mali, although parents were generally knowledgeable about ORS, they continued to seek a combination of multiple ineffective treatments, such as antidiarrheals and antibiotics, to treat the illness [234].

The promotion of ORS with zinc has been shown to simultaneously increase ORS use and acceptability among caretakers and physicians. A cluster-randomized study in India comparing ORS to ORS with zinc showed that 6 months following the intervention in the communities that received ORS with zinc, unwarranted medical care seeking was reduced by 34% (95% CI: 0.43, 0.99), oral antibiotic use was decreased by 88% (95% CI: 0.05, 0.32), intravenous fluid use rates decreased by 50% (95% CI: 0.22, 1.14), and ORS prescription rates increased over 12-fold (95% CI: 8.95, 17.35) [174]. A six site multicountry study introducing zinc for diarrhea treatment showed similar decreases in antibiotic use, and in five of six sites, ORS use rates were the same or higher in intervention communities [175]. Additionally, use of smaller packets of ORS can enable caregivers to more accurately prepare ORS, thereby limiting wasted dry salts as well as the likelihood of the prepared ORS becoming contaminated through the day. For example, the National Control of Diarrheal Diseases Program in Egypt used 200 ml packets effectively in the 1980s [235].

For many countries diarrhea treatment has not been a priority in the past 10–15 years. Funding for training and special promotional activities is minimal. If the coverage of key diarrhea prevention and treatment strategies is to increase (see supplementary figures for coverage of key interventions), national level governments as well as donors will need to recognize the dire need for additional resources to reinstitute proven treatment and prevention options to reduce mortality from diarrhea. A more recent approach, conceptualized by WHO and UNICEF, propose an integrated approach for pneumonia and diarrhea control, the two leading killers of children globally. Their efforts have led to the development of the integrated Global Action Plan for the Prevention and Control of Pneumonia and Diarrhea, which outlines the necessary actions for elimination of preventable child deaths from pneumonia and diarrhea by 2025. To reach this goal, national leaders and stakeholders must work together and apply lessons learned thus far to increase political will for eliminating pneumonia and diarrhea, prioritize interventions based on local context, increase investments in health policy research, and strengthen outcomes and accountability for results [236].

7.8 CONCLUSION

Infectious diarrhea caused by viral, bacterial, and parasitic pathogens is one of the leading causes of disease globally, causing billions of episodes a year. The dehydration and malnutrition caused by diarrhea result in hundreds of thousands of deaths each year, predominantly in children in developing countries. However, numerous well-proven treatment and prevention strategies exist that, when combined, have the potential to eliminate the mortality associated with infectious diarrhea. A renewed international commitment to control diarrheal diseases is necessary if the progress made against infectious diarrhea in the last few decades is to be maintained and strengthened.

REFERENCES

1. Collaborators (2016). Global, regional, and national life expectancy, all-cause mortality, and cause-specific mortality for 249 causes of death, 1980–2015: a systematic analysis for the Global Burden of Disease Study 2015. *Lancet* 388 (10053): 1459–1544.
2. de Wit, M.A., Koopmans, M.P., and van Duynhoven, Y.T. (2003). *Risk factors for norovirus, Sapporo-like virus, and group A rotavirus gastroenteritis. Emerg. Infect. Dis.* 9 (12): 1563–1570.
3. Malek, M.A. et al. (2006). Diarrhea- and rotavirus-associated hospitalizations among children less than 5 years of age: United States, 1997 and 2000. *Pediatrics* 117 (6): 1887–1892.
4. Black, R.E. et al. (1989). Incidence and etiology of infantile diarrhea and major routes of transmission in Huascar, Peru. *Am. J. Epidemiol.* 129 (4): 785–799.
5. Guerrant, R.L. et al. (1983). *Prospective study of diarrheal illnesses in northeastern Brazil: patterns of disease, nutritional impact, etiologies, and risk factors. J. Infect. Dis.* 148 (6): 986–997.
6. Singleton, R.J. et al. (2007). *Diarrhea-associated hospitalizations and outpatient visits among American Indian and Alaska native children younger than five years of age, 2000–2004. Pediatr. Infect. Dis. J.* 26 (11): 1006–1013.
7. Salama, P. et al. (2004). *Lessons learned from complex emergencies over past decade. Lancet* 364 (9447): 1801–1813.
8. Schwartz, B.S. et al. (2006). *Diarrheal epidemics in Dhaka, Bangladesh, during three consecutive floods: 1988, 1998, and 2004. Am. J. Trop. Med. Hyg.* 74 (6): 1067–1073.
9. Cronin, A.A. et al. (2009). *Quantifying the burden of disease associated with inadequate provision of water and sanitation in selected sub-Saharan refugee camps. J. Water Health* 7 (4): 557–568.
10. Kolstad, E.W. and Johansson, K.A. (2011). *Uncertainties associated with quantifying climate change impacts on human health: a case study for diarrhea. Environ. Health Perspect.* 119 (3): 299–305.
11. Checkley, W. et al. (2000). *Effect of El Niño and ambient temperature on hospital admissions for diarrhoeal diseases in Peruvian children. Lancet* 355 (9202): 442–450.
12. UNICEF (2009). *Diarrhoea: Why Children Are Still Dying and What Can Be Done*. Geneva: WHO/UN Children's Fund.
13. Baldi, F. et al. (2009). *Focus on acute diarrhoeal disease. World J. Gastroenterol.* 15 (27): 3341–3348.
14. Navaneethan, U. and Giannella, R.A. (2008). *Mechanisms of infectious diarrhea. Nat. Clin. Pract. Gastroenterol Hepatol.* 5 (11): 637–647.
15. Greenberg, H.B. and Estes, M.K. (2009). *Rotaviruses: from pathogenesis to vaccination. Gastroenterology* 136 (6): 1939–1951.
16. Huang, D.B., Chappell, C., and Okhuysen, P.C. (2004). *Cryptosporidiosis in children. Semin. Pediatr. Infect. Dis.* 15 (4): 253–259.
17. Laohachai, K.N. et al. (2003). *The role of bacterial and non-bacterial toxins in the induction of changes in membrane transport: implications for diarrhea. Toxicon* 42 (7): 687–707.

18. Lapointe, T.K., O'Connor, P.M., and Buret, A.G. (2009). *The role of epithelial malfunction in the pathogenesis of enteropathogenic E. coli-induced diarrhea. Lab. Investig.* 89 (9): 964–970.
19. Buret, A.G. (2008). *Pathophysiology of enteric infections with Giardia duodenalius. Parasite* 15 (3): 261–265.
20. Moon, H.W. (1997). *Comparative histopathology of intestinal infections. Adv. Exp. Med. Biol.* 412: 1–19.
21. Flores, J. and Okhuysen, P.C. (2009). *Enteroaggregative Escherichia coli infection. Curr. Opin. Gastroenterol.* 25 (1): 8–11.
22. Hookman, P. and Barkin, J.S. (2009). *Clostridium difficile associated infection, diarrhea and colitis. World J. Gastroenterol.* 15 (13): 1554–1580.
23. Guttman, J.A. and Finlay, B.B. (2009). *Tight junctions as targets of infectious agents. Biochim. Biophys. Acta* 1788 (4): 832–841.
24. Schroeder, G.N. and Hilbi, H. (2008). *Molecular pathogenesis of Shigella spp.: controlling host cell signaling, invasion, and death by type III secretion. Clin. Microbiol. Rev.* 21 (1): 134–156.
25. Janssen, R. et al. (2008). *Host-pathogen interactions in Campylobacter infections: the host perspective. Clin. Microbiol. Rev.* 21 (3): 505–518.
26. Haque, R. et al. (2003). Amebiasis. *N. Engl. J. Med.* 348 (16): 1565–1573.
27. Levine, M.M. (2006). *Enteric infections and the vaccines to counter them: future directions. Vaccine* 24 (18): 3865–3873.
28. Abba, K. et al. (2009). *Pathogens associated with persistent diarrhoea in children in low and middle income countries: systematic review. BMC Infect. Dis.* 9: 88.
29. WHO (1999). *The Evolution of Diarrhoeal and Acute Respiratory Disease Control at WHO. Achievements 1980–1995 in Research, Development, and Implementation.* Department of Child and Adolescent Health and Development (WHO/CHS/CAS/99.12).
30. Lima, A.A. and Guerrant, R.L. (1992). *Persistent diarrhea in children: epidemiology, risk factors, pathophysiology, nutritional impact, and management. Epidemiol. Rev.* 14: 222–242.
31. WHO (1995).The Treatment of Diarrhoea. A Manual for Physicians and Other Senior Health Workers. (WHO/CDR/95.3).
32. Bhandari, N., Bhan, M.K., and Sazawal, S. (1992). *Mortality associated with acute watery diarrhea, dysentery and persistent diarrhea in rural North India. Acta Paediatr.* 81 (Suppl .381)): 3 6.
33. Donowitz, M., Kokke, F.T., and Saidi, R. (1995). *Evaluation of patients with chronic diarrhea. N. Engl. J. Med.* 332 (11): 725–729.
34. Lanata, C.F. et al. (1991). *Epidemiologic, clinical, and laboratory characteristics of acute vs. persistent diarrhea in periurban Lima, Peru. J. Pediatr. Gastroenterol. Nutr.* 12 (1): 82–88.
35. Ahmed, F. et al. (2001). *Epidemiology of postshigellosis persistent diarrhea in young children. Pediatr. Infect. Dis. J.* 20 (5): 525–530.
36. Holtz, L.R., Neill, M.A., and Tarr, P.I. (2009). *Acute bloody diarrhea: a medical emergency for patients of all ages. Gastroenterology* 136 (6): 1887–1898.
37. O'Ryan, M., Prado, V., and Pickering, L.K. (2005). *A millennium update on pediatric diarrheal illness in the developing world. Semin. Pediatr. Infect. Dis.* 16 (2): 125–136.
38. DuPont, H.L. (2009). *Clinical practice Bacterial diarrhea. N. Engl. J. Med.* 361 (16): 1560–1569.
39. Gallay, A. et al. (2007). *Campylobacter antimicrobial drug resistance among humans, broiler chickens, and pigs, France. Emerg. Infect. Dis.* 13 (2): 259–266.
40. Parashar, U.D. et al. (2006). *Rotavirus and severe childhood diarrhea. Emerg. Infect. Dis.* 12 (2): 304–306.
41. Glass, R.I. et al. (1996). *The epidemiology of rotavirus diarrhea in the United States: surveillance and estimates of disease burden. J. Infect. Dis.* 174 (Suppl. 1)): S5–S11.
42. Parashar, U.D. et al. (2009). *Global mortality associated with rotavirus disease among children in 2004. J. Infect. Dis.* 200 (Suppl. 1)): S9–S15.
43. Goodgame, R.W. (2001). *Viral causes of diarrhea. Gastroenterol. Clin. N. Am.* 30 (3): 779–795.
44. Clark, B. and McKendrick, M. (2004). *A review of viral gastroenteritis. Curr. Opin. Infect. Dis.* 17 (5): 461–469.
45. Pires, S.M. et al. (2015). *Aetiology-specific estimates of the global and regional incidence and mortality of diarrhoeal diseases commonly transmitted through food. PLoS One* 10 (12): e0142927.
46. Liu, J. et al. (2016). *Use of quantitative molecular diagnostic methods to identify causes of diarrhoea in children: a reanalysis of the GEMS case-control study. Lancet* 388 (10051): 1291–1301.
47. Grais, R.F. et al. (2007). *Unacceptably high mortality related to measles epidemics in Niger, Nigeria, and Chad. PLoS Med.* 4 (1): e16.
48. Sniadack, D.H. et al. (1999). *Measles epidemiology and outbreak response immunization in a rural community in Peru. Bull. World Health Organ.* 77 (7): 545–552.
49. Thakur, J.S. et al. (2002). *Measles outbreak in a Periurban area of Chandigarh: need for improving vaccine coverage and strengthening surveillance. Indian J. Pediatr.* 69 (1): 33–37.
50. Sharma, M.K., Bhatia, V., and Swami, H.M. (2004). *Outbreak of measles amongst vaccinated children in a slum of Chandigarh. Indian J. Med. Sci.* 58 (2): 47–53.
51. Feachem, R.G. and Koblinsky, M.A. (1983). *Interventions for the control of diarrhoeal diseases among young children: measles immunization. Bull. World Health Organ.* 61 (4): 641–652.
52. Guerrant, R.L. et al. (1990). *Diarrhea in developed and developing countries: magnitude, special settings, and etiologies. Rev. Infect. Dis.* 12 (Suppl. 1)): S41–S50.
53. Pawlowski, S.W., Warren, C.A., and Guerrant, R. (2009). *Diagnosis and treatment of acute or persistent diarrhea. Gastroenterology* 136 (6): 1874–1886.
54. Majowicz, S.E. et al. (2014). *Global incidence of human Shiga toxin-producing Escherichia coli infections and deaths: a systematic review and knowledge synthesis. Foodborne Pathog. Dis.* 11 (6): 447–455.

55. Vieira, N. et al. (2007). *High prevalence of enteroinvasive Escherichia coli isolated in a remote region of northern coastal Ecuador. Am. J. Trop. Med. Hyg.* 76 (3): 528–533.
56. Niyogi, S.K. (2005). *Shigellosis. J. Microbiol.* 43 (2): 133–143.
57. Ali, M. et al. (2015). *Updated global burden of cholera in endemic countries. PLoS Negl. Trop. Dis.* 9 (6): e0003832.
58. Platts-Mills, J.A. et al. (2015). *Pathogen-specific burdens of community diarrhoea in developing countries: a multisite birth cohort study (MAL-ED). Lancet Glob. Health* 3 (9): e564–e575.
59. Podewils, L.J. et al. (2004). *Acute, infectious diarrhea among children in developing countries. Semin. Pediatr. Infect. Dis.* 15 (3): 155–168.
60. Leonard, E.M. et al. (1990). *An intrinsic pathogenicity index for microorganisms causing infection in a neonatal surgical unit. Microb. Ecol. Health Dis.* 3: 151–157.
61. Blackwelder, W.C. et al. (2012). *Statistical methods in the Global Enteric Multicenter Study (GEMS). Clin. Infect. Dis.* 55 (Suppl. 4): S246–S253.
62. Wilhelmi, I., Roman, E., and Sanchez-Fauquier, A. (2003). *Viruses causing gastroenteritis. Clin. Microbiol. Infect.* 9 (4): 247–262.
63. Mead, P.S. et al. (1999). *Food-related illness and death in the United States. Emerg. Infect. Dis.* 5 (5): 607–625.
64. Glass, R.I. et al. (2001). *Gastroenteritis viruses: an overview. Novartis Found. Symp.* 238: 5–19. discussion 19–25.
65. Kotloff, K.L. et al. (2013). *Burden and aetiology of diarrhoeal disease in infants and young children in developing countries (the Global Enteric Multicenter Study, GEMS): a prospective, case-control study. Lancet* 382 (9888): 209–222.
66. Gurgel, R.G. et al. (2009). *Incidence of rotavirus and all-cause diarrhea in northeast Brazil following the introduction of a national vaccination program. Gastroenterology* 137 (6): 1970–1975.
67. Walter, J.E. and Mitchell, D.K. (2000). *Role of astroviruses in childhood diarrhea. Curr. Opin. Pediatr.* 12 (3): 275–279.
68. Cheng, A.C., McDonald, J.R., and Thielman, N.M. (2005). *Infectious diarrhea in developed and developing countries. J. Clin. Gastroenterol.* 39 (9): 757–773.
69. Pang, X.L., Joensuu, J., and Vesikari, T. (1999). *Human calicivirus-associated sporadic gastroenteritis in Finnish children less than two years of age followed prospectively during a rotavirus vaccine trial. Pediatr. Infect. Dis. J.* 18 (5): 420–426.
70. Dennehy, P.H. (2005). *Acute diarrheal disease in children: epidemiology, prevention, and treatment. Infect. Dis. Clin. N. Am.* 19 (3): 585–602.
71. Marks, P.J. et al. (2003). *A school outbreak of Norwalk-like virus: evidence for airborne transmission. Epidemiol. Infect.* 131 (1): 727–736.
72. Martella, V. et al. (2009). Zoonotic aspects of rotaviruses. *Vet. Microbiol.* .
73. Bernstein, D.I. (2009). *Rotavirus overview. Pediatr. Infect. Dis. J.* 28 (3 Suppl): S50–S53.
74. Dennehy, P.H. (2008). *Rotavirus vaccines: an overview. Clin. Microbiol. Rev.* 21 (1): 198–208.
75. Lanata, C.F. et al. (2013). *Global causes of diarrheal disease mortality in children <5 years of age: a systematic review. PLoS One* 8 (9): e72788.
76. Santosham, M. et al. (1983). *Detection of rotavirus in respiratory secretions of children with pneumonia. J. Pediatr.* 103 (4): 583–585.
77. Santos, N. and Hoshino, Y. (2005). *Global distribution of rotavirus serotypes/genotypes and its implication for the development and implementation of an effective rotavirus vaccine. Rev. Med. Virol.* 15 (1): 29–56.
78. Gray, J. et al. (2008). *Rotavirus. J. Pediatr. Gastroenterol. Nutr.* 46 (Suppl. 2)): S24–S31.
79. Ahmed, S.M. et al. (2014). *Global prevalence of norovirus in cases of gastroenteritis: a systematic review and meta-analysis. Lancet Infect. Dis.* 14 (8): 725–730.
80. Lopman, B.A. et al. (2016). *The vast and varied global burden of norovirus: prospects for prevention and control. PLoS Med.* 13 (4): e1001999.
81. Dai, Y.C. et al. (2009). Surveillance and risk factors of norovirus gastroenteritis among children in a southern city of China in the fall-winter seasons of 2003-2006. *J. Paediatr. Child Health* 46 (1–2): 45–50.
82. Yori, P.P. et al. (2009). *Norovirus highly prevalent cause of endemic acute diarrhea in children in the peruvian Amazon. Pediatr. Infect. Dis. J.* 28 (9): 844–847.
83. Siebenga, J.J. et al. (2007). *Epochal evolution of GGII.4 norovirus capsid proteins from 1995 to 2006. J. Virol.* 81 (18): 9932–9941.
84. Payne, D.C., Parashar, U.D., and Lopman, B.A. (2015). *Developments in understanding acquired immunity and innate susceptibility to norovirus and rotavirus gastroenteritis in children. Curr. Opin. Pediatr.* 27 (1): 105–109.
85. Kim, Y. et al. (2015). *Recent advances in the discovery of norovirus therapeutics. J. Med. Chem.* 58 (24): 9438–9450.
86. Tan, M. and Jiang, X. (2014). *Vaccine against norovirus. Hum. Vaccine Immunother.* 10 (6): 1449–1456.
87. Walter, J.E. and Mitchell, D.K. (2003). *Astrovirus infection in children. Curr. Opin. Infect. Dis.* 16 (3): 247–253.
88. Guix, S. et al. (2002). *Molecular epidemiology of astrovirus infection in Barcelona, Spain. J. Clin. Microbiol.* 40 (1): 133–139.
89. Lenaerts, L., De Clercq, E., and Naesens, L. (2008). *Clinical features and treatment of adenovirus infections. Rev. Med. Virol.* 18 (6): 357–374.
90. Spitzer, M.D. (2002). *Viral causes of diarrhea. Pediatr. Rev.* 23 (7): 257–258.
91. Rima, B.K. and Duprex, W.P. (2006). *Morbilliviruses and human disease. J. Pathol.* 208 (2): 199–214.
92. Feachem, R.G., Hogan, R.C., and Merson, M.H. (1983). *Diarrhoeal disease control: reviews of potential interventions. Bull. World Health Organ.* 61 (4): 637–640.
93. Schneider-Schaulies, S. and Schneider-Schaulies, J. (2009). *Measles virus-induced immunosuppression. Curr. Top. Microbiol. Immunol.* 330: 243–269.

94. Hussey, G.D. and Clements, C.J. (1996). *Clinical problems in measles case management. Ann. Trop. Paediatr.* 16 (4): 307–317.
95. Carlos, C.C. et al. (1992). *Enteropathogens among measles patients with diarrhea in urban Filipino children. Phil. J. Microbiol. Infect. Dis.* 21 (2): 53–60.
96. Huilan, S. et al. (1991). *Etiology of acute diarrhoea among children in developing countries: a multicentre study in five countries. Bull. World Health Organ.* 69 (5): 549–555.
97. Nataro, J.P. and Kaper, J.B. (1998). *Diarrheagenic Escherichia coli. Clin. Microbiol. Rev.* 11 (1): 142–201.
98. Ochoa, T.J. et al. (2008). *New insights into the epidemiology of enteropathogenic Escherichia coli infection. Trans. R. Soc. Trop. Med. Hyg.* 102 (9): 852–856.
99. Callaway, T.R. et al. (2009). *Diet, Escherichia coli O157:H7, and cattle: a review after 10 years. Curr. Issues Mol. Biol.* 11 (2): 67–79.
100. Page, A.V. and Liles, W.C. (2013). *Enterohemorrhagic Escherichia coli infections and the hemolytic-uremic syndrome. Med. Clin. N. Am.* 97 (4): 681–695. xi.
101. Scallan, E. et al. (2011). *Foodborne illness acquired in the United States – major pathogens. Emerg. Infect. Dis.* 17 (1): 7–15.
102. German, E.-H.U.S.R. (2011). *The German 2011 epidemic of Shiga toxin-producing E. Coli – the nephrological view. Nephrol. Dial. Transplant.* 26 (9): 2723–2726.
103. Wong, C.S. et al. (2012). *Risk factors for the hemolytic uremic syndrome in children infected with Escherichia coli O157:H7: a multivariable analysis. Clin. Infect. Dis.* 55 (1): 33–41.
104. Ud-Din, A. and Wahid, S. (2014). *Relationship among Shigella spp. and enteroinvasive Escherichia coli (EIEC) and their differentiation. Braz. J. Microbiol.* 45 (4): 1131–1138.
105. Crump, J.A., Luby, S.P., and Mintz, E.D. (2004). *The global burden of typhoid fever. Bull. World Health Organ.* 82 (5): 346–353.
106. DuPont, H.L. (2007). *The growing threat of foodborne bacterial enteropathogens of animal origin. Clin. Infect. Dis.* 45 (10): 1353–1361.
107. Amour, C. et al. (2016). *Epidemiology and impact of campylobacter infection in children in 8 low-resource settings: results from the MAL-ED study. Clin. Infect. Dis.* 63 (9): 1171–1179.
108. (2001). Cholera vaccines. *Wkly. Epidemiol. Rec.* 76 (16): 117–124.
109. von Graevenitz, A. (2007). *The role of Aeromonas in diarrhea: a review. Infection* 35 (2): 59–64.
110. Qamar, F.N. et al. (2016). *Aeromonas-associated diarrhea in children under 5 years: the GEMS experience. Am. J. Trop. Med. Hyg.* 95 (4): 774–780.
111. Lessa, F.C. et al. (2015). *Burden of Clostridium difficile infection in the United States. N. Engl. J. Med.* 372 (24): 2369–2370.
112. Joshi, N.M., Macken, L., and Rampton, D.S. (2012). *Inpatient diarrhoea and Clostridium difficile infection. Clin. Med. (Lond.)* 12 (6): 583–588.
113. Furuya-Kanamori, L. et al. (2015). *Asymptomatic Clostridium difficile colonization: epidemiology and clinical implications. BMC Infect. Dis.* 15: 516.
114. Kim, J. et al. (2012). *Risk factors and outcomes associated with severe clostridium difficile infection in children. Pediatr. Infect. Dis. J.* 31 (2): 134–138.
115. Kociolek, L.K. and Gerding, D.N. (2016). Breakthroughs in the treatment and prevention of *Clostridium difficile* infection. *Nat. Rev. Gastroenterol. Hepatol.* 13 (3): 150–160.
116. Griffiths, J.K. (1998). *Human cryptosporidiosis: epidemiology, transmission, clinical disease, treatment, and diagnosis. Adv. Parasitol.* 40: 37–85.
117. Escobedo, A.A. and Cimerman, S. (2007). *Giardiasis: a pharmacotherapy review. Expert. Opin. Pharmacother.* 8 (12): 1885–1902.
118. Pritt, B.S. and Clark, C.G. (2008). *Amebiasis. Mayo Clin. Proc.* 83 (10): 1154–1159. quiz 1159–60.
119. Claeson, M. and Merson, M.H. (1990). *Global progress in the control of diarrheal diseases. Pediatr. Infect. Dis. J.* 9 (5): 345–355.
120. Hahn, S., Kim, S., and Garner, P. (2002). *Reduced osmolarity oral rehydration solution for treating dehydration caused by acute diarrhoea in children. Cochrane Database Syst. Rev.* 1: CD002847.
121. Lazzerini, M. and Wanzira, H. (2016). *Oral zinc for treating diarrhoea in children. Cochrane Database Syst. Rev.* 12: CD005436.
122. Baqui, A.H. et al. (2002). *Effect of zinc supplementation started during diarrhoea on morbidity and mortality in Bangladeshi children: community randomised trial. BMJ* 325 (7372): 1059.
123. Thomas, D. and Greer, F. (2010). Probiotics and prebiotics in pediatrics. *Pediatrics* 126 (6): 1217–1231.
124. Tomasik, E. et al. (2016). *Systematic review with meta-analysis: ondansetron for vomiting in children with acute gastroenteritis. Aliment. Pharmacol. Ther.* 44 (5): 438–446.
125. Bhutta, Z.A. et al. (1999). *Prevention of diarrhea and pneumonia by zinc supplementation in children in developing countries: pooled analysis of randomized controlled trials. Zinc Investigators' Collaborative Group. J. Pediatr.* 135 (6): 689–697.
126. Beaton, G., Martorell, R., and Aronson, K.J. (1993). *Effectiveness of Vitamin A Supplementation in the Control of Young Child Morbidity and Mortality in Developing Countries*. Toronto: International Nutrition Program.
127. al-Ali, F.M., Hossain, M.M., and Pugh, R.N. (1997). *The associations between feeding modes and diarrhoea among urban children in a newly developed country. Public Health* 111 (4): 239–243.
128. Huffman, S.L. and Combest, C. (1990). *Role of breast-feeding in the prevention and treatment of diarrhoea. J. Diarrhoeal Dis. Res.* 8 (3): 68–81.
129. VanDerslice, J., Popkin, B., and Briscoe, J. (1994). *Drinking-water quality, sanitation, and breast-feeding: their interactive effects on infant health. Bull. World Health Organ.* 72 (4): 589–601.

130. Ahiadeke, C. (2000). *Breast-feeding, diarrhoea and sanitation as components of infant and child health: a study of large scale survey data from Ghana and Nigeria. J. Biosoc. Sci.* 32 (1): 47–61.
131. Ejemot, R.I. et al. (2008). *Hand washing for preventing diarrhoea. Cochrane Database Syst. Rev.* 1: CD004265.
132. Fewtrell, L. et al. (2005). *Water, sanitation, and hygiene interventions to reduce diarrhoea in less developed countries: a systematic review and meta-analysis. Lancet Infect. Dis.* 5 (1): 42–52.
133. Vesikari, T. et al. (2006). *Safety and efficacy of a pentavalent human-bovine (WC3) reassortant rotavirus vaccine. N. Engl. J. Med.* 354 (1): 23–33.
134. Ruiz-Palacios, G.M. et al. (2006). *Safety and efficacy of an attenuated vaccine against severe rotavirus gastroenteritis. N. Engl. J. Med.* 354 (1): 11–22.
135. Ryland, S. and Raggers, H. (1998). *Childhood Morbidity and Treatment Patterns. DHS Comparative Studies No. 27.* Claverton, MD: Macro International Inc.
136. Clemens, J.D. et al. (1986). *Field trial of oral cholera vaccines in Bangladesh. Lancet* 2 (8499): 124–127.
137. Sanchez, J.L. et al. (1994). *Protective efficacy of oral whole-cell/recombinant-B-subunit cholera vaccine in Peruvian military recruits. Lancet* 344 (8932): 1273–1276.
138. King, C.K. et al. (2003). *Managing acute gastroenteritis among children: oral rehydration, maintenance, and nutritional therapy. MMWR Recomm. Rep.* 52 (RR-16): 1–16.
139. Duggan, C., Santosham, M., and Glass, R.I. (1992). *The management of acute diarrhea in children: oral rehydration, maintenance, and nutritional therapy. Centers for Disease Control and Prevention. MMWR Recomm. Rep.* 41 (RR-16): 1–20.
140. Latta, T. (1831-1832). *Malignant cholera: comments contributed by the Central Board of Health, London, relative to the treatment of cholera by copious injection of aqueous and saline fluids into the veins. Lancet* 2: 274–277.
141. Santosham, M. et al. (1997). *Oral rehydration therapy for diarrhea: an example of reverse transfer of technology. Pediatrics* 100 (5): E10.
142. Darrow, D.C., Pratt, E.L. et al. (1949). *Disturbances of water and electrolytes in infantile diarrhea. Pediatrics* 3 (2): 129–156.
143. King, C.K. et al. (2003). *Managing acute gastroenteritis among children: oral rehydration, maintenance, and nutritional therapy. MMWR Recomm. Rep.* 52 (RR-16): 1–16.
144. Harrison, H.E. (1954). *The treatment of diarrhea in infancy. Pediatr. Clin. N. Am.* 335–348.
145. Paneth, N. (1980). *Hypernatremic dehydration of infancy: an epidemiologic review. Am. J. Dis. Child.* 134 (8): 785–792.
146. Mahalanabis, D. et al. (1973). *Oral fluid therapy of cholera among Bangladesh refugees. Johns Hopkins Med. J.* 132 (4): 197–205.
147. Ruxin, J.N. (1994). *Magic bullet: the history of oral rehydration therapy. Med. Hist.* 38 (4): 363–397.
148. al-Awqati, Q.S., Mekkiya, M., and Thamer, M. (1969). *Establishment of a cholera treatment unit under epidemic conditions in a developing country. Lancet* 1 (7588): 252–253.
149. Chatterjee, H.N. (1953). *Control of vomiting in cholera and oral replacement of fluid. Lancet* 265 (6795): 1063.
150. Nalin, D.R. et al. (1968). *Oral maintenance therapy for cholera in adults. Lancet* 2 (7564): 370–373.
151. Nalin, D.R., Cash, R.A., and Rahman, M. (1970). *Oral (or nasogastric) maintenance therapy for cholera patients in all age-groups. Bull. World Health Organ.* 43 (3): 361–363.
152. Cash, R.A. et al. (1970). *A clinical trial of oral therapy in a rural cholera-treatment center. Am. J. Trop. Med. Hyg.* 19 (4): 653–656.
153. Duggan, C. et al. (2004). *Scientific rationale for a change in the composition of oral rehydration solution. JAMA* 291 (21): 2628–2631.
154. Mahalanabis, D. et al. (1995). *Hypotonic oral rehydration solution in acute diarrhoea: a controlled clinical trial. Acta Paediatr.* 84 (3): 289–293.
155. Santosham, M. et al. (1996). *A double-blind clinical trial comparing World Health Organization oral rehydration solution with a reduced osmolarity solution containing equal amounts of sodium and glucose. J. Pediatr.* 128 (1): 45–51.
156. CHOICE Study Group (2001). Multicenter, randomized, double-blind clinical trial to evaluate the efficacy and safety of a reduced osmolarity oral rehydration salts solution in children with acute watery diarrhea. *Pediatrics* 107 (4): 613–618.
157. (1995). *Multicentre evaluation of reduced-osmolarity oral rehydration salts solution. International study group on reduced-osmolarity ORS solutions. Lancet* 345 (8945): 282–285.
158. Murphy, C., Hahn, S., and Volmink, J. (2004). *Reduced osmolarity oral rehydration solution for treating cholera. Cochrane Database Syst. Rev.* 4: CD003754.
159. Sarker, S.A., Majid, N., and Mahalanabis, D. (1995). *Alanine- and glucose-based hypo-osmolar oral rehydration solution in infants with persistent diarrhoea: a controlled trial. Acta Paediatr.* 84 (7): 775–780.
160. Bryce, J. et al. (2005). *WHO estimates of the causes of death in children. Lancet* 365 (9465): 1147–1152.
161. Bhandari, N. et al. (1989). *Association of antecedent malnutrition with persistent diarrhoea: a case-control study. BMJ* 298 (6683): 1284–1287.
162. Guerrant, R.L. et al. (2002). *Updating the DALYs for diarrhoeal disease. Trends Parasitol.* 18 (5): 191–193.
163. Niehaus, M.D. et al. (2002). *Early childhood diarrhea is associated with diminished cognitive function 4 to 7 years later in children in a northeast Brazilian shantytown. Am. J. Trop. Med. Hyg.* 66 (5): 590–593.
164. Lorntz, B. et al. (2006). *Early childhood diarrhea predicts impaired school performance. Pediatr. Infect. Dis. J.* 25 (6): 513–520.
165. Chung, A.W. and Viscorova, B. (1948). *The effect of early oral feeding versus early oral starvation on the course of infantile diarrhea. J. Pediatr.* 33 (1): 14–22.

166. Armitstead, J., Kelly, D., and Walker-Smith, J. (1989). *Evaluation of infant feeding in acute gastroenteritis. J. Pediatr. Gastroenterol. Nutr.* 8 (2): 240–244.

167. Hjelt, K. et al. (1989). *Rapid versus gradual refeeding in acute gastroenteritis in childhood: energy intake and weight gain. J. Pediatr. Gastroenterol. Nutr.* 8 (1): 75–80.

168. Sandhu, B.K. et al. (1997). *A multicentre study on behalf of the European Society of Paediatric Gastroenterology and Nutrition Working Group on Acute Diarrhoea. Early feeding in childhood gastroenteritis. J. Pediatr. Gastroenterol. Nutr.* 24 (5): 522–527.

169. Santosham, M., Brown, K.H., and Sack, R.B. (1987). *Oral rehydration therapy and dietary therapy for acute childhood diarrhea. Pediatr. Rev.* 8 (9): 273–278.

170. Santosham, M. et al. (1985). *Role of soy-based, lactose-free formula during treatment of acute diarrhea. Pediatrics* 76 (2): 292–298.

171. WHO (2004). Clinical management of acute diarrhoea. In: *WHO/UNICEF Joint Statement* (ed. WHO/UNICEF). New York: WHO.

172. Patro, B., Golicki, D., and Szajewska, H. (2008). *Meta-analysis: zinc supplementation for acute gastroenteritis in children. Aliment. Pharmacol. Ther.* 28 (6): 713–723.

173. Lukacik, M., Thomas, R.L., and Aranda, J.V. (2008). *A meta-analysis of the effects of oral zinc in the treatment of acute and persistent diarrhea. Pediatrics* 121 (2): 326–336.

174. Bhandari, N. et al. (2008). *Effectiveness of zinc supplementation plus oral rehydration salts compared with oral rehydration salts alone as a treatment for acute diarrhea in a primary care setting: a cluster randomized trial. Pediatrics* 121 (5): e1279–e1285.

175. Awasthi, S. (2006). *Zinc supplementation in acute diarrhea is acceptable, does not interfere with oral rehydration, and reduces the use of other medications: a randomized trial in five countries. J. Pediatr. Gastroenterol. Nutr.* 42 (3): 300–305.

176. Baqui, A.H. et al. (2004). *Zinc therapy for diarrhoea increased the use of oral rehydration therapy and reduced the use of antibiotics in Bangladeshi children. J. Health Popul. Nutr.* 22 (4): 440–442.

177. Bhutta, Z.A. et al. (2000). *Therapeutic effects of oral zinc in acute and persistent diarrhea in children in developing countries: pooled analysis of randomized controlled trials. Am. J. Clin. Nutr.* 72 (6): 1516–1522.

178. Negi, R. et al. (2015). *Oral zinc supplements are ineffective for treating acute dehydrating diarrhoea in 5-12-year-olds. Acta Paediatr.* 104 (8): e367–e371.

179. Kimberlin, D.W. et al. (eds.) (2015). *Red Book: Report of the Committee on Infectious Diseases*, 30e. American Academy of Pediatrics.

180. Freedman, S.B. et al. (2015). *Gastroenteritis therapies in developed countries: systematic review and meta-analysis. PLoS One* 10 (6): e0128754.

181. Allen, S. et al. (2010). *Probiotics for treating acute infectious diarrhoea. Cochrane Database Syst. Rev.* 10 (11): CD003048. doi: 10.1002/14651858.CD003048.pub3.

182. Pulling, M. and Surawicz, C.M. (2008). *Loperamide use for acute infectious diarrhea in children: safe and sound? Gastroenterology* 134 (4): 1260–1262.

183. Santos, M. et al. (2009). *Use of racecadotril as outpatient treatment for acute gastroenteritis: a prospective, randomized, parallel study. J. Pediatr.* 155 (1): 62–67.

184. Isolauri, E., Vahasarja, V., and Vesikari, T. (1986). *Effect of cholestyramine on acute diarrhoea in children receiving rapid oral rehydration and full feedings. Ann. Clin. Res.* 18 (2): 99–102.

185. Richards, L., Claeson, M., and Pierce, N.F. (1993). *Management of acute diarrhea in children: lessons learned. Pediatr. Infect. Dis. J.* 12 (1): 5–9.

186. Alam, N.H. and Ashraf, H. (2003). *Treatment of infectious diarrhea in children. Paediatr. Drugs* 5 (3): 151–165.

187. Koo, H.L. et al. (2009). *Antimotility agents for the treatment of Clostridium difficile diarrhea and colitis. Clin. Infect. Dis.* 48 (5): 598–605.

188. Bhattacharya, S.K. and Sur, D. (2003). *An evaluation of current shigellosis treatment. Expert. Opin. Pharmacother.* 4 (8): 1315–1320.

189. Chandran, A. et al. (2010). *Prevention of rotavirus gastroenteritis in infants and children: rotavirus vaccine safety, efficacy, and potential impact of vaccines. Biologics* 4: 213–229.

190. Bhandari, N. et al. (2014). *Efficacy of a monovalent human-bovine (116E) rotavirus vaccine in Indian infants: a randomised, double-blind, placebo-controlled trial. Lancet* 383 (9935): 2136–2143.

191. Van der Wielen, M. and Van Damme, P. (2008). *Pentavalent human-bovine (WC3) reassortant rotavirus vaccine in special populations: a review of data from the Rotavirus Efficacy and Safety Trial. Eur. J. Clin. Microbiol. Infect. Dis.* 27 (7): 495–501.

192. Goveia, M.G. et al. (2008). *Efficacy of pentavalent human-bovine (WC3) reassortant rotavirus vaccine based on breast-feeding frequency. Pediatr. Infect. Dis. J.* 27 (7): 656–658.

193. Dennehy, P.H. et al. (2005). *Comparative evaluation of safety and immunogenicity of two dosages of an oral live attenuated human rotavirus vaccine. Pediatr. Infect. Dis. J.* 24 (6): 481–488.

194. Zaman, K. et al. (2009). *Successful co-administration of a human rotavirus and oral poliovirus vaccines in Bangladeshi infants in a 2-dose schedule at 12 and 16 weeks of age. Vaccine* 27 (9): 1333–1339.

195. Ciarlet, M., et al. (2007). Safety and immunogenicity of concomitant use of the pentavalent rotavirus vaccine (RotaTeq) and oral poliovirus vaccine (OPV). Porto, Portugal, European Society for Pediatric Infectious Diseases (ESPID).

196. Rha, B. et al. (2014). *Intussusception following rotavirus vaccination: an updated review of the available evidence. Expert Rev. Vaccines* 13 (11): 1339–1348.

197. (2009). *Meeting of the immunization Strategic Advisory Group of Experts, April 2009--conclusions and recommendations. Wkly. Epidemiol. Rec.* 84 (23): 220–236.

198. Loharikar, A. et al. (2016). *Status of new vaccine introduction — Worldwide, September 2016. Morb. Mortal. Wkly. Rep.* 65 (41): 1136–1140.

199. (2009). *Reduction in rotavirus after vaccine introduction--United States, 2000-2009. MMWR Morb. Mortal. Wkly. Rep.* 58 (41): 1146–1149.

200. Groome, M. et al. (2016). *Temporal association of rotavirus vaccine introduction and reduction in all-cause childhood diarrheal hospitalizations in South Africa. Clin. Infect. Dis.* 62 (Suppl. 2): S188–S195.

201. Armah, G. et al. (2016). *Impact and effectiveness of monovalent rotavirus vaccine against severe rotavirus diarrhea in Ghana. Clin. Infect. Dis.* 62 (Suppl. 2): S200–S207.

202. Orozco, M. et al. (2009). *Uptake of rotavirus vaccine and national trends of acute gastroenteritis among children in Nicaragua. J. Infect. Dis.* 200 (Suppl. 1): S125–S130.

203. Richardson, V. et al. (2010). *Effect of rotavirus vaccination on death from childhood diarrhea in Mexico. N. Engl. J. Med.* 362 (4): 299–305.

204. Mpabalwani, E. et al. (2016). *Impact of rotavirus vaccination on diarrheal hospitalizations in children aged <5 years in Lusaka, Zambia. Clin. Infect. Dis.* 62 (Suppl. 2): S183–S187.

205. Dóró, R. et al. (2014). *Review of global rotavirus strain prevalence data from six years post vaccine licensure surveillance: is there evidence of strain selection from vaccine pressure? Infect. Genet. Evol.* 28: 446–461.

206. Martin, S. et al. (2014). *Post-licensure deployment of oral cholera vaccines: a systematic review. Bull. World Health Organ.* 92 (12): 849–924.

207. Trach, D.D. et al. (2002). *Investigations into the safety and immunogenicity of a killed oral cholera vaccine developed in Viet Nam. Bull. World Health Organ.* 80 (1): 2–8.

208. Mahalanabis, D. et al. (2008). *A randomized, placebo-controlled trial of the bivalent killed, whole-cell, oral cholera vaccine in adults and children in a cholera endemic area in Kolkata, India. PLoS One* 3 (6): e2323.

209. Ali, M. et al. (2016). *Potential for controlling cholera using a ring vaccination strategy: re-analysis of data from a cluster-randomized clinical trial. PLoS Med.* 13 (9): e1002120.

210. Clemens, J.D. et al. (1988). *Measles vaccination and childhood mortality in rural Bangladesh. Am. J. Epidemiol.* 128 (6): 1330–1339.

211. Moss, W.J. (2009). *Measles control and the prospect of eradication. Curr. Top. Microbiol. Immunol.* 330: 173–189.

212. Boschi-Pinto, C., Velebit, L., and Shibuya, K. (2008). *Estimating child mortality due to diarrhoea in developing countries. Bull. World Health Organ.* 86 (9): 710–717.

213. Brown, K.H. et al. (2009). *Preventive zinc supplementation among infants, preschoolers, and older prepubertal children. Food Nutr. Bull.* 30 (Suppl. 1): S12–S40.

214. Grotto, I. et al. (2003). *Vitamin A supplementation and childhood morbidity from diarrhea and respiratory infections: a meta-analysis. J. Pediatr.* 142 (3): 297–304.

215. Faruque, A.S. et al. (1999). *Double-blind, randomized, controlled trial of zinc or vitamin A supplementation in young children with acute diarrhoea. Acta Paediatr.* 88 (2): 154–160.

216. Henning, B. et al. (1992). *Lack of therapeutic efficacy of vitamin A for non-cholera, watery diarrhoea in Bangladeshi children. Eur. J. Clin. Nutr.* 46 (6): 437–443.

217. Hossain, S. et al. (1998). *Single dose vitamin A treatment in acute shigellosis in Bangladesh children: randomised double blind controlled trial. BMJ* 316 (7129): 422–426.

218. Khatun, U.H. et al. (2001). *A randomized controlled clinical trial of zinc, vitamin A or both in undernourished children with persistent diarrhea in Bangladesh. Acta Paediatr.* 90 (4): 376–380.

219. Barreto, M.L. et al. (1994). *Effect of vitamin A supplementation on diarrhoea and acute lower-respiratory-tract infections in young children in Brazil. Lancet* 344 (8917): 228–231.

220. WHO (2000). Effect of breastfeeding on infant and child mortality due to infectious diseases in less developed countries: a pooled analysis. WHO Collaborative Study Team on the role of breastfeeding on the prevention of infant mortality. *Lancet* 355 (9202): 451–455.

221. Victora, C.G. et al. (1987). *Evidence for protection by breastfeeding against infant deaths from infectious diseases in Brazil. Lancet* 2 (8554): 319–322.

222. World Health Organization (2003). *Emerging Issues in Water and Infectious Disease*. Geneva: World Health Organization.

223. Haggerty, P.A. et al. (1994). *Community-based hygiene education to reduce diarrhoeal disease in rural Zaire: impact of the intervention on diarrhoeal morbidity. Int. J. Epidemiol.* 23 (5): 1050–1059.

224. Pinfold, J.V. and Horan, N.J. (1996). *Measuring the effect of a hygiene behaviour intervention by indicators of behaviour and diarrhoeal disease. Trans. R. Soc. Trop. Med. Hyg.* 90 (4): 366–371.

225. Luby, S.P. et al. (2005). *Effect of handwashing on child health: a randomised controlled trial. Lancet* 366 (9481): 225–233.

226. WHO (1978). *Control of diarrhoeal diseases: WHO's programme takes shape. WHO Chron.* 32 (10): 369–372.

227. Liu, L., Johnson, H.L., and Cousens, S. (2012). Global, regional, and national causes of child mortality: an updated systematic analysis for 2010 with time trends since 2000 (vol. 379, pp. 2151, 2012). *Lancet* 380 (9850): 1308.

228. WHO (2014). *The Global Burden of Disease: 2012. Update*. Geneva: World Health Organization.

229. Forsberg, B.C. et al. (2007). *Diarrhoea case management in low- and middle-income countries–an unfinished agenda. Bull. World Health Organ.* 85 (1): 42–48.

230. Barros, F.C.,.V.C., Forsberg, B., AGK, M. et al. (1991). *Management of childhood diarrhoea at the household level: a population-based survey in north-east Brazil. Bull. World Health Organ.* 69 (1): 59–65.

231. McDivitt, J.A., Hornik, R.C., and Carr, C.D. (1994). *Quality of home use of oral rehydration solutions: results from seven HEALTHCOM sites. Soc. Sci. Med.* 38 (9): 1221–1234.

232. Dua, T., Bahl, R., and Bhan, M.K. (1999). *Lessons learnt from Diarrheal Diseases Control Program and implications for the future. Indian J. Pediatr.* 66 (1): 55–61.

233. (2007). *National Family Health Survey (NFHS-3), 2005–06: India: Volume I.* Mumbai: IIPS.

234. Ellis, A.A. et al. (2007). *Home management of childhood diarrhoea in southern Mali – implications for the introduction of zinc treatment. Soc. Sci. Med.* 64 (3): 701–712.

235. Miller, P. and Hirschhorn, N. (1995). *The effect of a national control of diarrheal diseases program on mortality: the case of Egypt. Soc. Sci. Med.* 40 (10): S1–S30.

236. Qazi, S. et al. (2015). *Ending preventable child deaths from pneumonia and diarrhoea by 2025. Development of the integrated Global Action Plan for the Prevention and Control of Pneumonia and Diarrhoea. Arch. Dis. Child.* 100 (Suppl. 1): S23–S28.

8

SOIL-TRANSMITTED HELMINTHS: *ASCARIS, TRICHURIS,* AND HOOKWORM INFECTIONS

ALEXANDER T. YU AND BRIAN G. BLACKBURN

Department of Internal Medicine, Division of Infectious Diseases and Geographic Medicine, Stanford University School of Medicine, Stanford, CA, USA

8.1 INTRODUCTION

Ascaris lumbricoides, *Trichuris trichiura*, and the human hookworms (*Necator americanus*, *Ancylostoma duodenale* and *Ancylostoma ceylanicum*) are the five main parasitic organisms collectively known as the soil-transmitted helminths (STHs). These parasites share the requirement that their eggs or larvae must develop on soil before they can become infectious to humans. Although other helminths are transmitted primarily through soil (e.g. *Strongyloides stercoralis*, *Toxocara* spp.), they are not grouped with the soil-transmitted helminths either because they lack this requirement or are parasites for which humans are not the definitive host.

Endemic primarily in developing regions, these infections are intimately associated with poverty, not only resulting from poverty but also contributing to the perpetuation of poverty. Despite the economic and public health importance of the STHs, medical and international communities still largely fail to address their impact. They are considered among the "neglected tropical diseases," or NTDs. The neglect arises because the people most affected by these diseases are the world's most impoverished [1]. In addition, the infections have mostly insidious effects on health, and quantification of the impact of STHs has been difficult. When measured in disability-adjusted life years (DALYs) lost, the global burden from hookworm alone exceeds all other tropical infectious diseases except malaria, leishmaniasis, and lymphatic filariasis [2].

Poor sanitation, which includes unclean water as well as inadequate sewage management, is a prime contributor to the transmission of these parasites in endemic areas. Control and elimination campaigns have focused on mass chemotherapy with albendazole or mebendazole, but environmental modification, improved sanitation, and especially economic development (which would facilitate better access to clean water and footwear) are challenging to implement, but integral to addressing the underlying root causes of these diseases.

8.2 THE ORGANISMS

The soil-transmitted helminths (also known as the geohelminths) are intestinal nematodes. They are round, nonsegmented worms with elongate, cylindrical bodies. The sexes are separate, with fertilized females producing eggs that subsequently give rise to larvae. After molting, larvae become adult worms. The life cycles of the five species are similar, although their means of infecting humans are different. *Ascaris* and *Trichuris* are primarily ingested orally, while hookworms usually penetrate the skin directly after contact with infested soil. The soil-transmitted helminths do not reproduce in humans.

8.2.1 Ascaris

Ascaris lumbricoides is the most common helminth infection of humans, with an estimated prevalence between 800 million and 1.2 billion persons globally (Table 8.1) [1, 3]. Ascariasis is globally distributed, but is most abundant in the

Water and Sanitation-Related Diseases and the Changing Environment: Challenges, Interventions, and Preventive Measures, Second Edition. Edited by Janine M. H. Selendy.

Companion website: https://www.wiley.com/legacy/wileychi/selendy/

warm and wet tropics. Although most people infected with *Ascaris* are asymptomatic, mortality attributed to *Ascaris* is estimated at > 10 000 people per year with another 1.3 million disability-adjusted life years (DALYs) lost and 60 million people a year suffering morbidity in the form of growth retardation, overt symptoms such as abdominal pain and more severe complications including bowel obstruction and peritonitis [3, 4]. Morbidity disproportionately affects children, and most estimates reflect the physical impact of ascariasis while neglecting to include developmental outcomes associated with the disease.

Adult *Ascaris* worms are typically creamy or pink-white, 15–45 cm long, and live in the small intestine, particularly the jejunum (Table 8.1, Figure 8.1). Eggs are thick-shelled, oval, and about 65 μm × 45 μm (Figure 8.2). Eggs are passed in the feces of infected persons; once they reach a favorable environment (warm, moist soil), they become infectious after developing on soil for an interval that is dependent mostly on temperature (e.g., 10–14 days at 30 °C or six weeks at 17 °C) [5]. Humans usually acquire the infection by ingestion of these eggs via contaminated food (or water); eggs then hatch in the small intestine and release larvae that penetrate the intestine and migrate to the lungs a few days later. Over the week that follows, they ascend the tracheobronchial tree, are swallowed, and return to the intestines, where they mature into adult worms. Egg production begins approximately two months after infection, and adult worms live one to two years. Each adult female worm can produce over 200 000 eggs per day [6].

8.2.2 *Trichuris*

Trichuris trichiura is also known as the whipworm because of its long, whip-like head, which embeds into the intestinal mucosa (Figure 8.3). It is estimated that between 460 and 600 million people are infected with *Trichuris*, with distribution similar to that of *Ascaris*, predominating in wet and warm tropical climates. While significant morbidity can occur with *Trichuris* infection with 640 000 attributed DALYs lost, global mortality is thought to be low [3].

Adult worms are gray-to-pink and 3–5 cm long. They primarily reside in the cecum and ascending colon; with heavy infections, the distribution can extend to the entire colon. Eggs have a distinctive barrel-shaped appearance with a thick shell, bipolar plugs, and average 50 μm × 20 μm in size (Figure 8.4). The life cycle of *Trichuris* is simple: eggs shed in the feces become infective under appropriate (moist and shady) conditions once on soil, within two to four weeks. After infectious eggs have been ingested by a human, larvae emerge and move to the cecum. Here they molt, embed in the epithelium, and mature into adults. Oviposition begins two to three months after infection, and adult worms live one to three years. Female worms can produce up to 20 000 eggs per day (Table 8.1) [5, 7].

8.2.3 Hookworms

The human hookworms *A. duodenale, A. ceylanicum*, and *N. americanus* are small, gray-white nematodes about 0.7–1.3 cm long that live in the upper small intestine (Table 8.1). Globally, around 500–600 million people are thought to be infected with hookworms with 5 billion people at risk. Hookworms are estimated to cause the loss of 4 million DALYs and are responsible for up to 140 billion USD annually in lost wages globally [3, 8]. *A. ceylanicum*, once thought of as primarily zoonotic and rarely affecting humans, has more recently been recognized as a common human hookworm. This appears to be especially true in Asia where, based on molecular-based epidemiological surveys, it may be the second most common human hookworm [9–12].

The ovoid, thin-shelled eggs of the three species are identical, and measure about 60 μm × 40 μm (Figure 8.5). As with the other soil-transmitted helminths, eggs passed in the stool of infected persons must develop on soil (ideally warm, moist, and shady); they hatch there within one to two days, becoming larvae (Figure 8.6). Larvae then molt over 5–10 days, subsequently becoming infectious. Following contact with human skin, larvae enter the body and are carried to the lungs. They then move up the trachea, are swallowed, and migrate to the small intestine. Females start egg deposition four to six weeks after infection. Although *N. americanus* can only be transmitted via this percutaneous route, *A. duodenale* and *ceylanicum* can be transmitted orally as well as percutaneously. The adult lifespan averages three to five years for *N. americanus* and one year for *A. duodenale* and *ceylanicum* (Table 8.1) [1].

8.3 EPIDEMIOLOGY

Soil-transmitted helminths are among the most common human parasites globally, with 1.5 billion people infected by at least one of these worms, and many harbor infection with multiple species. It is thought that up to a billion people are infected with *Ascaris*, at least 460 million with *Trichuris*, and 500 million with hookworm [1, 3, 5–8, 13, 14]. These numbers are lower than earlier estimates, in large part due to the decline in STHs in China, where prevalence of any STH fell from 57% in 1990 to 18% in 2010. This is thought likely due to the rapidly developing economy, urban migration, and improvement in sanitation systems combined with a national control program and mass drug administration programs directed at school-aged children [3, 15–17].

The geographic distribution of these parasites is determined primarily by sanitation and climate. Eggs and larvae become infective on warm, moist soil; the tropics are thus well suited for transmission, and because eggs are spread via feces, areas with poor hygiene have the highest transmission rates. Sanitation therefore plays an important role in determining the degree of environmental contamination with eggs. Many developing countries share both the environmental and

TABLE 8.1 Epidemiologic, parasitologic, and clinical characteristics of the soil-transmitted helminths.

Characteristic	*Ascaris lumbricoides*	*Trichuris trichiura*	*Necator americanus*	*Ancyclostoma duodenale*	*Ancyclostoma ceylanicum*
Global prevalence (millions)	800–1200	460–600	500–600[a]	500–600[a]	500–600[a]
Size (mm)	150–450	30–50	7–13	8–13	6–10
Daily egg output (per adult female)	200 000	3000–5000	5000–10 000	10 000–30 000	Undefined
Daily blood loss	N/A	Undefined	0.03 ml	0.2 ml	Undefined
Life expectancy of adult worms (years)	1–2	1–3	3–5	1	Undefined
Larval developmental arrest in humans	No	No	No	Yes	Yes
Oral transmission	Yes	Yes	No	Yes	Yes
Cutaneous transmission	No	No	Yes	Yes	Yes

[a] Total for all three hookworm species collectively.

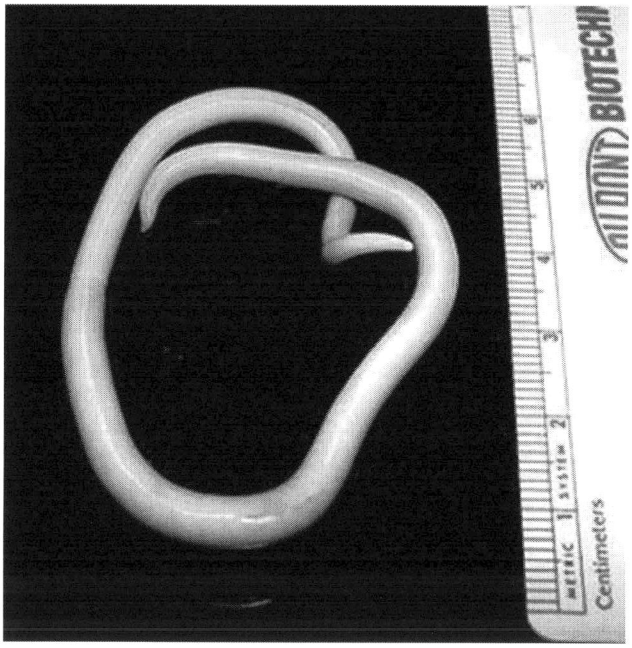

FIGURE 8.1 Adult *Ascaris lumbricoides* worm. (*Source*: Courtesy Division of Parasitic Diseases/Centers for Disease Control and Prevention).

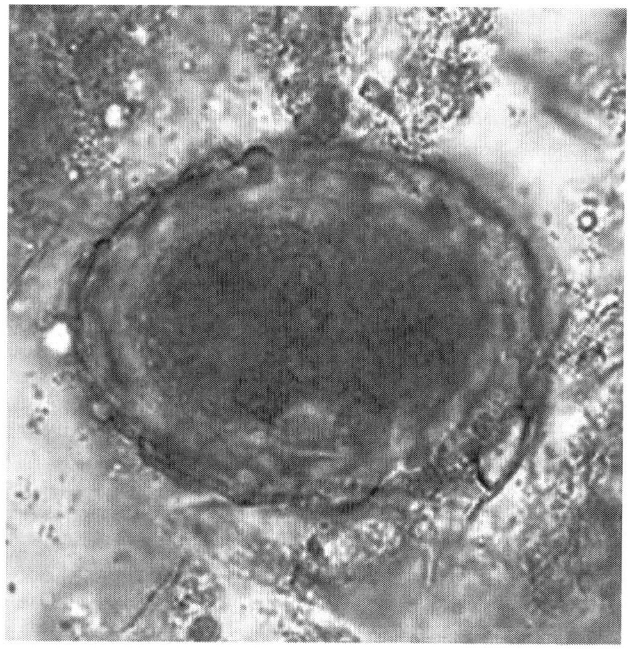

FIGURE 8.2 *Ascaris lumbricoides* egg. (*Source*: Courtesy Division of Parasitic Diseases/Centers for Disease Control and Prevention.) (*See insert for color representation of this figure.*)

climatic conditions necessary for transmission; although most are tropical, some temperate countries sustain transmission, albeit seasonally [1, 5–7, 13, 14].

Climate change will likely alter the range and intensity of STHs in the environment while also affecting parasite and human dynamics important for infectivity and transmission. By altering temperature, humidity, and soil conditions, climate change will likely extend the range of habitats suitable for parasite survival. Humidity impacts the duration and degree of STH infectivity, while higher temperatures accelerate egg development and hatching and also increases the survival of eggs and larvae [18, 19]. Climate change also alters important human host factors such as socioeconomics, infrastructure, behavior, and nutrition that directly affect human exposure and susceptibility to infectious diseases including soil-transmitted helminths [18, 20, 21].

While all soil-transmitted helminths predominate in tropical and subtropical areas, hookworm prevalence is not exaggerated in urban slums such as *Ascaris* and *Trichuris* are. Instead, hookworms occur mostly in poor, tropical, coastal, and agrarian communities [5, 13]. Larval development for hookworm is also more susceptible to climatic extremes (e.g., low rainfall or temperature) than it is for the eggs of *Ascaris*, and thus is more limited by the requirement for a warm, moist climate. Unlike *Ascaris* and *Trichuris*, hookworm prevalence most commonly increases throughout childhood and plateaus in young adulthood [14]. Overall, the highest prevalence rates (and intensity) of hookworm occurs in Sub-Saharan Africa, followed by China and Southeast Asia [13]. *N. americanus* is the predominant hookworm worldwide.

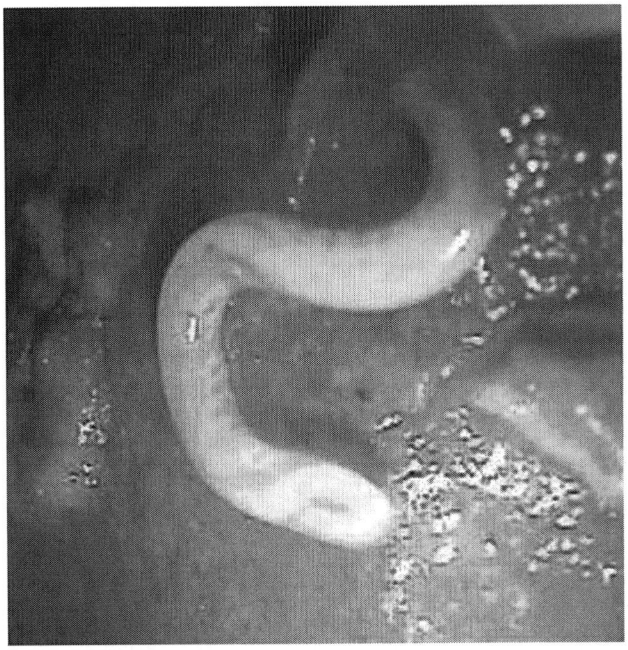

FIGURE 8.3 *Trichuris trichiura* adult worm (colonoscopy image). (*Source*: Courtesy Division of Parasitic Diseases/Centers for Disease Control and Prevention and Duke University Medical Center.) (*See insert for color representation of this figure.*)

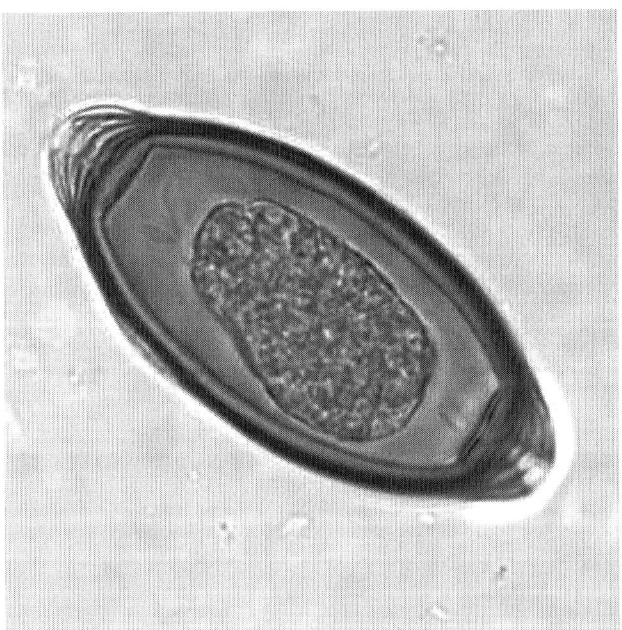

FIGURE 8.4 *Trichuris trichiura* egg. (*Source*: Courtesy Division of Parasitic Diseases/Centers for Disease Control and Prevention.)

Data corroborated from geographical information systems and satellite sensors have suggested that *A. lumbricoides* and *T. trichiura* do not occur in areas where temperatures exceed 37–40 °C. However, hookworm occurs throughout many such areas with these high temperatures, suggesting

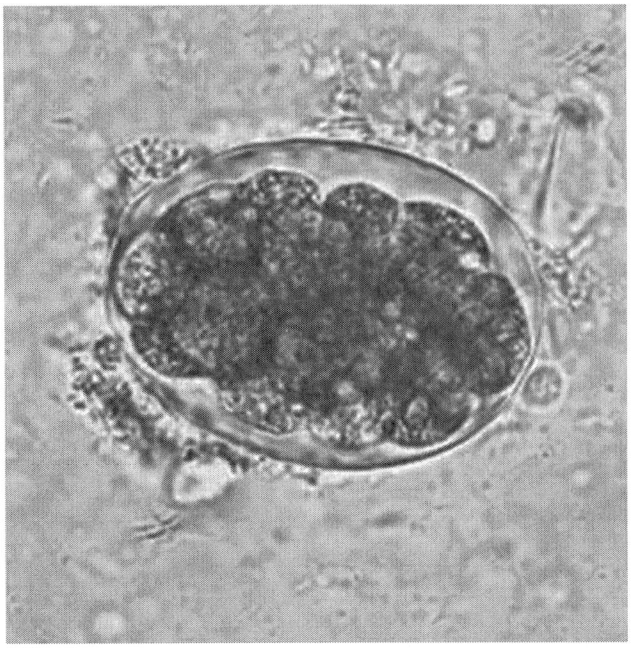

FIGURE 8.5 Hookworm egg. (*Source*: Courtesy Division of Parasitic Diseases/Centers for Disease Control and Prevention.) (*See insert for color representation of this figure.*)

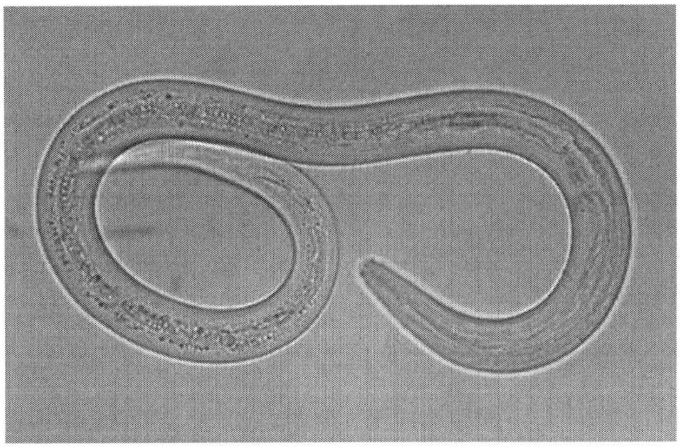

FIGURE 8.6 Infective hookworm larva. (*Source*: Courtesy Division of Parasitic Diseases/Centers for Disease Control and Prevention.)

that hookworm tolerates greater temperatures than *A. lumbricoides* and *T. trichiura* [22]. In the tropics, hookworm is generally confined to coastal plains below 150 m elevation. Above these altitudes, low temperatures (<20 °C) usually limit transmission [22]. Because *A. duodenale* larvae can undergo arrested development within humans, it can thrive in some areas where *N. americanus* cannot; this may allow *A. duodenale* to survive cold winter months. Hookworms are unable to survive desiccation and a minimum amount of rainfall is also an important determinant influencing hookworm transmission. In some endemic areas where hookworm transmission is seasonal, new infections with *A. duodenale* can appear 8–10 months after the rainy season, following emergence of larvae from this arrested development stage [13, 22, 23].

Soil type is another environmental factor important for hookworm transmission. Sandy soils are well suited for this because the small particle size and well-aerated texture allows infective larvae to migrate within the soil. Clay soils inhibit migration, which is another reason hookworm is most prevalent in coastal areas, where sandy soils predominate [22].

Most cases of soil-transmitted helminth infections in nonendemic regions occur among immigrants and travelers. Infections in such persons generally resolve within a few years even without treatment, when the adult worms die. Because of the requirement that geohelminth eggs or larvae develop on soil before becoming infectious, these parasites cannot be transmitted directly from person-to-person, and cannot multiply in the host. This contrasts, for example, with *Strongyloides stercoralis* (another nematode transmitted through soil contact), which can multiply in the host and persist for the life of the host, occasionally causing life-threatening disease, and which can rarely be transmitted from person-to-person.

8.4 PATHOGENESIS AND PATHOLOGY

The pathology stemming from *Ascaris* infection is mainly a result of the host response to the parasite. During larval migration, cells suffer mechanical trauma and lysis due to larval enzymes. Larvae also induce granuloma formation, and both *Ascaris* and hookworm larvae cause a hypersensitivity reaction in pulmonary parenchyma (Figure 8.7) [5]. The pathophysiologic consequences of *Ascaris* and *Trichuris* in the gastrointestinal tract result from the presence of worms in the lumen. Although the severity of symptoms is usually proportional to the worm burden, a single worm can (e.g., in the case of *Ascaris*) obstruct the common bile duct, resulting in severe symptoms. With *Trichuris*, heavier infections result in expansion of the ecological niche from the right colon to the entire colon. Production of bloody mucus from the mucosa occurs, with anemia and impaired growth as possible sequelae.

Hookworms, by contrast, exert their primary pathologic effect via blood loss. They attach to the intestinal mucosa using their strong buccal capsules, cutting plates, and teeth and secrete anticoagulants and anti-inflammatory factors, allowing continuous blood ingestion (Figure 8.8a and b). Chronic iron deficiency is particularly detrimental in childhood and may directly impair cognitive and intellectual development [2, 13, 14].

Human infection with these parasites leads to a predominantly Th2 immune response. It is thought that this shift in immunological response may impact the manifestations of

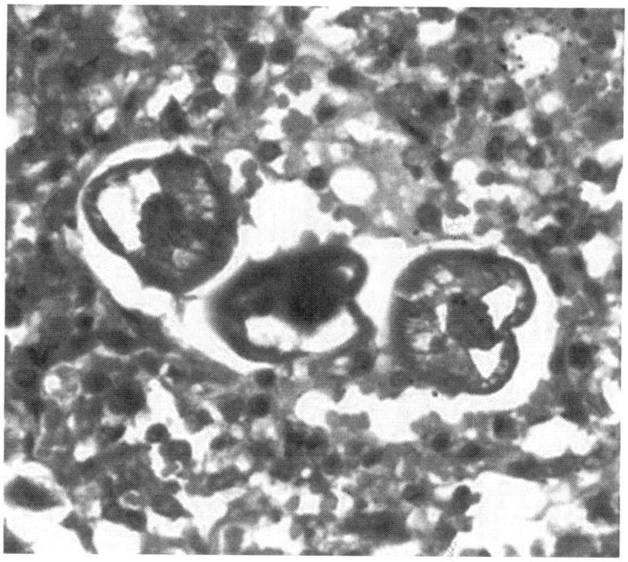

FIGURE 8.7 *Ascaris lumbricoides* larvae in lung tissue. (*Source*: Courtesy Division of Parasitic Diseases/Centers for Disease Control and Prevention.)

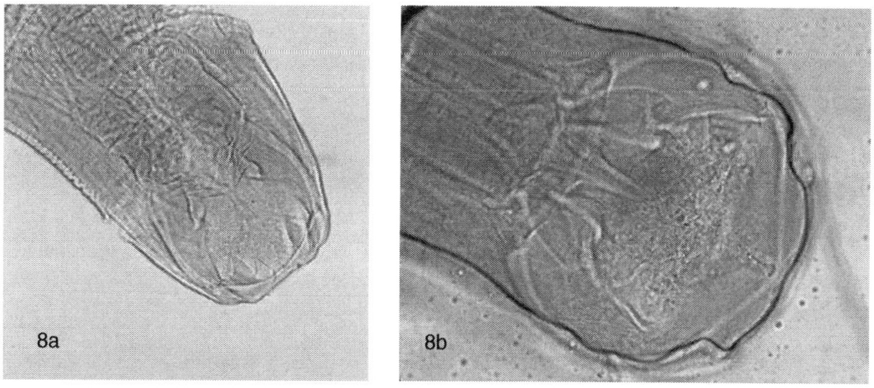

FIGURE 8.8 Adult hookworm mouthparts. (a) *Ancylostoma duodenale* cutting teeth; (b) *Necator americanus* cutting plates. (*Source*: Courtesy Division of Parasitic Diseases/Centers for Disease Control and Prevention.)

allergic and rheumatologic diseases as well as the response to infections typically controlled by the Th1 response (such as tuberculosis) [1]. There is controversy regarding whether antibodies produced in response to soil-transmitted helminth infections are protective or merely a marker of past or present infection [23].

8.5 CLINICAL MANIFESTATIONS

Most people infected with *Ascaris* are asymptomatic, although a small proportion develop pulmonary symptoms (such as a cough) while larvae migrate through the lungs (one to two weeks after infection). Eosinophilia may accompany this, as can eosinophilic pneumonia. With most established infections, adult worms in the small bowel lumen provoke no symptoms or produce only mild abdominal pain, nausea, or anorexia. However, heavier infections can adversely affect the nutritional status of children, especially in areas where malnutrition is otherwise common [24]. This can adversely impact intellectual development and growth; treatment of heavily infected children with antihelminthics improves these outcomes [24].

Other complications of chronic ascariasis are rare, largely due to mechanical issues related to the presence of the large adult worms in the gut, and can include intestinal, biliary, or pancreatic obstruction, appendicitis, and intestinal perforation. When these complications occur, they are more common in children from endemic areas [5]. Occasionally, adult worms can pass per rectum, through the mouth or nose, or tear ducts.

As with *Ascaris* infections, most people with trichuriasis are asymptomatic or have only eosinophilia. With heavy infections (particularly in children), chronic abdominal pain and diarrhea can result, as can anemia, growth retardation, and clubbing [7]. Stools may have an acrid smell and nocturnal stooling can occur [5, 7]. The *Trichuris* dysentery syndrome can result, characterized by tenesmus and frequent stools containing mucus and blood. Recurrent rectal prolapse occurs, with adult worms often visible [5, 7].

Similar to the other soil-transmitted helminths, most people infected with hookworms harbor light infections and are asymptomatic. Initial infection may be characterized by "ground itch," a pruritic maculopapular rash at the site of skin penetration, seen mostly in previously sensitized individuals. Serpiginous tracks (cutaneous larva migrans) on the skin occur with infection by animal hookworms (e.g., *Ancylostoma braziliense*) but not human hookworms [13]. Migration of larvae through the lungs may provoke pneumonitis, which is less common and less severe than it is with *Ascaris* [5]. Epigastric pain, diarrhea, and anorexia may occur about 6–12 weeks after penetration, as larvae begin attaching to the small bowel mucosa. When infection with *A. duodenale* occurs orally, the early migration of larvae can cause a syndrome known as Wakana disease, which is characterized by nausea, vomiting, pharyngeal irritation, cough, dyspnea, and hoarseness [13, 14]. There are important differences between the hookworm species: *N. americanus* is shorter, removes less blood (0.03 ml/day versus 0.20 ml/day), and produces fewer eggs (5000–10 000 versus 10 000–30 000 per day) than *A. duodenale* (Table 8.1) [13]; for *A. ceylanicum,* these parameters are not yet well defined. Larvae of *A. duodenale* can enter a stage of arrested development (including within humans), which may allow *A. duodenale* to persist in environmental conditions that would otherwise be too harsh [5, 13, 14]. Infants can also be infected by *A. duodenale*, most likely through breast milk, and possibly transplacentally [5, 13, 14].

Some who are infected with adult hookworms experience chronic abdominal pain and eosinophilia. However, the most important manifestations of hookworm disease are iron deficiency anemia and malnutrition. The anemia from hookworm infection develops gradually, but has especially severe consequences for children and pregnant women. It is a major health problem throughout the developing world and can impair physical, cognitive, and intellectual growth in children, diminish productivity of workers, and threaten the outcome of pregnancy for mother and child [24]. Anemia caused by hookworms is responsible for 60 000 deaths annually worldwide, and hookworms cause the greatest morbidity and mortality of the soil-transmitted helminths globally [2, 13, 14, 24]. Although malaria is co-endemic to many areas with high hookworm prevalence, it is clear that increasing egg burdens are associated with lower hemoglobin levels, independent of other factors such as malaria infection [14]. However, since much of the morbidity associated with both diseases results from anemia where the two are co-endemic, it is possible that hookworm disease exacerbates malarial anemia and vice versa.

8.6 DIAGNOSIS

The diagnosis of soil-transmitted helminth infections is straightforward and can be made by identifying eggs in stool samples. The high daily egg output of *Ascaris* worms means that infections are easily detected by stool exams and concentration techniques are unnecessary [1]. Trichuriasis and hookworm infections can also usually be diagnosed by simple stool smears, although concentration techniques may be necessary for light infections because the egg output is not as high as with *Ascaris* [1, 5]. Stool samples may be negative until two to three months after infection occurs. Larvae can sometimes be found in respiratory secretions of patients infected with *Ascaris* or hookworm during the early migratory phase of infection. At times, *Ascaris* worms may be seen radiographically (given their large size); all STHs may

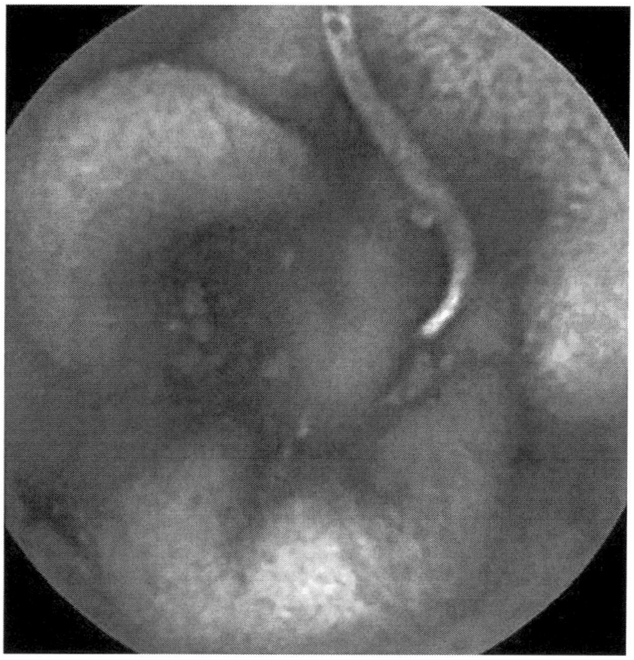

FIGURE 8.9 Hookworm in intestine (video capsule endoscopy image). (*Source*: Courtesy Dr Lauren Gerson, Stanford University School of Medicine.)

also be diagnosed by endoscopy (Figures 8.3 and 8.9). Serology is rarely needed to diagnose these infections and is used more frequently in epidemiologic studies.

8.7 TREATMENT

Soil-transmitted helminth infections should generally be treated with the oral benzimidazole drugs, albendazole or mebendazole. Although alternative antiparasitics are available, none are currently as favored for use against all five soil-transmitted helminth species as albendazole or mebendazole. Single-dose or short-course regimens with these agents result in high cure rates for *Ascaris*; a recent meta-analysis demonstrated cure rates of 88–95% with a single dose of these agents [25]. For *Trichuris* infections, single-dose cure rates with albendazole or mebendazole are low (28–47%), this parasite should therefore be treated with longer courses (at least three to seven days) or combinations of antiparasitics to achieve higher cure rates [5, 7, 26–28]. Low cure rates for *Trichuris* has led to investigations regarding the efficacy of combination therapy and extended duration therapies. Single-dose regimens of albendazole plus oxantel pamoate or albendazole plus ivermectin have been the most promising combination therapies; a randomized controlled trial (RCT) from Tanzania showed 68.5% and 27.5% cure rates, and 99.2% and 94.5% egg reduction rates for oxantel + albendazole and oxantel + ivermectin, respectively [29]. Daily dosing of either mebendazole or albendazole for three days also had higher efficacy than single-dose therapy of the same drugs in an RCT from China (71% and 56% versus 40% and 34% respectively) [17]. For hookworm, albendazole is preferred over mebendazole, with 69–88% versus 15–31% cure rates using a single-dose regimen [17, 26]. Daily dosing for three days yielded cure rates of 92% for albendazole and 54% for mebendazole in another trial, though current recommendations still largely call for single-dose albendazole [17]. This is because even in individuals not cured, a large reduction in parasite burden and egg output occurs [5, 28]. Although its use falls under pregnancy category C, the World Health Organization (WHO) has recommended the use of albendazole or mebendazole during pregnancy based on safety data, although they recommended deferring therapy during the first trimester; they also concluded that these drugs may be used in children as young as 12 months of age [30]. Replacement of iron and other micronutrients as an addition to antiparasitics is an important therapeutic adjunct. Some evidence suggests that resistance to albendazole and mebendazole may be developing although more research is needed in this regard [31].

8.8 CONTROL AND ELIMINATION

Soil-transmitted helminth infections are prevented primarily by provision of clean food and water, good community sanitation, and adequate footwear. As provision of safe water and adequate sanitation is expensive and logistically difficult for many developing countries to implement, WHO advocates treatment with antihelminthic drugs at regular intervals to populations at risk [30]. This is done primarily to reduce individual worm burdens below those that cause significant disease and to decrease the overall community worm burden. Albendazole is safe, inexpensive, and widely available; programs employing this agent (or mebendazole) are therefore feasible for most developing countries. Several studies have demonstrated that regular deworming of children (and pregnant/childbearing-age women) can prevent and reverse iron-deficiency anemia, impaired growth, malnutrition, poor school performance, and pregnancy complications [24].

A prominent example of successful soil-transmitted helminth elimination efforts is the southeastern United States, which was endemic for STHs through the early 1900s. An initiative, the Rockefeller Commission, was highly successful in achieving good control, and today transmission has been essentially eliminated [13, 23, 32, 33].

8.8.1 Mass Drug Administration

Current control initiatives focus largely on mass drug administration (MDA) with albendazole or mebendazole (Figure 8.10); these programs are also commonly referred to as preventive chemotherapy or deworming programs. Potential population-based approaches for drug distribution include

FIGURE 8.10 Mass drug administration, Central Nigeria. (*Source*: Photo courtesy Dr Brian Blackburn, Stanford University School of Medicine.) (*See insert for color representation of this figure.*)

directed administration (only to those known to be infected), universal administration (to the entire population), or targeted administration (to certain subsets of the population) [34]. Because of the cost and difficulty in establishing definitive diagnoses of STHs at large scale and the low-cost and good safety profile of antihelminthic medications, most programs do not pursue the directed approach. Most programs oriented solely at STHs have used the targeted approach, such as that of targeting schoolchildren for treatment with albendazole [34]. However, high re-infection rates and a desire to stop transmission rather than focus only on morbidity have led to the consideration of expanded universal coverage in certain settings since adolescents and adults can serve as a reservoir for community re-infection [35].

In 2001 the World Health Assembly passed Resolution 54.19, urging member states to control the morbidity of soil-transmitted helminths through mass drug administration programs (using albendazole or mebendazole) by targeting school-age children in developing countries. The established goal was to regularly treat >75% of school-age children at risk for infection globally by 2010 [36]. A WHO assessment of progress made between 2001 and 2010 found that only a third of children requiring treatment had access to antihelminthic medications, and only 200 of 600 million affected school-age children had been reached. The WHO therefore set a new deadline for 2020 to achieve the same goals, with substantial progress made since 2010. In 2015, 48 and 63% of pre-school-age and school-age children, respectively, were treated. [37]. This program has most commonly been operationalized through school-based drug administration and has become one of the largest public health programs worldwide.

The rationale for focusing on school-age children is based on this population having the highest infection intensity (defined as the number of worms infecting the human host) of *Ascaris* and *Trichuris* (Figure 8.11), and also because schools provide a cost-effective and logistically simple way to deliver antihelminthics. However, school-based programs may not accurately target vulnerable infected populations and also may not effectively reduce worm transmission or overall prevalence in the long term [38]. STHs are frequently prevalent in adults (including pregnant women), especially hookworm, and pre-school children are potentially vulnerable to permanent intellectual and cognitive deficits from infection [14, 34]. In addition, in areas with high-intensity transmission, infection remains endemic even when populations are regularly treated with antihelminthics.

Infection prevalence routinely rebounds, often returning to pre-treatment levels within months. After one community-wide treatment, rates of hookworm infection reached 80% of pre-treatment rates within 30–36 months, *A. lumbricoides* infection reached 55% of pre-treatment rates within 11 months, and *T. trichiura* infection reached 44% of pre-treatment rates within 17 months [39–42]. A 2012 meta-analysis of 24 studies found that at 3, 6, and 12 months after treatment, *A. lumbricoides* prevalence returned to 24%, 68%, and 94% of pre-treatment levels (respectively). For *T. trichiura*, respective post-treatment prevalence returned to 36%, 67%,

FIGURE 8.11 Typical transmission conditions for STHs, Central Nigeria. (*Source*: Photo curtesy Dr Brian Blackburn, Stanford University School of Medicine.) (*See insert for color representation of this figure.*)

and 82% of pre-treatment levels and hookworm prevalence returned to 30, 55, and 57% of pre-treatment levels [38]. In some cases, at least three treatments per year were required to improve iron status [14, 34].

High re-infection rates and a desire to stop transmission rather than focus only on morbidity have led to the consideration of expanded community-wide coverage in certain settings since adolescents and adults can serve as a reservoir for community re-infection [35, 43]. Emerging data suggest that the success of MDA programs may be higher when the entire population is treated rather than a targeted cohort such as schoolchildren, and there has been a reconsideration of global STH control policy with more emphasis on a community-wide approach [35, 43–45]. One model predicted that targeted treatment of school and pre-school children alone would be insufficient to eliminate STH transmission in all but low-intensity settings, at least in part due to the ability of adults to maintain transmission. Elimination required high levels of coverage of the entire community including adults, as well as repeat MDA at a frequency of every six months for at least 10 years and still risked an STH population rebound unless sanitation, water, and hygiene were simultaneously improved [44]. In addition to potentially being better able to achieve elimination, expanding to community-wide coverage from school-based programs is cost-effective. One model found integrated, community-wide MDA programs for STH and schistosomiasis to be highly cost-effective, even in low-burden settings [46].

In addition to deciding whom to target in MDA programs, defining the number of treatment days for each program is also important. The optimal number of treatment days differs between *Ascaris*, *Trichuris*, and hookworm. For *Ascaris*, single-dose albendazole regimens have cure and egg reduction rates over 85%. For hookworm, however, at least two treatment doses are required to reach similar rates. For whipworm, even three-day regimens are not able to achieve 85% cure rates although they can reduce the number of eggs shed by 90% [47]. Logistical constraints limit most programs to single doses; *Trichuris* control is thus achieved more slowly because of the low relative efficacy of treatment when monotherapy is used [34].

Another important programmatic question concerns the appropriate frequency of MDA. Although the most common approach is annual treatment, the optimal frequency should be determined in part by the local prevalence, transmission intensity, and programmatic goal. For *A. lumbricoides* several studies have shown that in both low- and high-transmission areas, the most cost-effective option is to treat every two years to reduce prevalence and parasite intensity. In contrast, when prevalence reduction is the measure of effectiveness, the most cost-effective option is to treat every four months in high-transmission areas and every year in low-transmission areas, since prevalence rebounds more rapidly than intensity in the absence of retreatment [34, 48]. Because of the non-linear relationship between prevalence and infection intensity, when prevalence rates are high, the initial impact of mass chemotherapy is a reduction of the worm burden, rather than a reduction in prevalence [49, 50]. As most complications of soil-transmitted helminth infections occur in the presence of heavy worm burdens, a decline in intensity (even in the absence of a decline in prevalence) serves to reduce morbidity and is the key goal when trying to decrease morbidity [49, 50].

Treatment two or three times a year is effective in reducing morbidity in areas of intense transmission (prevalence >70%, with >10% of infections being moderate or heavy), such as in Zanzibar and Myanmar [34]. In areas with a lower transmission intensity (prevalence 40–60% and <10% of infections moderate or heavy), such as in Oman, India, or Brazil, annual treatment was sufficient to reduce morbidity [34]. Notably, local prevalence can vary widely within a country, and aggregate generalizations at the country level may not be the most appropriate level of resolution for making policy. WHO guidelines suggest treatment twice yearly if the STH prevalence in school-age children is >50% and once yearly if it is 20–50%. Although school-age children are the priority in such programs, WHO suggests also treating pre-school children, women of childbearing age, pregnant women (in the second and third trimester), and lactating women [30]. However, the possibility that these guidelines may need to be revised is the subject of recent debate in light of the new data described above, and may reflect an approach that is not aggressive enough.

When implemented properly, the impact of MDA on short-term and long-term outcomes can be dramatic. Several studies have shown an impact on short-term nutritional and cognitive outcomes, decreased anemia, and improved school attendance. In India, when children one to three years of age received vitamin A and albendazole every six months, they gained 3.5 kg above expected in two years, and in Nepal twice-yearly distribution of vitamin A and albendazole to children under age five reduced anemia by 77% in one year [51]. In young children after 12 months of quarterly mebendazole administration in Zanzibar, moderate anemia was reduced by 59%, and growth and appetite improved [52]. Improvements in memory, language, problem-solving, and attention have been seen in school-age children as a result of deworming programs [34, 53]. Untreated Jamaican children with intense *T. trichiura* infections missed twice as many school days as their peers treated with albendazole (and those with lighter infection saw significantly more weight gain). In Kenya, school-based antiparasitics reduced school absenteeism by 25% [54, 55]. There, a single dose of albendazole or placebo was given to schoolchildren who had very high prevalence rates of hookworm, trichuriasis, and ascariasis; at follow-up six months later, the treated group had gained 1.3 kg in weight and 0.6 cm in height, significantly more than the placebo group [56]. The improvements were attributed more to the decrease in hookworm egg counts than to the other two parasites [56, 57]. In Nepal, albendazole treatment in the second trimester of pregnancy resulted in a significant decrease in severe anemia, the mean infant birthweight rose by 59 g, and the infant mortality rate fell by 41% [58]. In Uganda, two annual MDAs with albendazole resulted in a decrease in hookworm prevalence from 51% to 11% and a significant improvement in hemoglobin concentrations in just two years [59]. Long-term effects have also been observed. In Kenya, a study on males who participated in MDA had higher living standards and worked 17% more hours per week. The same study found that females were 25% more likely to pass their primary school exit exam and attend secondary school if they participated in MDA [60].

There has been debate about the robustness of these findings. The heterogeneity of the different study settings, community population, baseline morbidities, and the type and number of worms make these observational study results difficult to generalize. A 2007 meta-analysis of treatment of school-age children with antihelminthics showed a relatively consistent improvement in weight gain in those known to be infected, but little benefit for growth, school attendance, or cognition [61]. A 2017 meta-analysis attempted to address criticisms of the 2007 review including methodological and other issues surrounding re-infection, co-interventions, and baseline nutritional status. It also included data from the largest single trial regarding mass deworming, published in 2013, which included one million pre-school children in India who demonstrated no effect on weight gain or mortality. Notably, in that study the majority of children had a lower burden of infection [62]. The 2017 review showed, in contrast to the 2007 review, minimal benefit of mass drug administration for weight or height over a 12 month period. It also found no difference in stunting, cognition, school attendance, or mortality [63]. A 2015 Cochrane review found a possible increase in weight gain in targeted MDA programs but little evidence for benefits in nutritional status, hemoglobin levels, cognition, school performance, or mortality [64]. Criticisms remain that these studies are insufficiently lengthy to detect benefits in morbidity, that short-term symptoms such as diarrhea and abdominal pain are not considered, and that net individual benefits are too diluted when measured in populations that, as a whole, might have only light burdens of disease [65]. Thus, while the debate continues, mass drug administration programs are still recommended by WHO and considered effective. However, effectiveness likely varies depending on the local parasite species, existing water and sanitation infrastructure, burden of infection and patient comorbidities, and nutritional status [36, 66].

Several studies have demonstrated that hookworms contribute to 30–55% of moderate-to-severe anemia in pregnant women; it is estimated that 44 million pregnancies annually are affected by hookworm and 250 million women of childbearing age needed treatment [14, 37]. Treatment of women in endemic areas with antihelminthics once or twice during pregnancy appears to substantially improve maternal hemoglobin, birthweight, and infant mortality [13, 14, 34]. Again, the robustness of these findings has been questioned: a 2015 Cochrane review of four RCTs from three countries, Peru, Sierra Leone, and Uganda, showed no effect of treatment with antihelminthics during the second or third trimester on maternal anemia, low birthweight, or perinatal mortality [67]. Nevertheless, given the multiple studies that do suggest a benefit in pregnant women in hookworm-endemic areas, antihelminthic treatment is recommended during pregnancy (except during the first trimester) [36, 66].

Another problem is the potential emergence of drug resistance; this has been seen already in helminths affecting livestock being mass treated with antihelminthics in many areas, even threatening sustainability of livestock operations in some areas [68]. This has been thought due to the frequent and extended usage of a single drug from a narrow group of antihelminthic drugs often with suboptimal dosing in livestock populations [69]. Veterinary helminths are different from the ones affecting humans and carry different resistance genes, and it is unclear if antihelminthic resistance has become an issue in helminths affecting humans [26, 69]. While there is no conclusive evidence that resistance has developed in helminths affecting humans, a limited number of studies have suggested that, at the very least, decreased efficacy may be a problem [26]. In one example from Zanzibar, a population treated with mebendazole every four to six months for five years showed diminishing efficacy of mebendazole with repeated use. The cure rate (7.6%) and egg reduction rate (52.1%) for hookworm after the five-year mebendazole MDA was significantly lower than in the first years of the program (22.4 and 82.4%, respectively) [31].

8.8.2 Water, Sanitation, and Hygiene

Water, sanitation, and hygiene (WASH) are factors closely interrelated with transmission of the soil-transmitted helminths. Ascariasis was once highly prevalent in Europe, with elimination aided by the implementation of improved water, sanitation, and sewage systems [6]. Given the lack of animal reservoirs for *A. duodenale* and *N. americanus*, effective sanitation should interrupt hookworm transmission [13, 70]. While most WASH interventions for STHs have focused on sanitation, water and hygiene are also important factors for decreasing STH transmission [70]. Access to sufficient quantities of water enables people to practice improved hygiene behaviors, probably leading to reduced contamination by STHs; a systematic review found that use of treated water, handwashing after defecation, and sanitation reduced the chances of any STH infection by 54%, 53%, and 34%, respectively. Wearing shoes reduced odds of any STH infection by 70% [71]. Despite these impressive reductions in infection, the high environmental and community burden often means that the impact of WASH or improved footwear alone on reducing STH prevalence is often not realized for decades [34].

Sanitation consists of both improved personal hygiene (food preparation, handwashing, and toilet use) and access to the "hardware" necessary to invoke these behaviors (latrines or toilets, wells or other safe water supply, sewage treatment) [34]. Studies suggest that improved sanitary conditions can result in lower soil-transmitted helminth prevalence and reinfection rates [34]. Sanitation programs have been documented that successfully increase the number of latrines but have little impact on STH transmission, largely because the latrine usage rate in the population is low (especially among children) [72]. Besides low utilization, cost and sustainability remain important barriers to implementation of successful sanitation programs. While potentially prohibitive, these barriers can be overcome; a study in Vietnam assessed a market-based approach for sanitation items and found people were more willing to pay for services in an environment where they were considered customers rather than beneficiaries of a program. Proper sanitation, including decreased open defecation, increased 100% in just a year and was successful at improving STH transmission indices [73].

Such data are encouraging, as it is thought that the investment needed to provide access to adequate sanitation is currently beyond the resources of most low-income countries. In addition, the needed coverage rates of properly built, used, and maintained sanitation are likely to be >90% to decrease worm transmission [74]. Experience in Vietnam demonstrated that the per-child cost for latrines is 7.90 USD (about 200 times the cost per dose for deworming). In addition, this particular intervention only increased the latrine coverage in each community to less than 1% [34]. A study in Kenya demonstrated that *Ascaris* reinfection rates and infection intensity were significantly reduced by a school-based water treatment, hygiene, and sanitation program, yet this intervention had no impact on *Trichuris* nor hookworm [75]. In contrast, a study in Ethiopia showed that higher rates of sanitation usage did not appear to be a protective factor against STH infection [76]. A study from rural Bangladesh found that while latrine access was not associated with protection from STH infection, finished flooring was protective against *Ascaris* (but not whipworm nor hookworm) [77]. In an urban slum in Kenya, always treating water, having finished household floor material, and 10 m household elevation increases were protective against infection [78]. These mixed results show that the relationship between improved sanitation and STH control is complex and may depend on the specifics of the measures being implemented, the specific parasite in question, and the setting in which the measures will be employed. Poverty reduction and improved economic development have probably done more to eliminate hookworm in industrialized nations than any other single factor, including sanitation, antihelminthics, footwear, and health education. Economic development with improvements in living standards and the introduction of piped water, sewage systems, mechanized agriculture, and elimination of night soil are largely responsible for the control of hookworm infection in North America, Europe, Japan, and Korea [13]. These measures would most likely be effective in many developing countries. Education regarding the proper use of sanitation facilities and the avoidance of using night soil is also essential. However, achieving these goals in resource-limited settings may prove to be more challenging than the reachable goals of mass chemotherapy programs. In addition, it is not clear that any individual component (such as education, improved sanitation, footwear) alone would achieve lasting and significant reductions in the parasite burden; programs must instead integrate all of these components and must involve a high percentage of the population. Such programs can be difficult to successfully implement [34]. An important concept is that the soil-transmitted helminths not only stem from poverty but also cause poverty, perpetuating the cycle. In Kenya, for example, data suggest that deworming could raise per capita earning by 30% [79].

8.8.3 Program Integration

Integration of soil-transmitted helminth control and elimination programs with other such programs directed at parasites (such as lymphatic filariasis (LF), onchocerciasis, schistosomiasis, and malaria) may be a key path forward. Increasing public and private advocacy for this approach has been building and has the advantage of addressing multiple public health problems with fewer resources than each would require individually. Using existing infrastructure to deliver multiple interventions reduces costs and takes advantage of access to communities that is already in place. In 2015, for instance, 32 million school-aged children living in areas co-endemic for STH and schistosomiasis were treated for both diseases. However, an opportunity to treat even more children was lost, with 54 million children living in co-endemic areas treated for either STH or schistosomiasis alone [37]. Regardless, while integrated programs may lower the prevalence and intensity of STH infections, they are unlikely to eliminate transmission in the absence of the adjunctive factors described above, such as improved sanitation, clean water supplies, and economic development.

Other examples leveraging existing programs to add deworming medications include vitamin A or immunization campaigns. Integration of antihelminthics to vitamin A campaigns, for instance, takes advantage of the fact that over 167 million children are reached annually by vitamin A supplementation programs globally, and more than 50 countries report >70% coverage [80]. This approach is being used successfully in Cambodia, which in 2004 was the first country to reach the goal of 75% coverage of school-age children with albendazole [34]. The rationale for program integration with vitamin A supplementation seems particularly strong, given that ascariasis and trichuriasis in some studies have been associated with low serum retinol levels, as it is believed that these parasites interfere with vitamin A absorption [50, 81, 82].

Combination therapy for multiple NTDs is effective and safe, including one with three-drug combinations of antiparasitics (such as albendazole, ivermectin, and praziquantel) [83, 84]. Studies have also shown the addition of azithromycin for trachoma control with albendazole and ivermectin is likely safe, though may lead to increased ivermectin levels [85, 86]. Programs that target filariasis/onchocerciasis are particularly well suited for incorporation of soil-transmitted helminth control given that the drugs used are similar (most filariasis MDA programs already include albendazole) and involve MDA as a core control measure. Another overlap is with malaria control programs. Ivermectin acts as an *Anopheles* mosquito adulticide when ingested with a blood meal, and thus has been used for malaria MDA. The integration of albendazole for STHs with ivermectin MDA for malaria control is not only convenient but also potentially clinically synergistic. It is thought that STH and malaria co-infection may exacerbate anemia, especially in pregnant women, and that STH infection may facilitate malaria transmission [87]. Such combined therapies would improve impact on STH as well, especially trichuriasis, where combination therapy with albendazole or mebendazole plus ivermectin is more effective than albendazole or mebendazole alone [28, 88]. In other control programs, such as for LF and onchocerciasis, community-directed distribution of drugs has been the most cost-effective means of achieving high coverage in the population [27, 89].

Cost-effectiveness is a major principle in the planning of integrated campaigns. MDA with albendazole is inexpensive, on the order of 0.03–0.04 USD per person treated per year [89, 90]. It has been estimated that expanding coverage to include filariasis, onchocerciasis, schistosomiasis, vitamin A deficiency, and trachoma, in addition to the soil-transmitted helminths, could be done for as little as 0.40–0.50 USD per person per year [89, 91]. Similarly, when organized as an integrated campaign, these programs would save from 25 to 50% compared to the costs that each control effort would cost individually [91, 92]. Since these diseases are all presently best controlled by MDA campaigns, it is sensible to integrate them to save resources while controlling multiple public health issues simultaneously. It is even possible that combining campaigns increases compliance with all components of the program among targeted populations; the cost per DALY saved would be only 2–9 USD, indicating this approach would be highly cost-effective [91, 92].

8.9 VACCINATION

In the face of increasing drug resistance, a need for multiple re-treatments, challenges in implementing WASH, and logistical difficulties, vaccination remains the most desirable preventative measure for STHs by providing a simple and cost-effective means for control or elimination in a single step. Unfortunately, the lack of good animal models and a poor understanding of how soil-transmitted helminths persist in humans in the face of a potent immune response have hindered the development of an effective vaccine [93]. Nevertheless, a bivalent hookworm vaccine consisting of the recombinant larval antigens Na-APR-1 and Na-GST-1 is effective in animal models and has shown a protective association in immunoepidemiology studies [93, 94]. The Na-APR-1 and Na-GST-1 hookworm vaccine is now undergoing clinical development in humans and is in phase 1 clinical trials to determine safety and antigenicity [93, 94] This vaccine targets the larval forms of hookworms, but not adult worms. There is also effort towards development of a multivalent, pan-antihelminthic vaccine that builds off the bivalent hookworm antigens under investigation while adding a larval *A. lumbricoides* and an adult *T. trichiura* antigen as well [95].

8.10 CONCLUSIONS

The soil-transmitted helminths remain a major global public health problem. The most important impact globally is the association between soil-transmitted helminth infection and the adverse outcomes in infected young children, who suffer anemia, impaired growth, and impaired cognitive development. Infected pregnant women also suffer higher rates of anemia, resulting in poor pregnancy outcomes. Mass drug administration, endorsed by WHO and other agencies, has been the most widely accepted cost-effective approach to control these parasites to date. With increased global recognition and public and private advocacy for control of these neglected tropical diseases, expansion of these successful programs is ongoing in many areas. Integration of mass treatment campaigns with interventions involving other diseases and expansion to community-wide treatment may be the best way forward to expand these programs and maintain sustainability over the long term. Improved sanitation, water, hygiene, and prevention measures are also critical long-term elements of control and elimination campaigns for these diseases, although the best means of implementing these complex and resource-intensive measures has yet to be defined. Most importantly, improved economic development in affected areas will be needed to fully break the cycle of poverty and infection associated with these parasites. Breaking the cycle of poverty that is both a cause and effect of infections such as the soil-transmitted helminths is a fundamental aspect to eventual control and elimination of these diseases. While progress has been made in the past 10–15 years on achieving the goal of treating at least 75% of children in areas with endemic STH, continued vigilance, resources, and efforts are needed to maintain progress and eliminate morbidity from STH globally.

REFERENCES

1. Bethony, J., Brooker, S., Albonico, M. et al. (2006). Soil-transmitted helminth infections: ascariasis, trichuriasis, and hookworm. *Lancet* 367: 1521–1532.
2. Hotez, P.J., Bethony, J., Bottazzi, M.E. et al. (2005). Hookworm: "the great infection of mankind". *PLoS Med.* 2: e67.
3. Pullan, R.L., Smith, J.L., Jasrasaria, R., and Brooker, S.J. (2014). Global numbers of infection and disease burden of soil transmitted helminth infections in 2010. *Parasit. Vectors* 7: 37.
4. de Silva, N.R., Chan, M.S., and Bundy, D.A. (1997). Morbidity and mortality due to ascariasis: re-estimation and sensitivity analysis of global numbers at risk. *Tropical Med. Int. Health* 2: 519–528.
5. Maguire, J. (2010). Intestinal nematodes (roundworms). In: *Principles and Practice of Infectious Diseases* (ed. G. Mandell, R. Dolin and J. Bennett), 3577–3586. Philadelphia, PA: Churchill Livingston Elsevier.
6. Seltzer, E., Barry, M., and Crompton, D. (2006). Ascariasis. In: *Tropical Infectious Diseases–Principles, Pathogens and Practice* (ed. R. Guerrant, D. Walker and P. Weller), 1257–1264. Philadelphia, PA: Churchill Livingston Elsevier.
7. Cooper, E. (2006). Trichuriasis. In: *Tropical Infectious Diseases–Principles, Pathogens, and Practice* (ed. R.L. Guerrant, D. Walker and P. Weller), 1252–1256. Philadelphia, PA: Churchill Livingston Elsevier.
8. Bartsch, S.M., Hotez, P.J., Asti, L. et al. (2016). The global economic and health burden of human hookworm infection. *PLoS Negl. Trop. Dis.* 10 (9): e0004922. Available at: http://www.ncbi.nlm.nih.gov/pmc/articles/PMC5015833/. Accessed 31 January 2017.
9. Traub, R.J. (2013). Ancylostoma ceylanicum, a re-emerging but neglected parasitic zoonosis. *Int. J. Parasitol.* 43: 1009–1015.
10. Inpankaew, T., Schär, F., Dalsgaard, A. et al. (2014). High prevalence of *Ancylostoma ceylanicum* hookworm infections in humans, Cambodia, 2012. *Emerg. Infect. Dis.* 20: 976–982.
11. Ngui, R., Lim, Y.A.L., Traub, R. et al. (2012). Epidemiological and genetic data supporting the transmission of *Ancylostoma ceylanicum* among human and domestic animals. *PLoS Negl. Trop. Dis.* 6: e1522.
12. Jiraanankul, V., Aphijirawat, W., Mungthin, M. et al. (2011). Incidence and risk factors of hookworm infection in a rural community of Central Thailand. *Am. J. Trop. Med. Hyg.* 84: 594–598.
13. Hotez, P.J. (2006). Hookworm infections. In: *Tropical Infectious Diseases–Principles, Pathogens, and Practice* (ed. R. Guerrant, D. Walker and P. Weller), 1265–1273. Philadelphia, PA: Churchill Livingston Elsevier.
14. Hotez, P.J., Brooker, S., Bethony, J.M. et al. (2004). Hookworm infection. *N. Engl. J. Med.* 351: 799–807.
15. Chen, Y.-D., Tang, L.-H., and Xu, L.-Q. (2008). Current status of soil-transmitted nematode infection in China. *Biomed. Environ. Sci.* 21: 173–179.
16. McManus, D.P., Bieri, F.A., Li, Y.-S. et al. (2014). Health education and the control of intestinal worm infections in China: a new vision. *Parasit. Vectors* 7: 344.

17. Steinmann, P., Utzinger, J., Du, Z.-W. et al. (2011). Efficacy of single-dose and triple-dose albendazole and mebendazole against soil-transmitted helminths and *Taenia* spp.: a randomized controlled trial. *PLoS One* 6: e25003.
18. Weaver, H., Hawdon, J., and Hoberg, E. (2010). Soil-transmitted Helminthiases: implications of climate change and human behavior. *Trends Parasitol.* 26 (12): 574–581. Available at: http://www.ncbi.nlm.nih.gov/pubmed/20580609.
19. Hernandez, A.D., Poole, A., and Cattadori, I.M. (2013). Climate changes influence free-living stages of soil-transmitted parasites of European rabbits. *Glob. Chang. Biol.* 19: 1028–1042.
20. Palinkas, L.A., Pickwell, S.M., Brandstein, K. et al. (2003). The journey to wellness: stages of refugee health promotion and disease prevention. *J. Immigr. Health* 5: 19–28.
21. McMichael, C. (2015). Climate change-related migration and infectious disease. *Virulence* 6: 548–553.
22. Brooker, S., Clements, A.C.A., and Bundy, D.A.P. (2006). Global epidemiology, ecology and control of soil-transmitted helminth infections. *Adv. Parasitol.* 62: 221–261.
23. Brooker, S., Bethony, J., and Hotez, P.J. (2004). Human hookworm infection in the 21st century. *Adv. Parasitol.* 58: 197–288.
24. Hall, A., Hewitt, G., Tuffrey, V., and de Silva, N. (2008). A review and meta-analysis of the impact of intestinal worms on child growth and nutrition. *Matern. Child. Nutr.* 4 (Suppl. 1): 118–236.
25. Keiser, J. and Utzinger, J. (2008). Efficacy of current drugs against soil-transmitted Helminth infections: systematic review and meta-analysis. *JAMA* 299: 1937–1948.
26. Vercruysse, J., Behnke, J.M., Albonico, M. et al. (2011). Assessment of the anthelmintic efficacy of albendazole in school children in seven countries where soil-transmitted helminths are endemic. *PLoS Negl. Trop. Dis.* 5: e948.
27. Smits, H.L. (2009). Prospects for the control of neglected tropical diseases by mass drug administration. *Expert Rev. Anti-Infect. Ther.* 7: 37–56.
28. Knopp, S., Mohammed, K.A., Speich, B. et al. (2010). Albendazole and mebendazole administered alone or in combination with ivermectin against Trichuris trichiura: a randomized controlled trial. *Clin. Infect. Dis.* 51: 1420–1428.
29. Speich, B., Ali, S.M., Ame, S.M. et al. (2015). Efficacy and safety of albendazole plus ivermectin, albendazole plus mebendazole, albendazole plus oxantel pamoate, and mebendazole alone against *Trichuris trichiura* and concomitant soil-transmitted helminth infections: a four-arm, randomised controlled trial. *Lancet Infect. Dis.* 15: 277–284.
30. World Health Organization (2006). *Preventive Chemotherapy in Human Helminthiasis. Coordinated Use of Anthelminthic Drugs in Control Interventions. WHO Manual for Health Professionals and Programme Managers*, 1–63. World Health Organization.
31. Albonico, M., Bickle, Q., Ramsan, M. et al. (2003). Efficacy of mebendazole and levamisole alone or in combination against intestinal nematode infections after repeated targeted mebendazole treatment in Zanzibar. *Bull. World Health Organ.* 81: 343–352.
32. Ettling, J. (1990). The role of the Rockefeller Foundation in hookworm research and control. In: *Hookworm Disease* (ed. G. Schad and K. Warren), 3–14. London, Philadelphia: Taylor & Francis, Ltd.
33. Smith, F.H., Simmons, R.E., and Hatch, C.F. (1909). The Rockefeller Commission for the eradication of hookworm disease. *Science* 30: 635–636.
34. Albonico, M., Montresor, A., Crompton, D.W.T., and Savioli, L. (2006). Intervention for the control of soil-transmitted helminthiasis in the community. *Adv. Parasitol.* 61: 311–348.
35. Lo, N.C., Addiss, D.G., Hotez, P.J. et al. (2017). A call to strengthen the global strategy against schistosomiasis and soil-transmitted helminthiasis: the time is now. *Lancet Infect. Dis.* 17: e64–e69.
36. WHO/WHO Resolutions. Available at: http://www.who.int/tobacco/framework/wha_eb/wha_resolutions/en/. Accessed 22 July 2017.
37. WHO/Department of Control of Neglected Tropical Diseases (2016). Schistosomiasis and soil-transmitted helminthiases: number of people treated in 2015. *Wkly Epidemiol. Rec.* 91: 585–600.
38. Jia, T.-W., Melville, S., Utzinger, J. et al. (2012). Soil-transmitted Helminth reinfection after drug treatment: a systematic review and meta-analysis. *PLoS Negl. Trop. Dis.* 6: e1621.
39. Quinnell, R.J., Slater, A.F., Tighe, P. et al. (1993). Reinfection with hookworm after chemotherapy in Papua New Guinea. *Parasitology* 106 (Pt 4): 379–385.
40. Elkins, D.B., Haswell-Elkins, M., and Anderson, R.M. (1988). The importance of host age and sex to patterns of reinfection with *Ascaris lumbricoides* following mass anthelmintic treatment in a South Indian fishing community. *Parasitology* 96 (Pt 1): 171–184.
41. Chan, L., Bundy, D.A., and Kan, S.P. (1994). Aggregation and predisposition to *Ascaris lumbricoides* and *Trichuris trichiura* at the familial level. *Trans. R. Soc. Trop. Med. Hyg.* 88: 46–48.
42. Narain, K., Medhi, G.K., Rajguru, S.K., and Mahanta, J. (2004). Cure and reinfection patterns of geohelminthic infections after treatment in communities inhabiting the tropical rainforest of Assam, India. *Southeast Asian J. Trop. Med. Public Health* 35: 512–517.
43. Clarke, N.E., Clements, A.C.A., Doi, S.A. et al. (2017). Differential effect of mass deworming and targeted deworming for soil-transmitted helminth control in children: a systematic review and meta-analysis. *Lancet* 389: 287–297.
44. Anderson, R., Truscott, J., and Hollingsworth, T.D. (2014). The coverage and frequency of mass drug administration required to eliminate persistent transmission of soil-transmitted helminths. *Philos. Trans. R. Soc. Lond. B Biol. Sci.* 369: 20130435.
45. Anderson, R.M., Turner, H.C., Truscott, J.E. et al. (2015). Should the goal for the treatment of soil transmitted Helminth (STH) infections be changed from morbidity control in children to community-wide transmission elimination? *PLoS Negl. Trop. Dis.* 9: e0003897.
46. Lo, N.C., Bogoch, I.I., Blackburn, B.G. et al. (2015). Comparison of community-wide, integrated mass drug

administration strategies for schistosomiasis and soil-transmitted helminthiasis: a cost-effectiveness modelling study. *Lancet Glob. Health* 3: e629–e638.

47. Adegnika, A.A., Zinsou, J.F., Issifou, S. et al. (2014). Randomized, controlled, assessor-blind clinical trial to assess the efficacy of single- versus repeated-dose Albendazole to treat *Ascaris lumbricoides*, *Trichuris trichiura*, and hookworm infection. *Antimicrob. Agents Chemother.* 58: 2535–2540.
48. Guyatt, H.L., Bundy, D.A., and Evans, D. (1993). A population dynamic approach to the cost-effectiveness analysis of mass anthelmintic treatment: effects of treatment frequency on Ascaris infection. *Trans. R. Soc. Trop. Med. Hyg.* 87: 570–575.
49. Bundy, D.A., Hall, A., Medley, G.F., and Savioli, L. (1992). Evaluating measures to control intestinal parasitic infections. *World Health Stat. Q.* 45: 168–179.
50. de Silva, N.R. (2003). Impact of mass chemotherapy on the morbidity due to soil-transmitted nematodes. *Acta Trop.* 86: 197–214.
51. Awasthi, S., Pande, V.K., and Fletcher, R.H. (2000). Effectiveness and cost-effectiveness of albendazole in improving nutritional status of pre-school children in urban slums. *Indian Pediatr.* 37: 19–29.
52. Stoltzfus, R.J., Chway, H.M., Montresor, A. et al. (2004). Low dose daily iron supplementation improves iron status and appetite but not anemia, whereas quarterly anthelminthic treatment improves growth, appetite and anemia in Zanzibari preschool children. *J. Nutr.* 134: 348–356.
53. Watkins, W.E. and Pollitt, E. (1997). "Stupidity or worms": do intestinal worms impair mental performance? *Psychol. Bull.* 121: 171–191.
54. Simeon, D.T., Grantham-McGregor, S.M., Callender, J.E., and Wong, M.S. (1995). Treatment of *Trichuris trichiura* infections improves growth, spelling scores and school attendance in some children. *J. Nutr.* 125: 1875–1883.
55. Miguel, E. and Kremer, M. Worms: Education and Health Externalities in Kenya. National Bureau of Economic Research, 2001. Available at: http://www.nber.org/papers/w8481.
56. Stephenson, L.S., Latham, M.C., Kurz, K.M. et al. (1989). Treatment with a single dose of albendazole improves growth of Kenyan schoolchildren with hookworm, *Trichuris trichiura*, and *Ascaris lumbricoides* infections. *Am. J. Trop. Med. Hyg.* 41: 78–87.
57. Olsen, A. (2003). Experience with school-based interventions against soil-transmitted helminths and extension of coverage to non-enrolled children. *Acta Trop.* 86: 255–266.
58. Christian, P., Khatry, S.K., and West, K.P. (2004). Antenatal anthelmintic treatment, birthweight, and infant survival in rural Nepal. *Lancet* 364: 981–983.
59. Kabatereine, N.B., Brooker, S., Koukounari, A. et al. (2007). Impact of a national helminth control programme on infection and morbidity in Ugandan schoolchildren. *Bull. World Health Organ.* 85: 91–99.
60. Baird, S., Hicks, J.H., Kremer, M., and Miguel, E. (2016). Worms at work: long-run impacts of a child health investment. *Q. J. Econ.* 131: 1637–1680.
61. Taylor-Robinson, D.C., Jones, A.P., and Garner, P. (2007). Deworming drugs for treating soil-transmitted intestinal worms in children: effects on growth and school performance. *Cochrane Database of Syst. Rev.* (4): CD000371.
62. Awasthi, S., Peto, R., Read, S. et al. (2013). Population deworming every 6 months with albendazole in 1 million pre-school children in North India: DEVTA, a cluster-randomised trial. *Lancet* 381: 1478–1486.
63. Welch, V.A., Ghogomu, E., Hossain, A. et al. (2017). Mass deworming to improve developmental health and wellbeing of children in low-income and middle-income countries: a systematic review and network meta-analysis. *Lancet Glob. Health* 5: e40–e50.
64. Taylor-Robinson, D.C., Maayan, N., Soares-Weiser, K. et al. (2015). Deworming drugs for soil-transmitted intestinal worms in children: effects on nutritional indicators, haemoglobin, and school performance. *Cochrane Database of Syst Rev.* (7): CD000371. Available at: http://onlinelibrary.wiley.com/doi/10.1002/14651858.CD000371.pub6/abstract.
65. Andrews, J.R., Bogoch, I.I., and Utzinger, J. (2017). The benefits of mass deworming on health outcomes: new evidence synthesis, the debate persists. *Lancet Glob. Health* 5: e4–e5.
66. (2008). Soil-transmitted helminthiasis. Progress report on number of children treated with anthelminthic drugs: an update towards the 2010 global target. *Wkly Epidemiol. Rec.* 82: 237–252.
67. Salam, R.A., Haider, B.A., Humayun, Q., and Bhutta, Z.A. (2015). Effect of administration of antihelminthics for soil-transmitted helminths during pregnancy. *Cochrane Database of Syst. Rev.* (6): CD005547.
68. Albonico, M. (2003). Methods to sustain drug efficacy in helminth control programmes. *Acta Trop.* 86: 233–242.
69. Shalaby, H.A. (2013). Anthelmintics resistance; how to overcome it? *Iran. J. Parasitol.* 8: 18–32.
70. Campbell, S.J., Nery, S.V., McCarthy, J.S. et al. (2016). A critical appraisal of control strategies for soil-transmitted Helminths. *Trends Parasitol.* 32: 97–107.
71. Strunz, E.C., Addiss, D.G., Stocks, M.E. et al. (2014). Water, sanitation, hygiene, and soil-transmitted helminth infection: a systematic review and meta-analysis. *PLoS Med.* 11: e1001620.
72. Sow, S., de Vlas, S.J., Polman, K., and Gryseels, B. (2004). Hygiene practices and contamination risks of surface waters by schistosome eggs: the case of an infested village in Northern Senegal. *Bull. Soc. Pathol. Exot.* 97: 12–14.
73. Mukherjee, J. (2005). *Harnessing Market Power for Rural Sanitation*. Jakarta, Indonesia: Water and Sanitation Program.
74. Esrey, S.A., Potash, J.B., Roberts, L., and Shiff, C. (1991). Effects of improved water supply and sanitation on ascariasis, diarrhoea, dracunculiasis, hookworm infection, schistosomiasis, and trachoma. *Bull. World Health Organ.* 69: 609–621.
75. Freeman, M.C., Clasen, T., Brooker, S.J. et al. (2013). The impact of a school-based hygiene, water quality and sanitation intervention on soil-transmitted helminth reinfection: a cluster-randomized trial. *Am. J. Trop. Med. Hyg.* 89: 875–883.
76. Oswald, W.E., Stewart, A.E.P., Kramer, M.R. et al. (2017). Association of community sanitation usage with soil-transmitted helminth infections among school-aged children in Amhara Region, Ethiopia. *Parasit. Vectors* 10: 91.

77. Benjamin-Chung, J., Nazneen, A., Halder, A.K. et al. (2015). The interaction of deworming, improved sanitation, and household flooring with soil-transmitted Helminth infection in rural Bangladesh. *PLoS Negl. Trop. Dis.* 9: e0004256.
78. Worrell, C.M., Wiegand, R.E., Davis, S.M. et al. (2016). A cross-sectional study of water, sanitation, and hygiene-related risk factors for soil-transmitted Helminth infection in urban school- and preschool-aged children in Kibera, Nairobi. *PLoS One* 11: e0150744.
79. Yamey, G. and Hotez, P. (2007). Neglected tropical diseases. *BMJ* 335: 269–270.
80. The State of the World's Children 2005 – Childhood under Threat. Available at: https://www.unicef.org/publications/index_24432.html. Accessed 22 July 2017.
81. Sivakumar, B. and Reddy, V. (1975). Absorption of vitamin A in children with ascariasis. *J. Trop. Med. Hyg.* 78: 114–115.
82. Mahalanabis, D., Simpson, T.W., Chakraborty, M.L. et al. (1979). Malabsorption of water miscible vitamin A in children with giardiasis and ascariasis. *Am. J. Clin. Nutr.* 32: 313–318.
83. Reddy, M., Gill, S.S., Kalkar, S.R. et al. (2007). Oral drug therapy for multiple neglected tropical diseases: a systematic review. *JAMA* 298: 1911–1924.
84. Olsen, A. (2007). Efficacy and safety of drug combinations in the treatment of schistosomiasis, soil-transmitted helminthiasis, lymphatic filariasis and onchocerciasis. *Trans. R. Soc. Trop. Med. Hyg.* 101: 747–758.
85. Coulibaly, Y.I., Dicko, I., Keita, M. et al. (2013). A cluster randomized study of the safety of integrated treatment of trachoma and lymphatic filariasis in children and adults in Sikasso, Mali. *PLoS Negl. Trop. Dis.* 7: e2221.
86. Amsden, G.W., Gregory, T.B., Michalak, C.A. et al. (2007). Pharmacokinetics of azithromycin and the combination of ivermectin and albendazole when administered alone and concurrently in healthy volunteers. *Am. J. Trop. Med. Hyg.* 76: 1153–1157.
87. Kobylinski, K.C., Alout, H., Foy, B.D. et al. (2014). Rationale for the Coadministration of Albendazole and Ivermectin to humans for malaria parasite transmission control. *Am. J. Trop. Med. Hyg.* 91: 655–662.
88. Beach, M.J., Streit, T.G., Addiss, D.G. et al. (1999). Assessment of combined ivermectin and albendazole for treatment of intestinal helminth and *Wuchereria bancrofti* infections in Haitian schoolchildren. *Am. J. Trop. Med. Hyg.* 60: 479–486.
89. Hotez, P.J. (2009). Mass drug administration and integrated control for the world's high-prevalence neglected tropical diseases. *Clin. Pharmacol. Ther.* 85: 659–664.
90. Kabatereine, N.B., Tukahebwa, E.M., Kazibwe, F. et al. (2005). Soil-transmitted helminthiasis in Uganda: epidemiology and cost of control. *Tropical Med. Int. Health* 10: 1187–1189.
91. Hotez, P.J., Fenwick, A., Savioli, L., and Molyneux, D.H. (2009). Rescuing the bottom billion through control of neglected tropical diseases. *Lancet* 373: 1570–1575.
92. Hotez, P.J., Molyneux, D.H., Fenwick, A. et al. (2007). Control of neglected tropical diseases. *N. Engl. J. Med.* 357: 1018–1027.
93. Diemert, D.J., Bethony, J.M., and Hotez, P.J. (2008). Hookworm vaccines. *Clin. Infect. Dis.* 46: 282–288.
94. Hotez, P.J., Beaumier, C.M., Gillespie, P.M. et al. (2016). Advancing a vaccine to prevent hookworm disease and anemia. *Vaccine* 34: 3001–3005.
95. Zhan, B., Beaumier, C.M., Briggs, N. et al. (2014). Advancing a multivalent "pan-anthelmintic" vaccine against soil-transmitted nematode infections. *Expert Rev. Vaccines* 13: 321–331.

9

FOOD SYSTEMS AND NUTRITION IN THE CONTEXT OF CLIMATE CHANGE

José Graziano da Silva
Food and Agriculture Organization of the United Nations, Rome, Italy

9.1 INTRODUCTION

Ensuring food security and good nutrition for everyone in a sustainable manner is one of the main challenges of our time. According to the 2017 State of Food Insecurity and Nutrition in the World, 815 million people in the world were undernourished in 2016[1] [2]. However, ensuring sufficient energy intake is not enough: good nutrition requires a healthy diet with sufficient quantities of safe, diverse, and nutritious foods. Lack of proper nutrition continues to pose the most serious threat to global health, with nearly half of children's deaths on an annual basis linked to malnutrition. Global food demand is projected to increase, driven by population and income growth, as well as rapid urbanization [3]. Limited natural resources combined with effects of climate change will make the challenge of ending hunger and malnutrition even more difficult. The issues of ending hunger, achieving food security and improved nutrition and promoting sustainable agriculture, as well as taking urgent action to combat climate change and its impacts are directly addressed in the Sustainable Development Goals (SGDs), which aim to end poverty, protect the planet, and ensure prosperity for all.[2]

Efforts to ensure good nutrition include many entry points throughout the food system (see Box 9.1), such as improving the availability and access to nutritious foods, biofortification, and fortifying staple foods with essential micronutrients, but also actions and policies in other sectors, e.g., enhancing food safety, water, sanitation, and hygiene (WASH), access to health services, supplying micronutrient supplements, women's empowerment, education, employment, social protection, and safety nets.

This chapter starts off by briefly looking at the current malnutrition situation in the world. The effects of climate change and climate variability on nutrition are presented by considering their effects on food security (including on the nutrient content of foods and biodiversity) and on social, health, and living environments. Both direct impacts (e.g., those caused by changes in temperature, carbon dioxide concentration, and precipitation) as well as indirect impacts (e.g., via changes in diseases, pests, etc.) are included. The chapter also looks at how food systems impact climate

[1] Refers to the proportion of the population whose dietary energy consumption is less than a predetermined threshold. This threshold is country-specific and is measured in terms of the number of kilocalories required to conduct sedentary or light activities [1].

[2] SDG 2 is to "End hunger, achieve food security and improved nutrition and promote sustainable agriculture," while SDG 13 is to "Take urgent action to combat climate change and its impacts." Available from: www.un.org/sustainabledevelopment/sustainable-development-goals/.

Water and Sanitation-Related Diseases and the Changing Environment: Challenges, Interventions, and Preventive Measures, Second Edition. Edited by Janine M. H. Selendy.

Companion website: https://www.wiley.com/legacy/wileychi/selendy/

BOX 9.1 FOOD SYSTEMS

A food system can broadly be defined as gathering all the elements (environment, people, inputs, processes, infrastructures, institutions, etc.) and activities that relate to the production, processing, distribution, preparation and consumption of food, and the outputs of these activities, including socioeconomic and environmental outcomes [4]. Broad domains of the food system include: food production; food handling, storage, and transport; food processing and packaging; food trade, retail and marketing; and consumer demand, food preparation, and preferences.

A sustainable food system is a food system that delivers food security and nutrition for all in such a way that the economic, social, and environmental bases to generate food security and nutrition for future generations are not compromised [4].

change (touching on agriculture[3] and food production, food consumption patterns, and food loss and waste). Key opportunities are identified that can improve the sustainability of food systems and ensure good nutrition in the context of climate change, by focusing on interventions at production and post-harvest stages. Finally, policy recommendations are proposed.

9.2 CURRENT FOOD AND NUTRITION SITUATION, AND TRENDS[4]

Most countries are currently faced with the multiple burdens of malnutrition, including undernutrition, micronutrient deficiencies, and overweight and obesity. Different forms of malnutrition can co-exist within the same country, the same household, and even the same individual.

9.2.1 The State of Food Insecurity and Prevalence of Undernourishment

Hunger and extreme poverty have reduced globally in the last decade. In terms of caloric intake – the simplest form of measuring food deprivation – from 2000 to 2016, the total number of hungry people has been estimated to have declined by about 85 million [2]. However, in 2016 the number of undernourished people in the world increased to an estimated 815 million, up from 777 million in 2015 [2]. This is a cause for great concern, posing a significant challenge for international commitments to end hunger by 2030. This increase can be attributed to a variety of factors, including recent reductions in food availability and increases in food prices in regions affected by El Niño/La Niña-related phenomena – most notably in Eastern and Southern Africa and in Southeast Asia, as well as increasing conflicts, particularly in African and Near East nations [2]. These conflicts have led to food crisis situations, especially where compounded by droughts or other weather-related events and fragile response capacities [2]. Progress towards eradicating hunger is unevenly distributed worldwide (Table 9.1).

Some key points to note:

- In both Africa and the Near East and North Africa (NENA) the number of people undernourished has increased from 2000 to 2016. In Africa, about one in five people were undernourished in 2016. However, in terms of absolute numbers, the highest burden of undernourished people is found in Asia, almost 520 million people.
- The Europe and Central Asia region has managed to halve the number of undernourished people and the prevalence of undernourishment, while the Latin America and Caribbean (LAC) region has succeeded in almost halving the percentage of undernourished people. However, wide differences still exist across countries and subregions.

In addition to FAO's traditional indicator of the extent of hunger (the prevalence of undernourishment discussed above), the prevalence of severe food insecurity[5] can be used as a complementary measure to understand people's ability to obtain adequate food. Data collected by FAO in 2014, 2015, and 2016 in almost 150 countries reveal that nearly 1 in 10 people in the world (9.3%) suffered from severe food insecurity, corresponding to about 689 million people [2].

9.2.2 Prevalence of Stunting, Wasting, and Underweight

When we consider child malnutrition data, again inequalities are rife: less than half of all children under five live in lower-middle-income countries, but two-thirds of all stunted (low height-for-age) children live in these countries [6].

Childhood stunting is a largely irreversible outcome of inadequate nutrition and repeated bouts of infection during the first 1000 days of a child's life.[6] Stunting before the age of two years predicts poorer cognitive and educational

[3] The agriculture sectors include crops, livestock, forestry, fisheries, and aquaculture.

[4] Part of this text is adapted from reference [5].

[5] This is estimated based on data collected from adult individuals worldwide using the Food Insecurity Experience Scale (FIES).

[6] The 1000 days from a woman's pregnancy through her child's second birthday, when better nutrition can have a life-changing impact on a child's future; see http://www.thousanddays.org/.

TABLE 9.1 Summary of the undernourishment situation, based on FAO data.

	Africa[a]	Asia[b]	LAC[c]	NENA[d]	Europe and Central Asia[e]
Number of people undernourished (millions)					
2000	197.6	620.8	62.8	39.5	29.4
2016	243.2	519.6	42.5	50.7	13.6
% change from 2000 to 2016[f]	18.7	−16.3	−32.4	28.2	−53.8
Prevalence of undernourishment (PoU) (%)					
2000	24.3%	16.7%	12.0%	11.4%	3.4%
2016	20.0%	11.7%	6.6%	10.8%	1.5%
% change from 2000 to 2016[g]	−17.7	−29.9	−45.0	−5.2	−56.1

[a] Algeria, Angola, Benin, Botswana, Burkina Faso, Burundi, Cabo Verde, Cameroon, Central African Republic, Chad, Comoros, Congo, Côte d'Ivoire, Democratic Republic of the Congo, Djibouti, Egypt, Equatorial Guinea, Eritrea, Ethiopia, Gabon, Gambia, Ghana, Guinea, Guinea-Bissau, Kenya, Lesotho, Liberia, Libya, Madagascar, Malawi, Mali, Mauritania, Mauritius, Morocco, Mozambique, Namibia, Niger, Nigeria, Rwanda, Sao Tome and Principe, Senegal, Seychelles, Sierra Leone, Somalia, South Africa, South Sudan, Sudan, Swaziland, Togo, Tunisia, Uganda, United Republic of Tanzania, Western Sahara, Zambia, Zimbabwe.

[b] Afghanistan, Armenia, Azerbaijan, Bahrain, Bangladesh, Bhutan, Brunei Darussalam, Cambodia, Cyprus, Democratic People's Republic of Korea, Georgia, India, Indonesia, Iran (Islamic Republic of), Iraq, Israel, Japan, Jordan, Kazakhstan, Kuwait, Kyrgyzstan, Lao People's Democratic Republic, Lebanon, Malaysia, Maldives, Mongolia, Myanmar, Nepal, Oman, Pakistan, Palestine, Philippines, Qatar, Republic of Korea, Saudi Arabia, Singapore, Sri Lanka, Syrian Arab Republic, Tajikistan, Thailand, Timor-Leste, Turkey, Turkmenistan, United Arab Emirates, Uzbekistan, Viet Nam, Yemen.

[c] Latin America and the Caribbean (LAC). Antigua and Barbuda, Argentina, Bahamas, Barbados, Belize, Bolivia (Plurinational State of), Brazil, Chile, Colombia, Costa Rica, Cuba, Dominica, Dominican Republic, Ecuador, El Salvador, Grenada, Guatemala, Guyana, Haiti, Honduras, Jamaica, Mexico, Nicaragua, Panama, Paraguay, Peru, Saint Kitts and Nevis, Saint Lucia, Saint Vincent and the Grenadines, Suriname, Trinidad and Tobago, Uruguay, Venezuela (Bolivarian Republic of).

[d] Near East and North Africa (NENA), i.e., Algeria, Bahrain, Egypt, Iran (Islamic Republic of), Iraq, Jordan, Kuwait, Lebanon, Libya Mauritania, Morocco, Qatar, Saudi Arabia, Syrian Arab Republic, Oman, Sudan, Tunisia, United Arab Emirates and Yemen, plus Palestine. Please note that in the year 2016 South Sudan is not included (as it is considered an Eastern African country and therefore not part of NENA).

[e] Albania, Andorra, Armenia, Austria, Azerbaijan, Belarus, Belgium, Bosnia and Herzegovina, Bulgaria, Croatia, Cyprus, Czechia, Denmark, Estonia, Finland, France, Georgia, Germany, Greece, Hungary, Iceland, Ireland, Israel, Italy, Kazakhstan, Kyrgyzstan, Latvia, Lithuania, Luxembourg, Malta, Monaco, Montenegro, Netherlands, Norway, Poland, Portugal, Republic of Moldova, Romania, Russian Federation, San Marino, Serbia, Slovakia, Slovenia, Spain, Sweden, Switzerland, Tajikistan, The former Yugoslav Republic of Macedonia, Turkey, Turkmenistan, Ukraine, United Kingdom and Uzbekistan.

[f] The number of undernourished has reduced by 53.8% in Europe & Central Asia, 32.4% in LAC and 16.3% in Asia, but increased by 28.2% in NENA and 18.2% in Africa.

[g] PoU has fallen in all regions, most notably in Europe and Central Asia (by 56.1%) and LAC by 45%.

outcomes in later childhood and adolescence and has significant educational and economic consequences at the individual, household, and community levels [7]. According to the latest estimates available, 155 million children under five years of age across the world suffered from stunted growth in 2016 [2]. Globally, the prevalence of stunting fell from 29.5 to 22.9% between 2005 and 2016. However, at current trends, there would be 130 million stunted children by 2025. If current trends continue, the SDG target on reducing child stunting by 2030 will not be reached [2].

The majority of stunted children live in Asia (87 million) and Africa (59 million), while the prevalence of stunting is currently highest in Eastern Africa, Middle Africa, Western Africa, Southern Asia, and Oceania (excluding Australia and New Zealand), where more than 30% of children under five are stunted [2].

Wasting (low weight-for-height) occurs after a recent period of rapid weight loss due to starvation (e.g. following natural disasters like droughts, floods, and locusts) and/or severe disease. Wasting in either children or adults indicates a serious nutritional problem and requires urgent attention to prevent death. Wasting among under five year olds is mainly prevalent in Asia and Africa. In 2016, 7.7% of all children under five years globally suffered from severe wasting, which represents 51.7 million children [2]. Southern Asia stands out with a high prevalence of 15.4% – well above that of any other subregion [2].

9.2.3 Micronutrient Deficiencies

More than 2 billion people worldwide are affected by micronutrient deficiencies – also called "hidden hunger", as people who suffer may look healthy and not feel hunger [8]. Anemia affects 33% of women of reproductive age (about 613 million women between 15 and 49 years of age) globally, according to 2016 estimates [2]. The prevalence is highest in Africa and Asia, at over 35%. Anemia impairs health and well-being in women and increases the risk of maternal and neonatal adverse outcomes. It contributes to 20% of all maternal deaths [9] and about 50 000 women die in childbirth each year due to a lack of iron [8]. Vitamin A deficiency, a public health problem in more than 100 countries, causes an estimated 250 000–500 000 children to go blind every year, and half of them die within a year of losing their sight [10].

An estimated 38 million babies are born with iodine deficiency, which is the most common cause of preventable brain damage [11].

9.2.4 Overweight and Obesity, and Diet-Related Non-communicable Diseases

Overweight and obesity, which are increasing worldwide, increases the risk of non-communicable diseases (NCDs). Globally, 44% of diabetes, 23% of ischemic heart disease, and 7–41% of certain cancers are attributable to overweight and obesity [12]. In 2014, more than 39% of people aged 18 years and over were overweight and, of these, 13% were obese [13]. Childhood obesity increases the risk of early onset of obesity-related health complications and NCDs, which were once thought to be associated only with adult obesity [14]. An estimated 41 million children under five (about 6%) worldwide were considered overweight in 2016, up from 5% in 2005 [2]. In 2016, the prevalence of childhood overweight reached almost 12% in Southern Africa, 11% in Central Asia, 10% in Northern Africa, 8% in Northern America, and 7% in South-Eastern Asia and South America [2]. The number of overweight children in lower-middle-income countries has increased more rapidly than in other countries [15].

9.2.5 The Nutrition Transition, Changing Diets, and Demands on Food Systems

Although the nature and underlying causes of the various forms of malnutrition are complex and multidimensional, the common denominator is poor diets. Consuming a healthy diet throughout the life-course helps prevent malnutrition as well as a range of NCDs.

The nutrition transition is a marked shift in diets from traditional diets based on staples, legumes, fruits, and vegetables to diets that involve a high consumption of energy-dense foods with low nutritional value (high in fat, sugar, and salt), such as some snack foods, sweets, and sugar sweetened beverages; excessive consumption of some animal-source foods (especially processed meats); and lower consumption of grains, legumes, fruits, and vegetables. This shift in diets is common in countries as economic conditions improve and the urban population increases, and is generally accompanied by a simultaneous shift towards sedentary lifestyles.

With the shift in dietary patterns, the demand for processed food has grown, resulting in different demands on food systems. Between 1990 and 2013, the global consumption of some nutritious foods increased (e.g., fruit consumption increased in all regions, while vegetable consumption increased in some regions) (Master, 2016, cited in reference [5]). An increase was noted in the consumption of sugar-sweetened beverages in four out of seven regions, with the largest increase being in North America. In all regions, consumption of processed meat increased.

9.2.6 The Escalating Costs of Malnutrition

The costs of malnutrition are not limited to the human impacts of increased ill health and mortality, and impaired physical and mental growth. There are very real economic costs associated with malnutrition, experienced not only by the affected individuals and their immediate families but also by societies and states. In low-income countries, 54% of the cost of cardiovascular disease is met directly from household expenses [16]. No comprehensive global estimates exist that combines the productivity losses and health costs associated with all types of malnutrition and related diseases. However, a rough estimate suggests that malnutrition in all its forms may impose a cost of US$2.8–3.5 trillion, equivalent to 4–5% of global gross domestic product (GDP), or US$400–500 per person [17]. This value is made up of estimates of US$1.4–2.1 trillion for undernutrition and micronutrient deficiencies plus US$1.4 trillion for non-communicable diseases [17].

9.3 CLIMATE CHANGE AND CLIMATE-VARIABILITY RELATED IMPACTS ON MALNUTRITION

The risks of food insecurity and malnutrition can be exacerbated through short- and medium-term events such as extreme weather events (e.g., floods and droughts), the seasonality of the climate, and natural variations in the conditions of the atmosphere and ocean (like the El Niño Southern Oscillation), as well as by longer-term trends in the average climate, such as a rise in temperatures and changes in precipitation patterns. Climate change can also intensify the impacts of the short- and medium-term events.

Climate change and variability is expected to affect all four dimensions of food security (availability, access, stability, and utilization). Household food insecurity, together with inadequate care, and unhealthy environments and insufficient health services, constitute the underlying causes of malnutrition. As seen from Figure 9.1, climate change has an effect on agricultural production and post-harvest activities, which will have economic and social consequences, finally impacting on households' food security and nutrition [3]. The impacts will be transmitted through different channels and will affect food security in all four of its dimensions. At each stage of the transmission chain, the severity of impact will be determined by both the shock itself and by the vulnerability[7] of the system or population group under stress [3]. Not only the food security and nutrition of those

[7] FAO [3] uses the IPCC definition of vulnerability, i.e., "the extent to which a natural or social system is susceptible to sustaining damage from climate change impacts, and is a function of exposure, sensitivity and adaptive capacity."

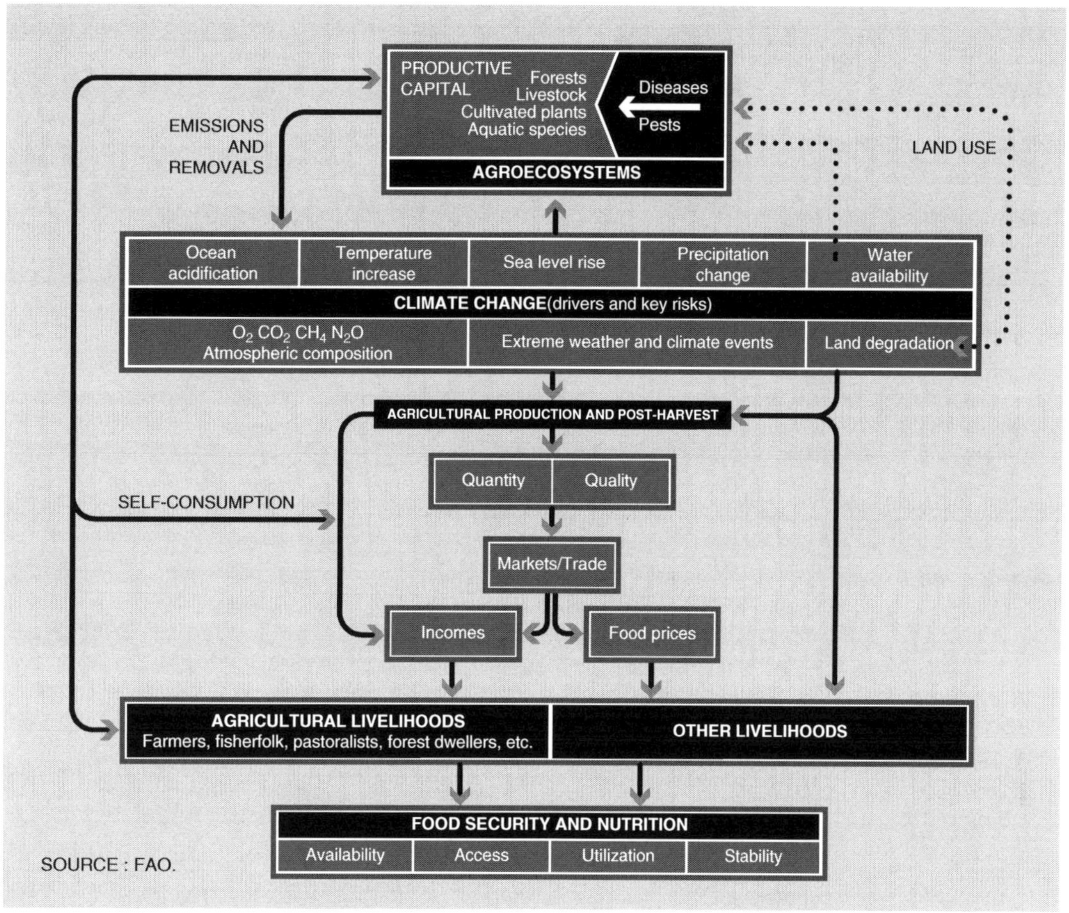

FIGURE 9.1 Impact pathways: from climate change to food security. *Source*: FAO [3].

who are directly dependent on agriculture for their food and livelihoods will be affected but also those of others who do not depend directly on agriculture, because price volatility and disrupted trade can have far-reaching effects.

The following section starts by summarizing the available estimates of how climate change and variability are expected to affect food security and nutrition. This will be followed by a more in-depth look at how nutrition will be affected through impacts on food security, diseases, water safety, sanitation, livelihoods, and care practices, which occur during extreme events, seasonal events, and long-term climate change.

Table 9.2 gives a non-exhaustive summary of the impacts of climate change on food security. A few of these aspects will be looked at in more depth later in this chapter.

9.3.1 How Is Climate Change and Variability Expected to Further Exacerbate the Existing Burden of Food Insecurity and Malnutrition?

The Fifth Assessment Report of the Intergovernmental Panel on Climate Change (IPCC) reported with high confidence that risks of undernutrition in poor regions will increase with climate change. Health impacts identified with high confidence included increased risks from foodborne and waterborne diseases and loss of work capacity and reduced labor productivity in vulnerable populations. Risks from vectorborne diseases were projected to generally increase with warming, due to the extension of the infection area and season, despite reductions in some areas that become too hot for disease vectors (medium confidence) [18].

According to the IPCC report, all aspects of food security will be potentially affected by climate change, including food production, access, use, and price stability (high confidence). For wheat, rice, and maize in tropical and temperate regions, climate change without adaptation[8] is projected to negatively impact production for local temperature increases of 2 °C or more above late twentieth century

[8] Adaptation and mitigation are complementary strategies for reducing and managing the risks of climate change. Substantial emissions reductions over the next few decades can reduce climate risks in the twenty-first century and beyond, increase prospects for effective adaptation, reduce the costs and challenges of mitigation in the longer term and contribute to climate-resilient pathways for sustainable development [26].

TABLE 9.2 Main impacts of climate change and variation on food security and nutrition.

Food security dimension	Effects of climate change
Availability *(the availability of sufficient quantities of food of appropriate quality)*	• Decrease in food availability as rain-fed agriculture in some areas will be affected by changes in precipitation patterns, leading to greater frequency of crop failures. • Fisheries and aquaculture production affected by sea water warming, change of salinity and water acidification, coastal flooding, and more frequent weather events. • Dietary diversity affected by reduced agricultural production (especially at tropical latitudes), as well as by loss of biodiversity. • Animals' vulnerability to disease will be higher, reducing fertility and meat and milk production. • Increased crop, livestock, and fisheries losses (both pre- and post-harvest) due to more frequent weather variability, extreme events, pests, and diseases.
Access *(both physical and economic access)*	• Higher food prices resulting from lower yields in some areas. • Loss of income due to the potential increase in damage to agricultural production, affecting agricultural livelihoods but also rural economies dependent on agriculture. • Short-term price volatility due to production variability, and effect on future price trends, especially of staples (a price increase in staple foods often leads to a decreased intake of other essential foods by the poor). • Risk of market disruptions and effects on supply and storage systems.
Stability *(access to adequate food at all times)*	• Increased food market volatility from the supply side, due to production variability. • Instability of incomes from agriculture. • Reliability of drinkable water supply will be severely affected by droughts and floods. • Reduced private investment in agriculture due to increased uncertainty.
Utilization *(utilization of food through adequate diet, clean water, sanitation, and health care to reach a state of nutritional well-being where all physiological needs are met)*	• If diseases increase (e.g., diarrheal diseases) the ability to utilize nutrients will decrease. • Climate variability and change will have an impact on water resources and the availability of clean drinking water. • Impact on food safety due to water pollution and changes in pests and diseases, e.g., contamination of food by mycotoxins. • Hygiene is likely to be affected by extreme weather events, such as flooding in environments where sound sanitation is absent. • In some cases, the nutritional composition of foods themselves will change. • Food security and health impacts include increased malnutrition.

levels, although individual locations may benefit (medium confidence) [18].

Climate change is expected to cause approximately 250 000 additional deaths per year from malnutrition, malaria, diarrhea, and heat stress between 2030 and 2050 [19]. If current trends continue, it is estimated that an additional 1–3 billion people will be affected by water scarcity and 200–600 million will suffer from hunger by 2080, particularly in Sub-Saharan African countries [20]. The likely impacts will differ by region, country, and location and will affect different population groups according to their vulnerability [3].

Climate change is expected to result in an additional 24 million undernourished children, 21% more than in a world with no climate change, almost half of whom would be living in Sub-Saharan Africa [21, 22]. Furthermore, it has been projected that climate change will lead to a relative increase in moderate stunting of 1–29% in 2050 compared with a future without climate change [23]. Climate change could have an even greater impact on rates of severe stunting, which are estimated to increase by 23% (in central sub-Saharan Africa) to 62% (in South Asia) [23].

Limited literature is available on how the prevalence of micronutrient deficiencies may change with climate change and variability. A recent modeling study considered the global distributions of selenium, an essential trace element that is required for humans [24]. Using moderate climate change projections, the study predicted future (2080–2099) soil selenium losses from 58% of modeled areas (mean loss = 8.4%). Predicted losses from croplands were even higher, with 66% of croplands predicted to lose 8.7% selenium. The authors conclude that these losses could increase the worldwide prevalence of selenium deficiency [24]. Even less information is available on the potential impact of climate change and variability on overweight, obesity, and related NCDs. A recent World Bank publication on the Pacific Island countries states that "growing global temperatures, combined with the Pacific's humid, tropical environment, will escalate the impacts of obesity in the Pacific Islands" [25]. The report mentions that heat exhaustion is one of the main climate change contributors to the burden of NCDs and that obese individuals reach higher core body temperatures more rapidly than their non-obese counterparts, initiating the associated symptoms of cardiovascular diseases [25].

9.3.2 Impacts of Extreme Weather Events on Food Security and Nutrition

Climate-related effects on nutrition are well documented for extreme weather events such as droughts, floods, and storms. Climate change can affect the frequency, intensity, and duration of extreme events. According to the IPCC Report, changes in many extreme weather and climate events have been observed since about 1950; some of these changes have been linked to human influences, including a decrease in cold temperature extremes, an increase in warm temperature extremes, an increase in extreme high sea levels, and an increase in the number of heavy precipitation events in a number of regions [26].

An FAO study found that between 2003 and 2013, natural hazards[9] in the developing regions affected more than 1.9 billion people and resulted in nearly half a trillion US dollars in estimated damage [27]. For climate-related disasters such as floods, droughts, and tropical storms, 25% of all damage and losses were on the agriculture sector – including crops, livestock, forestry, and fisheries [27]. Production losses from natural hazards had the greatest impact in Asia, followed by Africa. Production losses translate to reduced food availability for the affected populations, both directly, and via reduced grazing and fodder for animals, which leads to lower animal-source foods (ASFs) in the diet. In low-income countries, young children's diets are already lacking in ASFs. Demonstrated benefits of ASF consumption by undernourished children include improved growth, improved micronutrient status, cognitive performance, and levels of physical activity [28].

Studies from Bangladesh have shown that wasting and stunting rates increased among pre-school children in the years following floods, due to reduced access to food, increased difficulties of providing proper care, and greater exposure to contaminants [29]. The adverse effects of weather events also affect children in utero. Studies in Ethiopia and Kenya have found that children aged five or less born during a drought were respectively 36% and 50% more likely to be malnourished than children not born during a drought. In Niger, children aged two or less born in a drought year were 72% more likely to be stunted [30].

Climate-related disasters are also a major cause of displacement. Climate change is projected to increase the displacement of people [26]. The latest statistics available show that 17.5 million people were displaced by climate or weather-related disasters in 2014; when data for the preceding seven years were considered, the average number of people displaced was 62 000 per day [31]. Very often, displacement results in loss of livelihoods and lack of access to sufficient amounts of nutritious foods.

[9] The study covered natural hazards, namely drought, floods, storms such as cyclones and hurricanes, earthquakes, tsunamis, and volcanic eruptions.

9.3.3 Seasonality and Undernutrition

Seasonality is a key element of food availability and access in developing countries, which affects the nutritional status of vulnerable groups (see Box 9.2). A classical seasonality-nutrition scenario is that at the end of the dry season and during the rainy season food stocks run low, food prices increase, and food consumption decreases. The rainy season is also a period of intensive agricultural work with higher energy needs, coupled with greater exposure to infectious diseases due to the wet conditions. Since women are actively involved in agriculture, particularly during the busy planting season, and children are most vulnerable to infections (especially diarrhea, malaria, and upper respiratory infections) during this period, the nutritional status particularly of these groups is negatively affected [32]. This pattern is not universal, but is common in rural areas of developing countries among poor smallholder farm households, who are too poor to buy the food that is needed when their own stocks are depleted.

At the seasonal to interannual time scale, the El Niño Southern Oscillation phenomenon is the most significant natural driver of variability in the climate system, impacting the food security and nutritional status of millions of people (see Box 9.3).

The seasonal presence of infectious diseases is another way in which climate-related effects on malnutrition are evident – the spatial and seasonal distributions of many human parasitic, viral, and bacterial diseases (such as malaria, dengue, and cholera) are determined by rainfall and

BOX 9.2 SEASONAL EFFECTS ON WOMEN AND YOUNG CHILDREN'S NUTRITION

FAO food security and nutrition projects in Afghanistan, Cambodia, Laos, and Zambia found that seasonal food shortages were common among poor households, especially those with smaller pieces of land [33]. During the lean season (the pre-harvest months) meal diversity decreased (reduced dietary diversity) and some households experienced reduced daily meal frequency. These effects were seen also in the complementary foods fed to young children, where meals consisted mainly of staple (rice or maize) and lacked micronutrients during these periods. Seasonal effects linked to women's workload were also evident, such as mothers' time constraints for preparing enriched complementary foods during the busy planting seasons and reduced meal frequency for children. Studies on women's seasonal work and pregnancy outcomes have suggested that low birth weights (an essential first step to address in order to reduce stunting at two years of age) are related to women's seasonal workload [34].

BOX 9.3 THE EL NIÑO SOUTHERN OSCILLATION (ENSO)

El Niño is the warming of sea surface temperatures in the tropical Pacific, which occurs roughly every two to seven years, lasting from 6 to 24 months. This oceanic event is associated with a fluctuation of a global-scale tropical and subtropical surface pressure pattern called the Southern Oscillation (together called the El Niño Southern Oscillation). While reduced rainfall and drought is a key outcome of El Niño, the phenomenon can also cause heavy rains and flooding. With climate change, greater frequency and intensity of extreme weather events, such as the ENSO, are expected [3].

The agriculture, food security, and nutritional status of more than 60 million people around the globe is affected by El Niño-related droughts, floods, and extreme hot and cold weather [36]. For example, based on a recent FAO assessment on the impact of El Niño in Sudan, 1.2 million vulnerable people were affected. More than 160 000 households produced less than 300 kg of cereals, almost 9000 households lost over 25% of their livestock assets (estimated at USD 6.8 million), and 130 000 households were affected by severe pasture and fodder shortage [36].

temperature [16]. For example, across the Sahel subregion, most childhood malaria mortality and morbidity occur during the short rainy season, and is referred to as seasonal malaria [35].

9.3.4 Impacts of Climate Change and Variability on Food Production, Nutrient Content of Foods and Biodiversity

9.3.4.1 Impacts on Food Production Climate change will fundamentally alter global food production patterns, with negative crop productivity impacts expected, especially in low latitude and tropical regions [37]. The negative impacts of climate change on the productivity of crops, livestock, fisheries and forestry are expected to become increasingly severe beyond 2030 in all regions [3]. Indeed, negative impacts of climate change on crop yields is already seen, based on many studies covering a wide range of regions and crops [26].

Productivity declines are expected to have serious implications for food security. Food supply shortfalls can lead to increases in food prices in local markets, while increased climate variability can accentuate price volatility. Since the areas most affected would be those with already high rates of hunger and poverty, food price increases will directly affect millions of low-income people. Among the most vulnerable will be those who depend on agriculture for their livelihood and income, particularly smallholder producers in developing countries [3]. In situations of food insecurity, poor people generally cope by spending the greater part of their income on staple foods and neglecting more nutritious foods, which can have a negative impact on diet quality.

9.3.4.2 Impacts on Nutrient Content of Food The nutrient content in foods can also be negatively affected by climate change. Higher carbon dioxide (CO_2) concentration is shown to lower concentrations of zinc, iron, and protein and raise starch and sugar content in crop plants that use three-carbon (C3) fixation pathway[10] such as wheat, rice and some legumes. This may contribute to worsening the prevalence of micronutrient deficiencies problems [38]. FAO food balance sheets showed that in 2010, roughly 667 million people were living in countries whose populations received at least 60% of their dietary zinc and iron from C3 grains or legumes. In India, where up to a third of the rural population is at risk of not meeting protein requirements, and who depend on C3 plants, decreased protein content due to higher CO_2 levels may have serious health consequences [39]. Food safety may be compromised by chemical changes that can increase the prevalence of toxic compounds in food. For example, cassava grown under high levels of CO_2 has been shown to have increased cyanide levels in leaves (cassava leaves are richer in protein and many minerals and vitamins than cassava roots, and are commonly consumed in many countries) [40].

9.3.4.3 Impacts on Biodiversity and Ecosystems The general global trend has been toward diet simplification, with a negative impact on food security, nutritional balance, and the growing incidence of chronic diseases. Biodiversity of terrestrial and aquatic ecosystems contributes to a more diverse diet that can reverse this unhealthy trend and improve dietary choices [41]. Genetic resources for food and agriculture help to maintain and improve agricultural productivity and rural livelihoods. Furthermore, they can play a key role in resilience and adaptability to climate change [42]. Climate change is threatening the reservoir of crop and livestock genetic resources from which to breed the varieties that will be needed to adapt production systems to future challenges. As conditions change, varieties and breeds may be abandoned by farmers and livestock keepers, and may be lost forever. Forest tree populations, especially in the tropics, are unlikely to be able to migrate (by pollen and seed) sufficiently quickly to keep pace with the changing climate [43].

Climate change can also lead to geographical shifts of pests and diseases, and changes in relations between species in ecosystems such as the balance between crops and weeds,

[10] C3 carbon fixation is one of three metabolic pathways for carbon fixation in photosynthesis.

pests, and pollinators. Desert locust habitats, breeding, migration, and plague dynamics in West Asia and North Africa are expected to be affected by changes in the climate during the remainder of this century, with greater risks of outbreaks. A single swarm of locusts can contain billions of insects and migrate across continents, consuming as much food in one day as 2500 people may need [44].

9.3.5 Impacts of Climate Change and Variability on Social, Health and Living Environments

Impacts of climate change on the health and living environments include reduced availability of clean drinking water, extreme weather events causing flooding or drought affecting hygiene, and prevalence of diarrheal diseases (strongly correlated with temperature), which adversely affect the uptake of micronutrients [45]. Collection of stagnant water can lead to an increase in the breeding of disease-carrying mosquitos [40, 46]. Contamination of foods such as groundnut and maize by mycotoxins is a major health and nutrition issue and changes in climate could increase human exposure to these toxins in the food chain. Mycotoxins, especially aflatoxins, have been implicated in the linear growth retardation in children [47]. Climate shocks further influence health by affecting the delivery of health care services and people's access to such services.

As previously mentioned, seasonal demands on women's time (due to periods of intense agriculture activity) can negatively impact cooking and caring practices and intrafamily food distribution, going on to affect the nutritional status of especially children and women. These will become more unpredictable as climate patterns change.

9.3.6 The Effects of Climate Change on Vulnerable People

Low-income and remotely located populations are more vulnerable to natural hazards, undernutrition, diarrhea and other infectious diseases. Climate change can lead to a reduction of production and consumption of some foods that play a critical role in the diets of vulnerable rural and indigenous populations, such as fish, fruits and vegetables, and wild foods [48].

In regions with high food insecurity and high inequality, increased frequency of extreme events will particularly affect poorer households, which have no means and assets to cope [26]. Women may also be disproportionately affected, given their vulnerability and restricted access to resources (see Box 9.4). Climate change will put at high risk indigenous people, who depend on the environment and its biodiversity for their food security and nutrition – specifically those living in areas where significant climate change impacts are expected, such as mountain regions, the Pacific islands, coastal and other low-lying areas, and in the Arctic [26].

BOX 9.4 EXTREME WEATHER IN GHANA AFFECTS WOMEN DISPROPORTIONATELY

"How is climate change different for women? ....Because women do different work, they have access to different resources and have different priorities in their lives in most societies" [49].

Women farmers in Ghana suffer the "gender gap" in development. Ghanaian women face persistent social inequalities. They have a 15–20% longer workday than men. Only 19% complete formal education, compared to 37% of men, and only 42% are literate, compared to 66% of men. Women farmers face the further challenges of limited access to credit, machinery, labor, fertilizer, and agricultural extension services, as well as exclusion from more revenue-generating industries like agroforestry. They have weak land-tenure rights.

A study in northeast Ghana showed that women subsistence farmers were disproportionately affected by drought and floods. Particularly vulnerable were single women who lacked the household labor to plant a labor-intensive crop like rice. They also could not harness the community support that married women could to help undertake house building and repairs. These women do not grow cash crops but farm for the immediate subsistence of their dependents, who are their children, the children of deceased relatives, and elderly, sick, or handicapped relatives incapable of supporting themselves. The women had on average 6.2 dependents; the greatest number supported was 17 [50].

9.4 THE IMPACTS OF FOOD SYSTEMS ON CLIMATE CHANGE

It is clear that climate change will have serious impacts on agricultural production and food security and nutrition. However, food systems are themselves major contributors to climate change, and need to be considered in the response to address climate change.

9.4.1 The Agriculture Sectors, Food Systems and Climate Change

Projections show that feeding a world population of 9 billion people in 2050 would require raising overall food production by some 60% [3]. High food output is often achieved at the cost of great stress on natural resources. The agriculture sectors specifically are a major source of greenhouse gas (GHG) emissions. Taken together, the agriculture, forestry, and land-use change (AFOLU) sector accounts for about one-fifth of global GHG emissions [3]. CO_2 emissions from agriculture are mainly attributable to changes in land use, such as conversion of forests to pasture or cropland, and land degradation such as that caused by overgrazing. Direct emissions of methane and nitrous oxide are the result of enteric fermentation in livestock, rice production in flooded fields, and the application of nitrogen fertilizer and manure [3]. The share of the food system as a whole in total global GHG emissions is even greater (estimated to be 19–29% of total global anthropogenic GHG emission) [51], as further emissions are generated especially from post-harvest processes such as transportation, processing, storage, and retailing, as well as from food loss and waste.

9.4.2 Contribution of Food Loss and Waste to Climate Change

Globally about one-third of food produced for human consumption per year is estimated to be lost or wasted [52]. This constitutes a waste of all the natural resources used for growing, processing, packaging, transporting, and marketing this food [53]. Global food loss and waste is a major contributor to climate change and is estimated to contribute to 3.6 Gt of CO_2-eq/year, with an additional 0.8 Gt of CO_2-eq/year from land use change, altogether adding up to about 8% of total anthropogenic GHG emissions [53]. Food losses occur due to inefficiencies in the food system such as insufficient access to energy and technologies, poor infrastructure and logistics, limited access to markets, etc. [54]. Food waste is highly related to consumer behavior, especially in wealthier countries.

9.4.3 Impacts of Global Food Demand and Dietary Pattern Changes on Climate Change

Dietary patterns strongly influence some of the factors that drive climate change. In countries where food consumption is increasing, diets generally include more livestock products, vegetable oils, sugar, and processed foods [3]. The production of certain kinds of food is more greenhouse gas-intensive per unit than cultivation of other foods. A recent meta-analysis of greenhouse gas emissions for different foods from life-cycle assessment[11] studies indicates a clear greenhouse gas emissions hierarchy across food categories, with grains having the lowest impact and meat from ruminants having the highest impact [55].

9.5 ENSURING HEALTHY FOOD SYSTEMS AND NUTRITION IN A CHANGING CLIMATE: RESPONDING TO THE CHALLENGES

As the previous sections have shown, climate change and variability can exacerbate vulnerability, the breakdown of food systems, and aggravate rural poverty, hunger, and malnutrition [18, 26, 56]. This section will present some of the key areas that hold the potential to significantly improve food security and nutrition, and enhance the resilience of food systems to the effects of climate change and variability, while reducing the associated GHG emissions. It will also explore the enabling factors to overcome barriers related to the adoption of practices and technologies that are critical to a transformational change in global food systems.

Actions towards reducing global GHG emissions are urgently needed to ensure that future food production is not increasingly compromised and that the challenge of adaptation to climate change and the aggregated costs do not accelerate further. However, transforming food systems towards a more sustainable, climate resilient and low-emission path presents major challenges. Urbanization will prolong the food supply chain due to increased distance between food production sites and urban consumers, which will have implications on nutrition. Population increases are expected to be concentrated in areas that are the most vulnerable to climate change and where the prevalence of undernourishment is the highest [3]. An increase in food production will therefore be needed to match the growing demand for food. At the same time, drastically reducing GHG emissions arising from global food systems will be crucial in order to maintain the global average temperature increase to below 2°C, which is the main goal of the Paris Agreement, which entered into force in 2016 and recognizes "the fundamental priority of safeguarding food security and ending hunger, and the particular vulnerabilities of food production systems to the adverse effects of climate change." This will require fundamental changes and readjustments of global production and consumption patterns in order to address the competing demands for land and natural resources and to increase production without compromising natural ecosystems.

Responding to these challenges calls for multiple interventions and requires collective actions towards an integrated food system approach that considers all risks, challenges, opportunities, and potential trade-offs. Many of the responses to climate change related to food systems present cross-cutting opportunities to address adaptation and mitigation objectives while simultaneously delivering other social, economic, and environmental co-benefits and goals of the 2030 Agenda for Sustainable Development.

9.5.1 Strengthening the Natural Resource Base and the Ecological Foundation of Food Production

Increasing the diversity within production systems greatly improves the ability of an ecological resource base to adapt to climate change and variability. As mentioned earlier, biodiversity can play a key role in strengthening the resilience of terrestrial and aquatic ecosystems that support the productive capacity of food systems. Approaches to increase ecosystem, species, and genetic diversity include

[11] Life Cycle Assessment (LCA) is a tool for the systematic evaluation of the environmental aspects of a product or service system through all stages of its life cycle.

the combination of different types of production (crops, forestry, livestock, and fish) and/or increasing the number of varieties or breeds.

Diversifying crop varieties or cropping systems increases resilience of production to a number of climate-related challenges. Furthermore, new varieties can be selected that offer improved yields per unit and nutritional content, thus contributing to more nutritious diets and the elimination of micronutrient deficiencies and undernourishment. Promotion of diverse, nutrient-rich foods such as vegetables, fruits, legumes, ASFs, and sustainable managed fish will require agricultural policies that go beyond staples.

Maintaining healthy terrestrial and aquatic ecosystems is also key to strengthening resilience and to ensuring that the processes that deliver ecosystem goods and services are not impaired. These goods and services include, but are not limited to, pollination, watershed protection, soil retention, and enrichment along with natural pest and disease resistance, which are all directly linked to the sustained delivery of healthy food systems and nutrition. Addressing transboundary risks through integrated pest and disease management and prevention will also play an essential role.

In order to safeguard the well-being of terrestrial and aquatic ecosystems it is vital to enforce good governance, management, and practices to prevent ecosystem degradation and to ensure that sustainable harvesting is monitored and controlled. Additionally, forests (including mangroves) are vital for their role in sequestering and storing CO_2, thus acting as a key carbon sink.

Water is another essential resource to food systems that is under threat from climate change and variability. Adapting to more frequent and widespread drought conditions and water scarcity is therefore critical and can be enabled through improved water management techniques and technologies, along with a shift to more drought resistant (and nitrogen-efficient crops), heat-tolerant varieties and breeds. Agricultural practices that enhance soil water retention and prevent erosion should also be considered as part of overcoming challenges, and many of these practices and techniques such as minimum tillage, agroforestry, etc., will have a positive impact on soil carbon sequestration.[12]

Putting in place systems to assess risks along with adaptation and mitigation options must be enhanced to better understand and enable climate change adaptation while simultaneously harnessing the mitigation benefits that many actions have to offer. This includes the application of assessments and tools to evaluate risks, challenges, and opportunities. Monitoring systems such as Early Warning Systems along with appropriate training, stakeholder consultations, and extension services (through, for instance, farmer field schools) are critical to ensure that information is conveyed to farming, pastoralists, and forest and fishing communities in timely, accessible, and understandable ways to increase their adaptive capacities and involvement [48].

[12] Atmospheric concentrations of carbon dioxide can be lowered either by reducing emissions or by taking carbon dioxide out of the atmosphere and storing in terrestrial, oceanic, or freshwater aquatic ecosystems. This is known as soil carbon sequestration. Available from: http://www.fao.org/soils-portal/soil-management/soil-carbon-sequestration/en/.

9.5.2 Post-harvest Systems

Approaches to address climate change must also consider the challenges and opportunities beyond the farm gate. For example, improving efficiency in post-harvest systems contributes to climate-smart food systems that can also deliver safe and nutritious food.

Insufficient access or lack of access to grid-energy[13] is prevalent in areas with high poverty and malnutrition. In poor rural off-grid areas, smallholders often do not have access to energy sources and technologies to preserve their produce or process it into higher value products, and market access is often limited. Cooking is often done using wood fuel and besides adding pressures on natural forest lands, the stoves are often highly inefficient and release smoke that has serious health impacts on women and children. Improving access to reliable and sustainable energy, technologies, and infrastructure are key aspects of strengthening participation in post-harvest systems and mitigate impacts from wood fuel use and deforestation. This would reduce the time spent on collecting wood, mainly conducted by women and children along with dangers sometimes associated with this. However, many of these technologies rely on fossil fuels to generate electricity or heat for the different stages of the supply chain. Energy usage in food systems is significant: food systems are estimated to use about 30% of the world's available energy, with more than 70% of that share being utilized beyond the farm gate [3]. Therefore, in order to make the transition toward climate-smart food systems, increased deployment of climate-smart energy and technologies is required.

9.5.3 Mobilizing Global Climate Action Through Food Loss and Waste Reduction

Food loss and waste reduction presents a cross-cutting opportunity for climate action as it addresses both adaptation and mitigation, with its collective contributions toward more productive, resilient, and low-emission food systems. Increasing food availability through food loss and waste reduction can be important for ensuring food security for the efficient use of natural resources and hence as an important mitigation strategy throughout the food system. Scaling-up renewable energy and energy efficiency measures and

[13] The power grid is the network for delivering electricity from suppliers to consumers.

optimizing resource efficiency along with clean technology transfer and infrastructure to effectively reduce food losses presents a key opportunity to enhance the emission reduction potential and boost resilience in food systems [57]. A combination of technological solutions and appropriate incentives are necessary in this context.

9.5.4 Enabling Frameworks for Action

Enabling the technical options described above will depend on whether they are supported by coherent policies that provide appropriate incentives and by institutional frameworks and finance mechanisms necessary to fund investment costs. Climate change policies and strategies are in place in most countries. Such policies generally take into consideration the agricultural and health sectors. However, there is a strong need to formulate detailed action plans to achieve climate objectives related to food security and nutrition. Further, although there is strong interconnection, nutrition considerations are generally absent in national climate change plans and strategies related to agriculture and health. Ensuring a better integration of nutrition considerations in climate change action plans and an equal consideration of climate change impacts (along with dietary impacts) on nutrition as part in the formulation of health and nutrition policies will be essential for future planning processes. What follows will briefly present the recent advances under the United Nations Framework Convention on Climate Change (UNFCCC) and how to ensure that future policy frameworks are aligning objectives related to agricultural development, food and nutrition security, and climate change adaptation and mitigation.

9.5.5 The Agriculture Sector and Food Systems in Countries' Intended Nationally Determined Contributions (INDCs)

Ahead of the 21st Conference of the Parties (COP 21) of the UNFCCC, countries prepared and submitted their national objectives on how they will address climate change in terms of mitigation, adaptation, and finance aspects, referred to as "Intended Nationally Determined Contributions (INDCs)." The INDCs served as a basis for the negotiations at COP 21 and provided the foundation for the Paris Agreement. While the INDCs are not binding, they are expected to act as a roadmap in driving national aims for post-2020 climate action. More importantly, this bottom-up process helped to ensure that national mitigation and adaptation priority measures were selected by countries themselves. An analysis of the INDCs [58] shows that the agriculture sector featured prominently in the INDCs, with more than 90% of all countries including them in their mitigation and/or adaptation contributions [3]; 39% of the developing countries have explicitly highlighted the importance of food security and nutrition as part of their adaptation priorities. There is, however, scope for better integration of nutrition considerations in future cycles of the Nationally Determined Contributions[14] (NDCs) as part of the plans for agriculture sector, food security, and health sectors.

Given the relevance of the agriculture sector in the INDCs, there is a strong imperative to support further integration of food systems along with food security and nutrition into national climate change planning processes and strategies. This includes the mitigation and adaptation instruments under the UNFCCC, such as Nationally Appropriate Mitigation Actions (NAMAs) and National Adaptation Plans (NAPs), which are important tools for defining actions that address climate change in food systems (and all other sectors). Here it is important to not only consider actions related to production or along the food value chain but also to take into account measures to ensure the alignment of dietary choices with climate targets.

Strengthening policy frameworks and institutional arrangements while ensuring coherence at a cross-sectoral and interministerial/interinstitutional level will be critical to leveraging both public and particularly private sector investments to enable a transformational change in food systems. Support must prioritize policies and institutions that can facilitate this transformation through inclusive and integrated approaches that consider all food value chain actors. The emphasis that countries, in their INDCs, have placed on food systems as part of their adaptation, mitigation, and technology priorities underscores the need to support developing countries, least developed countries (LDCs), and small island developing states (SIDS) in particular, to appropriately respond to climate change within food systems. While climate finance[15] is expected to remain a fraction of the overall investment necessary in food systems, it could potentially help to leverage additional public and private sector capital flows needed to implement change along food systems. In this context, the proportion of international climate finance allocated to the agriculture sector continues to remain modest compared to its potential contribution vis-a-vis that of other sectors. Scaling up finance to fund the investment costs needed to enable more productive, resilient, and low-emission food systems must be prioritized by the international community, if food and nutrition security under a changing climate is to be achieved.

[14] INDCs become NDCs when Parties submit their instrument of ratification, acceptance, or approval of the Paris Agreement.

[15] There is no single definition of "climate finance," but it may be loosely defined as all finance that, regardless of origin, contributes to climate change adaptation and/or mitigation objectives [3].

9.5.6 Promoting Healthy Dietary Patterns and Consumption

In the response to climate change, information, education, and communication (IEC) can be an important means of helping people change behaviors, in order to adopt more environment friendly lifestyles that can still provide nutritious diets that are sustainable. National food-based dietary guidelines (FBDGs)[16] that embed both health and sustainability objectives can form the basis of policies (including agriculture policies) seeking to foster consumption patterns that are beneficial to both the environment and consumers.

9.6 CONCLUSIONS AND POLICY RECOMMENDATIONS

The adverse effects of climate change pose serious threats to food systems and their current and future ability to deliver safe, adequate, and nutritious food to a growing population. Without concerted and coordinated actions to address these vulnerabilities, climate change could undermine or potentially reverse the progress achieved over the past decades toward eradicating hunger, malnutrition, and poverty. At the same time, more ambitious efforts to reduce emissions and increase efficiency throughout the food systems will be a key part of the global response to climate change. Reshaping the global food system calls for multiple interventions and coordination at various stages of the system, involving all actors. Enabling this transformation, however, requires increased investment flows, which in turn will depend on whether the supporting policies and institutional frameworks are in place and that they are complementary. Considering the challenges and opportunities highlighted in this chapter, key concluding recommendations are presented as priority areas for strengthening the linkages between food systems, nutrition, and climate action:

1. Improving nutrition through sustainable food systems is key to delivering mitigation and adaptation objectives along with other social, economic, and environmental co-benefits. However, in order to realize these objectives, there is a need for better dialogue between the climate change and nutrition community – at policy level, nutrition considerations must be increasingly mainstreamed into climate change planning processes, especially for the areas concerned with food, agriculture, and health. Climate change, climate variability, and related risks should equally be considered and addressed in nutrition policies and planning.
2. More holistic approaches are needed to reshape the global food system to a changing climate. While the focus tends to be on practices and activities at farm level, post-harvest processes are equally important for poverty alleviation, food security, and nutrition. There are multiple interventions that can enhance climate action in post-harvest systems that should be prioritized. This includes increased dissemination of climate-smart technologies (for storage, processing, cooling, etc.), reducing food loss and waste, and improving infrastructure and access to modern and affordable energy, which can altogether help to increase market linkages, value addition, and food security and nutrition.
3. There is also a need to consider and better value the cultural dimension and geographical origin of foods and diets that are based on long-established knowledge and practices. Not only do indigenous and local communities often conserve and use domestic and wild species sustainably in unique combinations of foods, thereby helping to conserve biodiversity of ecosystems and dietary diversity, but they also have traditional knowledge, expertise, and strategies for coping with climate variability and extreme weather events, which can provide a strong basis for adaptation planning. The promotion and protection of traditional and local food and agricultural knowledge require interventions at national and international level with intercultural and interdisciplinary approaches, communication, and cooperation, such as collaborations with the Commission on Genetic Resources for Food and Agriculture (CGRFA) for issues relating to biodiversity, or with the United Nations Educational, Scientific and Cultural Organization (UNESCO) and the World Intellectual Property Organization (WIPO) for promoting the use of Geographical Indications as a way to protect these types of traditional knowledge.
4. There is a general need for better integration and consideration of the multiple factors influencing the food system, as reflected in its cross-sectoral dimension. Promoting multistakeholder approaches that bring together actors from across sectors can help to strengthen linkages between various areas and improve policy coherence. This could also help to foster partnerships between public, private, and civil society entities and unlock finance to fund investment costs.

[16] FBDGs are a set of easily understood messages for the general public that are country-specific and take into account each country's food availability, diet patterns, and nutrition-related health issues. They provide advice on foods, food groups, and dietary patterns to provide the required nutrients to the general public to promote overall health and prevent chronic diseases.

5. Considering equity issues is also fundamental when formulating integrated strategies to strengthen food systems, address climate change, and improve nutrition. Such integration at the policy level presents multiple trade-offs and requires policy-makers to prioritize the socioeconomic status and well-being of poor and marginalized people. Social policy options, social protection in particular, must be aligned with actions throughout the food systems to address vulnerability to malnutrition and climate change. For poor households facing shocks, social protection measures can help sustain consumption levels, reduce negative risk-coping strategies and encourage higher-risk but higher-return livelihood strategies, altogether ensuring better food security and nutrition.
6. There is also a need to incorporate explicit considerations on gender equality and women's empowerment during the design of interventions. Men and women have different priorities and capacities within the food systems and in the response to climate change, and due attention needs to be paid to gender roles. Interventions that reduce gender inequalities and promote women's empowerment hold potential for improving the sustainability of food systems while also reducing vulnerability to the adverse effects of climate change.
7. Sustainable food systems and nutrition approaches should be further integrated into future NDC cycles and climate change agendas in general. International organizations such as FAO should continue to increasingly emphasize the role of food systems and nutrition in the climate change negotiations under the UNFCCC, including through inputs to the Subsidiary Body for Scientific and Technological Advice (SBSTA). Broadening the dialogue on climate action through food system interventions could also help to improve climate finance opportunities as an additional means of attracting finance. Furthermore, international donors and suppliers of climate finance should also consider increasing support to activities that benefit nutrition.
8. Capacity building to address the research and data gaps, particularly those related to nutrition in the context of climate change and variability, will be critical to enhance the evidence base for informed decision-making. Improving the assessments of vulnerable and marginalized populations (including indigenous people) and their adaptive capacities must be prioritized to ensure that risks and options are properly analyzed. Furthermore, improving data collection for GHG inventories will also be required as part of the enhanced transparency framework of the Paris Agreement, in which developing countries will need support for implementing appropriate reporting mechanisms.

REFERENCES

1. FAO (2008). An Introduction to the Basic Concepts of Food Security. Food Security Information for Action Practical Guides. Rome, Italy: FAO Available at: http://www.fao.org/docrep/013/al936e/al936e00.pdf.
2. FAO, IFAD, UNICEF et al. (2017). The State of Food Security and Nutrition in the World 2017. Building Resilience for Peace and Food Security. Rome, Italy: FAO Available at: http://www.fao.org/3/a-I7695e.pdf.
3. FAO (2016). The state of food and agriculture: climate change, agriculture and food security. Available at: http://www.fao.org/3/a-i6030e.pdf.
4. HLPE (2014). Food Losses and Waste in the Context of Sustainable Food Systems. A report by the High Level Panel of Experts on Food Security and Nutrition of the Committee on World Food Security, Rome, Italy, 2014.
5. FAO (2017). The Future of Food and Agriculture: Trends and Challenges. Rome: Food and Agriculture Organization of the United States Available at: http://www.fao.org/3/a-i6583e.pdf.
6. UNICEF, WHO, and World Bank (2015). Levels and Trends in Child Malnutrition. UNICEF–WHO–World Bank Group Joint Child Malnutrition Estimates. Key findings of the 2015 edition. Available at: http://www.who.int/nutgrowthdb/jme_brochure2015.pdf.
7. WHO (2014). Global Nutrition Targets 2025: Stunting Policy Brief. Available at: http://apps.who.int/iris/bitstream/10665/149019/1/WHO_NMH_NHD_14.3_eng.pdf.
8. Nutrition International (formally Micronutrient Initiative) (2015). Available at: http://www.micronutrient.org/.
9. WHO (2016). Micronutrient Deficiencies: Iron Deficiency Anaemia [website]. Geneva: World Health Organization. Available at: http://www.who.int/nutrition/topics/ida/en/.
10. WHO (2016). Micronutrient Deficiencies: Vitamin A Deficiency [website]. Geneva: World Health Organization. Available at: http://www.who.int/nutrition/topics/vad/en/.
11. WHO (2016). Micronutrient Deficiencies: Iodine Deficiency Disorders [website]. Geneva: World Health Organization. Available at: http://www.who.int/nutrition/topics/idd/en/.
12. WHO (2009). Global Health Risks: Mortality and Burden of Disease Attributable to Selected Major Risks. Geneva: World Health Organization Available at: http://www.who.int/healthinfo/global_burden_disease/GlobalHealthRisks_report_full.pdf.
13. WHO (2016). Obesity and Overweight. Factsheet No. 311. Geneva: World Health Organization. Available at: http://who.int/mediacentre/factsheets/fs311/en/.
14. Park, M.H., Falconer, C., Viner, R.M., and Kinra, S. (2012). The impact of childhood obesity on morbidity and mortality in adulthood: a systematic review. *Obes. Rev.* 13 (11): 985–1000.
15. UNICEF, WHO, and World Bank (2016). Levels and trends in child malnutrition: UNICEF/WHO/World Bank Group Joint Child Malnutrition Estimates. Key Findings of the 2016 edition. Available at: http://www.who.int/nutrition/publications/jointchildmalnutrition_2016_estimates.pdf?ua=1.

16. IFPRI (2016). Global Nutrition Report 2016: From Promise to Impact: Ending Malnutrition by 2030. Washington, DC: International Food Policy Research Institute Available at: https://www.ifpri.org/publication/global-nutrition-report-2016-promise-impact-ending-malnutrition-2030.
17. FAO (2013). The State of Food and Agriculture: Food Systems for Better Nutrition. Rome: FAO Available at: http://www.fao.org/publications/sofa/2013/en/.
18. IPCC (2014). Climate Change 2014: Synthesis Report. Contribution of Working Groups I, II and III to the Fifth Assessment Report of the Intergovernmental Panel on Climate Change. Geneva: IPCC.
19. WHO (2016). Climate Change and Health. Fact sheet. Available at: http://www.who.int/mediacentre/factsheets/fs266/en/.
20. IUNS, FAO, UNSCN, and WHO (2015). Sustainable Food Systems and Health: The Convenient Truth of Addressing Climate Change While Promoting Health. Brief written by Cristina Tirado, UCLA as background information for the side event organized by IUNS, FAO, UNSCN, and WHO at COP21, on 7 December 2015. Available at: https://www.unscn.org/files/Announcements/EXE_2_Sustainable_Food_systems_and_health.pdf.
21. Nelson, G.C., Rosegrant, M.W., Koo, J. et al. (2009). Climate Change: Impact on Agriculture and Costs of Adaptation. Washington, DC: International Food Policy Research Institute (IFPRI).
22. Parry, M., Evans, A., Rosegrant, M.W., and Wheeler, T. (2009). Climate Change and Hunger: Responding to the Challenge. Washington, DC: International Food Policy Research Institute (IFPRI).
23. Lloyd, S.J., Kovats, S., and Chalabi, Z. (2011). Climate change, crop yields, and undernutrition: development of a model to quantify the impact of climate scenarios on child undernutrition. *Environ. Health Perspect.* 119: 1817–1823.
24. Jones, G.D., Droz, B., Greve, P. et al. (2017). Selenium deficiency risk predicted to increase under future climate change. *Proc. Natl Acad. Sci. USA* 114 (11): 2848–2853.
25. Hou, X., Anderson, I., and Burton-Mckenzie, E.-J. (2016). Background paper. Pacific Possible: Health and Non-Communicable Diseases. World Bank. . Available at: http://pubdocs.worldbank.org/en/942781466064200339/pacific-possible-health.pdf.
26. IPCC (2014). Climate Change 2014: Impacts, Adaptation, and Vulnerability. Part A: Global and Sectoral Aspects. Contribution of Working Group II to the Fifth Assessment Report of the Intergovernmental Panel on Climate Change (ed. C.B. Field, V.R. Barros, D.J. Dokken, et al.). Cambridge and New York: Cambridge University Press.
27. FAO (2015). The Impact of Natural Hazards and Disasters on Agriculture and Food Security and Nutrition – A call for Action to Build Resilient Livelihoods. Rome, Italy: Food and Agriculture Organization of the United Nations Updated May 2015. Available at: http://www.fao.org/3/a-i4434e.pdf.
28. Dror, D.K. and Allen, L.H. (2011). The importance of milk and other animal-source foods for children in low-income countries. *Food Nutr. Bull.* 32 (3): 227–243.
29. Del Ninno, C., Dorosh, P.A., and Smith, L.C. (2003). Public policy, markets and household coping strategies in Bangladesh: avoiding a food security crisis following the 1998 floods. *World Dev.* 31: 1221–1238.
30. Watkins, K. (2007). Human Development Report 2007/2008. Fighting Climate Change: Human Solidarity in a Divided World. New York: United Nations Development Programme.
31. IDMC and NRC (2015). Internal displacement monitoring centre and norwegian refugee council. Displacement Due to Natural Hazard-Induced Disasters. Global estimates 2015. Available at: http://www.internal-displacement.org/assets/library/Media/201507-globalEstimates-2015/20150713-global-estimates-2015-en-v1.pdf.
32. Hoorweg, J., Foeken, D., and Klaver, W. (1995). Seasons and Nutrition at the Kenya Coast. African Studies Centre Research Series 7/1995. Avebury.
33. Wijesinha-Bettoni, R., Kennedy, G., Dirorimwe, C., and Muehlhoff, E. (2013). Considering seasonal variations in food availability and caring capacity when planning complementary feeding interventions in developing countries. *Int. J. Child Health Nutr.* 2: 335–352.
34. ACC/SCN (1989). UN Administrative Committee on Coordination/Subcommittee on Nutrition. Women's Role in Food Chain Activities and the Implications for Nutrition – Nutrition Policy Discussion Paper No. 4. Available at: https://www.unscn.org/web/archives_resources/files/Policy_paper_No_4.pdf.
35. WHO (2012). WHO Global Malaria Programme. WHO Policy Recommendation: Seasonal Malaria Chemoprevention (SMC) for *Plasmodium falciparum* malaria control in highly season transmission areas of the Sahel sub-region in Africa. Available at: http://www.who.int/malaria/publications/atoz/smc_policy_recommendation_en_032012.pdf.
36. FAO (2016). 2015–2016 El Niño: Early Action and Response for Agriculture, Food Security and Nutrition. Report working draft. Update #10. Available at: http://www.fao.org/3/a-i6049e.pdf.
37. FAO (2015). Climate Change and Food Systems: Global Assessments and Implications for Food Security and Trade. Rome: Food and Agriculture Organization of the United Nations Available at: http://www.fao.org/3/a-i4332e.pdf.
38. Loladze, I. (2014). Hidden shift of the ionome of plants exposed to elevated CO_2 depletes minerals at the base of human nutrition. *eLifee* 3: e02245. doi: 107554/Elife.02245.
39. Myers, S., Zanobetti, A., Kloog, I. et al. (2014). Increasing CO_2 threatens human nutrition. *Nature* 510 (7503): 139–142. doi: 10.1038/nature13179 Epub 7 May 2014.
40. UNSCN (2010). SCN News No 38. Climate Change – Food and Nutrition Security Implications, early-2010. ISSN 1564–3743. Available at: http://www.unscn.org/files/Publications/SCN_News/SCN_NEWS_38_03_06_10.pdf.
41. Medhammar, E., Wijesinha-Bettoni, R., Stadlmayr, B. et al. (2012). Composition of milk from minor dairy animals and buffalo breeds: a biodiversity perspective. *J. Sci. Food Agric.* 92: 445–474.
42. FAO (2015). Coping with Climate Change – The Roles of Genetic Resources for Food and Agriculture. Rome: FAO Available at: http://www.fao.org/3/a-i3866e.pdf.

43. Loo, J., Fady, B., Dawson, I., Vinceti, B., and Baldinelli, G. (2011). Climate change and forest genetic resources: state of knowledge, risks and opportunities. Commission on Genetic Resources for Food and Agriculture. Background Study Paper No. 56. Rome, FAO. Available at: http://www.fao.org/docrep/meeting/023/mb696e.pdf.
44. Cressman, K. (2013). Climate change and locusts in the WANA Region. In: Climate Change and Food Security in West Asia and North Africa (ed. M.V.K. Sivakumar, R. Lal, R. Selvaraju and I. Hamdan)), 131–143. Springer. doi: 10.1007/978-94-007-6751-5_7.
45. Wheeler, T. and von Braun, J. (2013). Climate change impacts on global food security. 341 (6145): 508–513. doi: 10.1126/science.1239402.
46. IFPRI (2015). Global Nutrition Report 2015: Actions and Accountability to Advance Nutrition and Sustainable Development. Chapter 6 Climate Change and Nutrition. Available at: https://www.ifpri.org/publication/climate-change-and-nutrition-0.
47. Smith, L.E., Stoltzfus, R.J., and Prendergast, A. (2012). Food chain mycotoxin exposure, gut health, and impaired growth: a conceptual framework. *Adv. Nutr.* 3: 526–531.
48. FAO (2016). Climate Change and Food Security: Risks and Responses. Available at: http://www.fao.org/3/a-i5188e.pdf.
49. WE Speak (2007). Women and global climate change. *Women and Environments International* 74/75: p. 4.
50. Glazebrook, T. (2011). Women and climate change: a case-study from northeast Ghana. *Hypatia* 26 ((4)): –762, 782. (Fall, 2011).
51. Vermeulen, S.J., Campbell, B.M., and Ingram, J.S.I. (2012). Climate change and food systems. *Annu. Rev. Environ. Resour.* 37: 195–222.
52. FAO (2011). Global Food Losses and Food Waste – Extent, Causes and Prevention. Rome.
53. FAO (2015). Food Wastage Footprint and Climate Change. Rome. Available at: http://www.fao.org/fileadmin/templates/nr/sustainability_pathways/docs/FWF_and_climate_change.pdf.
54. FAO (2016). How Access to Energy Can Influence Food Losses. A Brief Overview. Available at: http://www.fao.org/3/a-i6626e.pdf.
55. Clune, S., Crossin, E., and Verghese, K. (2016). Systematic review of greenhouse gas emissions for different fresh food categories. *J. Clean. Prod.* 140: 766–783. doi: 10.1016/j.jclepro.2016.04.082.
56. Krishnamurthy, P.K., Lewis, K., Choularton, R.J. (2012). Climate impacts on food security and nutrition. The Met Office and WFP.
57. Wieben, E. (2016). The Post-2015 Development Agenda: How Food Loss and Waste (FLW) Reduction Can Contribute Towards Environmental Sustainability and the Achievement of the Sustainable Development Goals. United Nations University Institute for Integrated Management of Material Fluxes and of Resources (UNU-FLORES), Dresden.
58. FAO (2016). The Agriculture Sectors in the Intended Nationally Determined Contributions: Analysis. Available at: http://www.fao.org/3/a-i5687e.pdf.

10

MALARIA IN THE BRAZILIAN AMAZON: NEW UNDERSTANDING AND DIRECTIONS FOR INTERVENTION

Marcia C. Castro[1] and Burton H. Singer[2]

[1] *Department of Global Health and Population, Harvard School of Public Health, Boston, MA, USA*

[2] *Emerging Pathogens Institute, University of Florida, Gainesville, FL, USA*

10.1 INTRODUCTION

Malaria transmission in Brazil is restricted to the Amazon region [1] and accounts for more than 50% of all cases in Latin America. It is characterized by a dynamic interplay of social, environmental, economic, political, and behavioral factors [2, 3], operating at multiple spatial and temporal scales [4]. Transmission patterns are likely to be impacted in the future by climate change, recently considered as the "biggest global health threat of the twenty-first century" [5–7]. In addition, continued changes to the local environment alter the nature and intensity of the human-vector contact [8]. Ongoing and future infrastructure projects, either aimed at promoting development of the region or to better respond and adapt to climate change, may exacerbate the burden of malaria if their negative impacts are not anticipated and mitigated [9–11]. Moreover, intense human mobility in the region contributes to the spread of malaria across the region [12–14].

Mitigation of malaria in the Amazon is no easy task. The Amazon Basin covers 7% of the Earth's surface and roughly half of South America. Extending over nine countries – Bolivia, Brazil, Colombia, Ecuador, French Guiana, Guyana, Peru, Suriname, and Venezuela, it has significant importance for the regional climate [15]. Brazil holds the largest fraction of the forest, approximately 70% [16]. Brazil itself is divided into five major administrative regions (North, Northeast, Center-West, Southeast, and South), and the legal Brazilian Amazon occupies 61% of the Brazilian territory. The Brazilian Amazon is comprised of the states of the North region, one state from the Center-West region (Mato Grosso), and most of Maranhão state (west of longitude 44°), located in the Northeast region (Figure 10.1). According to the last Brazilian Census (2000), 21.1 million people lived in the Amazon region, or 12.4% of the national population. In 2008, approximately 59 and 63% of the population in the Amazon had access to sanitation and clean water [17].

The Brazilian Amazon experienced its most significant population growth during the past four decades. Between 1970 and 2007, the region grew from 7.8 million people to 23.6 million, an average growth rate of 3% per year against 1.8% observed during the same period for the entire country [18]. This was mostly a result of in-migration in response to governmental incentives in support of agriculture, mineral extraction, and wide-ranging human settlement [3, 19–23]. During the same period, the Amazon also observed significant environmental change. In 1978, approximately 169.9 thousand km^2 of the forested area had been removed and by 2003 the cleared area amounted to 648.5 thousand km^2 (16.2% of the initially forested area of the Amazon) [24]. Much of the disturbances to the forest contributed to an increase in the number of water habitats suitable for *Anopheles* mosquitoes (the malaria vectors) breeding [25, 26]. Migratory movements in response to governmental incentives brought to the region a largely malaria naïve population [27, 28]. In such a dynamically changing

Water and Sanitation-Related Diseases and the Changing Environment: Challenges, Interventions, and Preventive Measures, Second Edition. Edited by Janine M. H. Selendy.

Companion website: https://www.wiley.com/legacy/wileychi/selendy/

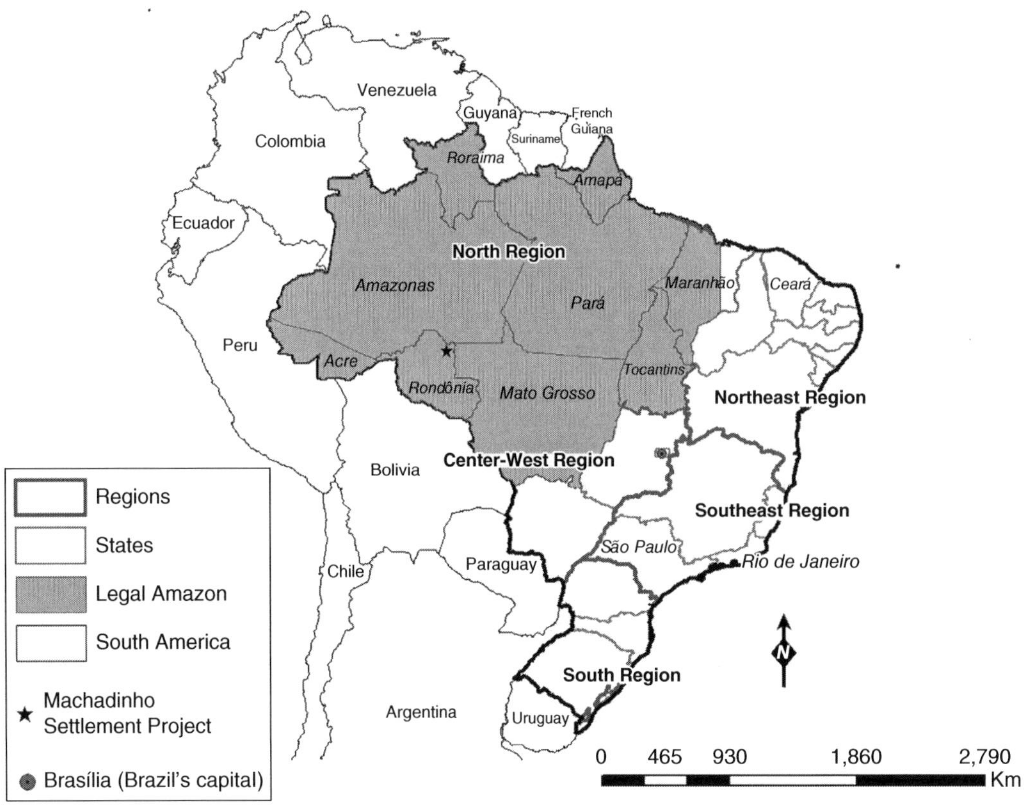

FIGURE 10.1 Major administrative divisions of Brazil (regions and states) and the location of the Brazilian Amazon.

environment, malaria took its toll. A steady increase in malaria cases was observed during the 1970s and 1980s, and the levels have fluctuated since the 1990s, most likely in response to new control strategies and, possibly, to the occurrence of extreme climatic events (Figure 10.2).

Since 1993, the Amazon region accounted for more than 99% of all malaria cases in Brazil. Cases are heterogeneously distributed in the region, mostly concentrated in agricultural settlement areas, mining camps, and areas of land invasion. A new concept of malaria (frontier malaria) was introduced to describe the complex factors, operating at multiple scales in space and time, which are associated with malaria risk in the Amazon. The formulation is multidisciplinary in nature and brings in the need to incorporate social sciences in the conceptual framework of transmission [2, 4, 29]. It is particularly suitable to describe the dynamics of malaria transmission in the rapidly transforming ecosystems of settlement areas in the Amazon, and has no parallel in other endemic areas.

In this chapter we discuss the process of malaria transmission in the Brazilian Amazon, characterizing its idiosyncrasies, framed within a historical context, highlighting current and future challenges for controlling the disease, and indicating important remaining knowledge gaps. Lessons taken from the Amazon case improve our understanding of the potential for malaria outbreaks consequential to environmental disturbance accompanied by inadequate planning for development of new settlement areas. Ultimately, the Amazon example sheds light on the choice of strategies that should be adopted in the future in order to minimize disease transmission, to prevent transmission following the construction of new infrastructure projects, and to mitigate potential changes in the pattern of malaria transmission due to extreme climatic events.

The remainder of this chapter is organized in six sections. We start with a brief review (Section 10.2) of important historical facts regarding malaria in Brazil generally and in the Amazon in particular. The objective is to facilitate understanding of current patterns of transmission. Section 10.3 contains a summary of evolutionary and behavioral aspects of *Anopheles* malaria vectors of importance in the Brazilian Amazon. Section 10.4 specifically addresses the drivers and characteristics of frontier malaria, and the following section provides an example of frontier malaria, based on a spatiotemporal analysis of a settlement project in the western Brazilian Amazon. Section 10.6 emphasizes some of the current and future challenges for malaria control in the Brazilian Amazon, highlighting areas where further research is urgently needed. Finally, we conclude with some remarks on lessons learned and on alternative strategies for mitigation of the current and likely future malaria transmission in the region.

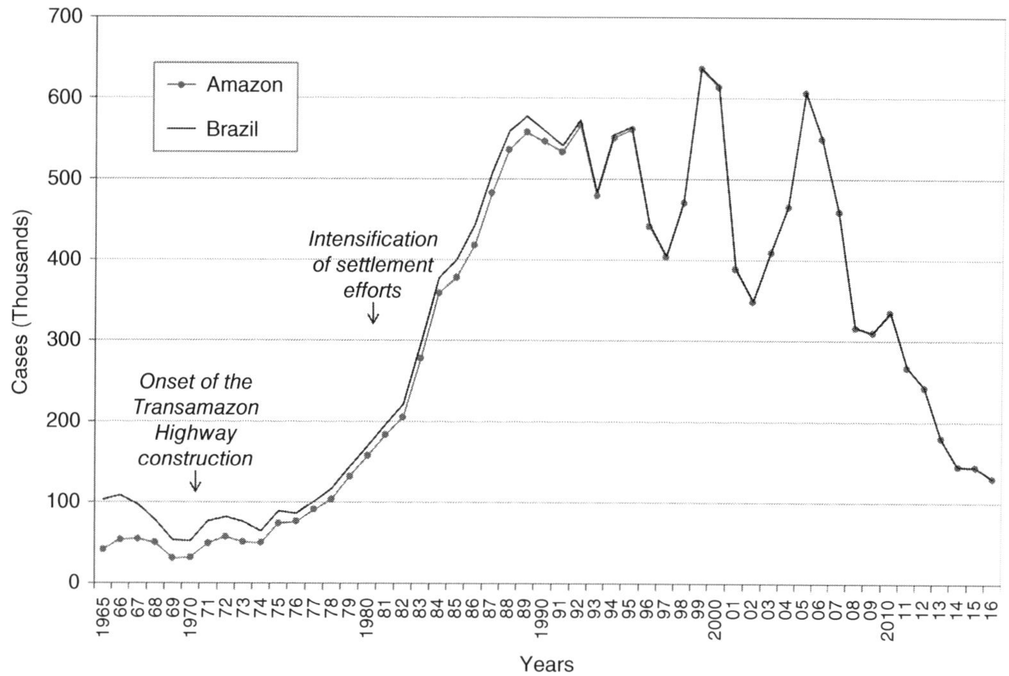

FIGURE 10.2 Number of malaria cases reported in Brazil and in the Amazon region, 1965–2016. *Source*: Brazilian Ministry of Health.

10.2 HISTORICAL PERSPECTIVE

Available evidence based on historical documents, studies in genetics and evolutionary biology, epidemiological issues, and archeological discoveries suggests that human malaria was not present in the New World prior to the European conquest [30]. The earliest references to "fevers" in Brazil are letters written by Jesuits in the sixteenth century [31–33]. One of the most detailed references dates back to 1587 [34], which provided evidence that at certain times and places, indigenous people suffered from tertian and quartan fevers. Usual treatment among the Portuguese colonizers was based on blood-letting and purging, and later on quinine [31]. The latter, however, was often overused [35], and the first cases of *Plasmodium falciparum* resistance to quinine were observed in Brazil as early as 1910 [36], including cases in the Amazon region [37].

The earliest map of the geographical distribution of malaria in Brazil was published in 1885 [38]. It indicated that the disease was present and endemic in areas along the major rivers of the country, particularly in the Amazon region, and less pronounced in the most southern portion of the country. This spatial pattern was corroborated by field observations collected along river valleys [39], and by observations of enlarged spleens among individuals living along the margins of major rivers from São Paulo state to the Amazon region [40]. In the early 1900s, Rio de Janeiro (the capital of Brazil at the time) and the states of the Amazon region were considered as areas where malaria transmission was severe [39].

In the case of the Amazon, a series of reports released in the early 1900s revealed the poor sanitary conditions of the region, and portrayed malaria as very severe in rubber plantations and proximal to rivers [37, 41–44]. Health assistance was precarious or totally absent, and malaria was considered to be the most severe disease prevailing in the region. A survey conducted among children from 6 months to 10 years of age in Manaus, the capital of Amazonas state, between 1905 and 1909, indicated that living in proximity of swamps considerably increased the risk of malaria. Blood examination through microscopy showed that 51.6 and 40.1% of children living near swamps in the suburbs and in the city, respectively, were infected with malaria parasites [44].

The sheer size of the Amazon region, local climate and hydrologic characteristics, and vector behavior are important challenges for controlling malaria. During the rainy season the water level of rivers increases dramatically, flooding the areas immediately proximal to the margins. As the rainy season ends, the water level decreases, and pools of water become widespread. Many of these offer ideal conditions for mosquito breeding. Larval control is not feasible given the size of the region, and the extremely difficult task of identifying and reaching all, or even a large majority of, breeding habitats. Hydraulic construction (ditches) near the margins of rivers, which would allow the accumulated water to flow back to the river, was proposed in 1916, but never implemented on a large scale [41]. Finally, an outdoor biting pattern has been observed for some Amazon malaria vectors since the mid-1980s and is likely to vary in intensity by location [45–49]. Although it is unclear which factor(s) triggered this exophilic behavior, some studies suggested a possible link with indoor residual spraying [50].

Nationwide antimalaria campaigns accompanied by engineering works (e.g. drainage and filling of marshes) were prominent in the first half of the 1900s, and facilitated the control of malaria in most regions of the country [51–53]. The Amazon, however, was an exception, and virtually all cases of malaria in Brazil are currently recorded in that region. In the remainder of this section we discuss multiple aspects that potentially led to this outcome.

10.2.1 The Construction of the Madeira–Mamoré Railway

The construction of the Madeira–Mamoré Railway, located in the state of Rondônia (Figure 10.1), is a good example of the diversity and magnitude of health challenges that forest disturbance can bring about [54]. The railway was planned to facilitate the flow of Brazilian and Bolivian products for export (mainly rubber). The closest city to the construction area was initially founded as a Jesuit mission in 1737, but later abandoned due to the prevalence of fevers. In 1910, the area had no infrastructure and during the rainy season was often transformed into a swamp [55].

After four failed attempts to construct the railway by different companies, mainly due to loss of workforce consequential to malaria and logistic difficulties [55–57], the final construction phase took place between 1907 and 1912. During that time, approximately 21 000 employees worked on the railway and probably more than 3000 died. Malaria was the major cause of incapacity for work in 87.5% of the cases. However, these numbers are likely to be underreported, since only workers in a critical condition would go to a hospital [55, 58, 59]. In 1910, a detailed set of recommendations for malaria control and prevention were proposed for the construction area, and this represented the first antimalaria campaign in the state of Rondônia [60, 61]. Among the recommendations were: the mandatory distribution of quinine, sanitary improvement, construction of screened houses for workers, avoidance of outdoors at dusk, and mandatory treatment of all sick people [60]. The Madeira–Mamoré Railway operated for 60 years, and the main cause of its deactivation was the decline in rubber export activity.

10.2.2 Control Efforts

The first official antimalaria campaign in Brazil[1] was implemented in the state of São Paulo in 1905, under the direction of the Oswaldo Cruz Institute,[2] in order to improve the health of employees of the port of Santos [64]. Although the campaign started with the use of quinine, the concept of malaria as a "household infection" (a disease acquired inside the houses and rarely outside) was proposed for the first time,[3] leading to novel approaches of malaria control [60, 65]. Mosquito screening inside the houses was done every eight days, nets installed in all windows, and in-house fumigation with sulfur, pyrethrum or tobacco adopted [62, 64]. It was the first time that a measure aimed at killing the adult mosquito was used [66], and the concept of malaria as a household infection became the basis for campaigns relying on indoor residual spraying. In 1910, Brazilian physicians introduced the concept of malaria as a local disease, suggesting that the success of interventions depended on local characteristics and therefore antimalaria campaigns should be developed in situ [67, 68]. Also, the idea of malaria as a disease confined to the tropics was refuted by Brazilian scientists, who claimed that the health status of countries across the globe were not dependent on latitude but on knowledge and local efforts [54].

In 1930, the highly effective African malaria vector, *Anopheles gambiae*, was discovered in Ceará state (Northeast region of Brazil; see Figure 10.1), most likely brought to the area by a French ship coming from Dakar, Senegal, in 1929 [69]. Precise identification of the invading mosquito was not possible at that time, since *An. gambiae* was recognized as a complex of seven species only in the 1960s. Recent DNA analysis of preserved museum species collected between 1932 and 1940 in Ceará suggest that *An. arabiensis* (a mosquito of the *An. gambiae* complex) is likely to have been the invading species [70].

In 1938–1939, as a direct consequence of the introduction of *An. gambiae*, there were 150 thousand malaria cases and 14 thousand deaths in only eight months [69, 71]. With the support of the Rockefeller Foundation, control efforts started at the end of 1938. Approximately 4 thousand workers were employed during 19 months, and measures of control included monthly house fumigation with pyrethrum, early case detection and rapid treatment, spraying of cars and trucks leaving or entering the endemic area, elimination of breeding sites, spreading Paris green on mosquito breeding habitats, and house capture. In 1940, *An. gambiae* had been eradicated from Brazil [72–74]. A similar campaign was repeated in Egypt in 1943–1944, inspiring the eradication strategy that marked the World Health Organization (WHO) agenda between 1955 and 1969 [75, 76].

[1] In 1898, shortly before Ronald Ross's discovery that *Anopheles* transmitted malaria, Adolfo Lutz (1855–1940), a Brazilian scientist and naturalist, suspected that malaria could be associated with mosquitoes, and recommended that houses for employees working on the construction of the São Paulo–Santos railway should be built far from the forest, where mosquito breeding sites were located [62].

[2] The Oswaldo Cruz Institute, founded in 1900, was the first research institute in Brazil's history. Research results from the Institute led to Brazil's international reputation in science [63].

[3] The concept was proposed by Carlos Chagas (1878–1934), a Brazilian physician who coordinated many successful antimalaria campaigns in Brazil, internationally known for the discovery of Chagas disease.

In the early 1940s, a successful environmental management strategy eradicated malaria transmitted by a bromeliad-breeding mosquito from the southern states of São Paulo, Paraná, Santa Catarina, and Rio Grande do Sul. Bromeliads were gradually removed from the urban areas and eucalyptus trees (on which bromeliads do not grow) were planted in forested areas [72, 77]. This intervention was based on the findings of Adolpho Lutz, who first discovered and described forest malaria and its vectors, as reported in 1903 [78, 79].

Nationwide antimalaria campaigns started in 1940, when the number of malaria cases was on the order of 4 to 5 million (out of a population slightly larger than 41 million) [72]. In 1945, DDT (dichloro diphenyl trichloroethane) was first utilized as an adulticide [80], and in 1947 it was used nationwide to spray the interior of houses every 6 months [69, 72, 81].[4] In the Amazon, DDT spraying started in 1945 in one municipality of Pará state. It was expanded to 27 and 91 municipalities by 1947 and 1948, respectively, located in the states of Amazonas, Pará, Amapá, and Rondônia (former Guaporé) [83, 84]. Efforts were initially mostly concentrated in capitals and major cities, progressively targeted to areas along streams (a common breeding ground for the anopheline) [84]. The DDT program continued to expand, favored by an adequate number of well-paid personnel [85], and in 1959 it formally became a component of the national control effort [86].

In 1953, six years after the onset of DDT use, the number of confirmed malaria cases in 11 localities in the South, Southeast, and Northeast regions of Brazil was reduced by 99% [87]. In an effort to tackle the malaria problem in the Amazon more efficiently, an unconventional method was introduced in 1953: distribution of chloroquinated salt to the population [72, 88]. The method was incorporated by the WHO as part of the strategies for malaria eradication, and 13 countries in South America, Africa, and Asia used medicated salt between 1952 and 1978 [89]. However, difficulties in the distribution and storage of the salt, as well as evidence of *P. falciparum* resistance to chloroquine both in Brazil and other countries, led to an interruption of the intervention [85, 89–91].

The total number of malaria cases in the country continued to decrease after the introduction of DDT, reaching its minimum level in 1970 when 52 371 cases were registered, 31 733 of them in the Amazon. That time, however, coincided with the commencement of important changes in the geopolitical strategy adopted by the Brazilian government [92, 93], which would bring about some of the most significant transformations of the Amazon region, both socially and environmentally. Migration to the Amazon, and most important, the human mobility inside the Amazon region, became intense. Financial and human resources were scarce, and the type of housing (e.g. with partial walls or walls and roof made of tree leaves) in some areas imposed serious constraints on indoor spraying [69, 85].

[4] DDT was prohibited for agricultural use in 1985, but continued to be used in antimalaria campaigns until 1997 [82].

10.2.3 Opening of the Amazon Frontier

Historical occupation of the Amazon rarely promoted settlement in the area [94]. Instead, economic booms (e.g., rubber, gold, minerals) often brought massive numbers of temporary migrants, which invariantly resulted in malaria epidemics [21, 58, 95, 96]. Although previous attempts to integrate the Amazon with the rest of country took place before 1960, it was not until that year, when the federal capital was moved from Rio de Janeiro to the interior of Brazil (Figure 10.1) and the first road into the Amazon was opened, that the modern Amazon frontier expansion was commenced [21, 97]. The major goal was to build roads linking the coast with the interior of Brazil, promoting national development and increasing the industrial power of the country [98].

After 1964, during the military dictatorial government, geopolitical strategies for integration, and ultimately for the strengthening of national security, included a phase of intense occupation of the Amazon [92, 99]. The construction of highways continued in the late 1960s and early 1970s [20, 100, 101]. Following the opening of roads, large numbers of migrants started to arrive in search of land and employment, putting pressure on services, infrastructure, and on the environment. The human population of the Amazon grew from 7.2 million in 1970 to 11 million in 1980 and then to 18.7 million by 1996. A major attempt to organize the occupation of the Amazon was the Northwest Region Integrated Development Program – POLONOROESTE [102].

Focusing on the area occupied by Rondônia and part of Mato Grosso states, POLONOROESTE was established in 1981, partly financed by the World Bank. It was a complex and ambitious project, designed to avoid problems faced by previous development efforts in the Amazon. However, a combination of factors contributed to early failures, including massive migration, lack of infrastructure, economic recession and fiscal crisis in the country, inability of government agencies to coordinate and execute the different phases of the project, lack of adequate information regarding soil quality of planned settlement areas, and absence of a detailed study regarding the sustainable carrying capacity of the project [102–104]. Among the most serious consequences of POLONOROESTE were very high rates of deforestation, conflicts with indigenous populations, and severe outbreaks of malaria [105]. These outbreaks resulted from a combination of factors that included human-made transformations to the environment, lack of acquired immunity among most of the settlers, precarious habitat conditions that offered no protection against mosquitoes, and modified vector behavior that included outdoor biting and peak biting times at dawn and dusk [3, 26].

After malaria reached its lowest level in Brazil in 1970, an increasing trend was observed following the rise of malaria in the Amazon region (Figure 10.2). In 1986, when approximately 443 thousand cases of malaria were reported in Brazil, Rondônia state was considered the "capital" of malaria in Brazil, containing 43% of all cases registered in the nation, and 46% of all cases in the Amazon. In 1989, more than 577 thousand cases were observed in the Amazon and a new control strategy was adopted (as detailed later).

10.3 *ANOPHELES* VECTORS IN THE AMAZON

There is considerable variation among current estimates of the origin of *An. darlingi*, the primary malaria vector in the Amazon basin of Brazil. Recent work places the expansion out of the central area at 304300 years BP, and out of the southeast and south areas at 121492 years BP [106]. This places expansion out of their ancestral groups in the late Pleistocene. In contrast, Mirabello and Conn [107] estimated the expansion of *An. darlingi* throughout South America at 25311 years BP. These estimates for the vector have corresponding widely varying counterparts for expansions of *Plasmodium* in South America [108]. The vector and *Plasmodium* timings for expansions are both estimated to be in the thousands of years, in contrast to the limits of actual documentation of human malaria in the region to hundreds of years [30]. The direct human evidence, however, does not include DNA-based studies and this gap is likely responsible for the substantial, qualitative disparity between vector, *Plasmodium*, and human estimates. This point notwithstanding, we are still left with great uncertainty about the evolution of the components of the malaria transmission system in the Brazilian Amazon. Much more fine-grained geographic and molecular-level studies than heretofore will be required to reduce the extant levels of uncertainty.

Given the continental scale distribution of *An. darlingi*, there is also significant additional variability due to biting preferences or resistance to control measures. Geographical barriers such as rivers and mountain ranges have contributed to substantially diversify *An. darlingi* in different areas of its range [106, 109]. An evaluation of the population genetics and historical expansion of *An. darlingi* in South America identified six population groups: Colombia, central Amazonia (geographically located around the Amazon River, covering most of the central and western Amazon basin, extending to Iquitos in Peru), southern Brazil (that actually spans from Pará to Paraná states), southeastern Brazil (Rio de Janeiro and Espírito Santo states), and two distinct groups in the northeast of Brazil. Based on orderings (in time) of *An. darlingi* haplotypes, the two oldest haplotypes are only distributed in Amazonian populations on either bank of the Amazon River (central group) [106]. More recently, three genetically diverse clusters were detected: (i) Atlantic Forest province, (ii) Parana Forest province, and (iii) Brazilian dominion population. Dispersal among these clusters was limited by the coastal mountain range of the Atlantic Forest and by the Chacoan dominion vegetation region (known as the Chaco, Cerrado, and Caatinga biomes), and it is possible that these three genetic clusters represent three putative species [110].

10.3.1 Classification, Behavior, and Genetic Structure

In a review of research on vectors of malaria in Brazil, Deanne [111] listed more than 50 species of anophelines (Table 10.1) that had been described and classified, mostly on the basis of morphological features, from 1900 until the early 1980s. Of those species of anophelines, all known vectors of human malaria belong to the subgenera *Nyssorhynchus* and *Kerteszia*. These are: *An. darlingi*, *An. aquasalis*, *An. albatarsis*, *An. triannulatus*, *An. nuneztovari*, *An. oswaldi*, *An. strodei*, and *An. evansae* from the *Nyssorhynchus* subgenus, and *An. cruzi*, *An. bellator*, and *An. homunculus* from the *Kerteszia* subgenus. As indicated previously, the most important primary vector is *An. darlingi*, found throughout the inland malarious areas of the legal Amazon and southward to the northern part of Paraná state (South of Brazil).

An. darlingi can maintain endemicity even at low densities [45] and is the only vector that can transmit all three malaria parasites found in the Amazon [112]. Surveys conducted in the Amazon between July 1942 and June 1946, before DDT was introduced in the region, indicated that *An. darlingi* represented 21% of all anophelines captured and 66% of all anophelines captured inside the houses. In addition, among all *An. darlingi* captured during the survey period, 88% were found indoors [113]. Therefore, before the large-scale indoor residual spraying with DDT (as described in Subsection 10.2.2), *An. darlingi* was mainly an endophilic vector, in contrast to the prevalent outdoor pattern observed since the mid-1980s [45–49]. Although it is unclear which factor(s) triggered this exophilic behavior, a few studies suggested a possible link with the use of DDT [50]. This could be a result of mosquito behavior change or of a selection process favoring mosquitoes that bite and rest outdoors. On the one hand, *An. darlingi* has been considered as a truly anthropophilic and endophilic mosquito, based on observations from several areas in South America. On the other hand, when analyzed in interior regions, where human population density is low and relatively mobile, the mosquito reveals a zoophilic and exophilic behavior [46, 85, 114].

Microsatellite analyses of Colombian *An. darlingi* populations indicated low genetic differentiation among samples collected in five geographical areas [115]. However, high-resolution molecular typing techniques, such as ddRADseq-derived single nucleotide polymorphisms (SNPs) are needed to capture microgeographic genetic variations [110, 116]. Using SNPs on *An. darlingi* populations from different locations in Brazil, ranging

TABLE 10.1 Anophelines found in Brazil.

Genus *Anopheles*	Subgenus *Nyssorhynchus*	*An. darlingi, An. argyritarsis argyritarsis, An. argyritarsis sawyeri, An. albitarsis, An. braziliensis, An. lanei, An. aquasalis, An. oswaldoi, An. ininii, An. dunhami, An. evansae, An. rangeli, An. nuneztovari, An. benarrochi, An. galvãoi, An. strodei, An. triannulatus triannulatus, An. triannulatus davisi, An. rondoni, An. lutzi, An. parvus, An. antunesi, An. nigritarsis.*
	Subgenus *Kerteszia*	*An. cruzi cruzi, An. cruzi laneanus, An. homunculus, An. bellator, An. bambusicolus, An. neivai.*
	Subgenus *Anopheles*	*An. eiseni, An. peryassui, An. mattogrossensis, An. tibiamaculatus, An. maculipes, An. mediopunctatus, An. punctimacula, An. intermedius, An. fluminensis, An. minor, An. shannoni, An. neomaculipalpus, An. evandroi, An. bustamantei, An. rachoui, An. anchietai.*
	Subgenus *Lophopodomyia*	*An. squamifemur, An. gilesi, An. pseudotibiamaculatus.*
	Subgenus *Stethomyia*	*An. nimbus, An. kompi, An. thomasi.*
Genus *Chagasia*		*Ch. fajardoi, Ch. bonneae, Ch. rozeboomi.*

Source: Deane [111].

from rural to urban settings, Campos et al. found genetic differentiation and population divergence at the microgeographical scale [116]. Specifically, analyses of *An. darlingi* populations collected outdoors in rural settlement areas 60 km apart and in an urban area about 700 km from the rural areas indicated that: (i) populations from rural areas had higher genetic diversity than the urban one and (ii) genetic diversity was higher in the younger rural population, which has larger patches of forest still standing. These results corroborate the hypotheses that deforestation and loss of genetic diversity are associated [116].

10.4 AGRICULTURAL SETTLEMENT AND ENVIRONMENTAL CHANGE

As of 2017, about 2900 agricultural settlements had been opened in the Brazilian Amazon, reaching more than 520 thousand families [117]. The settlements varied in terms of available resources, average size of land, and soil quality. Impacts were equally varied, such as rate of turnover, health outcomes, and patterns of deforestation and land use driven by a multitude of factors operating at varied temporal and spatial scales [118–120].

In any settlement project, the first environmental transformation is the process of clearing a small area for house construction and preparation for cultivation. The most common agricultural practice is slash-and-burn, which increases levels of soil pH, phosphorus, calcium, and magnesium, and decreases levels of toxic aluminum ions. A poor clearing and burning compromises the crop yields, can cause the obstruction of steams, and is likely to leave taller trees standing, providing the necessary partial shade for *An. darlingi* breeding [121]. This brings up the notion of forest fringe, a frontier between the forest and the property, where anopheline density is very high and the risk of malaria transmission is augmented [26].

In this rapidly transforming Amazon ecosystem, characterizing malaria risk requires a new concept of malaria that considers biological and ecological phenomena acting at multiple spatial scales, juxtaposed with behavioral and economic conditions. This concept is defined as frontier malaria [2, 122], and operates at three spatial scales [4]. First, at a micro/individual level, vector densities are high, as a consequence of ecosystem transformations that promote *An. darlingi* larval habitats (partial shade near the forest fringe and along river edges, clear standing water of high pH) [49, 123, 124]. Human exposure is intense, reflecting limited knowledge of transmission among settlers. *An. darlingi* has a bimodal biting pattern [47] – at dawn and dusk – just when settlers are going to and returning from their fields. *Plasmodium falciparum* is initially the primary parasite, although progressively *P. vivax* becomes more common. Immunity is low among new settlers (who mostly come from malaria-free areas), morbidity is high, but *P. falciparum* mortality is low. The relatively low virulence of *P. falciparum* in the Amazon suggests that it had a quite different evolutionary path compared to African strains [125]. Housing quality is poor, thereby rendering indoor residual spraying ineffective. Curative health services are sparsely available, thus limiting antimalarial drug distribution.

Second, at a community level, frontier malaria is characterized by weak institutions, minimal community cohesion, political marginality of the settlers, and high rates of both in- and out-migration. This combination of conditions severely limits organized attempts at ecosystem management to minimize malaria risk and development of health clinics. The human mobility ensures proliferation of parasites. Third, at a state and national level, frontier malaria is characterized by unplanned development of new settlement areas, stimulated by agricultural failures at previous settlement localities and by a desire of people to avoid further malaria episodes. This process, however, only serves to promote further transmission.

Frontier malaria was also conjectured to follow a distinctive time path [29]. At the opening of a settlement area, malaria rates rise rapidly, and the first two levels of the above spatial characterization are fully operative. After six to eight years, the unstable human migration (both in and out) and the highly variable ecological transformations (driven by variation in land clearance practices and local ecology) is replaced by a more organized process of urbanization and development of community cohesion. Deforestation, which is extremely intense in the initial years of occupation, reduces over time since fewer patches of the forest remain intact. This process likely reduces mosquito density. Thus, frontier malaria is gradually replaced by more stable low levels of transmission and lower malaria rates. The process of urbanization itself (especially the introduction of impervious surfaces and drains) is an important intervention, as it creates environments that are inhospitable to *An. darlingi* larvae, and that are increasingly remote from forest fringes, thereby substantially reducing human exposure.

10.5 FRONTIER MALARIA: AN EXAMPLE

To illustrate the context in which frontier malaria emerges, as well as its spatial and temporal characteristics, we use the Machadinho settlement project (Figure 10.1) as a case study. Started in late 1984, Machadinho was one of the settlement areas promoted by POLONOROESTE [102]. The area was primarily jungle before the settlement started, but sparsely populated by rubber tappers. It was the first settlement project that incorporated a plan of action to prevent the most harmful consequences of frontier expansion observed previously in the Amazon [27]. Instead of adopting the traditional fishbone pattern frequently used in most settlement projects in the Amazon, Machadinho had an original and carefully planned plot design (Figure 10.3). The shape of the plots was irregular, following the course of rivers and streams, so that every plot would have a natural source of water in its rear. Roads were planned in such a way that during the rainy season the increased water volume of rivers and streams would not leave them impassable. Areas with very irregular elevation were assigned as protected forest, and therefore not subjected to settlement or forest clearance. In total, the project had an area of 209 960 ha, of which approximately 68 000 ha were assigned as forest reserves. Such a design can only be implemented in areas of jungle, where there is no previous construction or agricultural practice in place [103].

In addition, the project included the construction of auxiliary roads, schools, health units, governmental agencies involved in rural development, forestry control posts, commercial and recreational areas; land use planning; execution of complementary soil analyses; provision of financial support for the acquisition of seeds and initial equipment; buildings for crop drying and storage in accordance with the local climatic and agricultural conditions; protection of natural parks and reserves, water sources, and consideration of endangered species; acquisition of all necessary equipment for schools; and establishment of community organizations. Finally, the plans included a health project, which had three main goals: strengthen malaria control, set up a network of health care facilities, and stimulate research [104].

However, a combination of flaws in the planning and implementation processes, and adverse economic conditions resulted in major social, environmental, and health problems in Machadinho [13]. By the time POLONOROESTE was approved (December 1981), Brazil was experiencing serious economic constraints: rising inflation, a critical balance of payments, and trade balance deficits. In 1982, Brazil started an economic stabilization program resulting in resource cutbacks. As a consequence, the federal government could not provide its share of financial resources for POLONOROESTE. Technical and financial support to new settlers was compromised [104]. Assistance with housing construction was limited due to scarce resources and only 30% of the families benefited from the distribution of wood to build a house [126]. The majority of settlers who did not receive assistance relied on plastic, palm thatch, cardboard, and other precarious materials to build a temporary house, which had extremely poor protection against mosquitoes, if any at all (Figure 10.4). Construction of schools had serious delays in the early years of the project and there was a shortage of teachers and furniture, all resulting in constraints on the provision of education for all children [127].

Regarding healthcare services, in 1985 there was only one health post and a mobile unit that was not sufficient to cover the entire area [128]. The activities of health agents were also compromised: in 1986 only one car and five bicycles were available [127]. Training was deficient and health professionals were scarce. Finally, soil quality was much lower than initially anticipated: 95% of plots assigned to settlers lacked adequate soil fertility at some level, and 58% of the area had good suitability for agricultural practice only if mechanization and intensive investment to improve the land were implemented [103, 129], inputs that poor settlers could not afford.

These problems were augmented by the local context. On average, settlers moving to Machadinho were low-income migrants (mainly from the southern portion of Brazil), with some agricultural experience (but little knowledge of agricultural potential or the techniques necessary for farming in a tropical rain forest area), and with low education. Most had no previous exposure to malaria, and consequently no acquired immunity against it, and very little knowledge of malaria transmission, prevention, and treatment [103]. Regarding the use of inputs in agriculture, only 28.6 and 33.5% of the settlers had a chainsaw in 1985 and 1986, respectively. As a result, the speed and quality of land clearance near the house was far from ideal. The number of

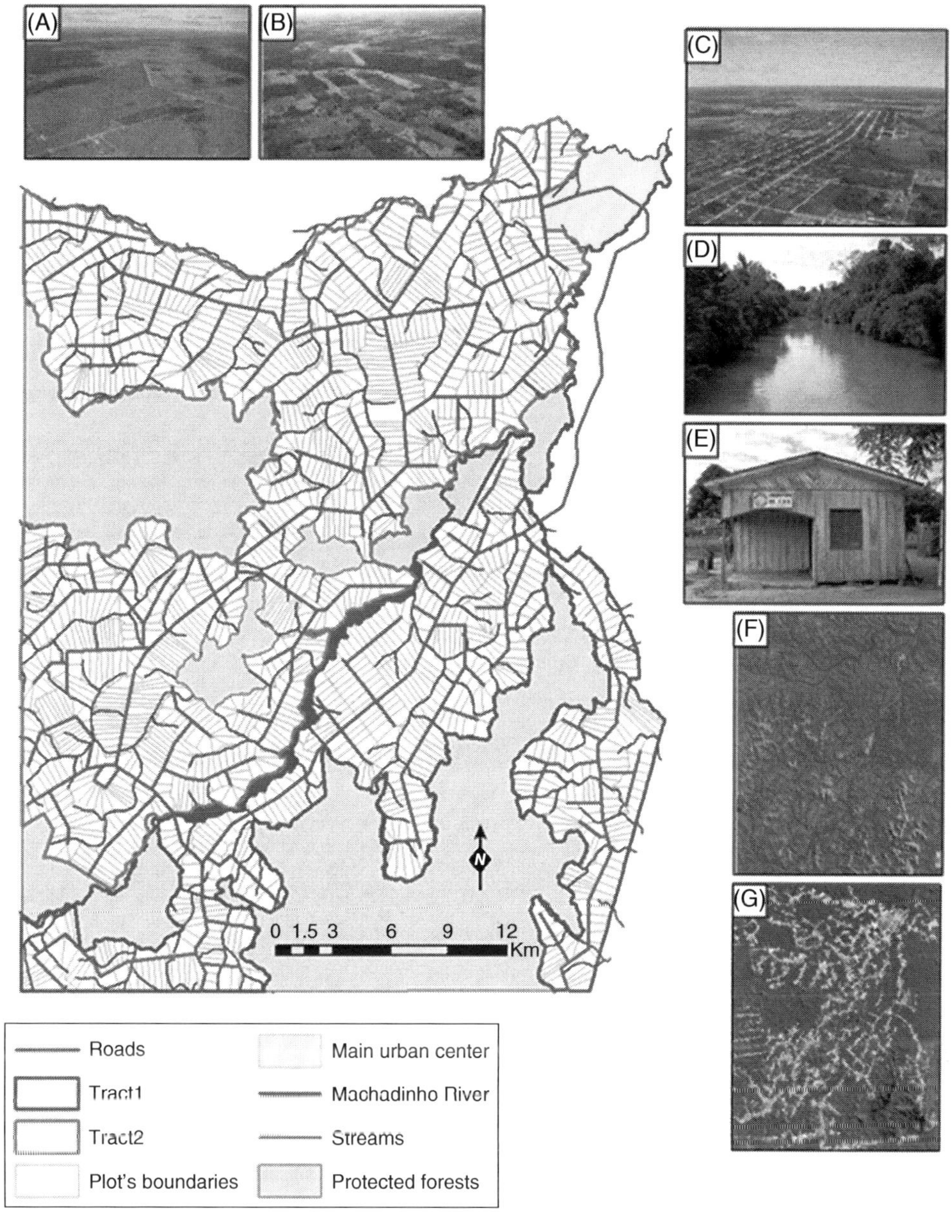

FIGURE 10.3 Location and design of the Machadinho settlement project. (A) Area with pasture. (B) Area with diversified crop production and patches of original forest. (C) Aerial view of the main urban center. (D) Machadinho River. (E) Laboratory where blood slides are prepared and analyzed. (F) and (G) Satellite images taken in 1985 and 1995, respectively (Landsat TM, 30 m of resolution).

partially shaded water bodies suitable for *Anopheles* breeding proliferated, and settlers became highly exposed to malaria transmission due to proximity to the forest fringe (Figure 10.5). Rubber tappers living in the protected forest reserves of Machadinho were most likely asymptomatic to malaria and may have served as reservoirs of malaria parasites [103].

An outbreak of malaria was soon observed, following the arrival of the first settlers in late 1984. In 1985, the Annual Parasite Index (API, the number of positive blood slides per total population) reached 3400 positive slides per thousand people, 65.7% of the population had malaria at least once, and this number jumped to 90.1% in the next year. Also in 1986, 55.9% of people had malaria episodes in more than five months of the year [27, 127]. By 1995, however, the level of transmission had been reduced, following the consolidation of the settlement project and corroborating the transition pattern of frontier malaria previously proposed [29].

Four household surveys conducted in 1985–1987 and 1995 appraised levels of malaria infection through self-report information and gathered data that allowed the identification of spatiotemporal malaria risk profiles based

FIGURE 10.4 Examples of housing in Machadinho. (A) House constructed with wood distributed by INCRA in 1984, photographed in 2001, when it was abandoned, without windows and without the original door. However, the overall sealing of the walls was of superior quality in comparison with houses that had plastic or thatch on the walls. (B) House built in 1985, all made of palm thatch. (C) House built in 1985; there is no door closing the house. (D) Rubber tappers housing, located in the border between forest reserves and plots, near streams (the canoe is used as a means of transportation).

FIGURE 10.5 Forest clearance and potential breeding habitats. (A) Forest fringe in 1986, the limit between the forest and the cleared area, located a very short distance from the house. (B) Water body ideal for *Anopheles* breeding, common after forest clearance, which leaves small streams exposed to partial shade and partially clogged by fallen trees.

on environmental, behavioral, and economic characteristics [4]. Important findings included: (i) in the initial years of settlement, environmental transformations and conditions dominated the high malaria risk profiles; (ii) by 1995, risky conditions essentially reflected personal behavior and economic circumstances, as the substantial land-clearance process left much of Machadinho relatively inhospitable to *An. darlingi* larval development; (iii) low-risk conditions were initially dominated by behavioral and economic measures, particularly reflecting ownership of capital equipment (chainsaws) and agricultural expertise that led to rapid clearance of land and construction of good-quality housing; and (iv) by 1995, there was substantial urban development in Machadinho city, a decided cattle ranching activity, and good-quality housing, conditions that facilitated low malaria risk profiles.

Most importantly, malaria risk profiles changed over time and across space, revealing idiosyncrasies of settlement efforts that need to be considered in the selection of combinations of control strategies [4]. These findings suggest some interventions that could be adopted in new settlement areas in the Amazon, including rapid initial clearance of land for agriculture and house construction that is protective and facilitated by government support. Land clearance should be of good quality and extend through a buffer area around the house that offers protection from the high density of adult mosquitoes and larval habitats located at the forest fringe. Indeed, field observations in the early twentieth century revealed that anophelines captured far from houses were not infected with the malaria parasite, and therefore efforts to reduce malaria needed to incorporate recommendations for house construction, including clearance of forest within a 100 m radius of the houses and avoidance of their construction at a distance less than 500 m from rivers, streams, lakes, and swamps [130]. Studies conducted during the 1980s in the Amazon suggested that the distance between the house and the forest should ideally be between 1000 and 1500 m [45]. Recommendations along these lines, however, were never implemented in past settlement projects. It is notable that the average plot size in some settlement programs, such as Machadinho (frontage size along the road averages being 400–500 m and the depth ranging between 700 and 900 m [27]), would make it unfeasible to achieve recommendations of large distances.

FIGURE 10.6 Percentage of *Plasmodium falciparum* cases reported in Brazil, 1962–2017. *Source*: Brazilian Ministry of Health.

10.6 MALARIA PREVENTION AND MITIGATION: RECENT DEVELOPMENTS AND CURRENT/FUTURE CHALLENGES

More than 99% of malaria cases in Brazil are recorded in the Amazon region, almost 90% of them are caused by *P. vivax*. A few biological reasons may explain the higher prevalence of vivax cases (e.g., submicroscopic *P. vivax* infections are common; dormant liver-stage parasites can reactivate and cause disease relapses; a long treatment course, which may cause hemolysis) [131]. In addition, the species composition has changed significantly since 2006, when artemisinin based combination (ACT) therapy was introduced by the Ministry of Health as the first-line treatment for falciparum cases. In 2006, 17% of the malaria cases were caused by *P. falciparum*, a sharp contrast with the more than 50% recorded back in the 1970s and 1980s (Figure 10.6).

The virulence of *P. falciparum* in Brazil, however, is much lower than that recorded in Africa and recent studies are shedding light on possible parasite-derived factors that may explain this difference. A comprehensive analysis of the repertoire of antigenic variants using samples from Brazil, other South American countries, and several countries in Africa (collected over 20 years) indicated that the Brazilian Amazon has fewer variants than any other falciparum endemic area [125]. Also, the representation of some Distinct Sequence Identifiers (DSIDs) was different in isolates from Brazil and South America, with major groups of antigenic variant genes absent [125]. A remaining gap in knowledge concerns whether these unique characteristics found in the Amazon can explain the lower virulence of *P. falciparum* commonly observed in the region.

The spatial distribution of malaria is very irregular, highlighting the need to target malaria control interventions spatially [132, 133]. In fact, after witnessing the number of malaria cases surpassing the mark of 600 thousand in 1999, the Brazilian government implemented the Intensification Plan of Malaria Control Activities in the Legal Amazon (PIACM) in July 2000, which was expected to be operational until December 2002 [134]. The PIACM targeted 254 municipalities in the Amazon (32.1% of the total number of municipalities), which accounted for 93.6% of the malaria cases. The criteria for selection of those municipalities were: (i) an API equal or greater than 50 cases per 1000 people; (ii) *P. falciparum* malaria was responsible for 20% or more of the total number of cases; (iii) the capitals of the nine states that comprise the Amazon region; (iv) the set of municipalities that accounted for at least 80% of all malaria cases in each state; and (v) those where urban malaria was observed [135]. A reduction of 45% in the number of cases was registered between 1999 and 2002, but this gain was not homogeneous across the Amazon region [134]. The reduction ranged from 78 to 35% in different states, and Rondônia was the only state that registered an increase of 12% in the number of cases.

Spatially targeted interventions were a significant improvement in the use of financial resources for control policies. As an example, in 1986, 60% of all malaria cases in the Amazon were concentrated in 458 municipalities, but 70% of the budget for malaria control was being spent in municipalities with only 3% of cases [136]. Targeted interventions demand good quality health data, and important improvements have been implemented by the Brazilian government in the past two decades. A new epidemiological surveillance system is in place and drug resistance is constantly monitored in sentinel sites in the Amazon [137].

In 2003, the government launched the National Malaria Prevention and Control Program (NMPCP), which focused on improvement of local health services, surveillance, diagnosis, and treatment; building local capacity; targeted vector control; research; and community sensitization [134]. Its impacts, however, were not reflected in declines of malaria cases until 2006. After more than 607 thousand cases were recorded in 2005, the number dropped to 307 thousand in 2009, and continued on a declining trend until 2016 when almost 130 thousand microscopically confirmed malaria cases were recorded, the lowest incidence in 38 years. These achievements in malaria control were recognized by the Pan American Health Organization (PAHO) in November 2015, when Brazil received the Malaria Champions of the Americas Award. In addition, the Brazilian Ministry of Health launched, in the same month, the Plan for Elimination of Malaria in Brazil. The plan focuses on *P. falciparum* malaria, providing guidelines on diagnostics, treatment, vector control, and community sensitization and education [138]. At the time when the elimination plan was launched only five municipalities still presented high transmission of *P. falciparum*, concentrating about 50% of the falciparum cases [138]. Sustaining those gains and moving toward malaria elimination in the Amazonian context, however, is likely to face many challenges [131]. Next, we discuss some of these challenges.

10.6.1 Asymptomatic and Submicroscopic Infections

At a local scale, natives and long-term residents, who have been continually exposed to malaria and most likely developed acquired immunity, often do not present symptoms of the disease, but carry *plasmodium* in the blood [139]. These asymptomatic individuals are potentially able to infect mosquitoes after a blood meal [140], and therefore can act as reservoirs of malaria, facilitating the spread of the disease to non-immune populations [139]. Since asymptomatic individuals are unlikely to search for health care, they remain "silently" infectious; a study that followed up asymptomatic individuals for six weeks suggest that less than one-fifth end up developing clinical symptoms [141].

When vector control is not effective, either because it has low coverage or because the local conditions impose barriers to its successful implementation, the presence of asymptomatic infections can potentially contribute to sustained levels of transmission [142]. Given the importance of outdoor biting in the Amazon [47, 48], the occurrence of asymptomatic infections is likely to produce local epidemics and/or maintain local levels of transmission. Indeed, asymptomatic infections are commonly reported in the Amazon [142–149]. In a settlement area in Amazonas state, more than half of *P. vivax* infections were asymptomatic, and about a third were both asymptomatic and submicroscopic [150]. In fact, in hypo- and mesoendemic areas in Brazil, between 50 and 75% of malaria infections can be submicroscopic [141, 143, 149, 151]. In addition, about 90% of submicroscopic and asymptomatic *P. vivax* infections were found to have gametocytes detectable by molecular techniques [141].

The high prevalence of asymptomatic malaria infections raises two issues. The first is related to diagnosis. An aggressive active case detection (AACD) strategy, in which members of a locality are tested monthly and treated for malaria [142], did show success in smaller and more isolated areas, but its widespread use in the Amazon region is likely to face significant logistic and financial difficulties [142, 149]. A possible alternative to AACD is reactive case detection (RCD), when a confirmed malaria infection triggers the active screening of individuals in a defined buffer area around the confirmed case [152]. This approach was applied to a setting of low/declining transmission in the Amazon, with individuals detected positive by passive surveillance followed-up four times over a period of six months, triggering RCD among residents of their houses, nearby houses (within 3 km), but also houses located more than 5 km from the initial case, which were considered as control houses. Samples were analyzed both by microscopy and quantitative real-time polymerase chain reaction (qPCR). Microscopy-based RCD missed 50.5% of the infections detected by qPCR, but identified 76.8% of the parasite biomass in the neighborhood of the initial case, suggesting that microscopy-based RCD could be an additional tool of the ongoing control efforts in Brazil [152]. However, the high prevalence of asymptomatic and submicroscopic infections imposes major limitations to this approach. First, submicroscopic infections would be completely missed and the importance of this parasite reservoir is unknown. Second, asymptomatic cases would only be identified if an infection is passively detected in the vicinity. In both cases, only AACD with highly sensitive diagnostic techniques would be able to identify those infections. However, the feasibility to implementing and regularly maintaining AACD in the Amazon setting, as well as its cost-effectiveness, is currently unknown.

The second issue is related to the infectiousness of asymptomatic individuals, which remains largely unknown (especially when the infections are also submicroscopic), as highlighted by the WHO [153]. In Africa, submicroscopic falciparum malaria represents 20–50% of mosquito infections [154]. In Brazil, one study showed that the mosquito infection rate after feeding on symptomatic *P. vivax* carriers was almost 20 times larger than that observed when carriers were asymptomatic [140]. This topic represents a gap in knowledge that has implications for effective control.

An additional challenge to malaria elimination efforts is the high *P. vivax* diversity in areas of declining transmission, as indicated by molecular genotyping [152]. This genetic diversity suggests complex networks of transmission, with continuous introduction of new strains, most likely influenced by migratory movements, but without a clear temporal pattern. Almost two-thirds of infections had unique genotypes, and thus were genetically unrelated [152]. As previously suggested, this diversity may impact the time it takes to acquire immunity [155], and mathematical models should account for genetic diversity to better capture its impact on control efforts [156].

10.6.2 Planned Large-Scale Infrastructure Projects

At a regional scale, development and integration plans (e.g., dams, roads, hydroways) that include or specifically target the Amazon region will bring about further changes to the social and environmental landscape. Large-scale infrastructure projects, although often justified by their perceived contribution to economic and regional development, commonly result in population displacements (voluntary and involuntary), increase in the incidence of vector- and water-borne diseases, and significant environmental disturbances [157, 158]. These and other negative impacts can potentially be avoided or minimized if comprehensive impact assessments are conducted, and mitigation plans proposed, implemented, and carefully monitored [9, 10, 159]. Yet, the implementation of these plans demands financial and human resources beyond the capacity of local health sectors and therefore should not solely rest under their responsibility [160].

10.6.3 Extreme Climatic Events

The Amazon's fragile ecosystem can be significantly impacted by climate change. The effects can occur directly, through extreme climate events (e.g., drought, flooding, excessive heat). Also, the region can be indirectly impacted by climate-driven population displacement, reduced crop production, and threatened water and food security, all of which impact an individual's vulnerability to infections. Future scenarios put forth by the 4th Assessment Report of the Intergovernmental Panel on Climate Change (IPCC) suggested that the Amazon region is likely to experience reduced rainfall due to climate change, followed by decreases in malaria transmission [161]. This scenario, however, carries much uncertainty. The IPCC report indicates a trend of drier dry seasons in the eastern portion and wetter wet seasons in the western side [15]. That variability could play a major role in the future spatial distribution and intensity of malaria transmission in the Amazon.

Although droughts are believed to reduce malaria incidence [161], this relationship is not straightforward and is modified by several confounding factors. At least six dry periods have been observed in the Amazon in the past 50 years. Three were a result of the El Niño Southern Oscillation phenomenon and tended to have a short duration. They have been connected to increased malaria during the second half of the year and during the following year in other parts of South America [162, 163]; no evaluation has been done for the Brazilian Amazon. A recent and prolonged drought was observed in 2005, not a result of El Niño, but related to a circulation pattern powered by warm seas in the Atlantic [164–166]; more than 600 thousand malaria cases were reported that year, although there has been no assessment of a connection with climate.

Extreme weather events are expected to become more numerous, despite the uncertainty of climate change scenarios. The likely impacts on the geographical pattern and intensity of malaria transmission is likely to vary within and between regions [167], and depends on factors such as: the variation in temperature [168]; the extent to which countries can anticipate and respond to climate change, and if adaptation strategies are already in place; the combination of malaria control interventions; the extent to which local risk factors can be maximized by climate variations; and local levels of population immunity [169, 170]. Currently, the extent to which coping strategies are adopted by local residents in response to these extreme weather events is unknown. Improved understanding of malaria–climate change dynamics in the Amazon region, with special attention to characteristics of reservoir of infection, parasite antigenic diversity, and extreme weather events, is needed.

10.6.4 Political and Economic Instability

Brazil experienced a major economic recession by the end of 2014, and in 2015 fiscal and monetary austerity policies were introduced resulting in severe cuts in public investments and in health and education budgets [171, 172]. Changes to national health policies that directly affect primary healthcare are expected to compromise services provided by municipalities, particularly smaller ones subjected to much financial instability [173]. With limited resources, municipalities may lay off critical personnel that work on surveillance and other crucial activities. The expected consequences of these measures for the health status of the Brazilian population are worrisome. With regards to malaria, after reaching the lowest incidence in 38 years, the number of malaria cases reported in the Amazon increased 47% in 2017 when compared with the previous year. The number of falciparum cases increased by 46% between 2016 and 2017, seriously threatening the goals set forth in the 2015 Elimination Plan [138].

Instability in neighboring countries is also a matter for concern. Between 2010 and 2015, malaria incidence has tripled in Venezuela and the country recorded 30% of all malaria cases in the Americas in 2015, almost one-fifth due to *P. falciparum* [174]. The highest transmission is observed in states bordering Brazil, mainly due to illegal gold mining activities. In fact, about 80% of malaria cases in Venezuela are observed among gold miners, who live in precarious conditions, with limited access to rapid malaria diagnosis and treatment [175]. A possibility of importation of malaria cases from Venezuela and a scenario of a weakening health system due to austerity measures recently adopted in Brazil, represent serious threats to malaria elimination attempts.

10.7 CONCLUSIONS

Malaria transmission in Brazil is mostly restricted to the Amazon, and this pattern is not expected to change in the near future. The declining trend in malaria incidence observed since 2005 has reversed in 2017, with an increase of 47% in the number of cases. Although Brazil launched a malaria elimination plan in 2015, initially focused on falciparum malaria, recurrent and new challenges put to question the feasibility of such a plan, unless intense surveillance, diagnosis, and treatment of infections is in place, including those that are asymptomatic and submicroscopic. Such an initiative would demand financial and human resources beyond what is currently available to the national malaria control program.

In the initial section of this chapter we reviewed important facts related to malaria endemicity and attempts at control over the past 100+ years in Brazil. These experiences hold lessons for contemporary efforts at malaria control in the Amazon region, including the planning of large-scale development projects; the recent International Finance Corporation (IFC) guidelines for HIAs [176] represent an important step forward toward prevention and mitigation of malaria problems associated with large-scale projects in the Amazon region and elsewhere in the tropics.

A common theme of agricultural settlement initiatives (planned and unplanned), mineral extraction projects, and large dams is that there is substantial ecosystem transformation associated with them [120, 177]. An immediate corollary of these transformations has frequently been a considerable expansion of desirable breeding and larval habitats for the primary malaria vector in the Amazon region, *An. darlingi*. In particular, cutting into primary forest for agriculture and mining with limited early availability of mechanized equipment and precarious establishment of cohesive communities have led to the phenomenon of frontier malaria previously described. This process continues to the present time in the formation of new settlements, some created as a way to "legalize" land invasions (and therefore allow these areas to receive basic infrastructure). This process of invasion, and later legalization, highlights problems in the current Brazilian legislation regarding land ownership, which contribute to increased deforestation, land conflicts, and, to a large extent, malaria outbreaks [178, 179]. Lessons from the past – e.g., the Machadinho settlement project discussed in Section 10.5′– are not always taken into account, and the frontier malaria phenomenon is a direct consequence.

In Section 10.3, we gave a brief overview of evolutionary and changing behavioral features of *Anopheles* vectors in the Amazon, emphasizing *An. darlingi*. Considerable uncertainty prevails about the timing and geographical details of *Anopheles* expansions in the region. Regarding the origin and timing of expansions of malaria vectors throughout Brazil and other parts of South America, much more refined DNA sequencing and intensive geographical sampling of *Anopheles* species than heretofore will be necessary to narrow the range of uncertainty – currently tens of thousands of years – about these population dynamics. Of necessity, such vector studies should be linked with much more refined evolutionary studies of human and animal plasmodia in the Amazon, so that a defensible historical picture of malaria in the Amazon can be put forward. This should be augmented by shorter-term studies to assess the root causes of behavioral changes in circulating *Anopheles* that is consequential to the major human migrations and ecosystem transformations initiated in the 1960s.

Finally, Brazil has been cited in recent years [180] as a country where malaria eradication should be feasible. Such commentary is based on coarse-level data focused on children and weighted on malaria rates from health posts/clinics. We view this as particularly problematic, since the substantial number of asymptomatic and submicroscopic carriers in the Amazon region represents a limiting factor for eradication efforts. Without aggressive active surveillance and appropriate treatment of asymptomatic cases – assuming that this is even a practical option – more nuanced mitigation strategies will, of necessity, have to be part of the future story of malaria in the Brazilian Amazon.

REFERENCES

1. PNCM (2008). Casos confirmados de Malária, segundo mês de notificação. Brasil, Grandes Regiões e Unidades Federadas. 2008. Programa Nacional de Controle da Malária-PNCM, Ministério da Saúde, Sistema de Informação da Vigilância Epidemiológica - Malária (Sivep-Malária). Available from: http://portal.saude.gov.br/portal/arquivos/pdf/casos_conf_malaria_mes_notificacao_2008.pdf. Accessed May 2009.
2. Sawyer, D.R. (1988). Frontier Malaria in the Amazon Region of Brazil: Types of Malaria Situations and Some Implications for Control. Brasília: PAHO/WHO/TDR, April 1988,
3. Sawyer, D.R. (1986). Malaria on the Amazon frontier: economic and social aspects of transmission and control. *The Southeast Asian Journal of Tropical Medicine and Public Health* 17 (3): 342–345.
4. Castro, M.C., Monte-Mór, R.L., Sawyer, D.O., and Singer, B.H. (2006). Malaria risk on the Amazon frontier. *Proceedings of the National Academy of Sciences* 103 (7): 2452–2457.
5. Costello, A., Abbas, M., Allen, A. et al. (2009). Managing the health effects of climate change. *The Lancet* 373: 1693–1733.
6. WHO (2009). *Protecting Health from Climate Change: Global Research Priorities*, 32. Geneva: World Health Organization.

7. Githeko, A.K., Lindsay, S.W., Confalonieri, U.E., and Patz, J.A. (2000). Climate change and vector-borne diseases: a regional analysis. *Bulletin of the World Health Organization* 78 (9): 1136–1147.
8. Deane, L.M., Ribeiro, C.D., Oliveira, R.L. et al. (1988). Study on the natural history of malaria in areas of the Rondônia state – Brazil and problems related to its control. *Revista do Instituto de Medicina Tropical de São Paulo* 30 (3): 153–156.
9. Krieger, N., Northridge, M., Gruskin, S. et al. (2003). Assessing health impact assessment: multidisciplinary and international perspectives. *Journal of Epidemiology and Community Health* 57 (9): 659–662.
10. Wood, C. (2003).Environmental Impact Assessment in Developing Countries: an overview. In: *Manchester, UK Conference on New Directions in Impact Assessment for Development: Methods and Practice*, 24–25 November 2003.
11. World Health Organization (2000). *Human Health and Dams*. Geneva: The World Health Organization's submission to the World Commission on Dams (WCD.
12. Monte-Mór, R.L. (2004). Modernities in the jungle: extended urbanization in the Brazilian Amazonia. PhD Thesis, University of California, Los Angeles, CA, USA.
13. Monte-Mór, R.L. (1997). Urban and rural planning: impact on health and the environment. In: *International Perspectives on Environment, Development, and Health: Toward a Sustainable World* (ed. G.S. Shahi, B.S. Levy, A. Binger, et al.), 554–566. New York: Springer.
14. Sawyer, D.O., Monte-Mór, R.L. (1992). Malaria Risk Factors Assessment in Brazil. Interregional Meeting on Malaria, Brasília, 1992.
15. Malhi, Y., Roberts, J.T., Betts, R.A. et al. (2008). Climate change, deforestation, and the fate of the Amazon. *Science* 319: 169–172.
16. IDB (1992). *Amazonia Without Myths, p. 99*. Washington, DC: Inter-American Development Bank and Commission on Development and Environment for Amazonia.
17. IBGE (2008). Pesquisa Nacional por Amostra de Domicílios, Síntese de Indicadores. Available at: http://www.ibge.gov.br/home/estatistica/populacao/trabalhoerendimento/pnad2008/default.shtm. Accessed March 2010.
18. IBGE (2009). SIDRA – Sistema IBGE de Recuperação Automática.Available at: http://www.sidra.ibge.gov.br/. Accessed December 2009.
19. Benchimol, S. (1985). Population changes in the Brazilian Amazon. In: *Change in the Amazon Basin, vol. II* (ed. J. Hemming), 37–50. Manchester: Manchester University Press.
20. Moran, E.F. (1985). An assessment of a decade of colonization in the Amazon Basin. In: *Change in the Amazon Basin: The Frontier after a Decade of Colonization*, vol. II (ed. J. Hemming), 91–102. Manchester: Manchester University Press.
21. Browder, J.O. and Godfrey, B.J. (1997). *Rainforest Cities: Urbanization, Development, and Globalization of the Brazilian Amazon*, 429. New York: Columbia University Press.
22. Schmink, M. and Wood, C.H. (1992). *Contested Frontiers in Amazonia*, 387. New York: Columbia University Press.
23. Schmink, M. and Wood, C.H. (1984). *Frontier Expansion in Amazonia*, 502. Gainesville: University of Florida Press.
24. Fearnside, P.M. (2005). Deforestation in Brazilian Amazonia: history, rates, and consequences. *Conservation Biology* 19 (3): 680–688.
25. Coimbra, C.E.A. Jr. (1988). Human factors in the epidemiology of malaria in the Brazilian Amazon. *Human Organization* 47 (3): 254–260.
26. Sawyer, D.R. (1992). *Malaria and the Environment, Documento de Trabalho*. Brasília, Spain: Instituto SPN.
27. Sawyer, D.R. and Sawyer, D.O. (1987). *Malaria on the Amazon Frontier: Economic and Social Aspects of Transmission and Control. Relatório de Pesquisa*. Belo Horizonte: CEDEPLAR.
28. Singer, B.H. and Castro, M.C. (2001). Agricultural colonization and malaria on the Amazon frontier. Population health and aging: strengthening the dialogue between epidemiology and demography. *Annals of the New York Academy of Sciences* 954: 184–222.
29. Sawyer, D.R. and Sawyer, D.O. (1992). The malaria transition and the role of social science research. In: *Advancing the Health in Developing Countries: The Role of Social Research* (ed. L.C. Chen), 105–122. Westport: Auburn House.
30. Castro, M.C. and Singer, B.H. (2005). Was malaria present in the Amazon before the European conquest? Available evidence and future research agenda. *Journal of Archaeological Science* 32: 337–340.
31. Santos Filho, L.C. (1997). *História Geral da Medicina Brasileira*, 397. São Paulo: Editora Humanismo Ciência e Tecnologia.
32. Azpilcueta, N.J. (1988). *Cartas Avulsas*, 529. Belo Horizonte: Editora da Universidade de São Paulo.
33. Leite, S. (1954). *Cartas dos Primeiros Jesuítas do Brasil*, 1094. São Paulo: Comissão do IV Centenário da cidade de São Paulo.
34. Soares de Sousa, G. and Varnhagen, F.A. *Tratado Descritivo do Brasil em 1587*, 4e, 389. São Paulo: Companhia Editora Nacional.
35. Bomtempo, J.M. (1825). *Memória Sobre Algumas Enfermidades do Rio de Janeiro, e mui Particularmente Sobre o Abuso Geral, e Pernicioso Efeito da Aplicação da Preciosa Casca Peruviana, ou Quina*, 74. Rio de Janeiro: Typographia Nacional.
36. Neiva, A. (1910). Formação de raça do hematozoario do impaludismo resistente a quinina. *Memórias do Instituto Oswaldo Cruz* 2 (1): 131–140.
37. Cruz, O. (1913). *Relatório Sobre as Condições Médico-sanitárias do Valle do Amazonas*. Rio de Janeiro: Typ.do Jornal do Commercio, de Rodriques & Comp.
38. Costa, D.A.M. (1885). *A Malária e Suas Diversas Modalidades Clínicas*. Rio de Janeiro: Imprensa a Vapor Lombaerts & Comp.
39. Penna, B. (1918). *Saneamento do Brasil. Sanear o Brasil é Povoal-o; é Enriquecel-o; é Moralisal-o*, 114. Rio de Janeiro: Typ. Revista dos Tribunais.
40. von Spix, J.B., von Martius, K.F.P., and Lloyd, H.E. (1824). *Travels in Brazil, in the Years 1817–1820: Undertaken by Command of his Majesty the King of Bavaria*, 298. London: Longman.

41. Peixoto, A. (1917). *O Problema Sanitário da Amazônia*. Rio de Janeiro: Imprensa Nacional.
42. Chagas, C. (1913). *Notas Sobre a Epidemiologia do Amazonas*. Rio de Janeiro: Instituto Oswaldo Cruz.
43. Chagas, C. (1903). *Estudos Hematológicos no Impaludismo*. Rio de Janeiro: Typographia da Papelaria União.
44. Thomas, H.W. (1910). The sanitary conditions and diseases prevailing in Manaos, north Brazil, 1905–1909, with plan of Manaos and chart. Expedition to the Amazon, 1905–1909. *Annals of Tropical Medicine and Parasitology* 4 (1): 7–55.
45. Tadei, W.P., Santos, J.M.M., Costa, W.L.S., and Scarpassa, V.M. (1988). Biologia de Anofelinos Amazônicos. XII. Ocorrência de espécies de *Anopheles*, dinâmica da transmissão e controle da malária na zona urbana de Ariquemes (Rondônia). *Revista do Instituto de Medicina Tropical de São Paulo* 30 (3): 221–251.
46. Bustamante, F.M. (1951). Efeito das aplicações intradomiciliárias de DDT sobre a densidade do *Anopheles darlingi* em várias regiões do Brasil. *Revista Brasileira de Malariologia* III (4): 571–590.
47. Klein, T.A. and Lima, J.B.P. (1990). Seasonal distribution and biting patterns of *Anopheles* mosquitoes in Costa Marques, Rondônia, Brazil. *Journal of the American Mosquito Control Association* 6 (4): 700–707.
48. Lourenço-de-Oliveira, R., Guimarães, A.E.G., Arlé, M. et al. (1989). Anopheline species, some of their habits and relation to malaria in endemic areas of Rondônia state, Amazon region of Brazil. *Memórias do Instituto Oswaldo Cruz* 84 (4): 501–514.
49. Tadei, W.P., Thatcher, B.D., Santos, J.M.M. et al. (1998). Ecologic observations on anopheline vectors of malaria in the Brazilian Amazon. *American Journal of Tropical Medicine and Hygiene* 59 (2): 325–335.
50. Bustamante, F.M., Pinto, O.S., and Freitas, J.R. (1952). Observações sobre o comportamento do *Anopheles darlingi* em casas experimentais tratadas com DDT e BHC na área de Engenheiro Dolabela, Estado de Minas Gerais. *Revista Brasileira de Malariologia e Doenças Tropicais* IV (4): 347–360.
51. Hochman, G., Mello, M.T.B., and Santos, P.R.E. (2002). A malária em foto: imagens de campanhas e ações no Brasil da primeira metade do século XX. *História, Ciências, Saúde* 9: 233–273.
52. Neiva, A. (1940). Profilaxia da malaria e trabalhos de engenharia: notas, comentários, recordações. *Revista do Clube de Engenharia* VI (70): 60–75.
53. Mello, M.T.V.B. (2007). *Imagens da Memória: Uma História Visual da Malária (1910–1960)* (ed. R.J. Niterói). Universidade Federal Fluminense, Departamento de História.
54. Peixoto, A. (1938). *Clima e Saúde; Introdução Bio-Geográfica à Civilização Brasileira*, 295. Rio de Janeiro: Companhia Editora Nacional.
55. Cruz, O. (1910). *Madeira-Mamoré Railway Company - Considerações gerais sobre as condições sanitárias do Rio Madeira*. Rio de Janeiro: Papelaria Americana.
56. Craig, N.B. and Vasconcelos, M.N. (1947). *Estrada de Ferro Madeira-Momoré, História Trágica de Uma Expedição*. São Paulo: Companhia Editora Nacional.
57. Ferreira, M.R. (1981). *A Ferrovia do Diabo: História de Uma Estrada de Ferro na Amazônia*, 400. São Paulo: Melhoramentos: Secretaria de Estado da Cultura.
58. Perdigão, F. and Bassegio, L. (1992). *Migrantes Amazônicos - Rondônia: a trajetória da ilusão*, 221. São Paulo, SP: Edições Loyola.
59. Thomas, H.W. (1910). Malaria in the Amazon region, and the protection of ships. In: *The Prevention of Malaria* (ed. R. Ross), 382–389. New York: E.P. Dutton & Company.
60. Cruz, O. (1910). Prophylaxis of malaria in central and southern Brazil. In: *The Prevention of Malaria* (ed. R. Ross), 390–399. New York: E.P. Dutton & Company.
61. Coimbra, M.E.L.S. (1985). *SUCAM and Malaria Control*. Belo Horizonte: CEDEPLAR.
62. Silveira, A.C. and Rezende, D.F. (2001). *Avaliação da Estratégia Global de Controle Integrado da Malária no Brasil*. Brasília: Organização Pan-Americana da Saúde.
63. Stepan, N. (1976). *Beginnings of Brazilian Science: Oswaldo Cruz, Medical Research and Policy, 1890–1920*, 225. New York: Science History Publications.
64. Chagas, C.(1934). *Luta Contra a Malária*. Conferência proferida no Nucleo Colonial S. Bento, Abril 1933. Discursos e Conferências, Rio de Janeiro: A Noite, pp. 191–231.
65. Chagas, C. (1905). *Prophylaxia do Impaludismo, p. 48*. Rio de Janeiro: Typ. Besnard Frères. (Trabalho do Instituto de Manguinhos).
66. Deane, L.M. (1986). *A História da Evolução dos Conhecimentos Sobre a Malária*. Rio de Janeiro.
67. Cruz, O., Chagas, C., and Peixoto, A. (1972). *Sobre o Saneamento da Amazônia, p. 205*. Manaus: P. Daou.
68. Stepan, N.L. (2003). The only serious terror in these regions: malaria control in the Brazilian Amazon. In: *Disease in the History of Modern Latin America: From Malaria to AIDS* (ed. D. Armus), 25–50. Durham, NC: Duke University Press.
69. Tadei, W.P. (1991). *Considerações Sobre as Espécies de Anopheles e a Transmissão da Malária na Amazônia*. Manaus: INPA – Instituto Nacional de Pesquisas da Amazônia.
70. Parmakelis, A., Russello, M.A., Caccone, A. et al. (2008). Short report: historical analysis of a near disaster: *Anopheles gambiae* in Brazil. *American Journal of Tropical Medicine and Hygiene* 78 (1): 176–178.
71. Deane, L.M.1990). A história do *Anopheles gambiae* no Brasil. Rio de Janeiro. In: III Seminário de Vetores Urbanos e Animais Sinatrópicos – 4a. Reunião Brasileira sobre Simulídeos.
72. Deane, L.M. (1988). Malaria studies and control in Brazil. *American Journal of Tropical Medicine and Hygiene* 38 (2): 223–230.
73. Soper, F.L. and Wilson, D.B. (1943). *Anopheles gambiae in Brazil, 1930 to 1940*, 262. New York City: The Rockefeller Foundation.
74. Killeen, G.F., Fillinger, U., Kiche, I. et al. (2002). Eradication of *Anopheles gambiae* from Brazil: lessons for malaria control in Africa. *Lancet Infectious Diseases* 2: 618–627.
75. Litsios, S. (1997). *The Tomorrow of Malaria*, revised edn, 183. Karori, NZ: Pacific Press.

76. Bailey, N.T.J. (1982). *The Biomathematics of Malaria*, 210. London: C. Griffin.

77. Pinotti, M. (1951). The biological basis for the campaign against the malaria vectors in Brazil. *Transactions of the Royal Society of Tropical Medicine and Hygiene* 44 (6): 663–682.

78. Lutz, A. (1903). Waldmosquitos und Waldmalaria. *Centralblatt für Bakteriologie, Parasitenkunde und Infektionskrankheiten* 33 (4): 282–292.

79. Benchimol, J.L. and Sá, M.R. (2005). *Adolpho Lutz: Obra Completa. Febre Amarela, Malária e Protozoologia*, 956. Rio de Janeiro, RJ: Editora Fiocruz.

80. Deane, L.M. (1947). Observações sobre a malária na Amazônia Brasileira. *Revista do Serviço Especial de Saúde Pública* 1: 3–60.

81. Loiola, C.C.P. (ed.) (1998). The Use of DDT in Malaria Control Programs in Brazil. Subregional Awareness Raising Workshop on Persistent Organic Pollutants (POPs). Puerto Iguazú, Argentina.

82. Mendes, R.A., Jesus, I.M., Santos, E.C.O. et al. (2007). Níveis séricos de DDT total de trabalhadores expostos no programa de controle da malária no estado do Pará. *Brasil. Cadernos de Saúde Coletiva* 15 (4): 559–568.

83. Deane, L.M., Freire, E.P.S., Tabosa, W., and Ledo, J. (1948). A aplicação domiciliar de DDT no contrôle da malária em localidades da Amazônia. *Revista do Serviço Especial de Saúde Pública* 1 (4): 1121–1161.

84. Deane, L.M., Ledo, J.F., Freire, E.P.S. et al. (1948). Contrôle da Malária na Amazônia pela aplicação domiciliar de DDT e sua avaliação pela determinação do índice de transmissão. *Revista do Serviço Especial de Saúde Pública* 2 (2): 545–560.

85. Tauil, P., Deane, L., Sabroza, P., and Ribeiro, C. (1985). A malária no Brasil. *Cadernos de Saúde Pública* 1 (1): 71–111.

86. Roberts, D.R. and Alecrim, W.D. (1991). Behavioral response of *Anopheles darlingi* to DDT-sprayed house walls in Amazonia. *Bulletin of PAHO* 25 (3): 210–217.

87. SNM (1953). Casos de malária confirmados por exame de laboratório, 1945 e 1953. SNM – Serviço Nacional de Malária. Ministério da Saúde. Casa de Oswaldo Cruz (FIOCRUZ) collection. CC/DT/19252040.

88. Pinotti, M., Lôbo, A.G.S., Damasceno, G., and Soares, R. (1955). Experiências de campo com o sal cloroquinado. *Revista Brasileira de Malariologia e Doenças Tropicais* 7: 5–23.

89. Payne, D. (1988). Did medicated salt hasten the spread of chloroquine resistance in *Plasmodium falciparum*? *Parasitology Today* 4 (4): 112–115.

90. Coimbra, M.E.L.S. (1987). Política pública e saúde: o caso da malária no Brasil. *Análise e Conjuntura* 2 (1): 72–90.

91. Wernsdorfer, W.H.1998). Drug Resistance of Malaria Parasites. Geneva: 20th Expert Committee on Malaria. Working Paper MAL/ECM/20/98/15.

92. Silva, G.C. (1967). *Geopolítica do Brasil*, 275. Rio de Janeiro: José Olympio.

93. Mattos, C.M. (1980). *Uma Geopolítica Pan-Amazônica*, 215. Rio de Janeiro, RJ: Biblioteca do Exército Editora.

94. Becker, B.K. (2005). Geopolítica da Amazônia. *Estudos Avançados* 19 (53): 71–86.

95. Singer, B.H. and Sawyer, D.O. (1992). Perceived malaria illness reports in mobile populations. *Health Policy and Planning* 7 (1): 40–45.

96. Aron, J.L. and Patz, J. (2001). *Ecosystem Change and Public Health: A Global Perspective*, 480. Baltimore: Johns Hopkins University Press.

97. Wood, C.H. and Wilson, J. (1984). The magnitude of migration to the Brazilian frontier. In: *Frontier Expansion in Amazonia* (ed. M. Schmink and C.H. Wood), 142–152. Gainesville: University of Florida Press.

98. Diegues, A.C.S. and Millikan, B. (1993). *A Dinâmica Social do Desmatamento na Amazônia: Populações e Modos de Vida em Rondônia e Sudeste do Pará*, 155. São Paulo: United Nations Research Institute for Social Development: Núcleo de Apoio à Pesquisa sobre Populações Humanas e Áreas Úmidas Brasileiras.

99. Silva, G.C. (1957). *Aspectos Geopolíticos do Brasil*, 81. Rio de Janeiro: Biblioteca do Exército.

100. Mahar, D.J. (1979). *Frontier Development Policy in Brazil: A Study of Amazonia*, 182. New York: Praeger.

101. Moran, E.F. (1981). *Developing the Amazon*, 292. Bloomington: Indiana University Press.

102. World Bank (1981). *Brazil: Integrated Development of the Northwest Frontier*. Washington: The World Bank.

103. Castro, M.C. (2002). Spatial configuration of malaria risk on the Amazon frontier: the hidden reality behind global analysis. PhD Thesis. Princeton: Princeton University.

104. World Bank (1992). *World Bank Approaches to the Environment in Brazil: A Review of Selected Projects, vol. 5: The POLONOROESTE Program*. Washington: The World Bank.

105. BENFAM (1987). Colonização: um projeto que pode fracassar. *População & Desenvolvimento* 21 (144): 47–48.

106. Pedro, P.M. and Sallum, M.A.M. (2009). Spatial expansion and population structure of the neotropical malaria vector, *Anopheles darlingi* (Diptera: Culicidae). *Biological Journal of the Linnean Society* 97: 854–866.

107. Mirabello., L. and Conn, J.E. (2006). Molecular population genetics of the malaria vector *Anopheles darlingi* in Central and South America. *Heredity* 96: 311–321.

108. Escalante, A.A., Barris, E., and Ayala, F.J. (1995). Evolutionary origin of human and primate malarias: evidence from the circumsporozoite protein gene. *Molecular Biology and Evolution* 12 (4): 616–626.

109. Naranjo-Díaz, N., Conn, J.E., and Correa, M.M. (2016). Behavior and population structure of *Anopheles darlingi* in Colombia. *Infection, Genetics and Evolution* 39: 64–73. doi: 10.1016/j.meegid.2016.01.004.

110. Emerson, K.J., Conn, J.E., Bergo, E.S. et al. (2015). Brazilian *Anopheles darlingi* root (Diptera: Culicidae) clusters by major biogeographical region. *PLoS One* 10 (7): e0130773. doi: 10.1371/journal.pone.0130773.

111. Deane, L.M. (1986). Malaria vectors in Brazil. *Memórias do Instituto Oswaldo Cruz* 81 (Suppl. II): 5–14.

112. Consoli, R.A.G.B. and Oliveira, R.L. (1998). *Principais Mosquitos de Importância Sanitária no Brasil*, 228. Rio de Janeiro: Editora Fiocruz.

113. SESP (1946). Adultos de Anophelinos Examinados Pelo Laboratório Central e Divisão de Malária. Julho de 1942 – Junho de 1946. SESP – Programa da Amazônia. Casa de Oswaldo Cruz (FIOCRUZ) Collection. LD/TP/19420740.

114. Giglioli, G. (1956). Biological variations in *Anopheles darlingi* and *Anopheles gambiae*: their effect on practical malaria control in the neotropical region. *Bulletin of the World Health Organization* 15 (3, 4, 5): 461–471.

115. Altamiranda-Saavedra, M., Conn, J.E., and Correa, M.M. (2017). Genetic structure and phenotypic variation of *Anopheles darlingi* in Northwest Colombia. *Infection, Genetics and Evolution* 56: 143–151. doi: 10.1016/j.meegid.2017.11.011.

116. Campos, M., Conn, J.E., Alonso, D.P. et al. (2017). Microgeographical structure in the major Neotropical malaria vector *Anopheles darlingi* using microsatellites and SNP markers. *Parasites and Vectors* 10: 76. doi: 10.1186/s13071-017-2014-y PubMed PMID: PMC5307779.

117. Incra nos Estados (2018). Informações gerais sobre os assentamentos da Reforma Agrária [Internet]. Instituto Nacional de Colonização e Reforma Agrária (INCRA) [cited 15 January 2018].

118. Brondízio, E.S., McCracken, S.D., Moran, E.F. et al. (2002). The colonist footprint: toward a conceptual framework of land use and deforestation trajectories among small farmers in the Amazonian frontier. In: *Deforestation and Land Use in the Amazon* (ed. C.H. Wood and R. Porro), 133–161. Gainesville: University Press of Florida.

119. Browder, J.O. (2002). Reading colonist landscapes: social factors influencing land use decisions by small farmers in the Brazilian Amazon. In: *Deforestation and Land Use in the Amazon* (ed. C.H. Wood and R. Porro), 218–240. Gainesville: University Press of Florida.

120. Wood, C.H. and Porro, R. (2002). *Deforestation and Land Use in the Amazon*, 385. Gainesville: University Press of Florida.

121. Fearnside, P.M. (1986). *Human Carrying Capacity of the Brazilian Rainforest*, 293. New York: Columbia University Press.

122. Sawyer, D.R. (1992. Deforestation and Malaria on the Amazon Frontier. Campinas, Brazil. Seminar on Population and Deforestation in the Humid Tropics, International Union for the Scientific Study of Population.

123. Charlwood, J.D. (1996). Biological variation in *Anopheles darlingi* root. *Memórias do Instituto Oswaldo Cruz* 91 (4): 391–398.

124. Charlwood, J.D. (1980). Observations on the bionomics of *Anopheles darlingi* root (Diptera: Culicidae) from Brazil. *Bulletin of Entomological Research* 70: 685–692.

125. Albrecht, L., Castineiras, C., Carvalho, B.O. et al. (2010). The South American *Plasmodium falciparum* var gene repertoire is limited, highly shared and possibly lacks several antigenic types. *Gene* 453 (1–2): 37–44. Epub 2010/01/19. doi: 10.1016/j.gene.2010.01.001 PubMed PMID: 20079817.

126. INCRA (1991). *Exposição de Motivos, mimeo.* Porto Velho: INCRA INCRA/SE-17/G/No. 02/91.

127. Sydenstricker, J.M. (1992). *Parceleiros de Machadinho: História Migratória e as Interações Entre a Dinâmica Demográfica e o Ciclo Agrícola em Rondônia. Master degree Dissertation.* Campinas, SP: Universidade de Campinas.

128. Mahar, D.J. (1989). *Government Policies and Deforestation in Brazil's Amazon Region*, vol. V, 56. Washington, DC: World Bank.

129. Fearnside, P.M. (1986). Settlement in Rondônia and the token role of science and technology in Brazil's Amazonian development planning. *Interciencia* 11 (5): 229–236.

130. Leão, F.A. *Em Torno do Saneamento do Brasil.* Rio de Janeiro: Typographia Leuzinger.

131. Ferreira, M.U. and Castro, M.C. (2016). Challenges for malaria elimination in Brazil. *Malaria Journal* 15 (1): 284. Epub 2016/05/22. doi: 10.1186/s12936-016-1335-1 PubMed PMID: 27206924; PubMed Central PMCID: PMCPMC4875681.

132. Carter, R., Mendis, K.N., and Roberts, D. (2000). Spatial targeting of interventions against malaria. *Bulletin of the World Health Organization* 78 (12): 1401–1411.

133. Castro, M.C., Sawyer, D.O., and Singer, B.H. (2006). Spatial patterns of malaria in the Amazon: implications for surveillance and targeted interventions. *Health and Place* 13: 368–380.

134. MS (2003). *Programa Nacional de Prevenção e Controle da Malária PNCM*, 132. Brasília: Ministério da Saúde. Secretaria de Vigilância em Saúde.

135. FUNASA (2000). Plano de Intensificação das Ações de Controle da Malária na Amazônia Legal. Available from:http://www.funasa.gov.br/epi/malaria/pdfs/plano_malaria.PDF and http://www.funasa.gov.br/epi/malaria/pdfs/resumo_exec_pcim.PDF.

136. Akhavan, D., Musgrove, P., Abrantes, A., and Gusmão, R.A. (1999). Cost-effective malaria control in Brazil: cost-effectiveness of a malaria control program in the Amazon basin of Brazil, 1988–1996. *Social Science and Medicine* 49: 1385–1399.

137. MS (2005). *Situação Epidemiológica da Malária no Brasil 2005.* Brasília: Ministério da Saúde.

138. Santelli, A.C.F.S., Damasceno, C.P., Peterka, C.L., and Marchesini, P.B. (2016). Plano de Eliminação de Malária no Brasil. Fase 1: Malária Falciparum. Brasília, DF: Ministério da Saúde. Available from: http://portalarquivos2.saude.gov.br/images/pdf/2017/janeiro/04/Plano-eliminacao-malaria-pub.pdf.

139. Coura, J.R., Suárez-Mutis, M., and Ladeia-Andrade, S. (2006). A new challenge for malaria control in Brazil: asymptomatic *Plasmodium* infection – A Review. *Memórias do Instituto Oswaldo Cruz* 101 (3): 229–237.

140. Alves, F.P., Gil, L.H.S., Marrelli, M.T. et al. (2005). Asymptomatic carriers of *Plasmodium* spp. as infection source for malaria vector mosquitoes in the Brazilian Amazon. *Journal of Medical Entomology* 42 (5): 777–779.

141. Barbosa, S., Gozze, A.B., Lima, N.F. et al. (2014). Epidemiology of disappearing *Plasmodium vivax* malaria: a case study in rural Amazonia. *PLoS Neglected Tropical Diseases* 8 (8): e3109. doi: 10.1371/journal.pntd.0003109.

142. Macauley, C. (2005). Aggressive active case detection: a malaria control strategy based on the Brazilian model. *Social Science and Medicine* 60: 563–573.

143. Alves, F.P., Durlacher, R.R., Menezes, M.J. et al. (2002). High prevalence of asymptomatic *Plasmodium vivax* and *Plasmodium falciparum* infections in native Amazonian populations. *American Journal of Tropical Medicine and Hygiene* 66 (6): 641–648.
144. Camargo, E.P., Alves, F., and Pereira da Silva, L.H. (1999). Symptomless *Plasmodium vivax* infections in native Amazonians. *The Lancet* 353: 1415–1416.
145. Ladeia-Andrade, S., Ferreira, M.U., Scopel, K.K.G. et al. (2007). Naturally acquired antibodies to merozoite surface protein (MSP)-1_{19} and cumulative exposure to *Plasmodium falciparum* and *Plasmodium vivax* in remote populations of the Amazon Basin of Brazil. *Memórias do Instituto Oswaldo Cruz* 102 (8): 943–951.
146. Tada, M.S., Marques, R.P., Mesquita, E. et al. (2007). Urban malaria in the Brazilian Western Amazon Region. I. High prevalence of asymptomatic carriers in an urban riverside district is associated with a high level of clinical malaria. *Memórias do Instituto Oswaldo Cruz* 102 (3): 263–269.
147. Suárez-Mutis, M., Cuervo, P., Leoratti, F.M.S. et al. Cross sectional study reveals a high percentage of asymptomatic *Plasmodium vivax* infection in the Amazon Rio Negro area, Brazil. *Revista do Instituto de Medicina Tropical de São Paulo* 49 (3): 159–164.
148. Katsuragawa, T.H., Gil, L.H.S., Tada, M.S., and Silva, L.H.P. (2008). Endemias e epidemias na Amazônia. Malária e doenças emergentes em áreas ribeirinhas do Rio Madeira. Um caso de escola. *Estudos Avançados* 22 (64): 111–141.
149. Ladeia-Andrade, S., Ferreira, M.U., M.E., C. et al. (2009). Age-dependent acquisition of protective immunity to malaria in riverine populations of the Amazon Basin of Brazil. *The American Journal of Tropical Medicine and Hygiene* 80 (3): 452–459.
150. Duarte, E.C., Gyorkos, T.W., Pang, L., and Abrahamowicz, M. (2004). Epidemiology of malaria in a hypoendemic Brazilian Amazon migrant population: a cohort study. *The American Journal of Tropical Medicine and Hygiene* 70 (3): 229–237. Epub 2004/03/20. PubMed PMID: 15031509.
151. da Silva, N.S., da Silva-Nunes, M., Malafronte, R.S. et al. (2010). Epidemiology and control of frontier malaria in Brazil: lessons from community-based studies in rural Amazonia. *Transactions of the Royal Society of Tropical Medicine and Hygiene* 104: 343–350.
152. Fontoura, P.S., Finco, B.F., Lima, N.F. et al. (2016). Reactive case detection for *Plasmodium vivax* malaria elimination in rural Amazonia. *PLoS Neglected Tropical Diseases* 10 (12): e0005221. doi: 10.1371/journal.pntd.0005221.
153. World Health Organization (2014). *WHO Evidence Review Group on Malaria Diagnosis in Low Transmission Setting*. Meeting Report. Geneva: World Health Organization.
154. Okell, L.C., Bousema, T., Griffin, J.T. et al. (2012). Factors determining the occurrence of submicroscopic malaria infections and their relevance for control. *Nature Communications* 3: 1237. Epub 2012/12/06. doi: 10.1038/ncomms2241 PubMed PMID: 23212366; PubMed Central PMCID: PMCPMC3535331.
155. Day, K.P. and Marsh, K. (1991). Naturally acquired immunity to *Plasmodium falciparum*. *Immunology Today* 12 (3): A68–A71. Epub 1991/03/01. PubMed PMID: 2069680.
156. Day, K.P., Artzy-Randrup, Y., Tiedje, K.E. et al. (2017). Evidence of strain structure in *Plasmodium falciparum* var gene repertoires in children from Gabon, West Africa. *Proceedings of the National Academy of Sciences of the United States of America* 114 (20): E4103-e11. Epub 2017/05/04. doi: 10.1073/pnas.1613018114 PubMed PMID: 28461509; PubMed Central PMCID: PMCPMC5441825.
157. Couldrey, M. and Morris, T. (eds.) (2002). *Dilemmas of Development-Induced Displacement (Forced Migration Review)*, vol. 12. Oxford: Refugee Studies Center.
158. Lerer, L.B. and Scudder, T. (1999). Health impacts of large dams. *Environmental Impact Assessment* 19: 113–123.
159. Winkler, M.S., Divall, M.J., Krieger, G.R. et al. (2009). Assessing health impacts in complex eco-epidemiological settings in the humid tropics: advancing tools and methods. *Environmental Impact Assessment Review* 30 (1): 52–61.
160. Castro, M.C., Krieger, G.R., Balge, M.Z. et al. (2016). Examples of coupled human and environmental systems from the extractive industry and hydropower sector interfaces. *Proceedings of the National Academy of Sciences* 113 (51): 14528–14535. doi: 10.1073/pnas.1605678113.
161. Confalonieri, U., Menne, B., Akhtar, R. et al. (2007). Human health. In: *Climate Change 2007: Impacts, Adaptation and Vulnerability Contribution of Working Group II to the Fourth Assessment Report of the Intergovernmental Panel on Climate Change* (ed. M. Parry, O. Canziani, J. Palutikof, et al.), 391–431. Cambridge: Cambridge University Press.
162. Gagnon, A., Smoyer-Tomic, K., and Bush, A. (2002). The El Niño Southern Oscillation and malaria epidemics in South America. *International Journal of Biometeorology* 46: 81–89. doi: 10.1007/s00484-001-0119-6.
163. Poveda, G., Rojas, W., Quinones, M. et al. (2001). Coupling between annual and ENSO timescales in the malaria-climate association in Colombia. *Environmental Health Perspectives* 109: 489–493. doi: 10.2307/3454707.
164. Cox, P.M., Harris, P.P., Huntingford, C. et al. (2008). Increasing risk of Amazonian drought due to decreasing aerosol pollution. *Nature* 453: 212–215.
165. Marengo, J.A., Nobre, C.A., Tomasella, J. et al. (2008). The drought of Amazonia in 2005. *American Meteorological Society* 21: 495–516.
166. Zeng, N., Yoon, J.-H., Marengo, J.A. et al. (2008). Causes and impacts of the 2005 Amazon drought. *Environmental Research Letters* 3: 9. doi: 10.1088/748-9326/3/1/014002.
167. Takken, W., P.T.R., V., Schneider, P., and Santos, F. (2005). Effects of environmental change on malaria in the Amazon region of Brazil. In: *Environmental Change and Malaria Risk: Global and Local Implications* (ed. W. Takken, P. Martens and R. Bogers), 113–123. Wageningen: Springer Netherlands.
168. Paaijmans, K.P., Blanford, S., Bell, A.S. et al. (2010). Influence of climate on malaria transmission depends on daily temperature variation. *Proceedings of the National Academy of Sciences* 107 (34): 15135–15139. doi: 10.1073/pnas.1006422107.
169. Laneri, K., Paul, R.E., Tall, A. et al. (2015). Dynamical malaria models reveal how immunity buffers effect of climate variability. *Proceedings of the National Academy of Sciences* 112 (28): 8786–8791. doi: 10.1073/pnas.1419047112.

170. Pascual, M. (2015). Climate and population immunity in malaria dynamics: harnessing information from endemicity gradients. *Trends in Parasitology* 31 (11): 532–534. doi: 10.1016/j.pt.2015.08.009.

171. Lima, L.D., Travassos, C., Carvalho, M.S., and Coeli, C.M. (2016). Um ano de perplexidade. *Cadernos de Saúde Pública* 32 (12): Epub December 15, 2016.

172. Rossi, P. and Dweck, E. (2016). Impactos do novo regime fiscal na saúde e educação. *Cadernos de Saúde Pública* 32 (12): e001943. Epub December 15, 2016.

173. Andrade, M.V., Coelho, A.Q., Neto, M.X. et al. (2018). Brazil's family health strategy: factors associated with programme uptake and coverage expansion over 15 years (1998–2012). *Health Policy and Planning* 33 (3): 368–380. Epub 15 January 2018. doi: 10.1093/heapol/czx189.

174. Pan American Health Organization (2016). *Report on the Situation of Malaria in the Americas, 2000–2015*. Washington, DC: Regional Malaria Program, Pan American Health Organization.

175. Pan American Health Organization (2017). *Epidemiological Alert: Increase in Cases of Malaria*. Washington, DC: Pan American Health Organization Available from: http://www.paho.org/hq/index.php?option=com_docman&task=doc_download&Itemid=&gid=38146&lang=pt.

176. IFC (2009). *Introduction to Health Impact Assessment*, 67. Washington, DC: Environment and Social Development Department, International Finance Corporation (IFC).

177. Moran, E.F. (1996). Deforestation in the Brazilian Amazon. In: *Tropical Deforestation: The Human Dimension* (ed. L.E. Sponsel, T.N. Headland and R.C. Bailey), 149–164. New York: Columbia University Press.

178. Alston, L.J., Libecap, G.D., and Mueller, B. (2000). Land reform policies, the sources of violent conflict, and implications for deforestation in the Brazilian Amazon. *Journal of Environmental Economics and Management* 39: 162–188.

179. Kirby, K.R., Laurance, W.F., Albernaz, A.K. et al. (2006). The future of deforestation in the Brazilian Amazon. *Futures* 38: 432–453.

180. Guerra, C.A., Gikandi, P.W., Tatem, A.J. et al. (2008). The limits and intensity of *Plasmodium falciparum* transmission: implications for malaria control and elimination worldwide. *PLoS Medicine* 5 (2): e38. doi: 10.1371/journal.pmed.0050038.

11

SCHISTOSOMIASIS

Pascal Magnussen[1], Birgitte Jyding Vennervald[2], and Jens Aagaard-Hansen[3,4]

[1] *Department of Immunology and Microbiology, Centre for Medical Parasitology and Department of Veterinary and Animal Sciences, Section for Parasitology and Aquatic Pathobiology, Faculty of Health and Medical Sciences, University of Copenhagen, Copenhagen, Denmark*

[2] *Department of Veterinary and Animal Sciences, Section for Parasitology and Aquatic Pathobiology, Faculty of Health and Medical Sciences, University of Copenhagen, Copenhagen, Denmark*

[3] *Steno Diabetes Center Copenhagen, Copenhagen, Denmark*

[4] *MRC Developmental Pathways for Health Research Unit, Department of Paediatrics, School of Clinical Medicine, Faculty of Health Sciences, University of the Witwatersrand, Johannesburg, South Africa*

11.1 INTRODUCTION

Schistosomiasis, a freshwater-related helminth infection, is considered the second most important parasitic infection after malaria in terms of public health impact [1]. It is endemic in almost 80 countries in the Americas, Africa, and Asia, with the highest burden of infection and disease on the African continent, especially in sub-Saharan countries; 600–700 million people are estimated to be at risk of infection and approximately 160–200 million individuals to be infected. Mortality figures for schistosomiasis are difficult to estimate as death due to schistosomiasis rarely takes place in health facilities, but in sub-Saharan Africa figures between 20 000 and 200 000 per year have been suggested. [2] Morbidity figures have recently been reassessed and disability estimates are that the global cost to a *Schistosoma*-infected individual is at least 2–15% chronic disability compared to previous figures of around 0.5% [1]. The main schistosome species causing human infection are *Schistosoma mansoni*, *S. haematobium* and *S. intercalatum* in Africa and the Middle East, Americas and the Caribbean, and *S. japonicum* and *S. mekongi* in China, the Philippines, Laos, Kampuchea, and South Thailand. For the African schistosome species, humans are the most important definitive host. *S. japonicum* and *S. mekongi* are both zoonotic parasites, with a wide range of definitive hosts such as pigs, water buffaloes, and various rodents.

In the following sections various aspects of schistosomiasis transmission, pathogenesis, epidemiology, clinical manifestations, diagnosis, treatment, social dynamics influencing transmission, and various control measures will be described.

11.2 SCHISTOSOMIASIS TRANSMISSION CYCLE

The schistosome life cycle is shown in diagrammatic form in Figure 11.1. The parasite alternates between two hosts: the definitive human host, in whom pairing of adult worms and sexual replication occurs, and the intermediate snail host, in which asexual replication occurs [3].

Parasite eggs are excreted via the feces or urine (1). The eggs hatch in fresh water to release miracidia (2), which locate vector snails (3) (see Figure 11.2), where asexual multiplication with formation of sporocysts occurs (4). *S. mansoni* is transmitted by snails of the genus *Biomphalaria*, whereas *S. haematobium* and *S. intercalatum* by the genus *Bulinus*. *S. japonicum*, and *S. mekongi* are transmitted by amphibious snails of the species *Oncomelania* and *Tricula*, respectively.

Water and Sanitation-Related Diseases and the Changing Environment: Challenges, Interventions, and Preventive Measures, Second Edition. Edited by Janine M. H. Selendy.

Companion website: https://www.wiley.com/legacy/wileychi/selendy/

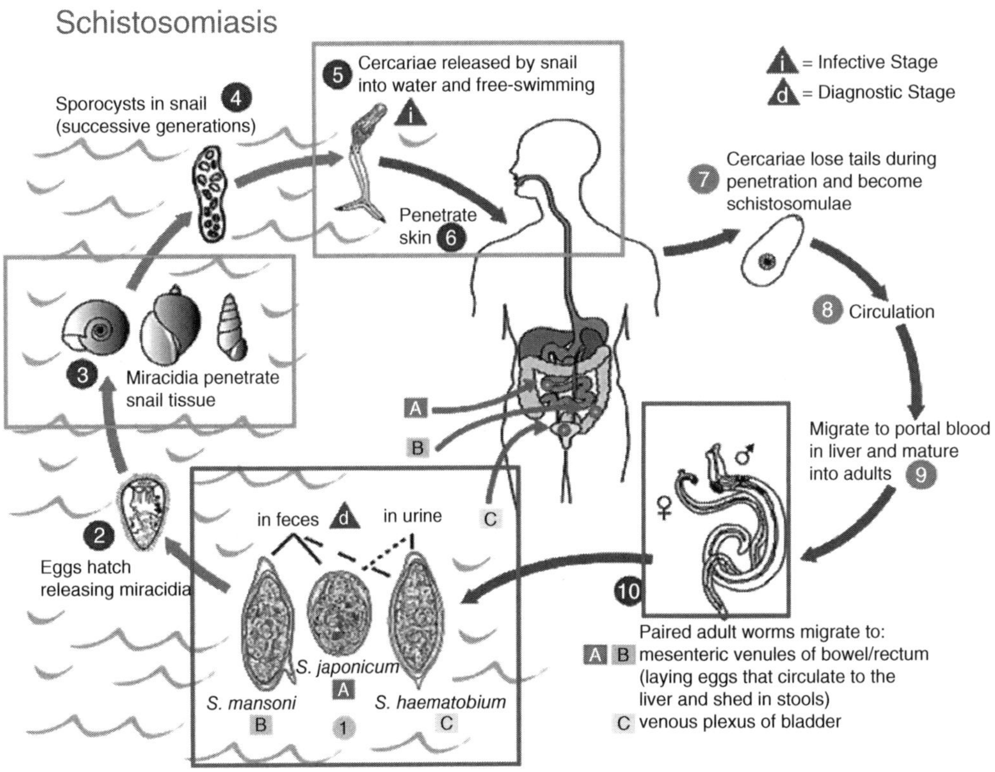

FIGURE 11.1 The transmission cycle of schistosomiasis. The numbers inserted in the text correspond to numbers indicated on the drawing (*Source*: With the courtesy of www. cdc.gov). (*See insert for color representation of this figure.*)

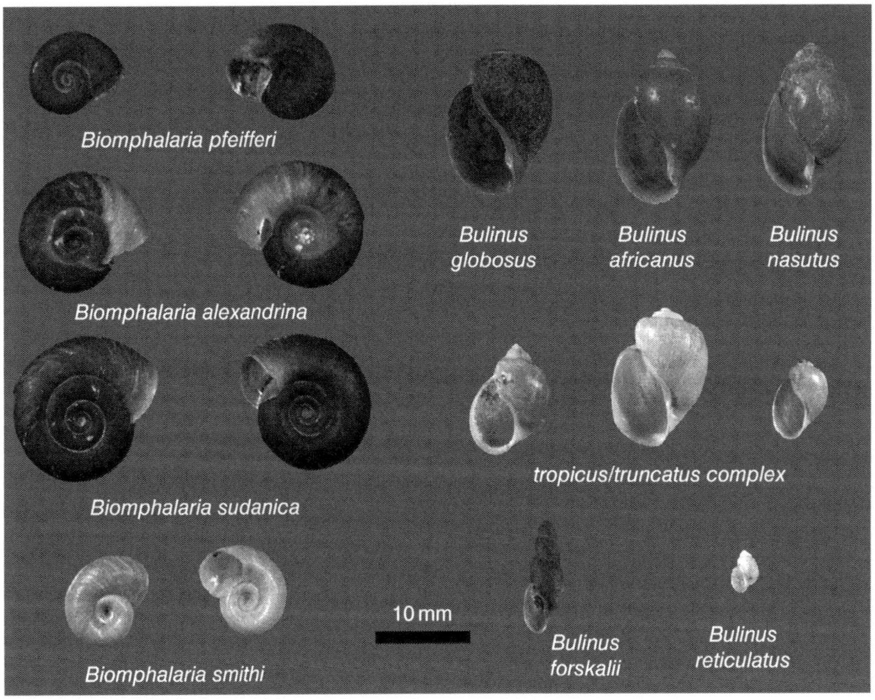

FIGURE 11.2 Schistosomiasis vector snails (*Source*: Photo: Henry Madsen). (*See insert for color representation of this figure.*)

FIGURE 11.3 Water contact in a branch of the Niger River at Diourou, Mali (*Source*: Photo: Henry Madsen). (*See insert for color representation of this figure.*)

FIGURE 11.4 Children playing in a small stream in Nigeria (*Source*: Photo: Henry Madsen). (*See insert for color representation of this figure.*)

After approximately five weeks, large numbers of cercariae (5), the infectious larval stage, are shed into the water and people are infected when the cercariae penetrate intact skin (6). People come into contact with water containing infectious larvae for several reasons, including washing themselves or their clothes and collecting water for drinking or cooking. Among children, playing or swimming in water is an important activity, while some adults, such as fishermen, car washers, canal cleaners, or people working in irrigation schemes or rice fields, may have a high occupational exposure [4] (Figures 11.3 to 11.6).

When the cercariae have penetrated the skin they transform into other larval forms, the schistosomula (7), which are either male or female and migrate in the body via the blood vessels (8). They pass through the lungs via the capillaries and reach the portal venous system where maturation takes place (9), and after pairing the worm pair travels to the small blood vessels around the intestines (10A, 10B) or urinary bladder (10C). Once they have established themselves, adult worms have an estimated mean lifespan of three years (*S. haematobium*) to seven years (*S. mansoni*). The female worms produce between 300 and 3000 eggs per day depending on the species. To reach the lumen of the intestine or urinary bladder and the exterior via the feces (1A, 1B) or urine (1C), eggs have to pass through the wall of the blood vessel as well as the intestinal or bladder wall, respectively [5] (Figure 11.7).

FIGURE 11.5 Canal cleaners removing aquatic plants from irrigation canals in the Sudan (*Source*: Photo: Danish Bilharziasis Laboratory). (*See insert for color representation of this figure.*)

FIGURE 11.6 Rice planting, Leyte, the Philippines (*Source*: Photo: Henry Madsen). (*See insert for color representation of this figure.*)

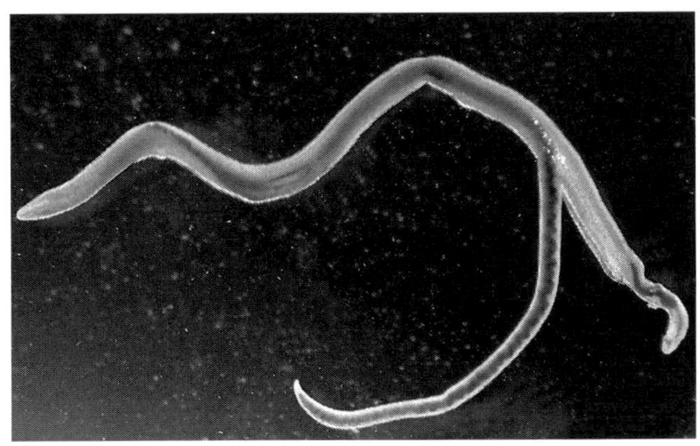

FIGURE 11.7 *Schistosoma* spp. worm pair. The female worm is located in the ventral canal of the male (*Source*: Photo: Maria Vang Johansen). (*See insert for color representation of this figure.*)

11.3 EPIDEMIOLOGY

In endemic countries children become infected when they start having contact with infected water. In many instances young children begin to acquire the infection from the age of three or four years but sometimes infections are found in even younger children. These very young children may sit in the shallow water where they are exposed to infection while the mother does her domestic chores such as washing clothes or utensils [6]. Both prevalence (the proportion of infected individuals) and intensity (the mean egg output) then rise sharply and in most endemic areas the proportion of infected individuals will be highest in the group aged 10–15 years, after which it reaches a plateau or declines among adults [7]. This pattern is caused by various factors. First, children generally have more frequent, prolonged, and extensive contact with water than adults. However, water contact patterns vary markedly between different locations and some communities show very high levels of occupational exposure among adults. Second, several studies have indicated that acquired immunity to schistosomiasis gradually develops among adults, so that they are reinfected to a much lesser extent than children [7]. Third, a progressive spontaneous death of adult worms from early infections acquired during childhood takes place over the years.

Local variations of infection patterns are pronounced and strongly influenced by gender roles [8]. Thus, in communities where women are secluded to the domestic area, for instance for religious reasons, they are relatively protected, as opposed to communities where they play an active role in agricultural work and fetching of water. Also, local perceptions of "appropriate behavior" for boys and girls may influence the likelihood that they swim in water bodies, thereby exposing themselves to the infection. Numerous anthropological studies have shown the many local varieties and adaptation patterns, which are important to know in order to plan appropriate control programs [8–10].

11.4 PATHOGENESIS

The eggs are the major cause of pathology in schistosomiasis. Approximately half of the eggs are not excreted with feces or urine but are retained in the tissue close to the adult worm pairs or swept into the portal circulation, where they lodge in the capillaries of the liver. A granulomatous inflammatory response is formed around the egg in the tissue. The inflammation is induced by antigens secreted by the miracidium inside the egg and this response is followed by fibrosis. Granuloma formation is a crucial feature in schistosomiasis pathology. Depending on the localization of the adult worms, granulomas will form in various organs, primarily the liver, large intestines including the rectum, and the genitourinary tract [11].

11.5 CLINICAL MANIFESTATIONS

Most infected people living in endemic areas have few overt symptoms but even a low intensity infection may be associated with anemia and have an impact on growth in children [1]. The clinical manifestations depend on the stage of infection. Cercarial penetration of the skin can provoke an immediate hypersensitivity reaction with intense itching and a rash, caused by an inflammatory reaction to cercaria. This is known as cercarial dermatitis or "swimmer's itch." A special and often severe type of cercarial dermatitis is seen after exposure to cercariae from bird schistosomes, e.g. *Tricobilharzia*, and this condition is regarded as a public health problem in areas where freshwater lakes are used for recreational purposes [12].

The migration of the young schistosomes does not normally give rise to symptoms, but when egg production starts an acute condition known as Katayama fever can be seen. It presents with high fever, eosinophilia, headache, and gastrointestinal symptoms, especially in people coming from non-endemic areas. It can resemble acute malaria, typhoid fever, or other febrile illnesses, but eosinophilia indicates Katayama fever [5].

It is the accumulation of eggs in the tissue and the resulting tissue inflammation and granuloma formation that causes the organ specific morbidity in schistosomiasis. The signs and symptoms seen in *S. haematobium* infection is the result of deposition of eggs in the urinary bladder wall, ureters, and genital organs. When the infection is established two stages can be recognized:

- An active stage with egg deposition in the genitourinary tract and egg excretion in urine. This is accompanied by blood (hematuria) and protein (proteinuria) in the urine. The blood in the urine is often clearly visible with the naked eye. This is mainly seen in children, adolescents, and younger adults.
- A chronic stage with sparse or absent urinary egg excretion but with urinary tract pathology and sometimes genital lesions.

Eggs are deposited in large conglomerates in the bladder wall from where they are sloughed off into the urine with ulceration and bleeding as a consequence. Hence, hematuria is often seen in *S. haematobium* infection, especially in the terminal portion of the urine voided. Eggs retained in the bladder wall cause inflammation and thickening of the wall. When the granulomas are situated in the ureters or near the ureteric openings into the urinary bladder, these can be constricted or even blocked, leading to obstructive uropathy with development of hydroureter and hydronephrosis. In some cases eggs can be found in other organs as well, such as the male and female genital organs, causing genital schistosomiasis, a condition that has received increasing attention because it has been hypothesized that genital lesions and inflammation caused by schistosome eggs may increase the transmission of HIV [13]. In rare cases eggs can lodge in the central nervous system or the spinal cord and give rise to severe neurological symptoms as a result of inflammation and edema. Infection with *S. haematobium* is associated with development of a specific type of cancer in the urinary bladder (squamous cell carcinoma) [14].

Intestinal schistosomiasis is associated with granuloma formation, inflammation, and fibrosis, primarily in the large intestine, and abdominal pain and bloody diarrhea can be seen, especially in high-intensity infections. Eggs that fail to traverse the wall of the blood vessel will be carried by the venous blood to the liver, and end in the small venuoles of the portal tract. Here they give rise to granulomas, which may result in enlargement of the liver and spleen. Granulomas can, in due course, lead to replacement by fibrous tissue (collagen) and result in periportal fibrosis (Symmers fibrosis), a typical lesion in schistosomiasis. Hepatosplenic schistosomiasis is a chronic manifestation of *S. mansoni* and *S. japonicum* infection. The term covers two clinical entities:

- Early inflammatory hepatosplenic disease with enlargement of the liver and spleen but generally no fibrosis. This is mainly seen in children and adolescents and the severity of organ enlargement is related to intensity of the infection. This type of hepatosplenic disease may also be associated with concomitant chronic exposure to malaria.

- Hepatosplenic disease with periportal fibrosis, which is mainly seen later in life, generally in young and middle-aged adults with long-standing intense exposure to infection. If organ enlargement and fibrosis is pronounced, portal hypertension may develop and in severe cases esophageal varices can occur. These can give rise to severe, life-threatening bleeding, where the infected person vomits large quantities of fresh blood (hematemesis). This is the most common cause of death in late-stage intestinal schistosomiasis. The frequency varies from area to area.

Besides the specific pathology and morbidity related to the species of schistosomes infecting humans (*S. haematobium* in the urinary tract, *S. mansoni, S. japonicum, S. mekongi*, and *S. intercalatum* in the gastrointestinal tract), there are more general and subtle effects on health such as anemia and growth retardation, which may lead to reduced school performance among children as well as impaired physical performance among adults [15].

11.6 DIAGNOSIS

Diagnosis of schistosomiasis is still made primarily by detection of eggs in urine or feces. The most widely used methods are urine filtration where 10 ml of urine is filtered where after the eggs trapped on the filter are counted or the Kato-Katz method for quantification of eggs in 25–50 mg of stool. The eggs are easily recognized and both methods are highly specific but they have a low sensitivity – especially the Kato-Katz method, which may easily underdiagnose low-intensity infections [16] (Figures 11.8 and 11.9).

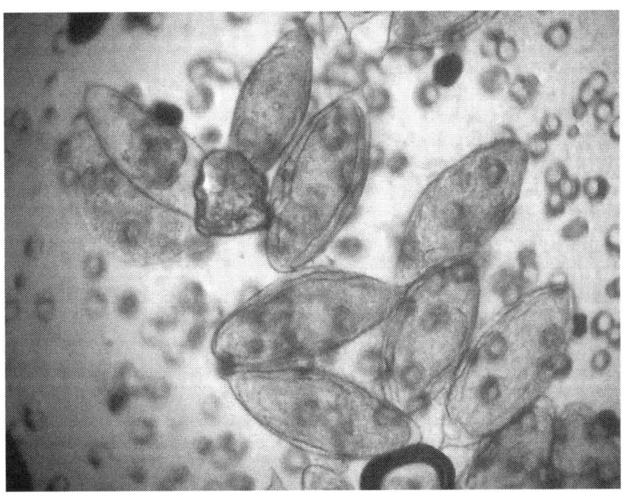

FIGURE 11.8 *Schistosoma haematobium* eggs on a Nucleopore filter after urine filtration (*Source*: Photo: Moussa Sacko). (*See insert for color representation of this figure.*)

Serological testing with detection of specific antibodies can be used in travelers from non-endemic areas returning with an infection, whereas antibody determination is of limited value in endemic areas with continued transmission. Another diagnostic approach is detection of circulating worm antigens in serum or urine and a simple Point-of Care test (POC-CCA test) for detection of worm antigens in urine has been developed. [17] (Figure 11.10).

Diagnosis is another issue, which is directly influenced by sociocultural and economic factors [8]. As the diagnosis is based on either urine or stool samples, local perceptions of privacy and embarrassment may in some cases lead to underdiagnosis simply because certain groups (e.g., women) shy away from tests. To the extent that the healthcare providers charge for the diagnostic procedures or the clinics are

FIGURE 11.9 Preparation of Kato-Katz microscopy slides from stool samples (*Source*: Photo: Birgitte J. Vennervald).

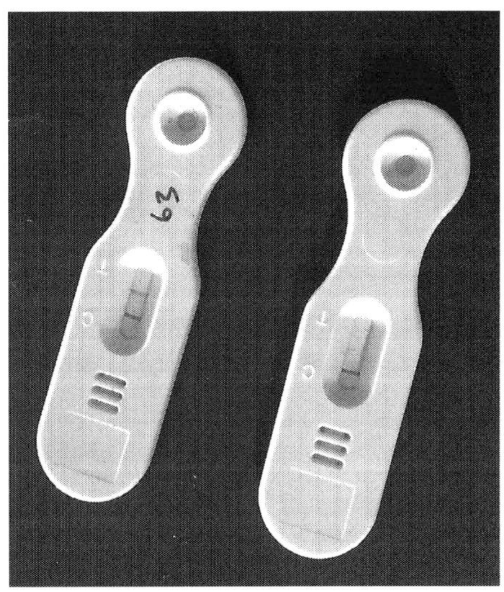

FIGURE 11.10 Point-of Care test (POC-CCA test) for detection of worm antigens in urine (two lines: positive test; one line: negative test) (*Source*: Photo: Henry Madsen).

situated far away, this may also inhibit diagnosis due to lack of funds. Thus, with respect to diagnosis, the local variations in contextual factors should be taken into consideration.

Detection of blood in urine can be used as an indirect indicator of *S. haematobium* infection either by direct visual inspection or by use of urine analysis reagent strips. This has been employed successfully as a means of cost-effective community diagnosis – especially in control programs directed toward schoolchildren [18]. No similar diagnostic procedure exists for intestinal schistosomiasis. The POC-CCA test which is based on urine samples may prove useful in this respect [19].

11.7 SOCIAL DYNAMICS INFLUENCING SCHISTOSOMIASIS TRANSMISSION

Population movements are relevant for the spread of schistosomiasis in many ways [20]. The main mechanism is that movement leads to fresh exposure to the infectious agents, either because the migrating population moves into new areas and gets in touch with infective populations or environments or because the migrating population carries the infection and brings it to hitherto unexposed populations at their point of arrival. People move for a multitude of different reasons, for example, pilgrimage, labor migration, pastoralism, trade, or tourism. It may also be caused by forced resettlement or refuge due to either war or natural disasters. All these aspects of population movements are directly linked to the spread of schistosomiasis as well as other infectious diseases. In some cases snails containing the parasites may also be transported from one place to another, thereby transferring the infection. This was seen during the outbreak of urogenital schistosomiasis in Corsica in 2013, where molecular data suggested that the parasite was imported by individuals infected in West Africa [21]. The global number of migrants is rapidly increasing due to wars and political unrest, and it is likely to increase further, when the consequences of cli mate change becomes more pronounced [22].

Urbanization is rapidly increasing also in schistosomiasis endemic countries. This phenomenon usually entails a combination of poor infrastructure in terms of water supply, sewage, and sanitation, and people often migrate to and from rural disease endemic areas. This combination of factors may create local environments suitable for transmission of schistosomiasis.

Large-scale irrigation and hydroelectric power schemes play an important role for establishment of new, or increase in existing, transmission [23]. To mitigate the negative effects of such water management projects it has become increasingly accepted that health impact assessments are needed as part of the project development process. Major health problems may ensue when people migrate to newly developed irrigations schemes, especially when coming from non-schistosomiasis endemic areas. Schistosomiasis with epidemic levels of transmission, high morbidity, and difficulties in treatment due to high infection intensities and rapid reinfection has occurred following such an influx of previously non-exposed populations [24, 25]. Sustaining control in such situations has also proven difficult due to inadequate health infrastructure and public health resources. However, once established, the transmission and continued exposure leads to fast development of partial immunity, thereby reducing the level of reinfection and intensity [26]. Another environmental aspect of schistosomiasis relates to climate change. The distribution of *Oncomelania hupensis*, the intermediate snail host for *S. japonicum*, is limited to areas where the mean January temperature in above 0 °C. Furthermore, the development of the schistosome stages in the snail is arrested at temperatures below 15 °C. Thus, the expected global warming may lead to an increase in the distribution of host snails and favor parasite development and result in significant increases in populations at risk, especially in China [27].

On a more general level, poverty constitutes an important social determinant for schistosomiasis in a multitude of ways. Poor people are likely to be more exposed due to more risky work, inadequate access to piped water, sanitary facilities, and medical services [28].

11.8 CONTROL

With the discovery of the complete life cycle of schistosomiasis during the period 1902–1915 [3] and the subsequent description of the various species of intermediate vector snails involved in the life cycle as well as the epidemiology and the pathogenesis, it became possible to put forward strategies for schistosomiasis control. The transmission cycle (Figure 11.1) illustrates the various places where the transmission can be interrupted and the disease thereby controlled.

Transmission of schistosomiasis is closely linked to human practices related to contact with freshwater bodies and sanitation practices. The following parameters are prerequisites for the transmission of schistosomiasis:

- Freshwater bodies harboring efficient snail intermediate hosts. It is the distribution of the snails that determines the distribution of schistosomiasis in any given area.
- A human population susceptible to infections with schistosomes.
- Poor sanitation leading to contamination of snail-infested water with feces or urine containing schistosome eggs.
- Existence of an infectious water sources leading the population to have direct contact with contaminated water for work, washing, and leisure purposes.

In principle, schistosomiasis should be a totally preventable infection. However, poor access to safe water in most rural areas of Egypt, Sudan, and sub-Saharan Africa, and elsewhere in low and middle income countries in South America and East Asia, combined with lack of adequate and appropriate sanitation, keep transmission at a high level in many endemic areas. It is, however, important to underline the often very focal nature of transmission, which has an impact on control strategies and implementation of control programs. As can be seen in Figure 11.1 (color plate), measures preventing transmission can basically be aimed at preventing or reducing contact with infested water (Figure 11.1, green boxes) and/or preventing or reducing contamination of water with schistosome eggs (Figure 11.1, blue boxes). Both aims can be obtained using disease-specific and non-disease specific control strategies and tools, as shown in Table 11.1.

TABLE 11.1 Classification of control strategies.

Strategy	Disease specific	Non-disease specific
Prevent or reduce contact with infected water	Control of intermediate vector	Adequate safe water supply for all domestic and leisure purposes and health education
Prevent or reduce contamination of water with schistosome eggs	Morbidity reduction by chemotherapy	Improve sanitation (latrines) and health education

Source: Adapted from Jordan et al. [3].

Initial attempts to control schistosomiasis focused on snail control, health education, and in some instances provision of a safe water supply and sanitation as part of more extensive development projects. Especially in Egypt and Sudan, snail control in irrigation canals remained the first-line strategy for decades. Major problems encountered were the high cost of molluscicides and the need for regular retreatment of water bodies due to reinvasion of snails. Furthermore, molluscicides were environmentally problematic as they also affected other species such as fish (Figure 11.11).

With the arrival during 1970–1980 of safe and effective antischistosomicidal drugs (metrifonate for *S. haematobium* and oxamniquine for *S. mansoni* and later praziquantel, which is effective against all species of schistosomes [29]), rational chemotherapy became possible at the individual level. With praziquantel, which can be given as a single dose, the concept of large-scale treatment targeted to at-risk populations or even whole populations developed. Due to the epidemiology of schistosomiasis the greatest risk of high intensity infection and pathology is found in school-age children and this group has thus routinely been targeted for mass drug administration programs. However, there are high-risk occupational groups among adults such as fishermen, car washers, and irrigation workers who should also be targeted for mass chemotherapy. It should be borne in mind that a control strategy, which depends mostly on one drug, praziquantel, is very vulnerable, because there is a risk that the parasites develop resistance toward the drug, although this has not been recorded during the last 10–15 years where large-scale preventive mass chemotherapy has been implemented.

Several studies have shown that repeated single-dose treatments with praziquantel once a year have an impact on

FIGURE 11.11 Molluscicidng (*Source*: Photo: Henry Madsen).

morbidity [30]. Due to reinfection, prevalence of infection remains high but intensity of infection is significantly reduced, and as morbidity is related to infection intensity, a prolonged effect on morbidity can be sustained with repeated annual treatments. Special considerations are needed regarding the zoonotic varieties of schistosomiasis caused by *S. japonicum* and *S. mekongi*, where other mammals such as pigs and water buffaloes are also definitive hosts and maintain a large reservoir of infection. In those endemic areas, pharmaceutical treatment of animals or the use of tractors instead of buffaloes to pull plows can be an integrated part of the control programs [31]. Hybridization between *S. haematobium* and the cattle schistosome *Streptococcus bovis* played a role in the outbreak of urogenital schistosomiasis in Corsica and illustrate how hybridization could increase the colonization potential of schistosomes [21].

There has been an increasing trend towards aiming at elimination and even eradication of schistosomiasis. However, in the short term, morbidity control will remain the cornerstone of schistosomiasis control often disseminated by mass drug administration programs. Though usually effective in the short term, they may entail long-term challenges. Partly, such programs depend on continued external funding. Changing priorities or financial difficulties of the donors may interrupt the program activities, thereby threatening their sustainability. Furthermore, there can be significant problems in securing high coverage if a strategy of mass drug administration is adopted unless the involved populations take ownership of the strategy and get involved in its implementation. WHO consider that in sub-Saharan Africa less than 45% of the school-aged populations in need access regular treatment with praziquantel, although the recommended strategy operates with a coverage of more than 75% [32]. Poor communication and insufficient sensitization of target populations as well as lack of understanding of local perceptions, attitudes, and practices may also lead to non-acceptance and resistance against a program [33–35]. In control programs with a more community directed approach, involving village health workers [36], it will be possible to improve existing strategies, taking factors into account such as timing of treatment according to the transmission season and including short interval (weeks) spaced treatments to ensure that juvenile worms get treated, thereby reducing morbidity [37, 38].

In the long term, schistosomiasis control should be based on the combination of regular treatment of high-risk population segments combined with an increasing effort to provide safe water and sanitation as well as appropriate health education approaches. Health impact assessments should be undertaken before major water management projects are implemented in order to provide information predicting changes in the epidemiology of existing schistosomiasis or provide bases for interventions to prevent introduction of schistosomiasis transmission [39].

Health education is a cross-cutting control activity, which underpins most of the other interventions. It is meant to motivate and guide the patients and targeted population segments in safe practices (avoiding risky water contact and using latrines) and the most appropriate ways of utilizing healthcare services with respect to diagnosis and treatment. However, numerous anthropological studies have shown that laypeople's perceptions of illness are often quite different from those promoted by the biomedical establishment and treatment-seeking practices often lead to self-treatment or take the patients to traditional healers. For instance, in many disease endemic countries hematuria (red urine) is so common among teenagers that it locally has been perceived as a normal sign of coming to age [40]. Consequently, it is mandatory that health education interventions be based on a thorough understanding of local perspectives, so that they can be formulated accordingly. Recently there has been a call to strengthen the global strategy against schistosomiasis and soil-transmitted helminthiasis, including refocusing on complementary interventions such as sanitation, safe water, snail control and information, education, and communication programs [41].

In order to ensure sustainability, schistosomiasis control should be undertaken within the framework of existing health services and integrated with other disease control programs. The importance of the way in which healthcare systems are organized is clearly illustrated by a case from China [42]. Whereas China achieved remarkable results in schistosomiasis control based on prevention and mass campaigns during the communist economic system, the neo-liberal reorganization has introduced an incentive-structure for the schistosomiasis control centers that encourages the health staff to emphasize curative services – often with unnecessarily expensive drugs for extended periods.

Several combinations of environmental and snail control, water, sanitation, and health education approaches with or without treatment have been tested in small scale with varying results. However, sustainability has been difficult to ensure in the long term in areas of high transmission.

Historically, schistosomiasis control has thus gone through a variety of different strategies. Initially it focused on vector snail control combined with health education and provision of safe water and sanitation. Later focus turned to morbidity reduction using regular administration of safe single-dose drugs [43–46]. Throughout the history of schistosomiasis control and up till now the implemented strategies have been focused on biomedical and technical aspects (vector control and morbidity control) while the social context of schistosomiasis has played a minor role for strategy and tools development [8].

Schistosomiasis is part of what is termed the Neglected Tropical Diseases (NTDs). Looking at this cluster of diseases as a whole provides new perspectives on control options and cost-effectiveness of interventions [47]. Integration of control programs for co-endemic NTDs such as soil-transmitted helminths, lymphatic filariasis, and onchocerciasis offers new ways of optimal use of sparse resources [48].

11.9 CONCLUSION

Schistosomiasis is considered to be the second most important parasitic infection after malaria in terms of public health impact and poor access to safe water combined with lack of adequate and appropriate sanitation keep transmission at a high level in many endemic areas. Based on recommendations from WHO, population-based treatment with praziquantel is now the main component in most national control programs. However, in the absence of ecological and behavioral changes this strategy has a minimal effect on transmission. This means that regular re-treatment of affected populations has to be continued for an unknown period and sustainability becomes a key element in treatment-based control. The transmission pattern and control of schistosomiasis is complex due to the intricate interaction of biological, environmental and sociocultural factors. Consequently, control measures should adapt to local conditions rather than applying simplistic, universal strategies. Controlling schistosomiasis is not just a matter of distributing drugs. Strengthening the local health systems so they are able to take care of patients and to integrate sustainable control measures are equally important and the social context of schistosomiasis should not be ignored. Schistosomiasis can in principle be controlled by behavioral changes, sanitation, and a safe water supply, but these measures will solve the problem only if poverty is eliminated.

REFERENCES

1. King, C.H. and Dangerfield-Cha, M. (2008). The unacknowledged impact of chronic schistosomiasis. *Chronic Illn.* 4: 65–79.
2. van der Werf, M.J., de Vlas, S.J., Brooker, S. et al. (2003). Quantification of clinical morbidity associated with schistosome infection in sub-Saharan Africa. *Acta Trop.* 86: 125–139.
3. Jordan, P., Webbe, G., and Sturrock, R.F. (eds.) (1993). *Human Schistosomiasis*. Wallingford: CAB International.
4. Tameim, O., Abdu, K.M., el Gaddal, A.A., and Jobin, W.R. (1985). Protection of Sudanese irrigation workers from schistosome infections by a shift to earlier working hours. *J. Trop. Med. Hyg.* 88: 125–130.
5. Gryseels, B., Polman, K., Clerinx, J., and Kestens, L. (2006). Human schistosomiasis. *Lancet* 368: 1106–1118.
6. Odogwu, S.E., Ramamurthy, N.K., Kabatereine, N.B. et al. (2006). *Schistosoma mansoni* in infants (aged < 3 years) along the Ugandan shoreline of Lake Victoria. *Ann. Trop. Med. Parasitol.* 100: 315–326.
7. Kabatereine, N.B., Vennervald, B.J., Ouma, J.H. et al. (1999). Adult resistance to schistosomiasis mansoni: age-dependence of reinfection remains constant in communities with diverse exposure patterns. *Parasitology* 118: 101–105.
8. Bruun, B. and Aagaard-Hansen, J. (2008). *The Social Context of Schistosomiasis and its Control. An Introduction and Annotated Bibliography*. Geneva: UNICEF/UNDP/World Bank/WHO Special Programme for Research and Training in Tropical Diseases (TDR).
9. Parker, M. (1993). Bilharzia and the boys: questioning common assumptions. *Soc. Sci. Med.* 37 (4): 481–492.
10. El Katsha, S. and Watts, S. (2002). *Gender, Behavior, and Health: Schistosomiasis Transmission and Control in Rural Egypt*. Cairo: American University in Cairo Press.
11. AAF, M. (ed.) (2001). *Schistosomiasis*, Tropical Medicine. Science and Practice, vol. 3, (eds G Pasvol, SL Hoffman). London: Imperial College Press.
12. Horák, P., Mikeš, L., Lichtenbergová, L. et al. (2015). Avian schistosomes and outbreaks of cercarial dermatitis. *Clin. Microbiol. Rev.* 1: 165–190.
13. Kjetland, E.F., Ndhlovu, P.D., Gomo, E. et al. (2006). Association between genital schistosomiasis and HIV in rural Zimbabwean women. *AIDS* 20 (4): 593–600.
14. IARC Working Group on the Evaluation of Carcinogenic Risks to Humans. Lyon, 24th February–3rd March, 2009. Schistosomes, liver flukes and *Heliobacter pylori*. IARC Monogr Eval Carcinog Risks Hum;X:x–xx.
15. World Bank (2006). *Disease Control Priorities in Developing Countries*, Chapter 24, 2e. New York: A co-publication of the World Bank and Oxford University Press.
16. Enk, M.J., Lima, A.C., Drummond, S.C. et al. (2008). The effect of the number of stool samples on the observed prevalence and the infection intensity with *Schistosoma mansoni* among a population in an area of low transmission. *Acta Trop.* 108 (2–3): 222–228.
17. van Dam, G.J., Wichers, J.H., Ferreira, T.M. et al. (2004). Diagnosis of schistosomiasis by reagent strip test for detection of circulating cathodic antigen. *J. Clin. Microbiol.* 42: 5458–5461.
18. Red Urine Study Group. Identification of high-risk communities for schistosomiasis in Africa: a multicountry study. Geneva: World Health Organization, Special Programme for Research and Training in Tropical Diseases (Social and Economic Research Projects Report, No. 15; TDR/SER/PRS/15);1995.
19. Adriko, M., Standley, C.J., Tinkitina, B. et al. (2014). Evaluation of circulating cathodic antigen (CCA) urine-cassette assay as a survey tool for in different transmission settings within Bugiri District, Uganda. *Acta tropica.* 136: 50–57.
20. Aagaard-Hansen, J., Nombela, N., and Alvar, J. (2010). Population movement: a key factor in the epidemiology of neglected tropical diseases. *Trop. Med. Int. Health* 15: 1281–1288.
21. Boissier, J., Grech-Angelini, S., Webster, B.L. et al. (2016). Outbreak of urogenital schistosomiasis in Corsica (France): an epidemiological case study. *Lancet Infect. Dis.* 16: 971–979.
22. International Organization for Migration. Available from: https://www.iom.int/world-migration. Accessed September 2017.

23. Steinmann, P., Keiser, J., Bos, R. et al. (2006). Schistosomiasis and water resources development: systematic review, meta-analysis, and estimates of people at risk. *Lancet Infect. Dis.* 6: 411–425.
24. Gryseels, B., Stelma, F.F., Talla, I. et al. (1994). Epidemiology, immunology and chemotherapy of *Schistosoma mansoni* infections in a recently exposed community in Senegal. *Trop. Geogr. Med.* 46 (4 Spec. No.)): 209–219.
25. Stelma, F.F., Talla, I., Sow, S. et al. (1995). Efficacy and side effects of praziquantel in an epidemic focus of *Schistosoma mansoni*. *Am. J. Trop. Med. Hyg.* 53 (2): 167–170.
26. Polman, K., Stelma, F.F., Le Cessie, S. et al. (2002). Evaluation of the patterns of *Schistosoma mansoni* infection and re-infection in Senegal, from faecal egg counts and serum concentrations of circulating anodic antigen. *Ann. Trop. Med. Parasitol.* 96 (7): 679–689.
27. Zhou, X.-N., Yang, G.-J., Yang, K. et al. (2008). Potential impact of climate change on schistosomiasis transmission in China. *Am. J. Trop. Med. Hyg.* 78 (2): 188–194.
28. Commission on Social Determinants of Health (2008). *Closing the Gap in a Generation: Health Equity through Action on the Social Determinants of Health*. Final Report. Geneva: World Health Organization.
29. WHO (1995). *Model Prescribing Information. Drugs Used in Parasitic Diseases*, 2e. Geneva: World Health Organization.
30. Magnussen, P. (2003). Treatment and re-treatment strategies for schistosomiasis control in different epidemiological settings: a review of 10 years' experiences. *Acta Trop.* 86: 243–254.
31. Jiang, Z., Zheng, Q., Wang, X., and Hua, Z. (1997). Influence of livestock husbandry on schistosomiasis transmission in mountainous regions of Yunnan province. *Southeast Asian J. Trop. Med. Public Health* 28 (2): 291–295.
32. WHO (2016). Schistosomiasis and soil-transmitted helminthiases: no. of people treated 2015. *Wkly. Epidemiol. Rec.* 91 (49/50): 585–595.
33. Mwanga, J.R., Magnussen, P., Mugashe, C.L. et al. (2004). Schistosomiasis related perceptions, attitudes and treatment seeking practices in Magu district, Tanzania: public health implications. *J. Biosoc. Sci.* 36: 63–81.
34. Parker, M., Allen, T., and Hastings, J. (2008). Resisting control of neglected tropical diseases: dilemmas in the mass treatment of schistosomiasis and soil- transmitted helminths in north-west Uganda. *J. Biosoc. Sci.* 40: 161–181.
35. Fleming, F.M., Fenwick, A., Tukahebwa, E.M. et al. (2009). Process evaluation of schistosomiasis control in Uganda, 2003–2007: perceptions, attitudes and constraints of a national programme. *Parasitology* 136 (13): 1759–1769.
36. Massa, K., Olsen, A., Sheshe, A. et al. (2009). Can coverage of schistosomiasis and soil transmitted helminthiasis control programmes targeting school-aged children be improved? New approaches. *Parasitology* 136: 1781–1788.
37. Augusto, G., Magnussen, P., Kristensen, T. et al. (2009). The influence of transmission season on parasitological cure rates and intensity of infection after praziquantel treatment of *Schistosoma haematobium* in Mozambique. *Parasitology* 136: 1771–1779.
38. Sacko, M., Magnussen, P., Traoré, M. et al. (2009). The effect of single dose versus two doses of praziquantel on *Schistosomiasis haematobium* infection and pathology among school-aged children in Mali. *Parasitology* 136: 1851–1857.
39. Lock, K. (2000). Health impact assessment. *BMJ* 320: 1395–1398.
40. Amazigo, U.O., Anago-Amanze, C.I., and Okeibunor, J.C. (1997). Urinary schistosomiasis among school children in Nigeria: consequences of indigenous beliefs and water contact activities. *J. Biosoc. Sci.* 1: 9–18.
41. Lo, N.C., Addiss, D.G., Hotez, P.J. et al. (2017). A call to strengthen the global strategy against schistosomiasis and soil-transmitted helminthiasis: the time is now. *Lancet Infect. Dis.* 2: e64–e69.
42. Bian, Y., Sun, Q., Zhao, Z., and Blas, E. (2004). Market reform: a challenge to public health – the case of schistosomiasis control in China. *Int. J. Health Plann. Manag.* 19: 79–94.
43. Jordan, P. and Rosenfield, P.L. (1983). Schistosomiasis control: past, present, and future. *Annu. Rev. Public Health* 4: 311–334.
44. WHO. Technical report Series. Epidemiology and Control of schistosomiasis. Technical report Series No. 643, Geneva: World Health Organization; 1980.
45. Report of a WHO Expert Committee. The Control of Schistosomiasis. Technical report Series No. 728, Geneva: World Health Organization; 1985.
46. Report of a WHO Expert Committee. Prevention and control of schistosomiasis and soil-transmitted helminthiasis. Technical report Series No. 912, Geneva: World Health Organization; 2001.
47. Aagaard-Hansen, J. and Chaignat, C.-L. (2010). Neglected tropical disease: equity and social determinants. In: *Equity, Social Determinants and Public Health Programmes* (ed. E. Blas and A.S. Kurup), 135–157. Geneva: WHO.
48. Deribe, K., Meribo, K., Gebre, T. et al. (2012). The burden of neglected tropical diseases in Ethiopia, and opportunities for integrated control and elimination. *Parasit. Vectors* 5: 240.

12

TRACHOMA

EMMA M. HARDING-ESCH[1], JOSEPH A. COOK[2], DAVID C. MABEY[1], AND ANTHONY W. SOLOMON[3]

[1] *London School of Hygiene & Tropical Medicine, University of London, London, UK*
[2] *University of North Carolina at Chapel Hill School of Public Health, Chapel Hill, NC, USA*
[3] *Department of Control of Neglected Tropical Diseases, World Health Organization, Geneva, Switzerland*

Trachoma is the world's leading infectious cause of blindness [1]. It spread throughout Europe during the Napoleonic Wars in the early nineteenth century, fuelled by returning British and French soldiers who had suffered an epidemic of "Egyptian ophthalmia" while fighting in Egypt [2], and then to North America predominantly via immigration. This led to the establishment of many now-venerable eye hospitals, and the establishment of ophthalmology as a medical discipline [3]. Trachoma disappeared from Europe and North America in the 1950s as a consequence of improved living and hygiene standards, pre-dating antibiotic treatment [3, 4]. However, it remains a significant cause of blindness worldwide, focalized in poor, dry, rural communities of developing countries and the Aboriginal communities of Australia [5]. The annual productivity losses due to trichiasis (in-turned eyelashes touching the eyeball) and trachoma-related reduced vision and blindness are up to US$8 billion [6]. In contrast, it has been estimated that trachoma elimination will cost approximately US$1 billion, with US$200–300 million already committed [7], and that investment in the SAFE strategy (described below) coordinated with other Neglected Tropical Diseases (NTDs) would prevent the loss of 4 million healthy life years by 2030 [8].

Population-based prevalence surveys, the quantity and quality of which have been considerably heightened through the Global Trachoma Mapping Project (GTMP) [9], indicate that 190 million people live in trachoma-endemic districts [10], 3.2 million need trichiasis surgery to avoid blindness due to trachoma and 1.9 million are blind or visually impaired due to trachoma [7, 11]. The current trachoma prevalence estimates are lower than those previously reported by the World Health Organization (WHO) [12]; this can be explained by the success in implementing a control strategy, by more accurate data, and by socioeconomic development in endemic countries [10]. Since 1998, great changes in the global trachoma burden have taken place as a result of rollout of the comprehensive, WHO-recommended strategy for elimination (SAFE, described below), the founding of the WHO Alliance for the Global Elimination of Trachoma by the year 2020 (GET2020), the GTMP, and its successor the Tropical Data service. With continued commitment by all GET2020 partners, global elimination of trachoma as a public health problem is within reach.

12.1 CAUSATIVE ORGANISM AND NATURAL HISTORY

Trachoma is caused by the Gram-negative bacterium *Chlamydia trachomatis*. Serovars A, B, Ba, and C are associated with trachoma, serovars D-K tend to infect the genital tract, and L1-L3 cause lymphogranuloma venereum (LGV). The non-LGV serovars target columnar and squamocolumnar epithelial cells, and are therefore infections of the conjunctivae and the linings of the genital, respiratory, and intestinal tracts [4]. *C. trachomatis* is an obligate intracellular bacterium and cannot replicate outside eukaryotic host cells. It has a unique 48–72 hours *in vitro* biphasic developmental cycle with dimorphic forms: the infectious elementary body (EB) and the non-infectious, metabolically active reticulate

Water and Sanitation-Related Diseases and the Changing Environment: Challenges, Interventions, and Preventive Measures,
Second Edition. Edited by Janine M. H. Selendy.

Companion website: https://www.wiley.com/legacy/wileychi/selendy/

body (RB). Within a host cell, the EB is encased in a membrane-bound vesicle that matures into an inclusion body. The EB differentiates into RBs that replicate rapidly and begin to transform into EBs. Following cell lysis and death, the released spore-like EBs attach to and enter other epithelial cells through endocytosis. The presence of these EBs in eye and nose secretions of infected persons (children especially) [13, 14] facilitate further transmission to family or contacts.

The intracellular location of the organism protects it from antibody and complement and there is a downregulation of the host major histocompatibility complex class I molecules by infected cells, thereby reducing killing by cytotoxic T cells. The frequent re-infection of individuals indicates that there is not a long-lasting protective immune response to infection. However, there is evidence of partial immune protection, with data demonstrating a reduction in the prevalence of active trachoma and ocular *C. trachomatis* duration with age [13]. While the immunopathology of trachoma is not well understood, it is accepted that scarring, trichiasis, and the resulting corneal damage occur after repeated infections. Detailed reviews of immunity and immunopathology are available in three Lancet Seminars on trachoma [15–17].

12.2 CLINICAL MANIFESTATIONS

After a brief incubation period of approximately 5–10 days, the initial infection will result in a mild mucopurulent conjunctivitis that is generally self-limiting and heals without permanent damage [15]. Repeated infections can result in intense inflammation and scarring of the subtarsal conjunctiva, causing distortion of the lid margin and resulting in shortening of the upper eyelid, pulling the eyelashes inwards to scratch against the eyeball (trichiasis). Abrasion of the cornea can eventually lead to corneal opacity and blindness. WHO recommends a simplified grading system for the assessment of trachoma in communities by non-specialist health personnel (Figure 12.1) [18], where 2.5× binocular magnifying loupes are used, with good lighting (torch or daylight). The upper eyelid is everted to examine for trachomatous inflammation—follicular (TF): the presence of five or more characteristic white or yellow follicles, each with a diameter ≥ 0.5 mm, in the central part of the upper tarsal conjunctiva. Some individuals develop intense inflammatory disease (trachomatous inflammation—intense (TI)), in which pronounced inflammatory thickening obscures more than half the normal deep tarsal vessels. People with TI have higher *C. trachomatis* bacterial loads and are more likely to develop conjunctival scarring [14, 19, 20]. Clinically apparent trachomatous scarring (TS) begins as small stellate scars that coalesce with time to form the dense scar tissue that distorts normal lid architecture. This is followed, in some individuals and usually after many years, by frank trachomatous trichiasis (TT) and corneal opacity (CO). The WHO simplified grading system shown in Figure 12.1 has allowed the trachoma burden to be assessed in prevalence surveys, population-level elimination efforts to be monitored and evaluated, and the changing pattern of trachoma prevalence over time to be analyzed.

12.3 DIAGNOSIS

Trachoma is generally diagnosed based on the clinical signs of the WHO simplified grading system. Laboratory techniques are available, but are not routinely used outside of research projects due to their expense and complexity. However, should it be feasible, detection of ocular *C. trachomatis* infection could help prioritize areas most in need of antibiotic, face cleaning, and environmental improvement interventions, the aim of which is to limit transmission of *C. trachomatis* to a point at which TT and blindness from trachoma do not ensue [21]. Although not generally needed for programmatic use in trachoma-endemic areas before mass drug administration (MDA) with antibiotics, where the prevalence of TF and ocular *C. trachomatis* infection is well correlated, such tests may have a future role to play post-MDA, when the positive predictive value of clinical diagnosis for ocular *C. trachomatis* infection falls [22]. With additional evidence and issuance of formal guidance, these assays could one day allow the discontinuation of antibiotic treatment sooner than reliance on clinical grading alone would otherwise dictate [23]. Serological assays also have a potential future role in guiding these decisions, including determining whether clinical signs ascribed to trachoma do, in fact, have a chlamydial etiology; a recent study in the Solomon Islands demonstrated that 80% of TF cases were serologically negative for prior *C. trachomatis* infection [24].

For detection of the presence of the organism itself, Giemsa staining, culture, and enzyme immunoassays have generally been superseded by nucleic acid amplification tests (NAATs), which have higher sensitivity and specificity. Although commercial NAATs have been developed primarily for diagnosis of urogenital chlamydial infections, a number have been successfully used for the detection of ocular *C. trachomatis* infections in the research context. The use of a commercial NAAT has been shown to be cost-saving relative to starting or continuing unnecessary community MDA [25], and reduced-cost NAATs for potential use in trachoma elimination programs have been developed [26]. No high-performing point-of-care tests (POCTs) for use in trachoma control programs currently exist [27]. Serological assays for detection of anti-*C. trachomatis* antibodies have also been developed and evaluated for use as a tool for post-elimination surveillance, but at present there is insufficient evidence to recommend the use of NAATs or serological assays within national trachoma elimination programs [23]. Therefore, for the time being, clinical

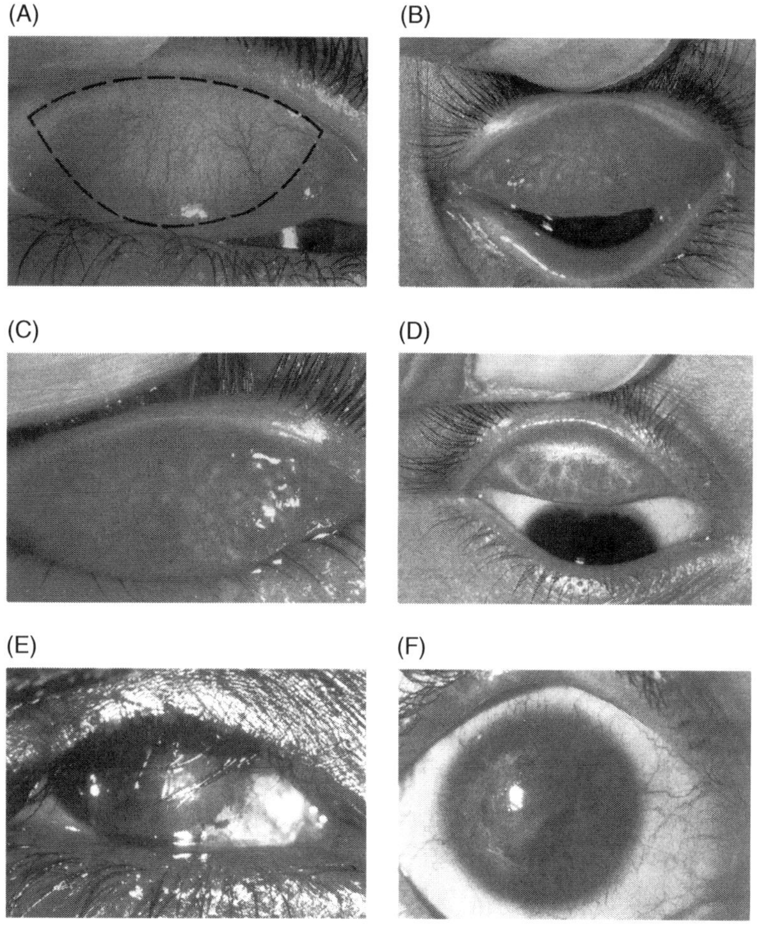

FIGURE 12.1 Clinical appearance of trachoma and the WHO simplified grading system [18]. (A) Normal everted upper tarsal conjunctiva; pink, smooth, thin, and transparent. Large deep-lying blood vessels that run vertically are present over the tarsal conjunctiva. The dotted line shows the area to be examined. (B) Trachomatous inflammation—follicular (TF): five or more follicles of >0.5 mm. (C) Trachomatous inflammation—follicular and intense (TF + TI): inflammatory thickening obscuring >one-half of the normal deep tarsal vessels. (D) Trachomatous scarring (TS). (E) Trachomatous trichiasis (TT). (F) Corneal opacity (CO). (These photographs are reproduced with permission from WHO.)

examination using the WHO simplified grading system remains the most practical and cost-effective diagnostic method for public health purposes.

12.4 EPIDEMIOLOGY

The Global Trachoma Mapping Project (GTMP) worked with health ministries to undertake standardized and robust baseline population-based prevalence surveys between December 2012 and January 2016 [28], resulting in over 2.6 million people being examined in 1546 districts across 29 countries [29]. The largest proportion (90%) of the estimated 190 million people living in trachoma-endemic areas [10] are in WHO's African Region, with the remainder in the WHO regions of the Eastern Mediterranean, the Western Pacific, South-East Asia, and the Americas [30]. Nearly half of the people at risk of trachoma live in three countries: Ethiopia, Malawi, and Nigeria. Ethiopia is the most affected country, with 75 million people at risk [6]. As of January 2018, five countries had officially been validated as having eliminated trachoma as a public health problem: Cambodia, Lao People's Democratic Republic, Mexico, Morocco, and Oman. A further six countries had reported achieving the elimination targets: China, Gambia, Ghana, Islamic Republic of Iran, Myanmar, and Nepal [10].

Children are the main reservoir of ocular *C. trachomatis* infection, with trachoma clustered at bedroom, household, compound, neighborhood, and community levels as a result of close contact easing transmission of infection via ocular and nasal secretions [17] (Figure 12.2). While children below 10 years of age account for the majority of infections, the late stages (scarring and TT) are predominantly seen in adults. Women have a 1.8-fold higher risk of TT compared with men

[31]; the cause of this is thought to be the frequent re-infection that many women experience while providing childcare, although other possible explanations, including biological susceptibility to the consequences of *C. trachomatis* infection, exist [32]. Figure 12.3 depicts the age-related distribution of active trachoma in childhood, with scarring, TT, and corneal opacity occurring in the older ages, drawn from data collected in South Sudan [33]. This pattern, to a greater or lesser extent, has been seen in all areas where trachoma is endemic.

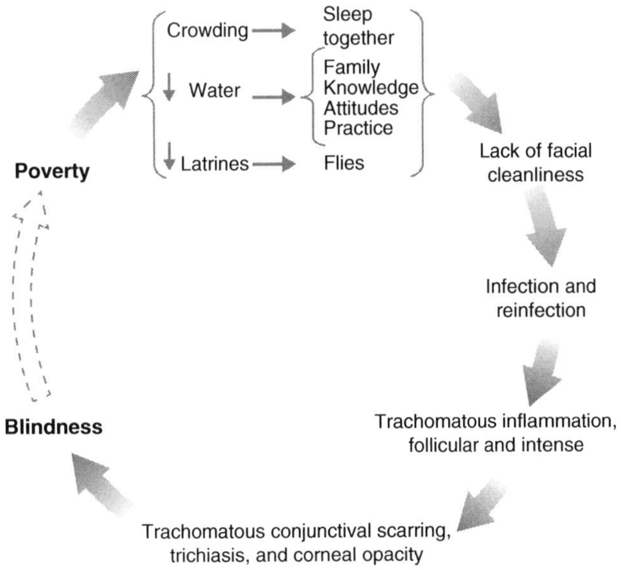

FIGURE 12.2 Interaction of risk factors for trachoma. (Reproduced with permission from Lancet [17].)

Transmission of ocular *C. trachomatis* is thought to occur via three main routes: fingers, fomites (shared cloths, bedding, and towels), and via the putative fly vector, *Musca sorbens*, but the relative importance of these transmission routes has not been ascertained [15]. The eye-seeking *M. sorbens* fly has been associated with trachoma in risk factor studies, and shown to carry *C. trachomatis* DNA externally [34], but trachoma transmission has also been found to exist in areas where muscid flies are rare or absent. Risk factor studies have repeatedly shown that children with ocular and/or nasal discharge are more likely to have trachoma. However, these data are difficult to interpret as discharge may be a consequence, as well as a cause, of ocular *C. trachomatis* infection [35]. A systematic review and meta-analysis conducted by Stocks et al. [36], updated by Freeman et al. [37], found that factors associated with lower odds of active trachoma (TF and/or TI) and ocular *C. trachomatis* infection were access to sanitation facilities and facial cleanliness indicators (lack of ocular or nasal discharge). Factors associated with reduced odds of active trachoma only were having a clean face and hygiene factors (for example, frequency of face washing, soap use, towel use, and daily bathing practices). However, data are inconsistent, with several risk factor studies finding no association between sanitation-related factors and trachoma. Furthermore, the majority of the studies included in the systematic review were observational, there was a dearth of studies including ocular *C. trachomatis* as an outcome measure, and varying definitions for the different putative explanatory variables associated with trachoma were used. There is a need for further randomized controlled trials with

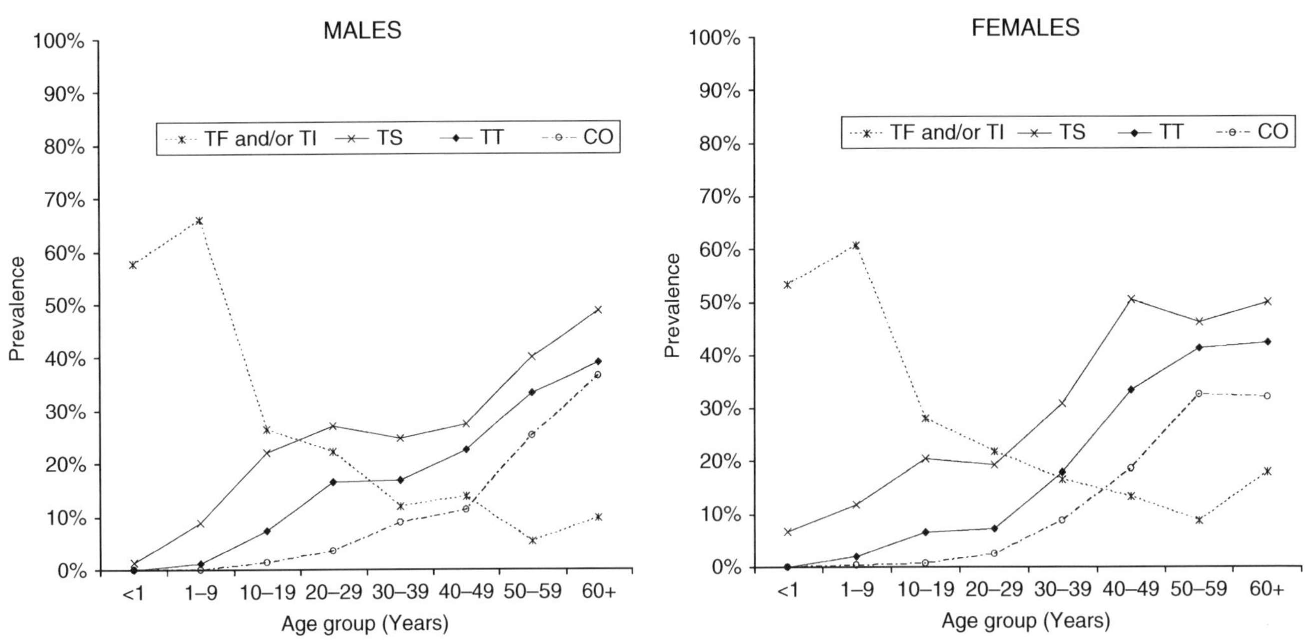

FIGURE 12.3 Age and sex distribution of active trachoma and cicatricial disease, based on data from South Sudan [33].

standardized methodologies to assess interventions to improve poor hygiene and environmental conditions in limiting ocular *C. trachomatis* transmission.

12.5 TRACHOMA—WATER AND SANITATION

The importance of water in the epidemiology of this disease seems intuitively obvious when one considers trachoma's modes of transmission. However, water's link to trachoma is perhaps more complex than it first appears. In a detailed review of available studies on water and trachoma, Prost and Negrel concluded that there would be a general reduction in trachoma prevalence as water access improved [38]. However, Cairncross has noted that a "water use plateau" exists in areas of restricted water availability. The amount used per person or per household may vary from one context to the next, but within a given community, a plateau of use exists as the time the nearest water source varies from a few minutes away from the house to 30 minutes away [39]. After 30 minutes, water use declines from the plateau and hence those living more than 30 minutes from a water source are likely to benefit most from the creation of new, more proximate, water sources. In Tanzania, active trachoma prevalence was substantially lower in children from households where more water was used for personal hygiene, independent of the total amount of water used [40]. The allocation of water to hygiene was directly associated with a shorter water collection time. Similarly, in The Gambia, trachoma-free households used more water for washing children than households with trachoma cases [41]. A systematic review of household water quantity found that the amount of water used for face washing, especially for children, led to better trachoma outcomes [42]. The key issue affecting trachoma in households is therefore water *use* for personal hygiene, as opposed to simply water access and quantity.

12.6 PREVENTION, CONTROL, AND ELIMINATION

It is water's link to poverty and poor hygiene that makes its role crucial in limiting trachoma transmission. Water is thought to be essential to trachoma elimination but not sufficient. That is, there is usually less secure access to water among poorer people, and poor personal hygiene cannot improve without water, but water alone will not prevent people from losing vision to trachoma. This fact underpinned WHO's endorsement of the strategy that it has recommended for trachoma control and elimination since 1993 [43], which incorporates primary, secondary, and tertiary preventive measures: the "SAFE" strategy [44]. The strategy entails: (S) surgery to halt corneal damage in individuals with TT (tertiary prevention); (A) antibiotic treatment (single-dose azithromycin, 20 mg per kilogram of body weight for children and 1 g for adults, given to residents of endemic districts as mass drug administration (MDA) (secondary prevention); (F) facial cleanliness (primary prevention); and (E) environmental improvement, including better access to water and sanitation (primary prevention). In 1996, WHO established an Alliance of health ministries, non-governmental organizations, academics, donors, and other interested parties committed to the GET2020 goal, and in 1998, a World Health Assembly resolution asked all countries where trachoma remained endemic to take steps to eliminate the disease by implementing SAFE. The trachoma elimination targets are a prevalence of TF $<5\%$ in one to nine year olds, a prevalence of TT unknown to the health system of $<0.2\%$ in ≥ 15 year-olds, and a system to identify and manage cases of incident TT [45, 46].

12.7 SURGERY FOR TRICHIASIS – S

If left unmanaged, TT may lead to corneal opacity, occurring in around one-third of individuals whose TT remains untreated for more than a year [47]. TT in itself is a cause of severe pain and reduced vision, even without corneal opacity, through the blepharospasm that it causes. Although epilation may have a role to play for minor TT, particularly for patients refusing surgery [48], surgical treatment (tertiary prevention) is thought to be the best means of definitively halting further damage, particularly for major TT. WHO recommends the use of either the bilamellar tarsal rotation (BLTR) or the posterior lamellar tarsal rotation (PLTR or "Trabut") procedure [49]. For BLTR, there is full-thickness incision of the upper eyelid and external rotation of the distal fragment using three to four sutures [50, 51]. This rotates the eyelashes up and away from the cornea. In PLTR, an approach via the conjunctiva is used and the anterior lamella (orbicularis muscle and skin) is not divided. A 2015 *Cochrane Review* concluded that the two methods are equally effective and safe [52]. The results of a subsequent randomized controlled trial, however, indicated that PLTR resulted in less TT recurrence at one year and had fewer intra-operative and immediate post-operative complications, indicating that it may be preferable [53]. Further trials are needed.

Post-surgical recurrence of TT is a major concern for patients, surgeons, and programs, with reports of 10% recurrence at three months, 20–40% within a year, and 60% at three years, using BLTR [52, 53]. This no doubt affects acceptability of surgery [54]. Multiple reasons for recurrence have been postulated, including the continuing process of lid shortening around scar coupled with the cicatrization from ongoing reinfection and variability in eye surgeon skill (reflective of training, innate ability, experience, and supervision) [55–57]. In a trial under field conditions, lid surgery by well-trained and well-equipped ophthalmic nurses has been shown to be as good as surgery provided by ophthalmologists [58] and has been successfully carried out in villages in national control programs in many countries. Moving this much needed surgery to the community setting improves acceptability, reduces costs, and is therefore more likely to halt further vision impairment. To help programs assure improved quality of surgery, WHO has produced a manual for teaching and evaluating performance of BLTR and PLTR [49].

12.8 ANTIBIOTICS – A

From 1952 until the early 1990s, the recommended treatment for trachoma was topical 1% tetracycline ointment, initially to be given four times a day, then later recommended as a twice-per-day treatment given over six to eight weeks. While results could be good in programs where therapy was observed or administered by healthcare personnel, when tubes of ointment were simply given to individuals in the community the results were often poor. Tetracycline ointment stings after application and disturbs vision for refractive reasons; in most cases, treatment is discontinued before the course is complete. Doxycycline, an oral tetracycline with a long half-life, was found to be effective if given once daily for two or three weeks. However, since the target population is young children, it was not a good candidate treatment because of the dental damage and photosensitivity it precipitates.

In the mid-1990s, three randomized trials showed that azithromycin in a single oral dose was at least as effective in treating active trachoma as topical tetracycline ointment [59–61]. Azithromycin is a broad-spectrum antibiotic, active against *C. trachomatis* as well as other Gram-positive and Gram-negative bacteria including *Staphylococcus aureus*, *Streptococcus pneumoniae*, *Haemophilus influenzae*, and *Moraxella* [62]. Studies in conjunctival biopsies of uninflamed eyes showed that the inhibitory concentrations were maintained in these tissues for at least 15 days after a single treatment [63]. Community MDA with three weekly doses of azithromycin was shown to be an effective control strategy in Egypt, Tanzania, and The Gambia [64].

Having an oral antibiotic that is effective in a single dose revolutionized trachoma control programs and has played an important role in progress toward elimination. In 1996, at a Global Scientific Meeting on Trachoma, WHO endorsed the use of azithromycin by trachoma control programs [65]. This endorsement led Pfizer Inc. to set up a generous azithromycin (Zithromax®) donation scheme, expressly for this purpose. Pfizer joined the Edna McConnell Clark Foundation in founding the International Trachoma Initiative (ITI), a non-governmental organization, to steward the donated drug and help advance the elimination agenda. WHO recommendations for MDA with azithromycin are based on the prevalence of TF in one to nine year olds at district level [45]. If the prevalence is 5–9.9%, a single round of MDA (plus "F" and "E") is recommended before reassessment. If the prevalence is 10–29.9%, "AFE" is recommended for three years before re-survey, whereas if the prevalence is ≥ 30%, five years of "AFE" are recommended. Between 1999 and December 2017, ITI had shipped over 700 million doses of azithromycin to programs in 36 countries [66].

Several research studies, as well as analysis of ITI's azithromycin treatments database, have assessed the impact of MDA on trachoma prevalence. The data suggest that in certain low prevalence settings, a single-round of MDA may be sufficient for achieving trachoma elimination [67–69]. In other studies, however, particularly in higher prevalence settings, infection prevalence is seen to fall immediately post-treatment and then increase again within 12 months of MDA [64, 70–72]. Mathematical modeling suggests that strategies such as more frequent (e.g. biannual) treatment would enable elimination to be achieved in high prevalence settings, such as Ethiopia [73–76]. Trials of biannual dosing, however, have shown a somewhat disappointing lack of additional effect.

As with the broad use of any antibiotic, the specter of antibiotic resistance must be considered. No convincing evidence of stable azithromycin resistance has been reported in *Chlamydia* species [77]. However, short-term azithromycin resistance has been observed in other pathogens (including *S. pneumoniae*, *S. aureus* and *Escherichia coli*) following azithromycin MDA given for trachoma elimination purposes. Fortunately, the prevalence of macrolide-resistant strains generally returns to pre-treatment levels after 12 months [78–84].

In addition to the impact on trachoma, MDA with oral azithromycin can have other community benefits. There is evidence that azithromycin helps in malaria control [85, 86] and in reducing morbidity due to acute lower respiratory infections [87] and diarrhea [88]. Furthermore, reduced mortality rates in under five year old children were observed in both a randomized controlled trial and a cohort study in Ethiopia, possibly due to beneficial effects on gastrointestinal and pulmonary disease [89, 90]. This potential additional benefit of MDA for trachoma elimination is being further investigated in the Mortality Reduction After Oral Azithromycin: Mortality Study (MORDOR) trial [91].

These important studies are indicative of both the power and limitations of effective antibiotic treatment. Mass antibiotic treatment alone is unlikely to eliminate ocular chlamydial infection in areas such as Ethiopia, which is the country with the largest number of people at risk of trachoma blindness [6, 30, 92]. For global elimination of trachoma as a public health problem to be reached and maintained, full implementation of the SAFE strategy is required, with good hygiene and improved environmental conditions enabling prevalence reductions induced through MDA to be sustained.

12.9 F AND E – FACIAL CLEANLINESS AND ENVIRONMENTAL IMPROVEMENT

The F and E components of the SAFE strategy involve the use of water and sanitation to curb transmission of ocular *C. trachomatis* infection. The efficacy of improved hygiene conditions for bringing about trachoma elimination, even in the absence of antibiotics, was demonstrated ecologically in Western Europe and North America by the mid-twentieth century. On the other hand, Rabiu and colleagues systematically reviewed the evidence for effectiveness of formally implemented environmental sanitary measures in reducing active trachoma prevalence and concluded that there was a dearth of acceptable data to determine the effectiveness of any aspects of F or E in the way these things are delivered by programs [93]. Another Cochrane review assessing the evidence for face washing promotion in preventing trachoma also concluded that there was inconclusive evidence (from the two eligible trials) for the effectiveness of face washing alone or in combination with tetracycline eye ointment for controlling trachoma [94]. The absence of conclusive randomized controlled trials in this area is not surprising as it involves the difficult-to-study topics of cultural/behavioral change, and requires whole communities to be assigned to various alternative interventions. Also, in the face of the relative ease of providing single-dose antibiotics, behavioral change, health education, and provision of water supplies and latrines are seen to be much more labor intensive and expensive. Still, it seems obvious that water and sanitation should have benefits far beyond trachoma elimination, in terms of disease control, quality of life, and expenditure of time and energy to obtain water. Thus, it is important that water and sanitation, education, and health promotion organizations and government agencies work in close collaboration to bring about improvements in these aspects of SAFE [95].

There is, however, some evidence that has been generated outwith randomized controlled trials that F and E interventions are associated with a reduced risk of trachoma. A case control study in almost 1000 households in eight Tanzanian villages examined the association between use and quality of latrines and the risk of trachoma. Use of latrines was significantly associated with a decreased risk of trachoma. The condition or type of latrine was not a determining factor. Latrines, then, need not be expensive or elaborate to provide health benefits [96]. In a community-based study of 507 children aged one to nine years in 232 Ethiopian households, lack of access to a latrine increased the risk of having active trachoma 4.4-fold, but the absence of a clean face (defined simply as only the absence of discharge from the eyes or nose) increased the odds of having trachoma more than seven-fold [97]. The prevalence of trachoma in this community was 53%. In Vietnam, a public health intervention study was conducted between 2002 and 2005 in two villages (separated by 10 km and five other villages) to assess the impact of improved water and sanitation. In one village, all four SAFE components were implemented, whereas in the other village only SA were implemented [98]. The "SA" village active trachoma prevalence fell from 10.2 to 6.7%, while in the "SAFE" village it fell from 13.3 to 1.4%. Also, 59% of the active trachoma decline in residents of all ages and 37% of the decline in children under the age of 15 years was attributed to the addition of health education, water, and sanitation. Another study in Ethiopia reported that the prevalence of active trachoma in intervention communities (woredas) fell from 26.6% at baseline to 18.3% after one year ($p<0.001$) of provision of water, hygiene, and sanitation education, compared with control woredas, where active trachoma prevalence was essentially unchanged [99].

Studies in The Gambia [100] and Ethiopia [101] have shown that chlamydial DNA can be found by polymerase chain reaction (PCR) on flies caught on children's faces. *M. sorbens* preferentially breeds in human feces left exposed on the soil, hence the link to latrines. A trial in The Gambia showed that fogging with insecticides reduced both the fly population and fly–eye contacts, and this effect persisted as long as insecticide use was continued. There was also a concomitant reduction in active trachoma prevalence [102]. However, this study only compared two pairs of villages. In a subsequent larger study in The Gambia, villages were randomized so that seven received fly control, seven received latrines, and seven had no additional interventions. Persons with intense trachoma were treated with either tetracycline or azithromycin. After six months, in the fly control clusters, the number of *M. sorbens* on children's faces had decreased by 88% (95% CI: 64–100) and there was a significant relative reduction (56%) in active trachoma. In the latrine provision clusters, a 30% decrease in *M. sorbens*–eye contacts was observed, and a concomitant but insignificant 30% reduction in trachoma prevalence [103]. However, a randomized controlled trial in a hyperendemic area of Tanzania incorporating eight clusters per arm did not show any benefit from fly control when used in conjunction with mass azithromycin treatment [104]. Similarly, studies in Ethiopia, where transmission intensity is thought to be extraordinarily high, imply that in the presence of mass treatment with azithromycin, fly control might not add much to trachoma elimination efforts. Flies were collected in villages that had received mass oral azithromycin distribution and were compared with flies in untreated villages. PCR was performed to detect chlamydial DNA on the flies. *C. trachomatis* was found on 23% of the flies in untreated villages but on only 0.3% of flies in treated villages. Therefore, mass treatment of children with antibiotics would seem to drastically reduce the likely role of flies as a vector [101].

12.10 THE PATH TOWARD ELIMINATION BY 2020

Between the founding of the WHO Alliance for GET2020 and January 2018, five countries were validated as having eliminated trachoma as a public health problem and a further six had notified WHO that they had achieved the epidemiological criteria and were preparing dossiers [6]. In 2016, health ministries reported the largest ever annual collective number of TT surgeries performed (260 759 surgeries, compared with 185 087 in 2015 and 139 441 in 2014), with over two-thirds of these having taken place in Ethiopia [10]. Between 1999 and December 2017, over 700 million doses of azithromycin had been shipped by ITI. In 2016 alone, over 85 million people were treated with antibiotics for trachoma elimination purposes, achieving global-level population coverage of 45% (compared to 30% in 2015 and 25% in 2014) [10]. In contrast, progress for the F and E components of the SAFE strategy, aimed at continued suppression of ocular *C. trachomatis* transmission, has lagged behind. It is estimated that an additional investment of US$258 million is needed from the WASH sector for environmental improvement [7]. The GTMP, and subsequently Tropical Data, have enabled the global burden of trachoma to be understood, interventions to be targeted and prioritized in the areas most in need, and the impact of control interventions to be assessed, in order for the GET2020 target to be reached.

Trachoma elimination may be aided through integration with other NTD control programs which, in particular, could provide greater advocacy for F and E interventions. NTDs are a group of 20 diseases, affecting over 1 billion of the world's poorest people [105]. Many are chronic and cause life-long disabilities, which are often stigmatizing [106]. The five strategies employed for their control, are: preventive chemotherapy, innovative and intensified disease management, vector control, veterinary public health measures, and improved water and sanitation [107]. In addition to the Pfizer donation of azithromycin for trachoma, there are drug donation programs by major pharmaceutical companies for four other NTDs: ivermectin (Merck & Co., Inc.) for onchoceriasis; albendazole (Glaxo Smith Kline) given along with ivermectin and/or diethyl carbamazine for lymphatic filariasis (LF); mebendazole (Johnson & Johnson) for soil transmitted helminths (STHs); and praziquantel (Merck Serono) for schistosomiasis. In 2012, WHO published its "roadmap for implementation" for the "the control, elimination or eradication of 17 Neglected Tropical Diseases (NTDs) by 2020" [105]. This was supported later that year by the London Declaration on NTDs, a partnership of organizations focusing on the elimination of ten NTDs, including the five with MDA as cornerstones of their control strategies [108]. A key aspect of this work is for countries to integrate control programs when these diseases overlap. This includes national NTD control programs actively seeking to integrate mapping, prevention and control efforts, such as in Burundi [109], and opportunistic synergy with other programs. For example, a program in the Amhara region of Ethiopia has combined treatment and health education in a joint trachoma and malaria control effort [110].

12.11 CONCLUSION

The evidence gathered from operational research studies and the experience of national control programs clearly supports the effectiveness of the SAFE strategy for the elimination of trachoma as a public health problem. The impact of mass treatment with azithromycin is remarkable, but alone it is unlikely to sustainably eliminate the irreversible consequences of the disease at scale. The benefits of access to water and hygiene (including the availability of appropriate methods for safely separating humans from their feces, normalization of their use, and behavioral changes for personal hygiene) for general health and quality of life make these critical elements in elimination programs for trachoma and other NTDs.

REFERENCES

1. WHO. 2017. *Trachoma fact sheet* [Online]. Available from: http://www.who.int/mediacentre/factsheets/fs382/en/.
2. Marmion, V.J. (2005). The origin of eye hospitals. *Br. J. Ophthalmol.* 89: 1396–1397.
3. Taylor, H.R. (2009). Doyne lecture: trachoma, is it history? *Eye* 23: 2007–2022.
4. West, S.K. (2004). Trachoma: new assault on an ancient disease. *Prog. Retin. Eye Res.* 23: 381–401.
5. Cowling, C., Liu, B., Snelling, T. et al. (2015). National trachoma surveillance annual report, 2012. *Commun. Dis. Intell.* 39: E146–E157.
6. ICTC 2016. *Eliminating Trachoma: Accelerating Towards 2020.* International Coalition for Trachoma Control, on behalf of the WHO Alliance for the Global Elimination of Trachoma by 2020. Available from: http://www.trachomacoalition.org/sites/all/themes/report-2016/PDF/GET2020_2016_EN.pdf.
7. ICTC. 2017. *Key trachoma facts and statistics* [Online]. Available from: http://www.trachomacoalition.org/trachomastatistics.
8. De Vlas, S.J., Stolk, W.A., Le Rutte, E.A. et al. (2016). Concerted efforts to control or eliminate neglected tropical diseases: how much health will be gained? *PLoS Negl. Trop. Dis.* 10: e0004386.
9. Engels, D. (2016). The global trachoma mapping project: a catalyst for progress against neglected tropical diseases. *Ophthalmic Epidemiol.* 23: 1–2.

10. WHO (2017). WHO alliance for the global elimination of trachoma by 2020: progress report on elimination of trachoma, 2014–2016. *Wkly. Epidemiol. Rec.* 92: 359–368.
11. Bourne, R.R., Stevens, G.A., White, R.A. et al. (2013). Causes of vision loss worldwide, 1990–2010: a systematic analysis. *Lancet Glob. Health* 1: e339–e349.
12. WHO (2012). Global WHO alliance for the elimination of blinding trachoma by 2020. Progress report on elimination of trachoma, 2010. *Wkly. Epidemiol. Rec.* 87: 161–168.
13. Bailey, R., Duong, T., Carpenter, R. et al. (1999). The duration of human ocular *Chlamydia trachomatis* infection is age dependent. *Epidemiol. Infect.* 123: 479–486.
14. Solomon, A.W., Holland, M.J., Burton, M.J. et al. (2003). Strategies for control of trachoma: observational study with quantitative PCR. *Lancet* 362: 198–204.
15. Mabey, D.C., Solomon, A.W., and Foster, A. (2003). Trachoma. *Lancet* 362: 223–229.
16. Taylor, H.R., Burton, M.J., Haddad, D. et al. (2014). Trachoma. *Lancet* 384: 2142–2152.
17. Wright, H.R., Turner, A., and Taylor, H.R. (2008). Trachoma. *Lancet* 371: 1945–1954.
18. Thylefors, B., Dawson, C.R., Jones, B.R. et al. (1987). A simple system for the assessment of trachoma and its complications. *Bull. World Health Organ.* 65: 477–483.
19. Burton, M.J., Holland, M.J., Faal, N. et al. (2003). Which members of a community need antibiotics to control trachoma? Conjunctival *Chlamydia trachomatis* infection load in Gambian villages. *Invest. Ophthalmol. Vis. Sci.* 44: 4215–4222.
20. Wolle, M.A., Munoz, B.E., Mkocha, H., and West, S.K. (2009). Constant ocular infection with *Chlamydia trachomatis* predicts risk of scarring in children in Tanzania. *Ophthalmology* 116: 243–247.
21. Gambhir, M., Basanez, M.G., Burton, M.J. et al. (2009). The development of an age-structured model for trachoma transmission dynamics, pathogenesis and control. *PLoS Negl. Trop. Dis.* 3: e462.
22. Ramadhani, A.M., Derrick, T., Macleod, D. et al. (2016). The relationship between active trachoma and ocular *Chlamydia trachomatis* infection before and after mass antibiotic treatment. *PLoS Negl. Trop. Dis.* 10: e0005080.
23. WHO 2016. Trachoma Alternative Indicators Study Data Review, 31 August 1 September 2016. World Health Organization Strategic and Technical Advisory Group for Neglected Tropical Diseases. Working Group on Monitoring and Evaluation. Available from: http://apps.who.int/iris/bitstream/10665/259397/1/WHO-HTM-NTD-PCT-2017.10-eng.pdf?ua=12016.
24. Butcher, R., Sokana, O., Jack, K. et al. (2017). Age-specific prevalence of anti-Pgp3 antibodies and severe conjunctival scarring in the Solomon Islands. *bioRxiv* 141135. doi: 10.1101/141135.
25. Harding-Esch, E., Jofre-Bonet, M., Dhanjal, J.K. et al. (2015). Costs of testing for ocular *Chlamydia trachomatis* infection compared to mass drug administration for trachoma in the Gambia: application of results from the PRET study. *PLoS Negl. Trop. Dis.* 9: e0003670.
26. Butcher, R., Houghton, J., Derrick, T. et al. (2017a). Reduced-cost *Chlamydia trachomatis*-specific multiplex real-time PCR diagnostic assay evaluated for ocular swabs and use by trachoma research programmes. *J. Microbiol. Methods* 139: 95–102.
27. Harding-Esch, E.M., Holland, M.J., Schemann, J.F. et al. (2011). Diagnostic accuracy of a prototype point-of-care test for ocular *Chlamydia trachomatis* under field conditions in the Gambia and Senegal. *PLoS Negl. Trop. Dis.* 5: e1234.
28. Solomon, A.W., Pavluck, A.L., Courtright, P. et al. (2015). The global trachoma mapping project: methodology of a 34-country population-based study. *Ophthalmic Epidemiol.* 22: 214–225.
29. Trotignon, G., Jones, E., Engels, T. et al. (2017). The cost of mapping trachoma: data from the global trachoma mapping project. *PLoS Negl. Trop. Dis.* 11: e0006023.
30. WHO (2017). Meeting of the international task force for disease eradication, November 2016. *Wkly Epidemiol. Rec.* 92: 106–116.
31. Cromwell, E.A., Courtright, P., King, J.D. et al. (2009). The excess burden of trachomatous trichiasis in women: a systematic review and meta-analysis. *Trans. R. Soc. Trop. Med. Hyg.* 103: 985–992.
32. Courtright, P. and West, S.K. (2004). Contribution of sex-linked biology and gender roles to disparities with trachoma. *Emerg. Infect. Dis.* 10: 2012–2016.
33. Ngondi, J., Ole-Sempele, F., Onsarigo, A. et al. (2006). Blinding trachoma in postconflict southern Sudan. *PLoS Med.* 3: e478.
34. Miller, K., Pakpour, N., Yi, E. et al. (2004). Pesky trachoma suspect finally caught. *Br. J. Ophthalmol.* 88: 750–751.
35. Hu, V.H., Harding-Esch, E.M., Burton, M.J. et al. (2010). Epidemiology and control of trachoma: systematic review. *Tropical Med. Int. Health* 15: 673–691.
36. Stocks, M.E., Ogden, S., Haddad, D. et al. (2014). Effect of water, sanitation, and hygiene on the prevention of trachoma: a systematic review and meta-analysis. *PLoS Med.* 11: e1001605.
37. Freeman, M.C., Garn, J.V., Sclar, G.D. et al. (2017). The impact of sanitation on infectious disease and nutritional status: a systematic review and meta-analysis. *Int. J. Hyg. Environ. Health* 220: 928–949.
38. Prost, A. and Negrel, A.D. (1989). Water, trachoma and conjunctivitis. *Bull. World Health Organ.* 67: 9–18.
39. Cairncross, S. (1999). Trachoma and water. *Community Eye Health* 12: 58–59.
40. Polack, S., Kuper, H., Solomon, A.W. et al. (2006). The relationship between prevalence of active trachoma, water availability and its use in a Tanzanian village. *Trans. R. Soc. Trop. Med. Hyg.* 100: 1075–1083.
41. Bailey, R., Downes, B., Downes, R., and Mabey, D. (1991). Trachoma and water use; a case control study in a Gambian village. *Trans. R. Soc. Trop. Med. Hyg.* 85: 824–828.
42. Stelmach, R.D. and Clasen, T. (2015). Household water quantity and health: a systematic review. *Int. J. Environ. Res. Public Health* 12: 5954–5974.
43. Francis, V. and Turner, V. (1993). *Achieving Community Support for Trachoma Control (WHO/PBL/93.36)*. Geneva: World Health Organization.

44. WHO. 1998. World Health Assembly. Global Elimination of Blinding Trachoma. 51st World Health Assembly, Geneva, 16 May 1998, Resolution WHA51.11. Geneva: World Health Organization.
45. WHO. 2010. Report of the Third Global Scientific Meeting on Trachoma. Available from: http://www.who.int/blindness/publications/WORLDHEALTHORGANIZATIONGSMmtgreportFINALVERSION.pdf?ua=1: Baltimore, 19–20 July 2010.
46. WHO. 2015. Technical Consultation on Trachoma Surveillance, September 11–12, 2014. Technical Advisory Group on Neglected Tropical Diseases. Task Force for Global Health, Decatur, USA. Available from: http://apps.who.int/iris/bitstream/10665/174085/1/WHO_HTM_NTD_2015.02_en.pdf.
47. Bowman, R.J., Faal, H., Myatt, M. et al. (2002). Longitudinal study of trachomatous trichiasis in the Gambia. *Br. J. Ophthalmol.* 86: 339–343.
48. Rajak, S.N., Habtamu, E., Weiss, H.A. et al. (2011). Surgery versus epilation for the treatment of minor trichiasis in Ethiopia: a randomised controlled noninferiority trial. *PLoS Med.* 8: e1001136.
49. WHO (2015). *Trichiasis Surgery for Trachoma*, 2e (ed. S. Merbs, S. Resnikoff, A. Kello, et al.). Geneva: Switzerland World Health Organization Available from: http://apps.who.int/iris/bitstream/10665/155227/1/9789241549011_eng.pdf.
50. Reacher, M., Foster, A., and Huber, M.J. (1993). *Trichiasis Surgery for Trachoma. The Bilamellar Tarsal Rotation Procedure*. Geneva: World Health Organisation.
51. Reacher, M.H., Munoz, B., Alghassany, A. et al. (1992). A controlled trial of surgery for trachomatous trichiasis of the upper lid. *Arch. Ophthalmol.* 110: 667–674.
52. Burton, M., Habtamu, E., Ho, D., and Gower, E.W. (2015). Interventions for trachoma trichiasis. *Cochrane Database Syst. Rev.* 13 (11): Cd004008.
53. Habtamu, E., Wondie, T., Aweke, S. et al. (2016). Posterior lamellar versus bilamellar tarsal rotation surgery for trachomatous trichiasis in Ethiopia: a randomised controlled trial. *Lancet Glob. Health* 4: e175–e184.
54. Solomon, A.W. (2016). Optimising the management of trachomatous trichiasis. *Lancet Glob. Health* 4: e140–e141.
55. Burton, M.J., Kinteh, F., Jallow, O. et al. (2005). A randomised controlled trial of azithromycin following surgery for trachomatous trichiasis in the Gambia. *Br. J. Ophthalmol.* 89: 1282–1288.
56. Rajak, S.N., Makalo, P., Sillah, A. et al. (2010). Trichiasis surgery in the Gambia: a 4-year prospective study. *Invest. Ophthalmol. Vis. Sci.* 51: 4996–5001.
57. West, E.S., Mkocha, H., Munoz, B. et al. (2005). Risk factors for postsurgical trichiasis recurrence in a trachoma-endemic area. *Invest. Ophthalmol. Vis. Sci.* 46: 447–453.
58. Alemayehu, W., Melese, M., Bejiga, A. et al. (2004). Surgery for trichiasis by ophthalmologists versus integrated eye care workers: a randomized trial. *Ophthalmology* 111: 578–584.
59. Bailey, R.L., Arullendran, P., Whittle, H.C., and Mabey, D.C. (1993). Randomised controlled trial of single-dose azithromycin in treatment of trachoma. *Lancet* 342: 453–456.
60. Dawson, C.R., Schachter, J., Sallam, S. et al. (1997). A comparison of oral azithromycin with topical oxytetracycline/polymyxin for the treatment of trachoma in children. *Clin. Infect. Dis.* 24: 363–368.
61. Tabbara, K.F., Abu-El-Asrar, A., Al-Omar, O. et al. (1996). Single-dose azithromycin in the treatment of trachoma. A randomized, controlled study. *Ophthalmology* 103: 842–846.
62. Baral, K., Osaki, S., Shreshta, B. et al. (1999). Reliability of clinical diagnosis in identifying infectious trachoma in a low-prevalence area of Nepal. *Bull. World Health Organ.* 77: 461–466.
63. Tabbara, K.F., Al-Kharashi, S.A., Al-Mansouri, S.M. et al. (1998). Ocular levels of azithromycin. *Arch. Ophthalmol.* 116: 1625–1628.
64. Schachter, J., West, S.K., Mabey, D. et al. (1999). Azithromycin in control of trachoma. *Lancet* 354: 630–635.
65. WHO. (1997). Planning Meeting for the Global Elimination of Trachoma, Geneva, 97–28 November 1996. WHO/PBL/97.60. Available from: http://whqlibdoc.who.int/hq/1997/WHO_PBL_97.60.pdf.
66. ITI. (2017). Available from: http://www.trachoma.org/.
67. Burton, M.J., Holland, M.J., Makalo, P. et al. (2005). Re-emergence of *Chlamydia trachomatis* infection after mass antibiotic treatment of a trachoma-endemic Gambian community: a longitudinal study. *Lancet* 365: 1321–1328.
68. Harding-Esch, E.M., Sillah, A., Edwards, T. et al. (2013). Mass treatment with azithromycin for trachoma: when is one round enough? Results from the PRET trial in the Gambia. *PLoS Negl. Trop. Dis.* 7: e2115.
69. Solomon, A.W., Harding-Esch, E., Alexander, N.D. et al. (2008). Two doses of azithromycin to eliminate trachoma in a Tanzanian community. *N. Engl. J. Med.* 358: 1870–1871.
70. Jimenez, V., Gelderblom, H.C., Mann Flueckiger, R. et al. (2015). Mass drug administration for trachoma: how long is not long enough? *PLoS Negl. Trop. Dis.* 9: e0003610.
71. Lakew, T., House, J., Hong, K.C. et al. (2009). Reduction and return of infectious trachoma in severely affected communities in Ethiopia. *PLoS Negl. Trop. Dis.* 3: e376.
72. West, S.K., Munoz, B., Mkocha, H. et al. (2005). Infection with *Chlamydia trachomatis* after mass treatment of a trachoma hyperendemic community in Tanzania: a longitudinal study. *Lancet* 366: 1296–1300.
73. Biebesheimer, J.B., House, J., Hong, K.C. et al. (2009). Complete local elimination of infectious trachoma from severely affected communities after six biannual mass azithromycin distributions. *Ophthalmology* 116 (11): 2047.
74. Lietman, T., Porco, T., Dawson, C., and Blower, S. (1999). Global elimination of trachoma: how frequently should we administer mass chemotherapy? *Nat. Med.* 5: 572–576.
75. Melese, M., Alemayehu, W., Lakew, T. et al. (2008). Comparison of annual and biannual mass antibiotic administration for elimination of infectious trachoma. *JAMA* 299: 778–784.

76. Ray, K.J., Porco, T.C., Hong, K.C. et al. (2007). A rationale for continuing mass antibiotic distributions for trachoma. *BMC Infect. Dis.* 7: 91.

77. West, S.K., Moncada, J., Munoz, B. et al. (2014). Is there evidence for resistance of ocular *Chlamydia trachomatis* to azithromycin after mass treatment for trachoma control? *J. Infect. Dis.* 210: 65–71.

78. Bojang, E., Jafali, J., Perreten, V. et al. (2017). Short-term increase in prevalence of nasopharyngeal carriage of macrolide-resistant *Staphylococcus aureus* following mass drug administration with azithromycin for trachoma control. *BMC Microbiol.* 17: 75.

79. Chern, K.C., Shrestha, S.K., Cevallos, V. et al. (1999). Alterations in the conjunctival bacterial flora following a single dose of azithromycin in a trachoma endemic area. *Br. J. Ophthalmol.* 83: 1332–1335.

80. Fry, A.M., Jha, H.C., Lietman, T.M. et al. (2002). Adverse and beneficial secondary effects of mass treatment with azithromycin to eliminate blindness due to trachoma in Nepal. *Clin. Infect. Dis.* 35: 395–402.

81. Gaynor, B.D., Chidambaram, J.D., Cevallos, V. et al. (2005). Topical ocular antibiotics induce bacterial resistance at extraocular sites. *Br. J. Ophthalmol.* 89: 1097–1099.

82. Gaynor, B.D., Holbrook, K.A., Whitcher, J.P. et al. (2003). Community treatment with azithromycin for trachoma is not associated with antibiotic resistance in *Streptococcus pneumoniae* at 1 year. *Br. J. Ophthalmol.* 87: 147–148.

83. Leach, A.J., Shelby-James, T.M., Mayo, M. et al. (1997). A prospective study of the impact of community-based azithromycin treatment of trachoma on carriage and resistance of *Streptococcus pneumoniae*. *Clin. Infect. Dis.* 24: 356–362.

84. Seidman, J.C., Coles, C.L., Silbergeld, E.K. et al. (2014). Increased carriage of macrolide-resistant fecal *E. coli* following mass distribution of azithromycin for trachoma control. *Int. J. Epidemiol.* 43: 1105–1113.

85. Chico, R.M., Pittrof, R., Greenwood, B., and Chandramohan, D. (2008). Azithromycin-chloroquine and the intermittent preventive treatment of malaria in pregnancy. *Malar. J.* 7: 255.

86. Sadiq, S.T., Glasgow, K.W., Drakeley, C.J. et al. (1995). Effects of azithromycin on malariometric indices in the Gambia. *Lancet* 346: 881–882.

87. Coles, C.L., Levens, J., Seidman, J.C. et al. (2012). Mass distribution of azithromycin for trachoma control is associated with short-term reduction in risk of acute lower respiratory infection in young children. *Pediatr. Infect. Dis. J.* 31: 341–346.

88. Coles, C.L., Seidman, J.C., Levens, J. et al. (2011). Association of mass treatment with azithromycin in trachoma-endemic communities with short-term reduced risk of diarrhea in young children. *Am. J. Trop. Med. Hyg.* 85: 691–696.

89. Keenan, J.D., Ayele, B., Gebre, T. et al. (2011). Childhood mortality in a cohort treated with mass azithromycin for trachoma. *Clin. Infect. Dis.* 52: 883–888.

90. Porco, T.C., Gebre, T., Ayele, B. et al. (2009). Effect of mass distribution of azithromycin for trachoma control on overall mortality in Ethiopian children: a randomized trial. *JAMA* 302: 962–968.

91. clinicaltrials.gov. 2017. Mortality Reduction After Oral Azithromycin: Mortality Study (MORDOR) [Online]. Available from: https://clinicaltrials.gov/ct2/show/NCT02047981. Accessed 14 December 2017.

92. Lietman, T.M., Gebre, T., Abdou, A. et al. (2015). The distribution of the prevalence of ocular chlamydial infection in communities where trachoma is disappearing. *Epidemics* 11: 85–91.

93. Rabiu, M., Alhassan, M.B., Ejere, H.O., and Evans, J.R. (2012). Environmental sanitary interventions for preventing active trachoma. *Cochrane Database Syst. Rev.* (2): Cd004003.

94. Ejere, H.O., Alhassan, M.B., and Rabiu, M. (2015). Face washing promotion for preventing active trachoma. *Cochrane Database Syst. Rev.* 20 (2): Cd003659.

95. ICTC. (2016). Sustaining impact, halting transmission: Focus on F&E. Eliminating Trachoma: Accelerating Towards 2020 [Online]. Available from: http://www.trachomacoalition.org/sites/all/themes/report-2016/PDF/Focus_on_F&E.pdf. Accessed 15 December 2017.

96. Montgomery, M.A., Desai, M.M., and Elimelech, M. (2010). Assessment of latrine use and quality and association with risk of trachoma in rural Tanzania. *Trans. R. Soc. Trop. Med. Hyg.* 104: 283–289.

97. Golovaty, I., Jones, L., Gelaye, B. et al. (2009). Access to water source, latrine facilities and other risk factors of active trachoma in Ankober, Ethiopia. *PLoS One* 4: e6702.

98. Khandekar, R., Ton, T.K., and Do Thi, P. (2006). Impact of face washing and environmental improvement on reduction of active trachoma in Vietnam – a public health intervention study. *Ophthalmic Epidemiol.* 13: 43–52.

99. Tadesse, B., Worku, A., Kumie, A., and Yimer, S.A. (2017). Effect of water, sanitation and hygiene interventions on active trachoma in north and south Wollo zones of Amhara Region, Ethiopia: a quasi-experimental study. *PLoS Negl. Trop. Dis.* 11: e0006080.

100. Emerson, P.M., Bailey, R.L., Mahdi, O.S. et al. (2000). Transmission ecology of the fly *Musca sorbens*, a putative vector of trachoma. *Trans. R. Soc. Trop. Med. Hyg.* 94: 28–32.

101. Lee, S., Alemayehu, W., Melese, M. et al. (2007). *Chlamydia* on children and flies after mass antibiotic treatment for trachoma. *Am. J. Trop. Med. Hyg.* 76: 129–131.

102. Emerson, P.M., Lindsay, S.W., Walraven, G.E. et al. (1999). Effect of fly control on trachoma and diarrhoea. *Lancet* 353: 1401–1403.

103. Emerson, P.M., Lindsay, S.W., Alexander, N. et al. (2004). Role of flies and provision of latrines in trachoma control: cluster-randomised controlled trial. *Lancet* 363: 1093–1098.

104. West, S.K., Emerson, P.M., Mkocha, H. et al. (2006). Intensive insecticide spraying for fly control after mass antibiotic treatment for trachoma in a hyperendemic setting: a randomised trial. *Lancet* 368: 596–600.

105. WHO 2012. Accelerating work to overcome the global impact of neglected tropical diseases. A roadmap for implementation. Available from: http://apps.who.int/iris/bitstream/10665/70809/1/WHO_HTM_NTD_2012.1_eng.pdf.

106. Molyneux, D. (2013). Neglected tropical diseases. *Community Eye Health* 26: 21–24.

107. Molyneux, D.H., Savioli, L., and Engels, D. (2017). Neglected tropical diseases: progress towards addressing the chronic pandemic. *Lancet* 389: 312–325.

108. Uniting to Combat NTDs. 2017. *Uniting to Combat Neglected Tropical Diseases* [Online]. Available from: http://unitingtocombatntds.org/about-us/. Accessed 15 December 2017.

109. Ndayishimiye, O., Ortu, G., Soares Magalhaes, R.J. et al. (2014). Control of neglected tropical diseases in Burundi: partnerships, achievements, challenges, and lessons learned after four years of programme implementation. *PLoS Negl. Trop. Dis.* 8: e2684.

110. ITI 2010. Zithromax® in the Elimination of Blinding Trachoma. A Program Manager's Guide. Available from: http://endinafrica.org/wp-content/uploads/2012/11/ITIZithromaxManagersGuide_en.pdf.

SECTION III

ANTHROPOGENIC AND NATURALLY OCCURRING POLLUTANTS

13

IMPACTS OF PHARMACEUTICALS AND PERSONAL CARE PRODUCTS IN THE ENVIRONMENT

M. DANIELLE MCDONALD

Department of Marine Biology and Ecology, Rosenstiel School of Marine and Atmospheric Science, University of Miami, Miami, FL, USA

It has been almost two decades since scientists first became aware of the presence of pharmaceuticals and the active ingredients in personal care products in the world's rivers, lakes, and marine coastal ecosystems [1–3]. While this discovery was mostly driven by increased detection efficiencies in environmental chemistry techniques, many pharmaceuticals and personal care products (PPCPs) are in high enough concentrations to cause harm to aquatic organisms. Compounds such as non-steroidal anti-inflammatory drugs (pain relievers), beta-blockers (blood pressure modulators), blood lipid lowering agents (cholesterol reducers), neuroactive compounds (e.g., antidepressants), steroidal hormones (e.g., contraceptives), antibiotics, surfactants (detergents), musks (perfumes), and UV filters (sunscreens) can enter the environment through the agricultural runoff from veterinary-treated livestock [4, 5], during manufacturing, by irrigation with reclaimed wastewater and by landfill leachates [1, 6–9]. However, the main route of entry is through human consumption, excretion, and the subsequent treatment, or lack thereof, and release by sewage treatment plants (STPs). PPCPs also enter STPs through bathing and by the improper disposal of unused and expired pharmaceuticals, either down the toilet or sink [10]. STP elimination efficiencies span a large range from completely ineffective (0% elimination) to relatively effective (99% elimination) in removing these compounds, depending on the STP treatment technology, hydraulic retention time, and season [3, 11]. Furthermore, while adsorption to suspended solids (sewage sludge), one of the two main elimination processes important in wastewater treatment (the other being biodegradation) removes PPCPs from wastewater, degradation of these compounds in sludge does not always occur and many compounds, such as steroid hormones, end up being present in sludge in measurable amounts. This can then contribute to further problems when the contaminated sludge is then used as a land fertilizer [12–14].

In terms of the toxicity of environmental PPCP contamination, aquatic organisms are continuously exposed to low levels of PPCPs over long periods of time or even over their entire life cycle, since PPCP concentrations are typically not high enough to be lethal. This chronic, sublethal exposure can result in changes in the reproductive success, behavior, and general fitness of aquatic organisms, leading to population level effects that could impact the ecosystem. Furthermore, long-term exposure of aquatic organisms to low concentrations of PPCPs can also have implications on human health. Because of their lipophilicity, many PPCPs are readily taken up by aquatic organisms, accumulating in their tissues and then moving through the aquatic and terrestrial food chain, with humans being one of the top consumers. While the direct risk to human health through exposure to or consumption of contaminated aquatic organisms or even polluted drinking water [15, 16] is not yet cause for concern in most cases (i.e., we would have to eat hundreds of thousands of kilograms of fish or drink an equivalent volume of water per day to get a daily dose of

Water and Sanitation-Related Diseases and the Changing Environment: Challenges, Interventions, and Preventive Measures, Second Edition. Edited by Janine M. H. Selendy.

Companion website: https://www.wiley.com/legacy/wileychi/selendy/

any given pharmaceutical), the potential additive or synergistic effects of low amounts of multiple compounds if bioaccumulated in fish and then consumed by humans is, at this time, unstudied.

Unlike the active ingredients in personal care products and many other environmental toxicants that may have unexpected mechanisms of toxicity (e.g., nonylphenol, an active ingredient in household detergents, disrupts reproduction (e.g., see [17]) and UV filters found in sunscreen can have central nervous system effects (e.g., see [18])), pharmaceuticals are designed to target specific pathways in humans. Thus, the impacts of these compounds on aquatic organisms should be predictable, since the target organs or receptors across organisms are oftentimes conserved. Indeed, recent work has put more emphasis on the idea of "read-across" or taking what is known regarding the therapeutic dose and targeted effect of a pharmaceutical in humans and applying it to predict endpoints and effective waterborne pharmaceutical concentrations in aquatic organisms [19–22]. An example of pharmaceutical compounds with predictable effects are the steroid hormone, estradiol (E2), and the synthetic estrogen, 17α-ethinylestradiol (EE2), found in birth control pills and known as endocrine disrupting compounds (EDCs). In humans, their target is the endocrine axis and receptors responsible for regulating reproductive endpoints. Because of the high conservation of this system across the animal kingdom, waterborne exposure to these estrogenic EDCs result in substantial reproduction-level effects in aquatic organisms, namely a decrease in reproductive behavior and fecundity [23, 24], an increase in feminization, often resulting in "intersex" species [25–27], or a decrease in the fertilization rate [28]. While the changes in reproductive abilities of aquatic biota are devastating ecologically and will ultimately have a global impact if left unchecked, evidence also suggests that environmental EDCs also pose a clear risk to human health [29, 30], which may include a role in the decline in the human sex ratio [31–34], several female reproductive disorders (reviewed by [35]), and obesity (reviewed by [36]).

In addition to environmental contamination with EDCs, another potentially serious problem to the aquatic ecosystem and human health is the indiscriminant use and misuse of antibiotics, which is the leading proposed cause of the increased and spreading resistance among bacteria [37]. Drug resistance in bacteria is not believed to develop in surface waters, since environmental concentrations of antibiotics are not high enough to be toxic and promote the selection of resistant bacteria. However, if our current habits do not change, antibiotics will reach levels in the environment that are toxic, the surviving bacteria will multiply, and resistance can develop.

Limiting the consumption of pharmaceuticals may be the most straightforward way to reduce their appearance in the environment [38]. This can be done in many ways. At the level of the healthcare professional, the correct medication and the lowest effective dosage for each patient should be selected. The pharmaceutical industry could work towards repackaging medicines into smaller quantities, especially those medicines that are prone to being thrown out because they have expired before being used in their entirety. The patient should be held responsible for completing treatments as prescribed so that illnesses do not reoccur and the risk of antibiotic resistance is avoided. Finally, the general public should avoid disposing of unused medications in the trash can or flushing them down the toilet or sink [10, 39, 40], as these methods do not only lead to detrimental effects on environmental health but also human health. A recent review details that only in Sweden and Germany does returning unused and expired pharmaceuticals to the pharmacy consistently occur [40]. Pharmaceutical take-back programs such as this not only reduce the amount of pharmaceutical waste introduced to the environment by keeping these compounds out of landfills and the water supply but they help to avoid dosing errors, drug abuse, and accidental poisonings that result in the accumulation of unused pharmaceuticals in the home. In the United States, pharmaceutical take-back programs are becoming more common and depend on local pharmacies and local government household trash and recycling services. The involvement of the US federal government is an important first step in getting on top of this problem in this country – it is now up to individuals to hold themselves and their community leaders accountable.

REFERENCES

1. Daughton, C.G. and Ternes, T.A. (1999). Pharmaceuticals and personal care products in the environment: agents of subtle change? *Environ. Health Perspect.* 107: 907–938.
2. Kolpin, D.W., Furlong, E.T., Meyer, M.T. et al. (2002). Pharmaceuticals, hormones, and other organic wastewater contaminants in US streams, 1999–2000: a national reconnaissance. *Environ. Sci. Technol.* 36: 1202–1211.
3. Ternes, T.A. (1998). Occurrence of drugs in German sewage treatment plants and rivers. *Water Res.* 32: 3245–3260.
4. Boxall, A.B., Kolpin, D.W., Halling-Sorensen, B., and Tolls, J. (2003). Are veterinary medicines causing environmental risks? *Environ. Sci. Technol.* 37: 286A–294A.
5. Boxall, A.B.A., Blackwell, P., Cavallo, R. et al. (2002). The sorption and transport of a sulphonamide antibiotic in soil systems. *Toxicol. Lett.* 131: 19–28.
6. Balcioglu, I.A. and Ötker, M. (2003). Treatment of pharmaceutical wastewater containing antibiotics by O-3 and O-3/H2O2 processes. *Chemosphere* 50: 85–95.
7. Cordy, G.E., Duran, N.L., Bouwer, H. et al. (2004). Do pharmaceuticals, pathogens, and other organic waste water compounds

persist when waste water is used for recharge? Ground water monitor. *Remed.* 24: 58–69.

8. Kinney, C.A., Furlong, E.T., Werner, S.L., and Cahill, J.D. (2006). Presence and distribution of wastewater-derived pharmaceuticals in soil irrigated with reclaimed water. *Environ. Toxicol. Chem.* 25: 317–326.
9. Kinney, C.A., Furlong, E.T., Zaugg, S.D. et al. (2006). Survey of organic wastewater contaminants in biosolids destined for land application. *Environ. Sci. Technol.* 40: 7207–7215.
10. Seehusen, D.A. and Edwards, J. (2006). Patient practices and beliefs concerning disposal of medications. *J. Am. Board Fam. Med.* 19: 542–547.
11. Fent, K., Weston, A.A., and Caminada, D. (2006). Ecotoxicology of human pharmaceuticals. *Aquat. Toxicol.* 76: 122–159.
12. Dizer, H., Fischer, B., Sepulveda, I. et al. (2002). Estrogenic effect of leachates and soil extracts from lysimeters spiked with sewage sludge and reference endocrine disrupters. *Environ. Toxicol.* 17: 105–112.
13. Golet, E.M., Xifra, I., Siegrist, H. et al. (2003). Environmental exposure assessment of fluoroquinolone antibacterial agents from sewage to soil. *Environ. Sci. Technol.* 37: 3243–3249.
14. Jones-Lepp, T.L. and Stevens, R. (2007). Pharmaceuticals and personal care products in biosolids/sewage sludge: the interface between analytical chemistry and regulation. *Anal. Bioanal. Chem.* 387: 1173–1183.
15. Focazio, M.J., Kolpin, D.W., Barnes, K.K. et al. (2008). A national reconnaissance for pharmaceuticals and other organic wastewater contaminants in the United States – II. Untreated drinking water sources. *Sci. Total Environ.* 402: 201–206.
16. Heberer, T. (2002). Tracking persistent pharmaceutical residues from municipal sewage to drinking water. *J. Hydrol.* 266: 175–189.
17. Watanabe, H., Horie, Y., Takanobu, H. et al. (2017). Medaka extended one-generation reproduction test evaluating 4-nonylphenol. *Environ. Toxicol. Chem.* 36: 3254–3266.
18. Ma, B., Lu, G., Liu, J. et al. (2017). Bioconcentration and multi-biomarkers of organic UV filters (BM-DBM and OD-PABA) in crucian carp. *Ecotoxicol. Environ. Saf.* 141: 178–187.
19. Berninger, J.P. and Brooks, B.W. (2010). Leveraging mammalian pharmaceutical toxicology and pharmacology data to predict chronic fish responses to pharmaceuticals. *Toxicol. Lett.* 193: 69–78.
20. Huggett, D.B., Cook, J.C., Ericson, J.F., and Williams, R.T. (2003). A theoretical model for utilizing mammalian pharmacology and safety data to prioritize potential impacts of human pharmaceuticals to fish. *Hum. Ecol. Risk. Assess.* 9: 1789–1799.
21. Rand-Weaver, M., Margiotta-Casaluci, L., Patel, A. et al. (2013). The read-across hypothesis and environmental risk assessment of pharmaceuticals. *Environ. Sci. Technol.* 47: 11384–11395.
22. Winter, M.J., Owen, S.F., Murray-Smith, R. et al. (2010). Using data from drug discovery and development to aid the aquatic environmental risk assessment of human pharmaceuticals: concepts, considerations, and challenges. *Integr. Environ. Assess. Manag.* 6: 38–51.
23. Brown, K.H., Schultz, I.R., Cloud, J.G., and Nagler, J.J. (2008). Ahneuploid sperm formation in rainbow trout exposed to the environmental estrogen 17α-ethynylestradiol. *Proc. Natl. Acad. Sci. USA* 105: 19786–19791.
24. Nash, J.P., Kime, D.E., Van der Ven, L.T. et al. (2004). Long-term exposure to environmental concentrations of the pharmaceutical ethinylestradiol causes reproductive failure in fish. *Environ. Health Perspect.* 112: 1725–1733.
25. Hirai, N., Nanba, A., Koshio, M. et al. (2006). Feminization of Japanese medaka (*Oryzias latipes*) exposed to 17β-estradiol: formation of testis-ova and sex-transformation during early ontogeny. *Aquat. Toxicol.* 77: 78–86.
26. Metcalfe, C.D., Metcalfe, T.L., Kiparissis, Y. et al. (2001). Estrogenic potency of chemicals detected in sewage treatment plant effluents as determined by in vivo assays with Japanese medaka (*Oryzias latipes*). *Environ. Toxicol. Chem.* 20: 297–308.
27. Salierno, J.D. and Kane, A.S. (2009). 17 alpha-ethinylestradiol alters reproductive behaviors, circulating hormones, and sexual morphology in male fathead minnows (*Pimephales promelas*). *Environ. Toxicol. Chem.* 28: 953–961.
28. Van den Belt, K., Verheyen, R., and Witters, H. (2001). Reproductive effects of ethynylestradiol and 4t-octylphenol on the zebrafish (*Danio rerio*). *Arch. Environ. Contam. Toxicol.* 41: 458–467.
29. Mclachlan, J.A. (2001). Environmental signaling: what embryos and evolution teach us about endocrine disrupting chemicals. *Endocr. Rev.* 22: 319–341.
30. Mclachlan, J.A., Newbold, R.R., Burow, M.E., and Li, S.F. (2001). From malformations to molecular mechanisms in the male: three decades of research on endocrine disrupters. *APMIS* 109: 263–272.
31. Hood, E. (2005). Are EDCs blurring issues of gender? *Environ. Health Perspect.* 113: A670–A677.
32. Jongbloet, P.H., Zielhuis, G.A., Groenewoud, H.M.M., and Pasker-de Jong, P.C.M. (2001). The secular trends in male: female ratio at birth in postwar industrialised countries. *Environ. Health Perspect.* 109: 749–752.
33. Mackenzie, C.A., Lockridge, A., and Keith, M. (2005). Declining sex ratio in a first nation community. *Environ. Health Perspect.* 113: 1295–1298.
34. WHO/IPCS, 2002. Global Assessment of the State-of-the-Science of Endocrine Disruptors. World Health Organization/ International Program on Chemical Safety. WHO/PCS/EDE/0. 02. Available at: www. who. int/ppcs/emerg_site/edc/global_ edc_ch5. pdf.
35. Crain, D., Janssen, S., Edwards, T. et al. (2008). Female reproductive disorders: the roles of endocrine-disrupting compounds and developmental timing. *Fertil. Steril.* 90: 911–940.
36. Newbold, R.R., Padilla-Banks, E., and Jefferson, W.N. (2009). Environmental estrogens and obesity. *Mol. Cell. Endocrinol.* 304: 84–89.

37. Gould, I.M. (2009). Antibiotic resistance: the perfect storm. *Int. J. Antimicrob. Agents* 34: S2–S5.

38. Daughton, L.G. (2003). Cradle-to-cradle stewardship of drugs minimizing their environmental disposition while promoting human health II. Drug disposal, waste reduction and future directions. *Environ. Health Perspect.* 111: 775–785.

39. Kuspis, D.A. and Krenzelok, E.P. (1996). What happens to expired medications? *Vet. Hum. Toxicol.* 38: 48–49.

40. Paut Kusturica, M., Tomas, A., and Sabo, A. (2017). Disposal of unused drugs: knowledge and behavior among people around the world. *Rev. Environ. Contam. Toxicol.* 240: 71–104.

FIGURE 1.1 A woman fetching water in India. *Source*: License information: McKay Savage from London, UK (https://commons.wikimedia.org/wiki/File:India_-_Sights_&_Culture_-_32_-_woman_fetching_water_(2458024353).jpg), “India – Sights & Culture – 32 – woman fetching water (2458024353),” https://creativecommons.org/licenses/by/2.0/legalcode.

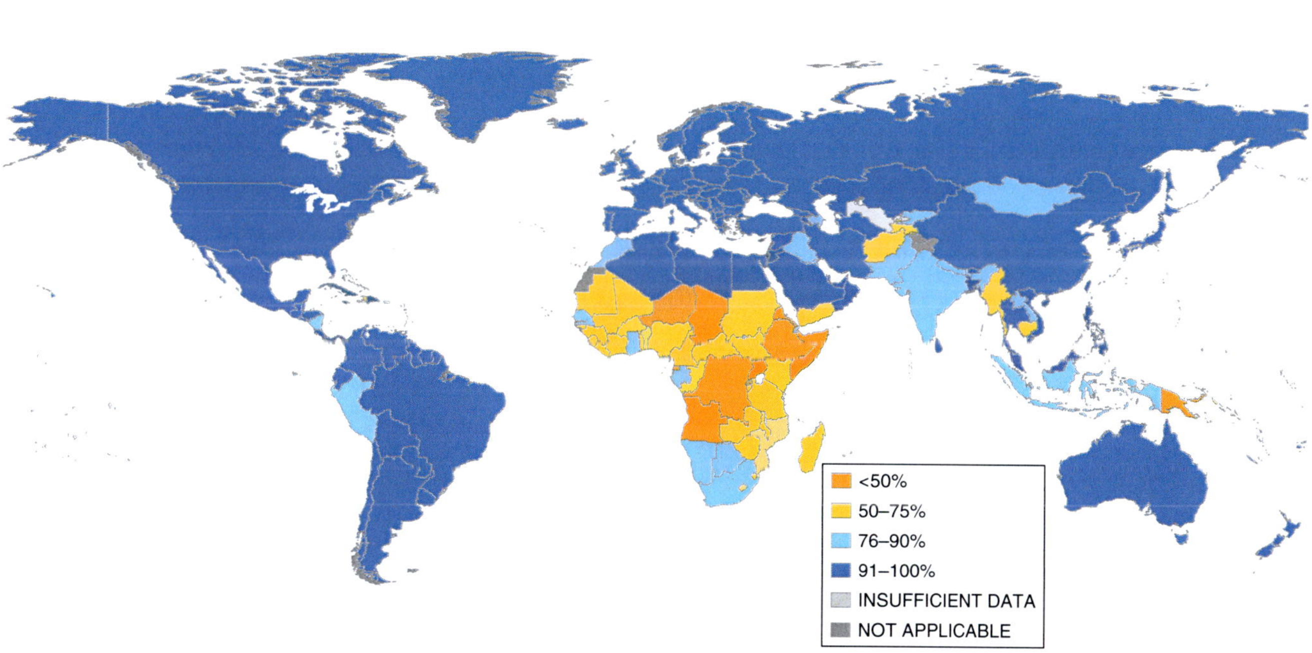

FIGURE 1.6 Countries represented by the percentage of population using at least basic drinking water services in 2015. *Source*: WHO/UNICEF [4].

Water and Sanitation-Related Diseases and the Changing Environment: Challenges, Interventions, and Preventive Measures, Second Edition. Edited by Janine M. H. Selendy.

Companion website: https://www.wiley.com/legacy/wileychi/selendy/

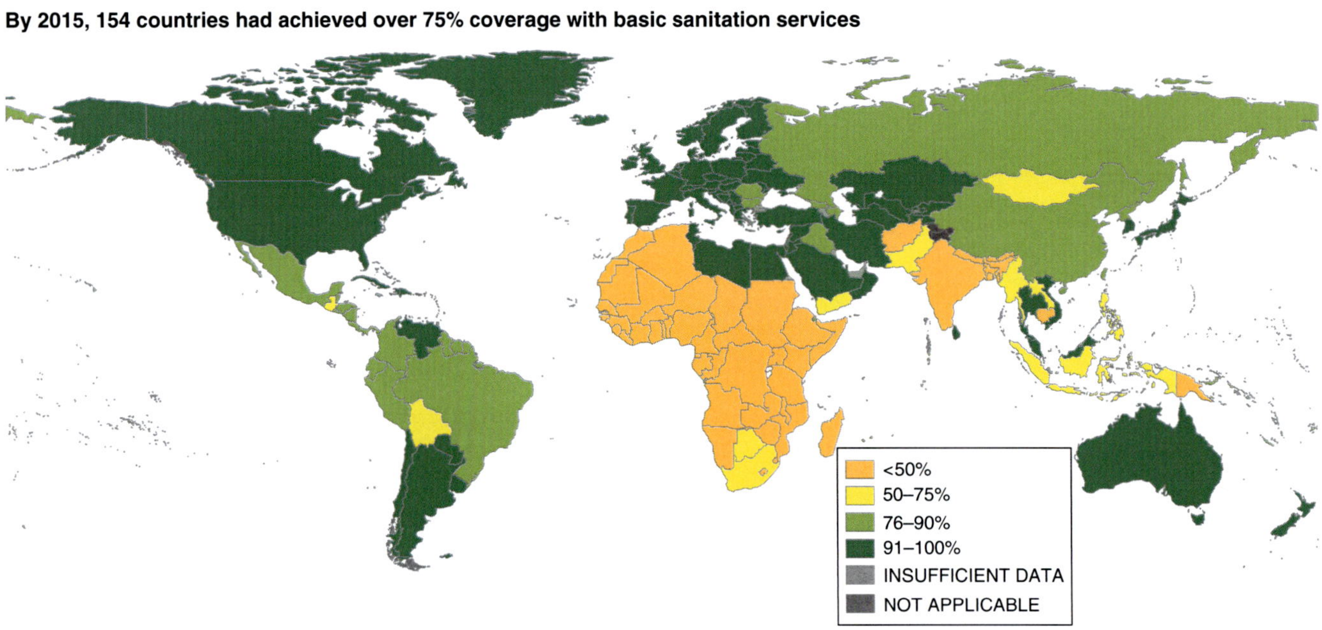

FIGURE 2.2 Global sanitation coverage as of 2015. *Source*: World Health Organization [3].

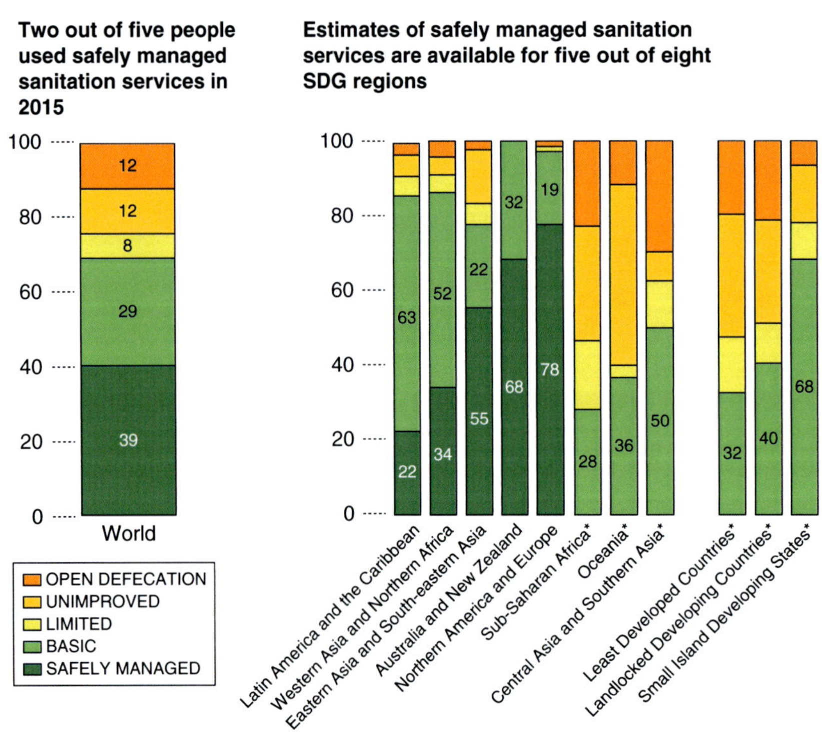

FIGURE 2.3 Global access to safely managed sanitation. *Source*: World Health Organization [3].

FIGURE 4.2 A farmer burned a harvested wheat field last month on the outskirts of Jalandhar, India. *Source*: Credit Shammi Mehra/ Agence France-Presse – Getty Images. Available from: http://www.nytimes.com/2016/11/03/world/asia/farmers-unchecked-crop-burning-fuels-indias-air-pollution.html?_r=0.

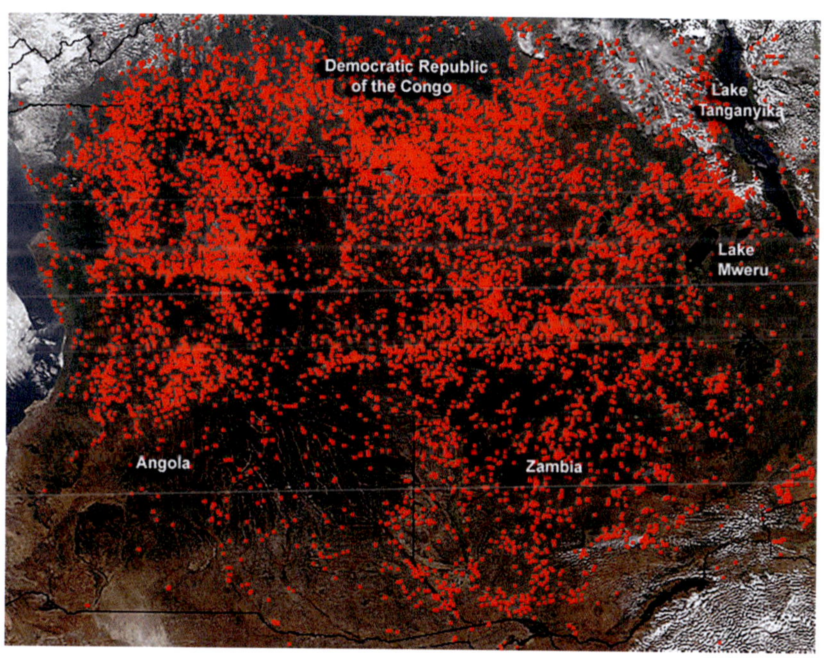

FIGURE 4.5 Widespread agricultural burning throughout central Africa. Fires burning in the southern region of the Democratic Republic of the Congo, Zambia, and Angola. Actively burning areas are outlined in red. *Source*: Photo from the NASA's Aqua satellite June 3, 2016. NASA, NOAA, and the US Department of Defense. Available from: https://www.nasa.gov/sites/default/files/styles/image_card_4×3_ratio/public/thumbnails/image/centralafricasnpp66_0.jp.

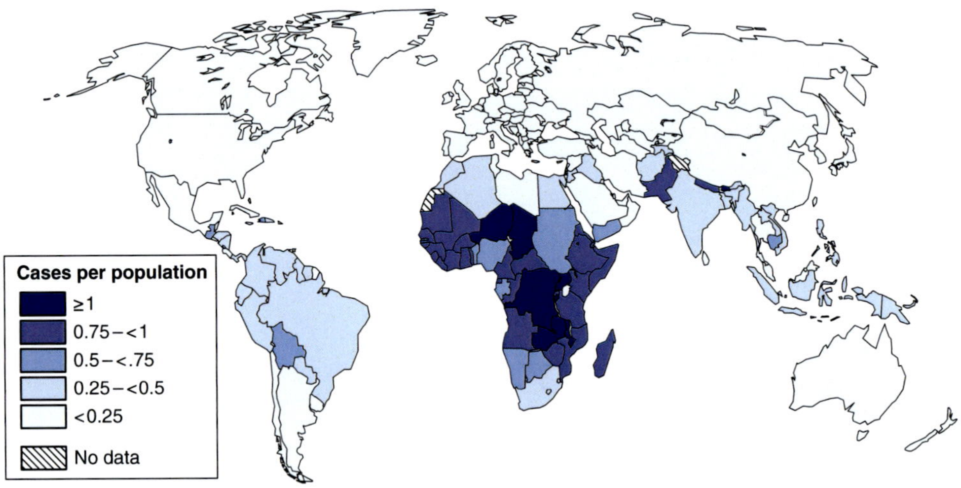

FIGURE 7.1 Incidence on diarrhea per population. *Source*: Adapted from The Global Burden of Disease: 2015 Update; WHO [1].

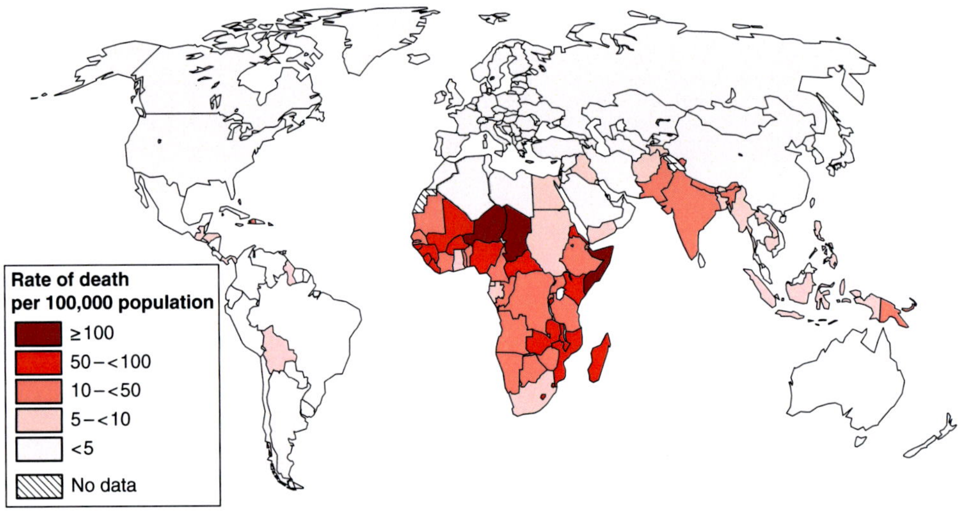

FIGURE 7.2 Rate of death due to diarrhea per 100 000 population. *Source*: Adapted from The Global Burden of Disease: 2015 Update; WHO [1]. Population estimates from World Bank database 2015.

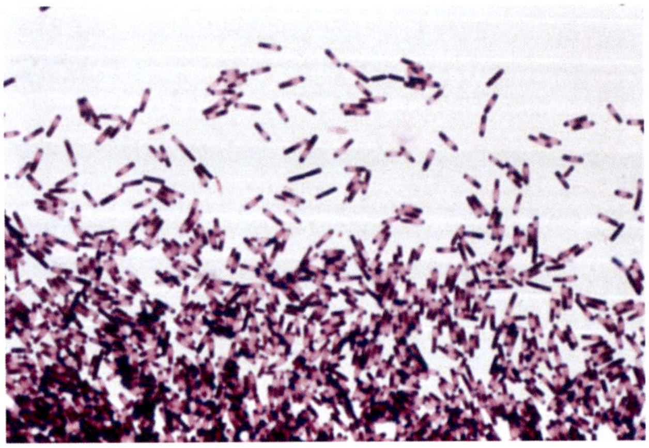

IMAGE 7.12 Gram-stained light micrograph of *Clostridium difficile*. (Content provider: CDC. *Source*: CDC, Public Health Image Library, Image ID # 3876.)

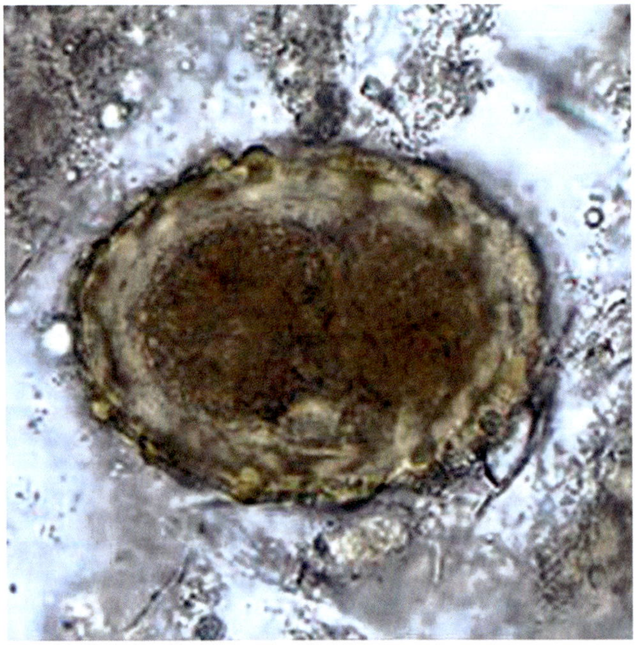

FIGURE 8.2 *Ascaris lumbricoides* egg. (*Source*: Courtesy Division of Parasitic Diseases/Centers for Disease Control and Prevention.)

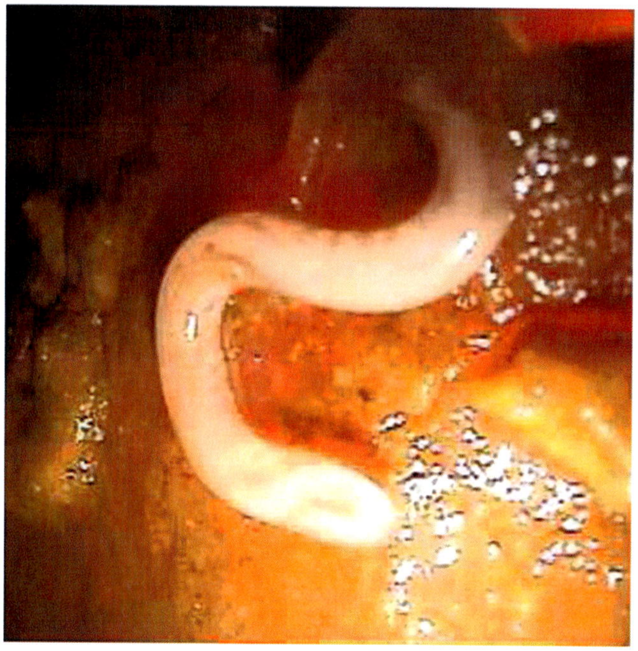

FIGURE 8.3 *Trichuris trichiura* adult worm (colonoscopy image). (*Source*: Courtesy Division of Parasitic Diseases/Centers for Disease Control and Prevention and Duke University Medical Center.)

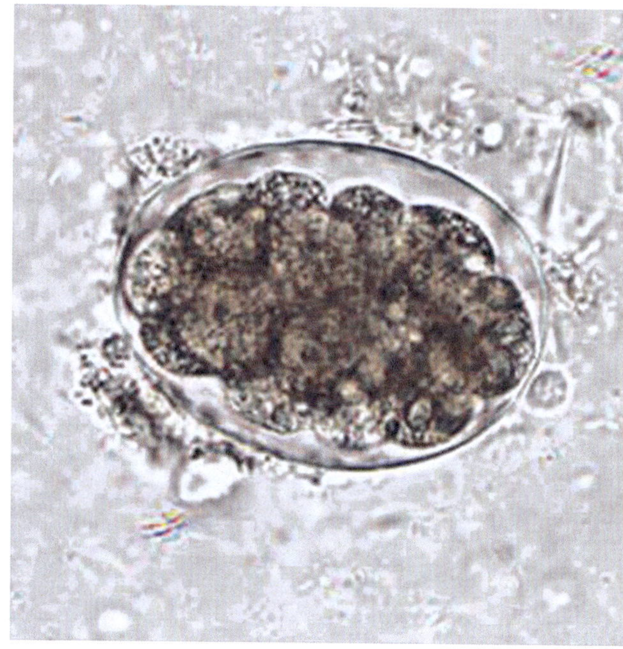

FIGURE 8.5 Hookworm egg. (*Source*: Courtesy Division of Parasitic Diseases/Centers for Disease Control and Prevention.)

FIGURE 8.10 Mass drug administration, Central Nigeria. (*Source*: Photo courtesy Dr Brian Blackburn, Stanford University School of Medicine.)

FIGURE 8.11 Typical transmission conditions for STHs, Central Nigeria. (*Source*: Photo curtesy Dr Brian Blackburn, Stanford University School of Medicine.)

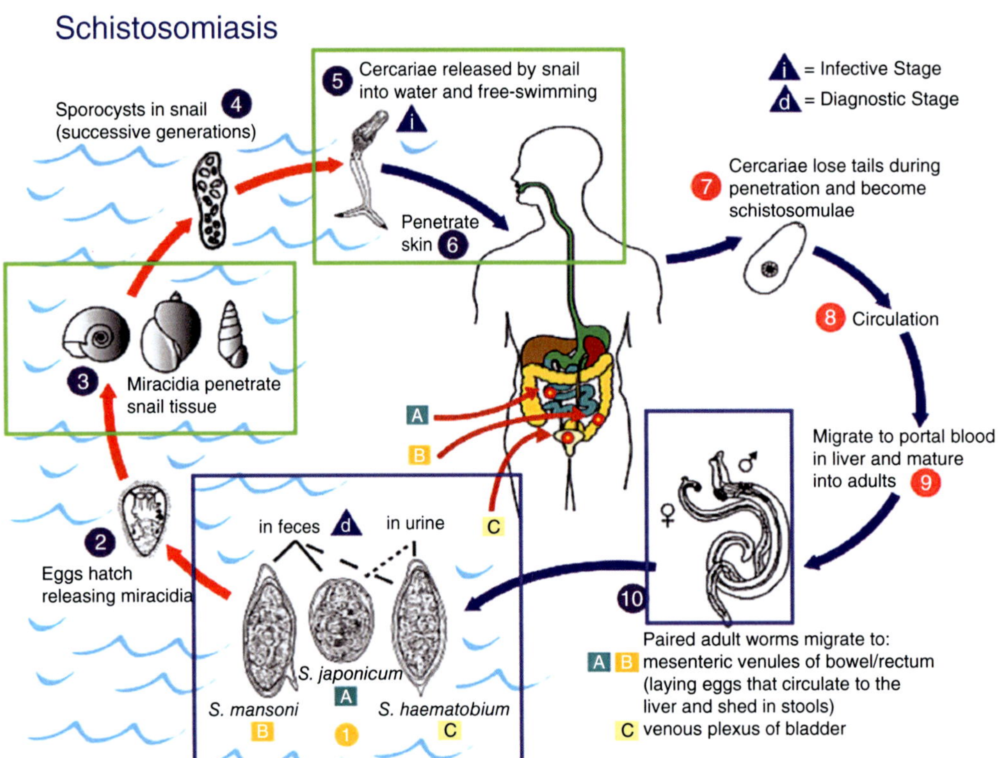

FIGURE 11.1 The transmission cycle of schistosomiasis. The numbers inserted in the text correspond to numbers indicated on the drawing (*Source*: With the courtesy of www. cdc.gov).

FIGURE 11.2 Schistosomiasis vector snails (*Source*: Photo: Henry Madsen).

FIGURE 11.3 Water contact in a branch of the Niger River at Diourou, Mali (*Source*: Photo: Henry Madsen).

FIGURE 11.4 Children playing in a small stream in Nigeria (*Source*: Photo: Henry Madsen).

FIGURE 11. 5 Canal cleaners removing aquatic plants from irrigation canals in the Sudan (*Source*: Photo: Danish Bilharziasis Laboratory).

FIGURE 11.6 Rice planting, Leyte, the Philippines (*Source*: Photo: Henry Madsen).

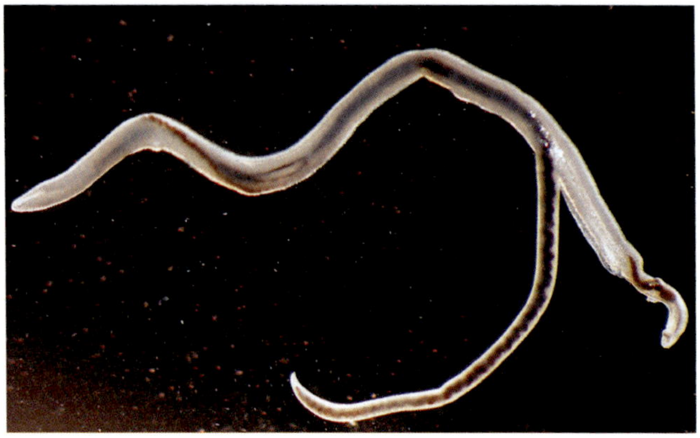

FIGURE 11.7 *Schistosoma* spp. worm pair. The female worm is located in the ventral canal of the male (*Source*: Photo: Maria Vang Johansen).

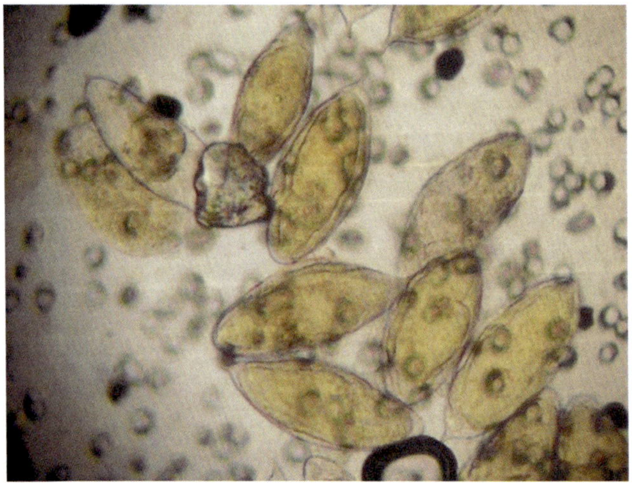

FIGURE 11.8 *Schistosoma haematobium* eggs on a Nucleopore filter after urine filtration (*Source*: Photo: Moussa Sacko).

FIGURE 14.4 Potential routes of transmission of antibiotic-resistant bacteria.

FIGURE 16.1 Vestergaard–Frandsen LifeStraw Filter 2.0.

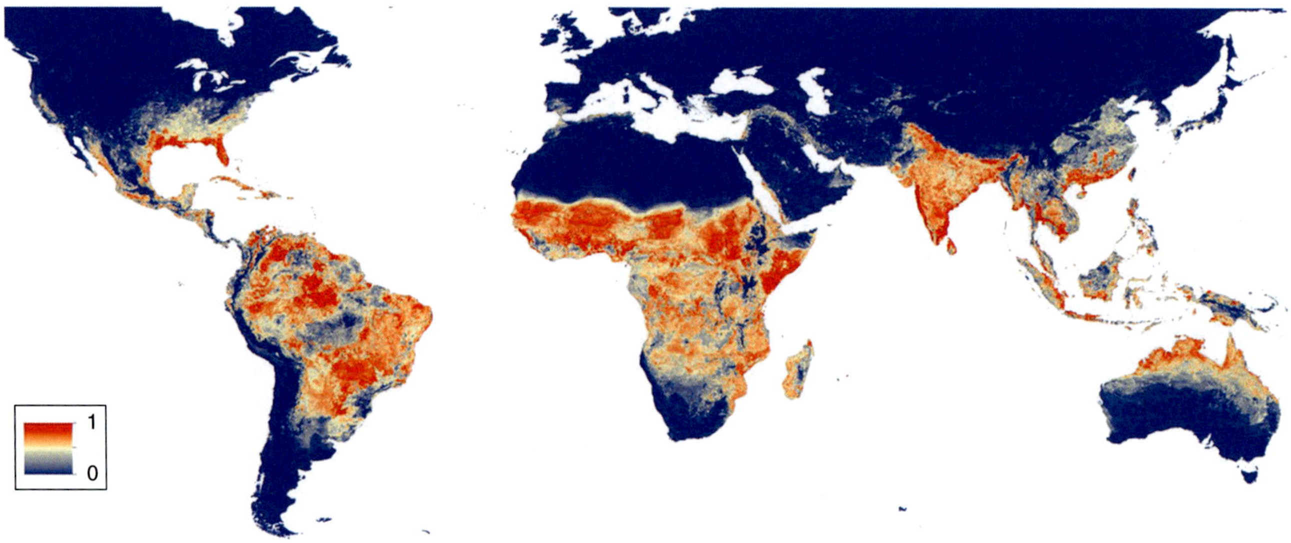

FIGURE 17.1 Distribution *Ae. aegypti*. (*Source*: Reproduced from [18].

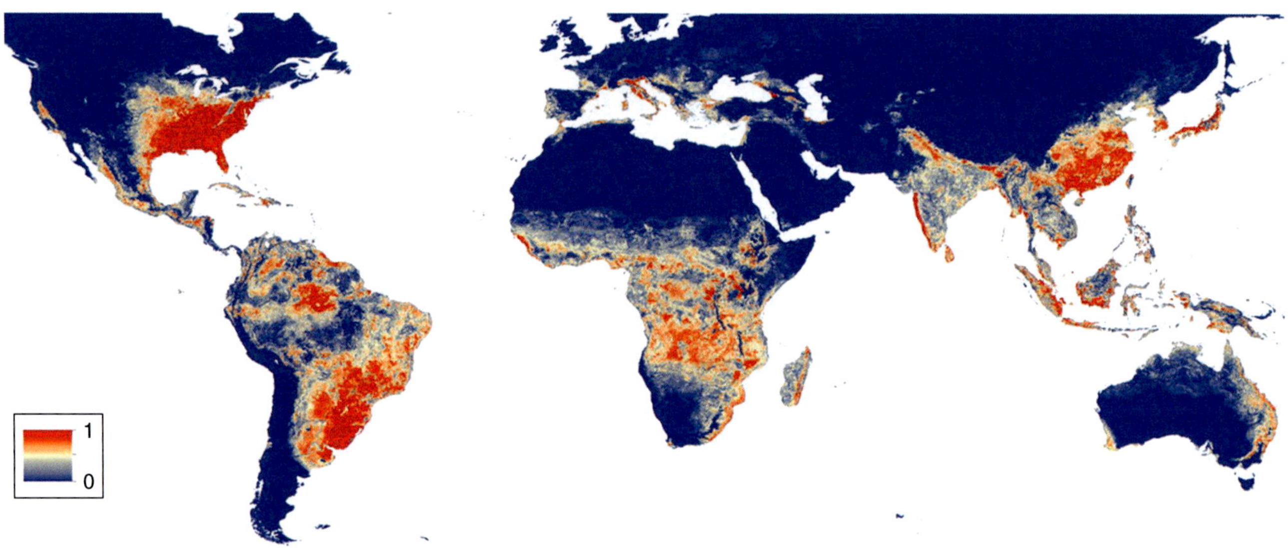

FIGURE 17.2 Distribution *Ae. albopictus* (globally) and areas with rapidly expanding range [18]. MAPS Figures 17.1 and 17.2: This is an open-access article free of all copyright and may be freely reproduced, distributed, transmitted, modified, built upon, or otherwise used by anyone for any lawful purpose. The work is made available under the Creative Commons CC0 public domain dedication.

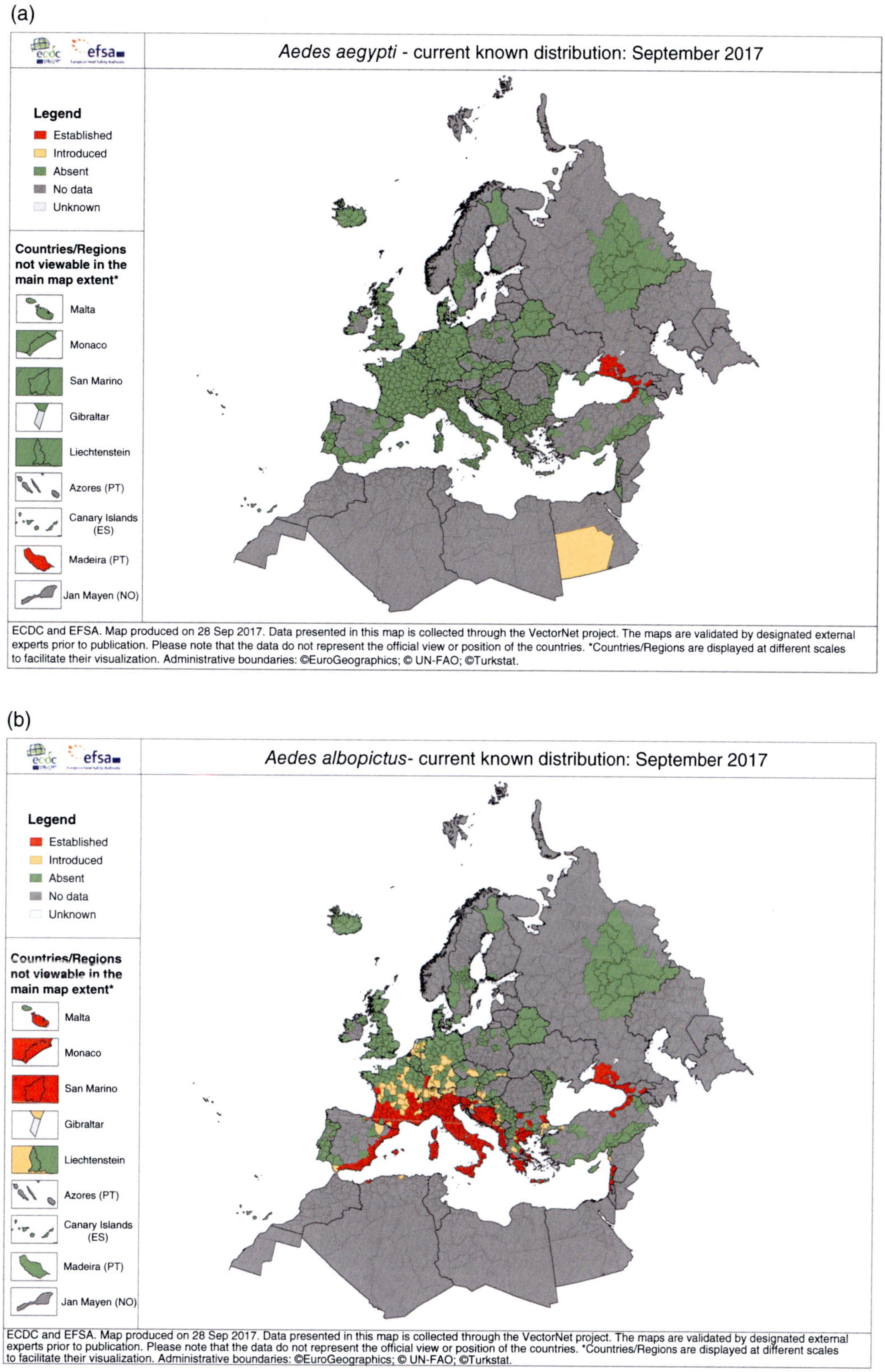

FIGURE 17.5 Maps: current distribution of *Ae. aegypti* (a) and *Ae. albopictus* (b) in Europe from ECDC. Available from: http://ecdc.europa.eu/en/activities/pages/programme_on_emerging_and_vector-borne_diseases_maps.aspx.

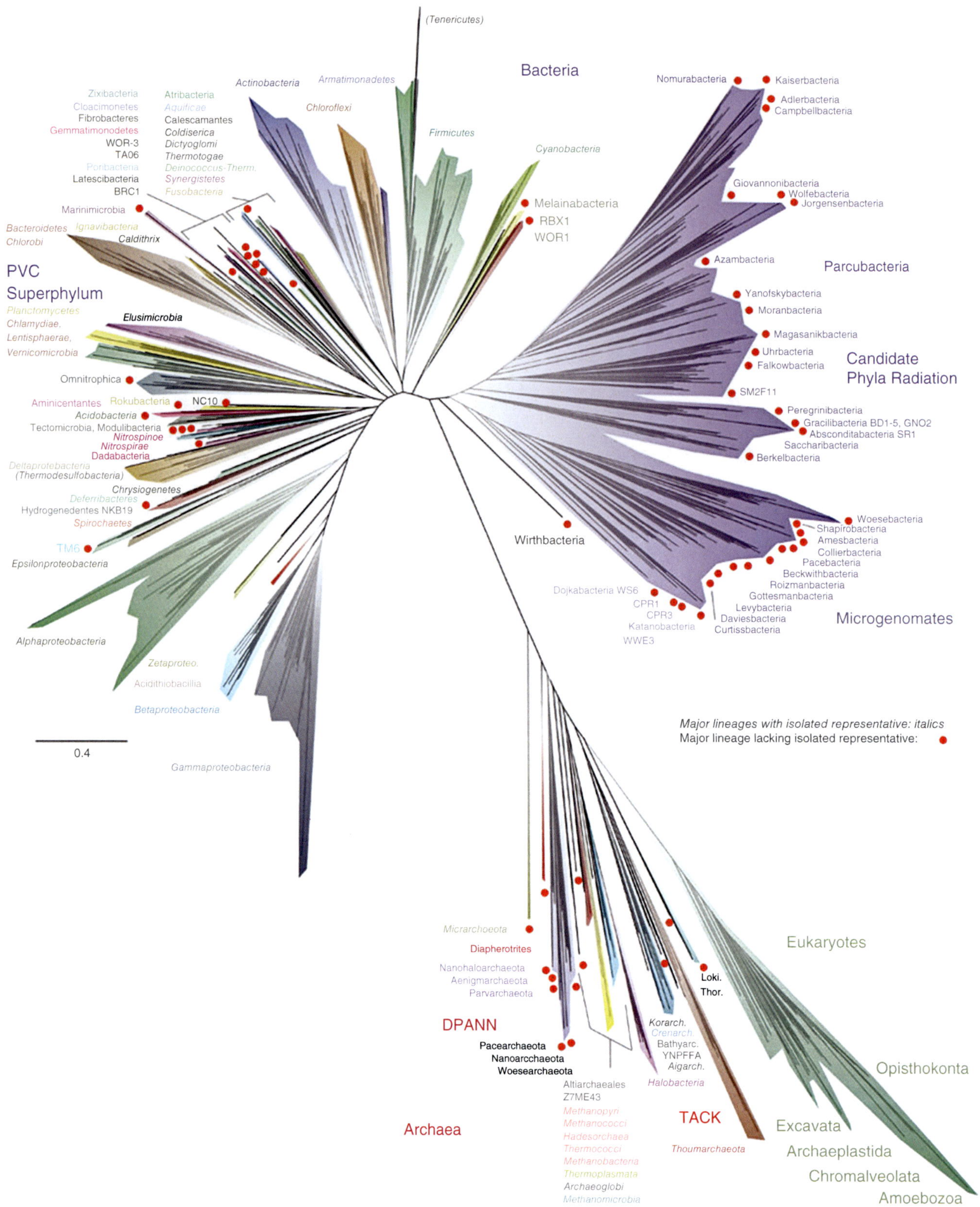

FIGURE 19.1 The new tree of life, based on 1000 newly sequenced genomes, places, all of the plants (the Archaeplastida, bottom right corner), animals and fungi (the Opisthokonta, bottom right corner) on an evolutionary edge, having emerged from the microbes that support our continued existence on Earth. For explanation of names and relationships, see the *source*: Hug et al. [8]. Available from: https://www.nature.com/articles/nmicrobiol201648#s6. Licensed under CC BY 4.0.

FIGURE 19.3 The rapid transition in a shallow lake from clear water dominated by one (a) or many species (b) of plant, supported mutually by various aquatic animals, to a turbid state (c) where large plants and the animals that depend on and support them are excluded in favor of microscopic algae and cyanobacteria. The turbid water shades the sediments so large plants cannot root and the loss of plants removes the refuge for zooplankton (from fish predators), so they can no longer sweep the water clear of phytoplankton. Shown are communities in water taken from Hickling Broad, Norfolk, UK, for a study of alternative stable states [47, 48]. (a) Shows clear water with *Chara intermedia*, (b) shows clear water with *Myriophyllum spicatum* L, *Ceratophyllum demersum* L, *Enteromorpha intenstinalis* L, *Potamogeton pectinatus* L, *Lemna trisulca* L, and *Elodea canadensis* Michx, (c) shows the turbid, green, water of the degraded state, empty of plants, and dominated by species of microscopic algae and cyanobacteria.

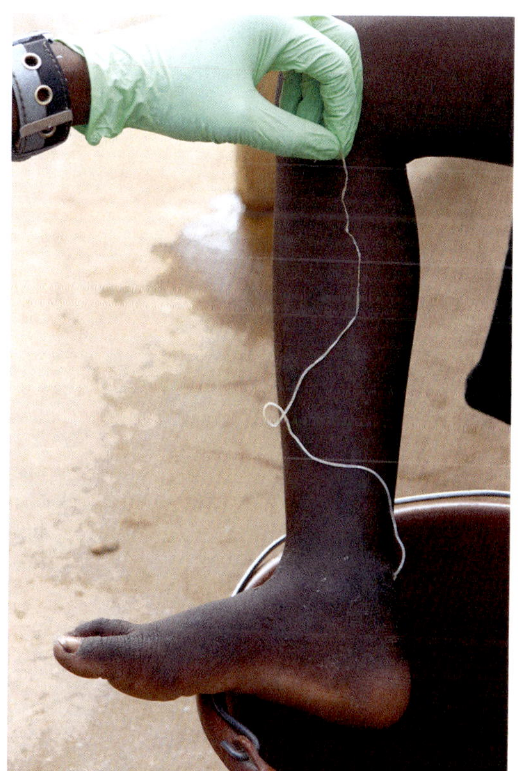

FIGURE 22.1 Emerging Guinea worm. *Source*: Photo credit: The Carter Center/Louise Gubb.

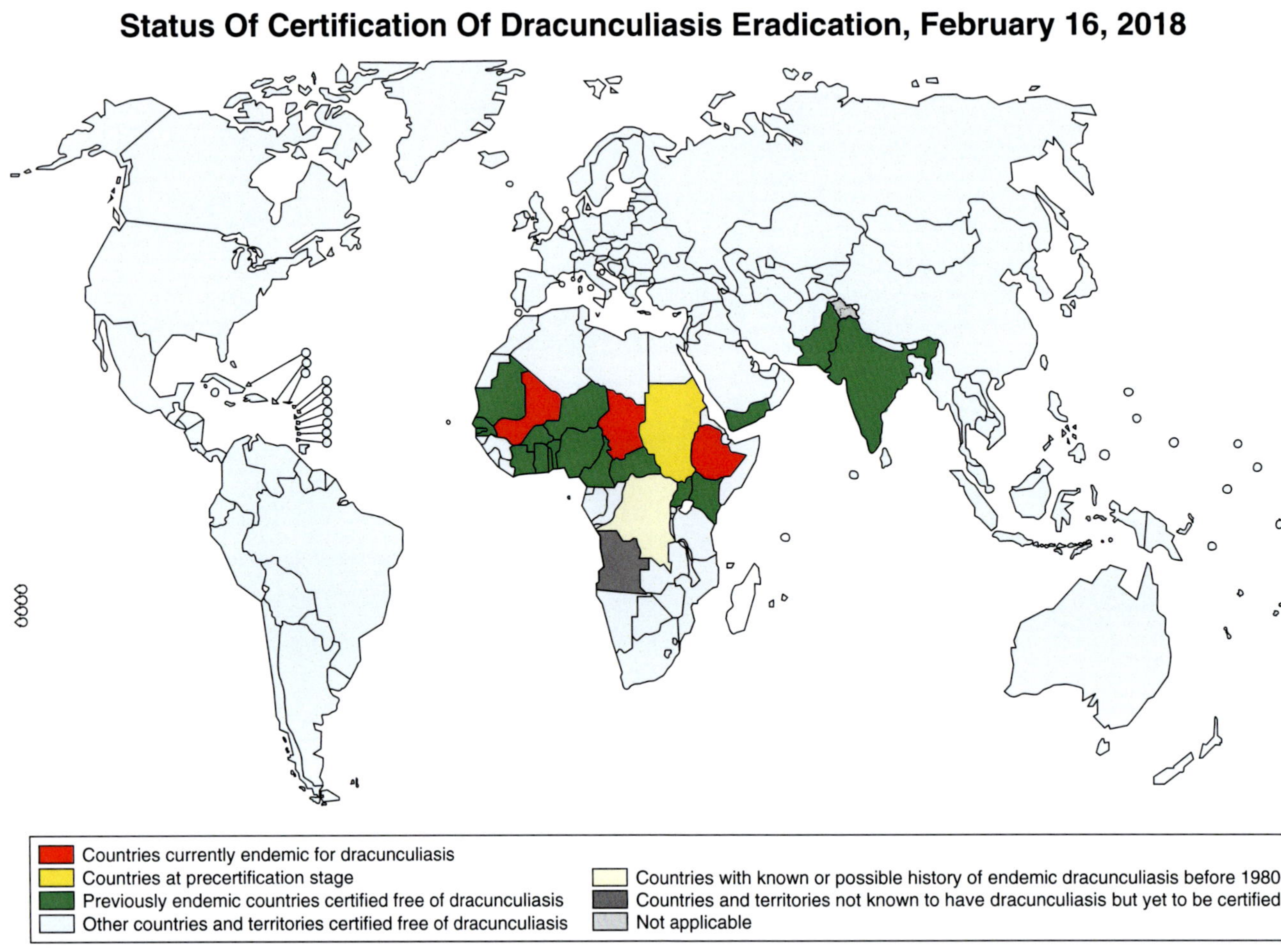

FIGURE 22.3 Distribution of 30 indigenous cases of Dracunculiasis reported during 2017. For the first time since the global Guinea Worm Eradication Program began, only two countries, Chad and Ethiopia, reported cases of Guinea worm disease (dracunculiasis) in humans during 2017 (Figure 22.1). *Source*: Reproduced from The Carter Center and the Centers for Disease Control and Prevention. (9 March 2018), "Guinea Worm Wrap-Up #253". Available from: https://www.cartercenter.org/resources/pdfs/news/health_publications/guinea_worm/wrap-up/253.pdf.

A community map being drawn in rural Madagascar as part of the "triggering" meeting. Participants identify key features in their community, where the main open defecation areas are, and how they inadvertently eat each other's feces. *Source*: Credit: WSSCC.

Support for making latrines more accessible often comes from family members. In Malawi, Dorothy's brother decided to build an accessible pit latrine and bathing shelter for her, with a small ramp at the entrance. Source: *Credit*: Sarah House.

Sita – a member of a Dalit community in Khana VDC, Nepal – stands outside her newly built latrine. *Source*: Credit: Sue Cavill.

Constructing a twin-pit latrine in India. *Source*: Credit: WSSCC.

A newly constructed pour-flush latrine with a handwashing facility in Cambodia. *Source*: Credit: WSSCC.

14

OTHER WATER POLLUTANTS: ANTIMICROBIAL RESISTANCE

ROCHELLE RAINEY

Environmental Health, USAID Global Health Bureau, Washington, DC, USA

14.1 INTRODUCTION

In addition to the general health and environmental risks from unsafely managed sanitation, as discussed in previous chapters, wastewater and other aquatic systems can also serve as a reservoir for antibiotics and antimicrobial-resistant microorganisms along with their genetic material.

Wastewater is the general term for fresh water that has passed through one or more human systems, with a subsequent decline in microbiological and/or chemical quality. Wastewater can come from a combination of domestic, industrial, commercial or agricultural activities, surface runoff, and from sewer inflows. Poor quality or lack of sanitation services including access to toilets and latrines, garbage collection and disposal, fecal sludge removal from latrines and septic tanks, and public drainage can also contribute to contamination of water bodies. Globally, it is estimated that over 80% of wastewater is released to the environment without adequate treatment [1]. Climate change is expected to increase the frequency and severity of extreme rainfall events, which can overload the capacity of existing functioning sewer systems, resulting in additional untreated sewage flowing into the environment.

14.2 ANTIMICROBIALS AND ANTIBIOTICS

Antimicrobials are a group of drugs that are designed to kill, prevent, or slow the growth of microorganisms – a category that includes bacteria, viruses, protozoa, and fungi. The best-known group of antimicrobials is antibiotics, which are designed to kill bacteria and treat bacterial infections. Antibiotics can occur naturally, be produced semi-synthetically from natural products, or chemically synthesized, based on the structure of natural antibiotics [2]. Antibiotics are grouped by chemical structure (class) and by the mechanism of inhibiting synthesis in bacterial cells. Antimicrobials, especially antibiotics, are widely used in human and veterinary health and in agriculture to cure infections and maintain health and productivity.

Global consumption of antibiotics for human health increased by 35% between 2000 and 2010 [3], influenced by an increasing and aging population, longer survival with complex illnesses, changes in food systems, and other social, cultural, and economic factors. Per capita consumption is generally higher in high-income countries, but the greatest increase in antibiotic use between 2000 and 2010 was in low- and middle-income countries [4]. Changes in per capita use of antibiotics for human health for 69 countries between 2000 and 2010 are shown in Figure 14.1 [3].

While an estimated 80–90% of the volume of human antibiotic use occurs in outpatient settings [5], inappropriate applications, including unnecessary use and inappropriate selection, dosing and duration, may approach 50% of all outpatient antibiotic use [6–8]. Thus, the social aspects of antimicrobial use and impacts are significant.

Disposal of unused antibiotics can also be a challenge, whether the prescribed course was completed correctly or

This reflects the views of the author and does not necessarily reflect the views of USAID or the US Government.

Water and Sanitation-Related Diseases and the Changing Environment: Challenges, Interventions, and Preventive Measures, Second Edition. Edited by Janine M. H. Selendy.

Companion website: https://www.wiley.com/legacy/wileychi/selendy/

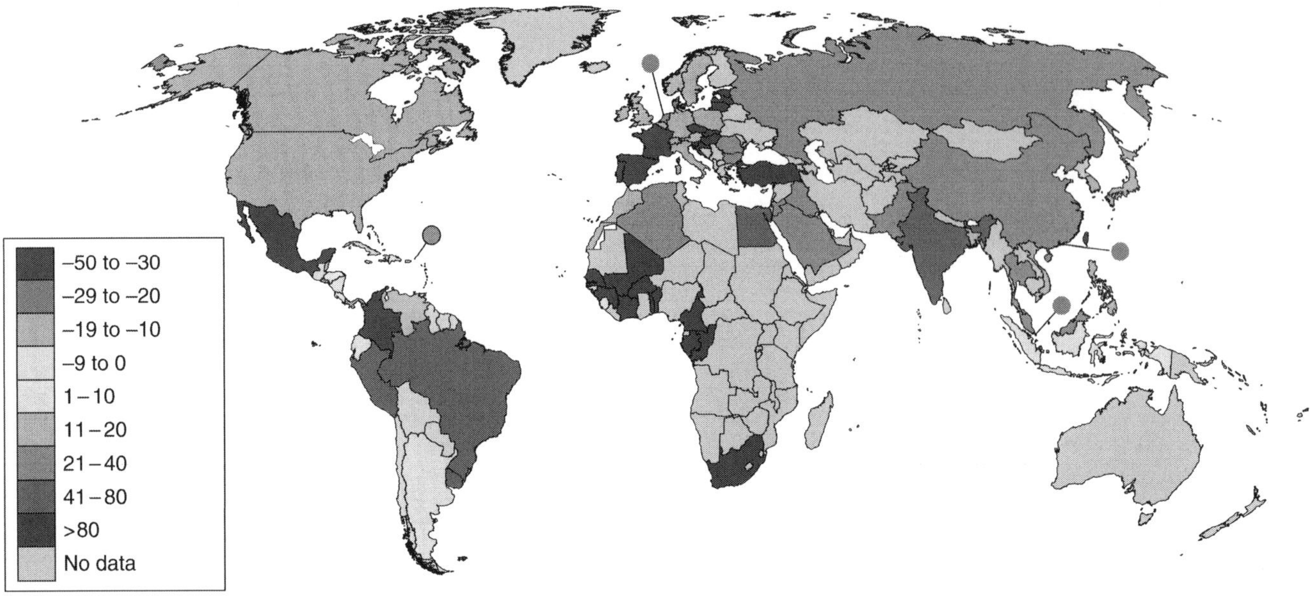

FIGURE 14.1 Percentage change in antibiotic consumption per capita 2000–2010.

not. A study of 250 individuals in Malaysia found that 36% of them had taken antibiotics in the year of the study and nearly 55% discontinued use once symptoms disappeared, while 79% reported disposing of leftover antibiotics and only 6% of participants returned leftover antibiotics to the pharmacist or doctor [9]. Drug donations, including antibiotics, are frequently part of international development health programs, as well as a component of humanitarian responses. Most of these are used properly and save lives, but in some cases the pharmaceuticals may arrive past or near their expiry date, or they may be inappropriate for the needs. In these cases, safe disposal of these drugs can create a major problem and can entail heavy expense. Inappropriate disposal into a water supply can affect the makeup of the bacterial community in the wastewater treatment plant and the water downstream, and also the soil ecology if sludge from the wastewater treatment plant containing it is applied as fertilizer.

The United States Food and Drug Administration provides guidance for disposal of unused medicines (including antimicrobials) in the United States [10], while the #MedsDisposal website [11] has an interactive map of Europe with a description of the regulatory framework in each country and links to national-level information on disposal. The World Health Organization (WHO) issued guidelines for safe disposal of unwanted pharmaceuticals in and after emergencies for countries without local, regional, or national frameworks [12]. Medicine take-back programs are the best way to safely dispose of most types of unneeded medicines where they exist. In the United States, consumers and nursing homes and other types of long-term care facilities can transfer unused medicines to collectors registered with the Drug Enforcement Administration, which may include retail pharmacies, hospital or clinic pharmacies, and law enforcement locations. If there are no take-back or authorized collectors available, the recommended steps for household disposal are: (i) mix medicines (do not crush tablets or capsules) with an unpalatable substance such as dirt, kitty litter, or used coffee grounds; (ii) place the mixture in a container such as a sealed plastic bag; and (iii) dispose of the container with other household solid waste [10].

There is an increased demand for animal protein leading to a shift towards large-scale farming and increased use of antimicrobials in animal production. Estimates of total global consumption in agriculture vary due to poor surveillance and data collection, but range from around 63 000 metric tons [3] to over 240 000 metric tons annually [13]. China is the largest user, with an estimated 23% of the global total, followed by the United States at 13%. Antimicrobial consumption is projected to rise by 67% by 2030, and nearly double in Brazil, Russia, India, China, and South Africa. This does not include use of antimicrobials in aquaculture, for example, salmon and shrimp farming [14]. It is estimated that more than 50% of the antibiotics that are deemed medically important for human health sold worldwide are used in livestock production [15].

The lack of appropriate regulations in the manufacture, sales, and use of antimicrobials enables misuse and poor-quality products. Voluntary and regulatory efforts to reduce use in human and animals in different contexts have had mixed results, with the most stringent restrictions in European countries [14, 16–20]. Figure 14.2 shows the percentages of WHO member states with a National Regulatory Authority, national quality standards, and ability to enforce standards for antibiotic manufacture and use [21].

In Europe, all use of antibiotics for promoting growth has been banned since 2006, and they are also banned in Mexico,

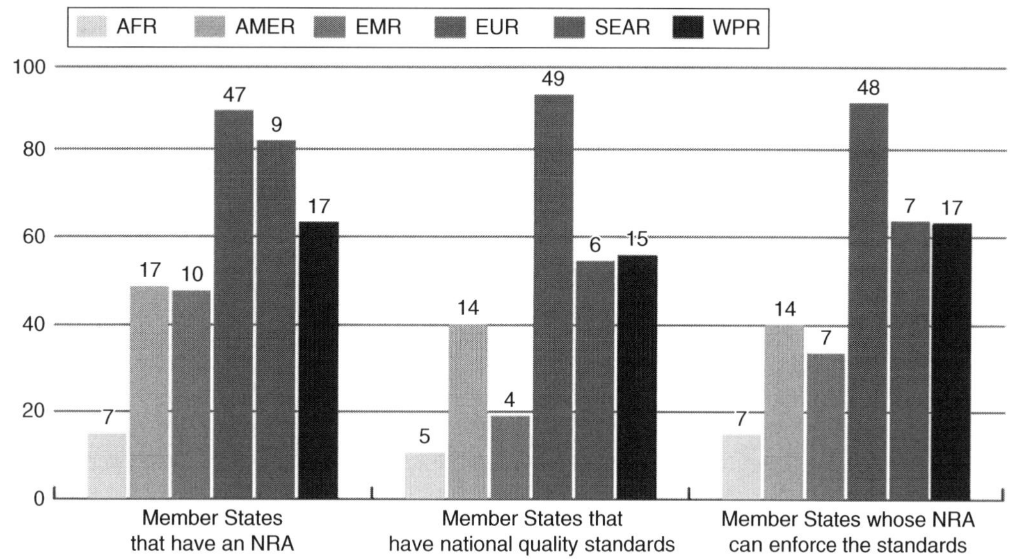

FIGURE 14.2 Percentages of UN member states with regulatory authority, quality standards and the capacity to enforce standards for manufacture and use of antibiotics, all regions.

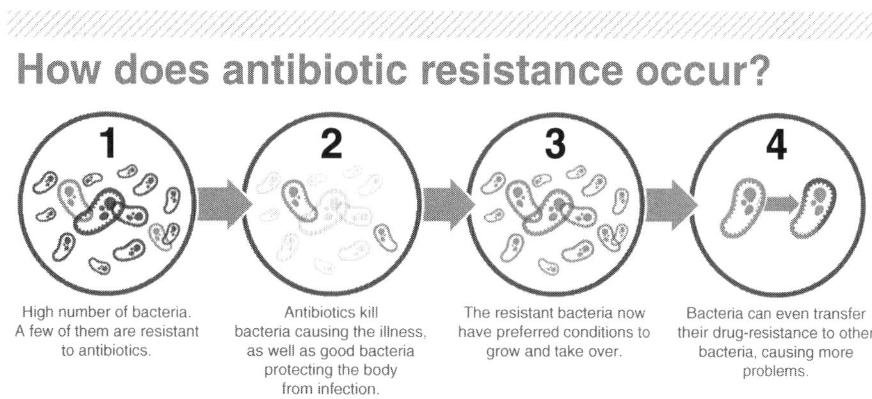

FIGURE 14.3 How does antibiotic resistance occur?

South Korea, New Zealand, and partially banned in Australia. In the United States, the use of antibiotic growth promoters is not banned, but in 2013 the Federal Drug Authority (FDA) issued voluntary guidelines for the industry to withdraw the use of medically important antibiotics as growth promoters, and Canada followed with a similar voluntary approach in 2014. Antibiotic growth promoters are not banned in many countries that are major meat producers, including China, Brazil, Russia Federation, Argentina, India, Indonesia, Philippines, and South Africa [22].

14.3 ANTIMICROBIAL RESISTANCE (AMR)

Antimicrobial Resistance (AMR) occurs when bacteria, viruses, protozoa, or fungi change their genetic makeup in response to exposure to an antimicrobial agent, developing resistance genes. These mutations can then be passed on to future generations, spreading the genetic resistance through reproduction or by sharing the genetic material with other bacteria through horizontal gene transfer of the resistance genes [23].

Emergence of resistance is unavoidable from an evolutionary perspective – any use of antimicrobials, whether in humans, animals, plants, or in the environment, creates evolutionary pressures that can allow the resistant microbes to thrive and the susceptible microbes to die off, as shown in Figure 14.3 [24].

This was graphically demonstrated in time-lapse videos showing bacteria mutating to overcome higher and higher concentrations of antibiotics meant to destroy them [25]. However, excessive use of antimicrobial drugs has resulted in wide dissemination of antimicrobials, AMR and associated resistance genes, in the community and the environment. Some factors that influence the development and spread of AMR include the antimicrobial compounds used, how they

FIGURE 14.4 Potential routes of transmission of antibiotic-resistant bacteria. (*See insert for color representation of this figure.*)

are used, microbial co-selection, fitness and persistence mechanisms, use of wastewater and sludge in agriculture, globalization and urbanization, host lifestyle, contact with wildlife, and treatment conditions [26]. Figure 14.4 and the explanatory text below (used with permission of bioMérieux) describe the potential routes of transmission of antibiotic-resistant bacteria to and through the environment [27].

Humans, pets, livestock, and fish farms rely on similar classes of antibiotics to fight infectious diseases. Both pathogenic and non-pathogenic bacteria evolve or exchange the ability to survive when exposed to these antibiotics. Then they can spread into the environment through different routes, for example, through water sanitation systems (1), as wastewater treatment facilities do not entirely remove antibiotic resistant bacteria before releasing water in waterways. Another common route is through the application of manure to fields with cultivated crops (2), where antibiotic-resistant bacteria can flourish on the plants (3). The uptake of these resistant bacteria can then happen through the food chain, when humans later consume these plants (4) or the contaminated flesh of animals and fish harboring resistant bacteria (5). As bacteria can easily reach water reserves, water distribution infrastructure is also a potential route for the spread of these bacteria (6). Even wildlife, insects, and other bugs are potential carriers of antimicrobial resistance (7). Tourism, migrations, and food imports (8) are nevertheless reported as the fastest way of spreading resistant strains of bacteria across borders. At the healthcare facility level, resistant bacteria can spread by contact between patients or with healthcare staff, or through contaminated surfaces and medical devices.

Antimicrobial resistance crosses geographic boundaries and is a serious threat to public and environmental health globally. Infection associated with antimicrobial resistance results in increased morbidity, mortality, and healthcare costs. For example, methicillin-resistant *Escherichia coli* (MRSA) is a type of bacteria that is resistant to certain methicillin and other common antibiotics such as oxacillin, penicillin, and amoxicillin. Serious or life-threatening occurrences of *Staphylococcus* infections, including MRSA, occur most frequently among persons in hospitals and healthcare facilities (such as nursing homes and dialysis centers). Infections with multidrug-resistant pathogens are responsible for at least 700 000 deaths every year globally, a number estimated to increase to 10 million by 2050 [28].

14.4 AMR IN THE ENVIRONMENT

Between 10 and 90% of the antibiotics administered to humans or animals are shed in urine or feces either in their original formulation, partly degraded, or as metabolites in forms that are still biologically active [29]. Therefore, antibiotics have been discharged into wastewater treatment plants since they were first developed. However, antibiotics are only partially eliminated during wastewater treatment

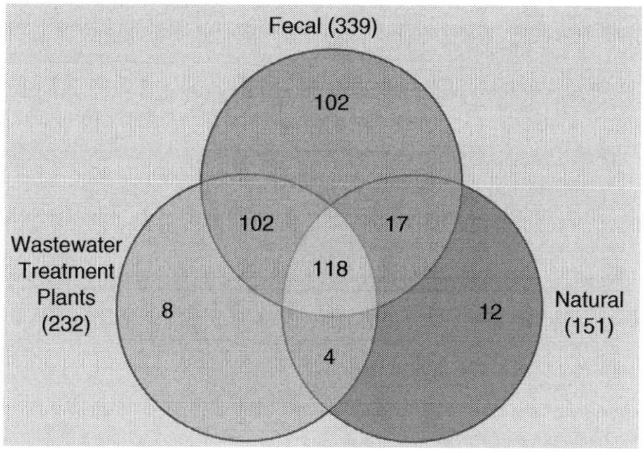

FIGURE 14.5 Shared ARGs in different environments (based on the data of Li et al. [39]).

[30–32]. As a result, antibiotics and other antimicrobial agents along with their metabolites, antimicrobial-resistant (AMR) bacteria, and their antimicrobial resistance genes (ARGs) are now common and widespread contaminants in water and other environmental media. This includes wastewater used for agricultural purposes or directly discharged to the environment [33]. AMR bacteria have been widely detected in wastewater, drinking water, soils, and ambient air, and in the human food chain [34–38].

Figure 14.5 shows a summary of antibiotic resistance genes (ARGs) unique to specific environments, and shared between and among the natural environment (mainly river water and sediments), fecal samples, and wastewater treatment plants, showing a large overlap of ARGs from both fecal and environmental sources also found in wastewater treatment plants [39]. The authors also found that the levels of ARGs in wastewater are comparable to the fecal sources, indicating a need for an integrated strategy to address the spread of ARGs.

Understanding the scope and magnitude of existing risks to human and ecosystem health is ongoing. Although the concentration of individual antibiotics is generally low in the aquatic environment [40], the presence of a broad range of antibiotics in some source waters suggests that the overall effect should not be underestimated, and that the aquatic environment can play multiple roles in the emergence, exposure, and spread of AMR [41, 42]. A study of ARG from environmental water samples in Germany and Australia found that the levels of antibiotic resistance genes for five classes of antibiotics (sulfonamide, trimethoprim, macroline, beta-lactams, and chloramphenicol) were significant enough to consider the examined ecosystems as reservoirs [43]. Therefore, in some cases, high background concentrations of ARG in the receiving water body overshadow any potential effects of wastewater treatment plant effluent discharge [44].

Antibiotics gravitate to different environmental compartments according to their physical and chemical properties, where they may undergo further physical or biological transformation. For example, wastewater treatment removes up to 90% of the antibiotic ciprofloxacin, but the mechanism of removal is sorption to sewage sludge, where the ciprofloxacin then accumulates [45]. When the sludge is applied to soil as fertilizer, the ciprofloxacin can persist for more than 90 days with only minimal transformation, where it can affect a range of soil microorganisms [29, 46].

A study in China showed that the long-term (10-month) application of sewage sludge and chicken manure can change bacterial communities, causing an increase in antibiotic-resistant forms. Two of these were important human pathogens: the causative agent of tuberculosis (*Mycobacterium tuberculosis*) and a common cause of hospital-associated infections (*Pseudomonas aeruginosa*) [47].

14.5 WASTEWATER TREATMENT

Conventional wastewater treatment consists of multiple processes. In primary treatment, heavy solids settle out, and fats, grease, and oils float to the surface and form a scum. These solids are removed and treated separately, while the remaining wastewater proceeds to secondary treatment. Here, biological contaminants are degraded using beneficial microorganisms, called "activated sludge," that feed on the contaminants in the wastewater. This process forms excess sludge that is removed from the treatment process in order to maintain the correct ratio of activated sludge to biological contaminants during secondary treatment. Tertiary treatment refers to any additional treatment, where water is disinfected or additional chemical contaminants are removed.

Wastewater treatment plants consolidate high loads of fecal and environmental bacteria along with residual concentrations of antibiotics. The mixing of different types of bacteria in wastewater treatment plants provides a conducive environment for horizontal gene transfer of ARGs. There are many studies that demonstrate that wastewater treatment plants and untreated sewage are major environmental reservoirs of antimicrobial-resistant bacteria and ARGs within their systems [48–54]. A meta-analysis indicates that wastewater treatment appears to increase the proportion of resistant bacteria, suggesting that antimicrobial-resistant bacteria are better able to survive the wastewater treatment process than non-resistant strains [55].

Wastewater treatment plant effluent also functions as a major point source that releases AMR into the environment, where they disperse [56–58]. Wastewater effluent can favor the persistence and spread of antimicrobial resistance in aquatic microbial communities [59] and has been shown to contribute to antibiotic resistance in downstream ecosystems [60, 61].

Researchers have begun to examine the fate and removal of various antibiotic bacteria and their ARGs during various processes of wastewater treatment [62–64] in order to characterize the risks and improve treatment processes to reduce the spread of antibiotic resistance. There is a lack of standardized methods to detect antimicrobials and to assess selective pressures [48], but it is known that the removal efficiency of antibiotics, antibiotic-resistant bacteria and associated ARGs during wastewater treatment processes varies, and is dependent on a combination of properties of the source water, the physical and chemical properties of the antibiotic, and the operating conditions of the treatment systems, including temperature, season, and rainfall [65–68]. Often researchers do not specify the exact type of wastewater treatment processes that were investigated in their studies, making comparison of treatment effects difficult. In order to allow comparison and evaluation of the removal efficiency of different layouts and concepts in wastewater treatment, additional information like the sludge retention time (SRT) and sludge load (food to microorganism, or the F/M ratio) are necessary [69].

The removal of environmentally relevant concentrations of 11 antibiotics from six different classes was investigated using batch processing of activated sludge in freshwater and saline sewage treatment plants. Biodegradation and adsorption were the major removal routes for the antibiotics. Cefalexin and two sulfonamides were mainly removed by biodegradation in both freshwater and saline sewage systems, while ampicillin, norfloxacin, ciprofloxacin, ofloxacin, tetracycline, roxithromycin, and trimethoprim were mainly removed by adsorption. Erythromycin was persistent in both saline and freshwater systems in this experiment and could not be removed [70].

Another study examined ARG composition and removal efficiencies, mobilization of ARGs, and co-occurrences of ARGs and antibiotic genes through two different full-scale wastewater treatment plants – an attached growth-trickling filter biofilm system and a suspended growth-activated sludge treatment system. The analysis included 19 different antibiotic resistance categories and over 175 different ARGs. The results showed that ARGs were slightly more persistent and abundant on mobile genetic elements throughout the biofilm system and were also more abundant in trickling filter clarifier sludge from the biofilm system compared to sludge from the suspended growth system. However, the researchers found large increases in mobilization of ARGs throughout both types of wastewater treatment systems [71].

In a large study in China, wastewater treatment reduced the overall concentration of ARGs by between one and two orders of magnitude. However, 56 out of 211 resistance genes were enriched in activated sludge and 71 were enriched in the effluents. The researchers found that the process of wastewater treatment changed the composition of both the resistome and its host bacterial community, with a strong correlation between the two communities of bacteria, indicating that host/gene interactions are probably important for persistence throughout the treatment process [72].

In a study comparing the ARG fate from raw water and sludge from five activated sludge wastewater treatment plants in California, researchers found that a gene responsible for resistance to tetracycline was reduced in the sludge from all five wastewater treatment plants, but that two other ARGs increased in four of the plants while being reduced in the fifth. Ongoing work is evaluating potential correlations between operational parameters and quantified ARGs at each wastewater treatment plant [73].

Chlorination has also been studied as an option for inactivating antibiotic-resistant bacteria during wastewater treatment. One study found that while most AMR bacteria were inactivated completely at a dose of 15 mg Cl_2 min/l, about 40% of erythromycin-resistance genes and 80% of tetracycline-resistance genes could not be removed by chlorination; thus chlorination is not considered to be effective in controlling antimicrobial resistance [65].

A study of tertiary treatment by ultraviolet light/ozonation of wastewater reduced the majority of clinically relevant bacteria, but antibiotic resistance genes were found to be present in the effluent and some exhibited an increased abundance within the surviving population [74], similar to results after activated sludge treatment.

Constructed wetlands that use ecosystem processes in natural wetlands are considered an alternative wastewater treatment option for removal of antibiotic-resistant bacteria and ARGs. A study compared removal rates in constructed wetlands at three hydraulic loading rates (of 10, 20, and 30 cm/day) and using four substrates (oyster shell, zeolite, medical stone, and ceramic). Seven target antibiotic compounds (erythromycin–H_2O, lincomycin, monensin, ofloxacin, sulfamerazine, sulfamethazine, and novobiocin) were detected, and all 18 resistance genes tested for were detected in the domestic wastewaters. The aqueous removal rates of the total antibiotics ranged from 18 to 98%, while those for the total ARGs varied between 50 and 86%. After considering their aqueous removal rates in combination with their mass removals, the constructed wetland using zeolite as the substrate and a hydraulic loading rate of 20 cm/day was selected as the best choice. The research of combined chemical and biological analyses indicates that both microbial degradation and physical sorption processes were responsible for the fate of antibiotics and ARGs in the wetlands [75].

A new combination of microbial fuel cell coupled with a constructed wetland (CW-MFC) has been designed to simultaneously treat wastewater and produce energy. A study was conducted to assess the removal efficiency of tetracycline and sulfamethoxazole and the development of associated resistance genes in this new treatment process. The results indicated that the CW-MFCs could significantly reduce the concentration of the two antibiotics in wastewater, with accumulation of antibiotics and the relative abundances of the resistance genes in different layers of the CW-MFCs. The researchers also found that the relative abundances of the resistance genes were significantly affected by the antibiotics in effluent of the CW-MFCs [76].

Further experiments to study the effects of variables such as wastewater disinfection methods, effluent dilution ratios, bacterial strains, and antibiotic types are currently underway. The application of alternative techniques including membrane processes, activated carbon adsorption, and advanced oxidation processes, which may lead to higher removals, may be necessary before final disposal of effluents or their reuse for irrigation or groundwater recharge [56].

14.6 THE WAY FORWARD

The Global Health Security Agenda (GHSA), a partnership of nations, international organizations, and non-governmental stakeholders, was launched in 2014 to help build capacity to create a world safe from infectious disease threats and to elevate global health security as a national and global priority. One of the risks identified for this Agenda is the rise of drug resistance [77].

Recognizing the pervasiveness of AMR in humans, animals, food, and the environment, in 2016 the United Nations General Assembly adopted a declaration and a commitment to action [78] based on the World Health Organization's (WHO) Global Action Plan on Antimicrobial Resistance [79]. The Global Action Plan was developed in collaboration with the Food and Agriculture Organization of the United Nations and the World Organization for Animal Health, to address the multisectoral nature of AMR. The overall goal of the Global Action Plan is "to ensure, for as long as possible, continuity of the ability to treat and prevent infectious diseases with effective and safe medicines that are quality-assured, used in a responsible way, and accessible to all who need them." The development of the Global Action Plan included a review of many existing plans, initiatives, and activities [80], but the Global Action Plan notes that progress on these proposals and initiatives has been slow due to inadequate monitoring and reporting, and inadequate recognition of the need for action across all stakeholders.

A focus only on developing new antibiotics assumes that the microorganisms are the enemy instead of a public global good, and downplays the significant roles of hygiene, sanitation, vaccines, and cultural changes in addressing the problem of AMR [81]. Therefore, the Global Action Plan has five overarching strategic objectives. The objectives, along with the means for achieving them, are shown in Table 14.1.

TABLE 14.1 Strategic objectives of WHO's Global Action Plan on antimicrobial resistance.

Objectives	Means
1. Improve awareness and understanding of AMR	Communication, education and training
2. Strengthen the knowledge and evidence base on AMR	Surveillance and research
3. Reduce the incidence of infections	Sanitation, hygiene, and infection prevention measures
4. Limit the emergence and spread of AMR	Optimal use of antimicrobial medicines in human and animal health
5. Develop new tools to fight AMR	New medicines, diagnostic tools, vaccines, and other interventions

The comprehensive "One Health" approach adopted by the American Medical Association in 2007, and subsequently endorsed by the World Health Organization and other national and global medical and veterinary groups, integrates multiple disciplines at different scales to attain optimal health for people, animals, and the environment. It is an approach that is well suited for addressing the complex and systemic issue of AMR [26, 82]. At the 2016 G7 meeting in Japan, the heads of the seven most industrialized countries agreed to strengthen the One Health approach and cooperation to address the cross-cutting issue of AMR in human and animal health [83].

A 2015 WHO survey on AMR found that only 21% of countries that responded have a financed national plan with multisectoral input for addressing antimicrobial resistance [21]. Therefore the Global Action Plan urges all member states and other stakeholders including FAO, the World Bank, and industry associations to align with the Global Action Plan and implement action plans in their respective fields of responsibility. The action plans should reflect the principles of: a whole-of-society engagement, including a One Health approach; prevention first, including good sanitation, hygiene, and other infection prevention measures to avoid the need for treatment; equitable access to, and appropriate use of, existing and new antimicrobial medicines; an assessment of resource needs that includes long-term investments in surveillance, operational research, laboratories, regulatory capacity, and training for both human and animal health sectors; and incremental targets for implementation recognizing that member states are at different stages in developing and implementing national plans.

The issue of AMR in wastewater and the environment is not specifically addressed in the WHO Action Plan. Current WHO guidelines for drinking water, recreational water, and use of wastewater and excreta in agriculture and aquaculture contain no information on antibiotics and other antimicrobial agents, their metabolites, antimicrobial-resistant bacteria or their AMR genes. There is a lack of specific guidance on how to assess, manage, or minimize the potential human health risks associated with antimicrobial agents and antimicrobial-resistant bacteria and ARGs through various exposure and transmission routes [84]. Although many individual surveys of antibiotics in wastewater treatment plants have been conducted, systematic monitoring is still limited. Monitoring resistance at wastewater treatment plants could contribute to a better understanding of the problem and inform improved control strategies, including tertiary treatment, that protect both human and environmental health. Water professionals will need to develop a new risk-assessment framework that considers the fate and transport of antimicrobials and their associated resistance genes throughout the drinking water and wastewater treatment process, and provide scientific input into the new policies, management systems, and guidance required to address this global challenge [85].

REFERENCES

1. Corcoran E, Nellemann C, Baker E, Bos R, Osborn D, Savelli H (eds). 2010. Sick Water? The Central Role of Wastewater Management in Sustainable Development – A Rapid Response Assessment. United Nations Environment Program (UNEP) and UN-HABITAT, Nairobi, Kenya [Online]. Available from: http://www.fao.org/sustainable-food-value-chains/library/details/en/c/266054/. Accessed 30 December 2017.
2. Demain, A.L. (2009). Antibiotics: natural products essential to human health. *Medicinal Research Reviews* 29 (6): 821–842. doi: 10.1002/med.20154.
3. Van Boeckel, T., Gandra, S., Ashok, A. et al. (2014). Global antibiotic consumption 2000 to 2010: an analysis of national pharmaceutical sales data. *The Lancet Infectious Diseases* 14 ((8): 742–750. doi: 10.1016/S1473-3099(14)70780-7.
4. State of the World's Antibiotics, 2015. Center for Disease Dynamics, Economics and Policy (CDDEP): Washington, DC [Online]. https://cddep.org/publications/state_worlds_antibiotics_2015/. Accessed December 29, 2017.
5. Goossens, H., Ferech, M., Vander Stichele, R. et al. (2005). Outpatient antibiotic use in Europe and association with resistance: a cross-national database study. *The Lancet Infectious Diseases* 365 (9459): 579–587. doi: 10.1016/S0140-6736(05)17907-0.
6. United States Centers for Disease Control and Prevention (CDC) (2011). Office-related antibiotic prescribing for persons aged ≤14 years -United States, 1993–1994 to 2007–2008. *MMWR Morbidity and Mortality Weekly Report* 60 (34): 1153–1156.
7. Saleh, N., Awada, S., Awwad, R. et al. (2015). Evaluation of antibiotic prescription in the Lebanese community: a pilot study. *Infection Ecology and Epidemiolgy* 5: 27094. doi: 10.3402/iee.v5.27094.
8. Shapiro, D.J., Hicks, L.A., Pavia, A.T., and Hersh, A.L. (2014). Antibiotic prescribing for adults in ambulatory care in the USA, 2007–09. *The Journal of Antimicrobial Chemotherapy* 69 (1): 234–240. doi: 10.1093/jac/dkt301.
9. Fatokun, O. (2014). Exploring antibiotic use and practices in a Malaysian community. *International Journal of Clinical Pharmacy* 36 (3): 564–569. doi: 10.1007/s11096-014-9937-6.
10. United States Food and Drug Administration. Disposal of Unused Medicines: What You Should Know. Last update 20 October 2017 [Online]. Available from: https://www.fda.gov/Drugs/ResourcesForYou/Consumers/BuyingUsingMedicineSafely/EnsuringSafeUseofMedicine/SafeDisposalofMedicines/ucm186187.htm. Accessed November 12, 2017.
11. #medsdisposal. 2017. Disposal of Medicines in Europe [Online]. Available from: http://medsdisposal.eu/. Accessed November 12, 2017.
12. World Health Organization and partners. 1999. Guidelines for Safe Disposal of Unwanted Pharmaceuticals In and After Emergencies. World Health Organization [Online]. Available from: http://www.who.int/immunization_safety/publications/waste_management/en/Guidelines_safe_disposal_unwanted_pharmaceuticals.pdf. Accessed December 29, 2017.
13. Delia G. 2015. Review of Evidence on Antimicrobial Resistance and Animal Agriculture in Developing Countries, Evidence on Demand, International Livestock Research Institute. doi: http://dx.doi.org/10.12774/eod_cr.june2015.graced., cited in: Antimicrobials in Agriculture and the Environment – Reducing Unnecessary Use and Waste. *The Review on Antimicrobial Resistance*. 2015. Jim O'Neill, Chair [Online]. Available from: https://ec.europa.eu/health/amr/sites/amr/files/amr_studies_2015_am-in-agri-and-env.pdf. Accessed November 12, 2017.
14. Van Boeckel, T., Brower, C., Gilbert, M. et al. (2015). Global trends in antimicrobial use in food animals. *Proceedings of the National Academy of Sciences of the United States of America* 112 (18): 5649–5654. doi: 10.1073/pnas.1503141112.
15. Antimicrobials in Agriculture and the Environment-Reducing Unnecessary Use and Waste. The Review on Antimicrobial Resistance. Jim O'Neill, Chair. 2015 [Online]. Available from: https://ec.europa.eu/health/amr/sites/amr/files/amr_studies_2015_am-in-agri-and-env.pdf. Accessed November 12, 2017.
16. Combating Antibiotic Resistance: A Policy Roadmap to Reduce Use of Medically Important Antibiotics in Livestock. Expert Commission on Addressing the Contribution of Livestock to the Antibiotic Resistance Crisis. 2017 [Online]. Available from: http://battlesuperbugs.com/sites/battlesuperbugs.com/files/Final%20Report%208.25.17.pdf#page=9. Accessed December 29, 2017.
17. Sales of Veterinary Antimicrobial Agents in 29 European Countries in 2014. European Medicines Agency (EMA/61769/2016) [Online]. Available from: http://www.ema.europa.eu/docs/en_GB/document_library/Report/2016/10/WC500214217.pdf. Accessed December 29, 2017.
18. Martin, M., Thottathil, S., and Newman, T. (2015). Antibiotics overuse in animal agriculture: a call to action for health care providers. *American Journal of Public Health* 105 (12): 2409–2410. doi: 10.2105/AJPH.2015.302870.
19. Guidance for Industry New Animal Drugs and New Animal Drug Combination Products Administered in or on Medicated Feed or Drinking Water of Food Producing Animals: Recommendations for Drug Sponsors for Voluntarily Aligning Product Use Conditions with GFI #209. United States Food and Drug Administration (FDA). 2017 [Online]. Available from: https://www.fda.gov/downloads/AnimalVeterinary/GuidanceComplianceEnforcement/GuidanceforIndustry/UCM299624.pdf. Accessed December 29, 2017.

20. Dall, C. 2017. Farm Antibiotics: Does New FDA Policy Go Far Enough? Center for Infectious Disease Research and Policy (CIDRAP) [Online]. Available from: http://www.cidrap.umn.edu/news-perspective/2017/01/farm-antibiotics-does-new-fda-policy-go-far-enough. Accessed December 29, 2017.

21. World Health Organization. 2015. Worldwide Country Situation Analysis: Response to Antimicrobial Resistance [Online]. Available from: http://apps.who.int/iris/bitstream/10665/163468/1/9789241564946_eng.pdf?ua=1&ua=1. Accessed December 29, 2017.

22. Laxminarayan R, Van Boeckel T., Teillant A. 2015. The Economic Costs of Withdrawing Antimicrobial Growth Promoters from the Livestock Sector. OECD Food, Agriculture and Fisheries Papers, No. 78. Paris: OECD Publishing. doi: 10.1787/5js64kst5wvl-en.

23. Salyers, A. and Whitt, D. (2002). *Bacterial Pathogenesis: A Molecular Approach*, 2e. Washington, DC: ASM Press.

24. Pharmaceutical Microbiology. Microbiology, Pharmaceuticals, Healthcare and Contamination Control News and Discussion Site. Established 2010. Edited by Dr Tim Sandle [Online]. Available from: http://www.pharmamicroresources.com/2017/06/antibiotics-are-potential-link-to-bowel.html. Accessed December 29, 2017.

25. Baym, M., Lieberman, T., Kelsic, E. et al. (2016). Spatiotemporal microbial evolution on antibiotic landscapes. *Science* 353 (6304): 1147–1151. doi: 10.1126/science.aag0822 Videos in Supplementary Materials at: http://science.sciencemag.org/content/suppl/2016/09/07/353.6304.1147.DC1?_ga=2.88673363.970618696.1507585734-1793372695.1507585734.

26. Harbarth, S., Balkhy, H.H., Goossens, H. et al. (2015). Antimicrobial resistance: one world, one fight! For the world healthcare-associated infections resistance forum participants. *Antimicrobial Resistance and Infection Control* 4: 49. doi: 10.1186/s13756-015-0091-2.

27. Biomerieux. 2016. Bacteria and the Environment website [Online]. Available from: https://www.antimicrobial-resistance.biomerieux.com/popup/bacteria-and-the-environment/. December 29, 2017.

28. Tackling Drug-Resistant Infections Globally: Final Report and Recommendations. 2016. The Review on Antimicrobial Resistance. Chaired by Jim O'Neill [Online]. Available from: https://amr-review.org/sites/default/files/160525_Final%20paper_with%20cover.pdf. Accessed December 29, 2017.

29. Berkner, S., Konradi, S., and Schönfeld, J. (2014). Antibiotic resistance and the environment – there and back again. *EMBO Reports* 15 (7): 740–744. doi: 10.15252/embr.201438978.

30. Watkinson, A., Murby, E., and Costanzo, S. (2007). Removal of antibiotics in conventional and advanced wastewater treatment: implications for environmental discharge and wastewater recycling. *Water Research* 41: 4164–4176.

31. Boxall, A. (2004). The environmental side effects of medication. *EMBO Reports* 5 (12): 1110–1116. doi: 10.1038/sj.embor.7400307.

32. Meena, V., Dotaniya, M., Saha, J., and Patra, A. (2015). Antibiotics and antibiotic resistant bacteria in wastewater: impact on environment, soil microbial activity and human health. *African Journal of Microbiology Research* 9 (14): 965–978. doi: 10.5897/AJMR2015.7195.

33. Briefing Note, Antimicrobial Resistance: An Emerging Water, Sanitation and Hygiene Issue. 2014. World Health Organization WHO/FWC/WSH/14.07 [Online]. Available from: www.who.int/water_sanitation_health/emerging/AMR_briefing_note.pdf. December 29, 2017.

34. Zhu, Y.G., Johnson, T., Su, J.Q. et al. (2013). Diverse and abundant antibiotic resistance genes in Chinese swine farms. *Proceedings of the National Academy of Sciences of the United States of America* 110 (9): 3435–3440. doi: 10.1073/pnas.1222743110.

35. Marti, R., Scott, A., Tien, Y.C. et al. (2013). Impact of manure fertilization on the abundance of antibiotic-resistant bacteria and frequency of detection of antibiotic resistance genes in soil and on vegetables at harvest. *Applied and Environmental Microbiology* 79 (18): 5701–5709. doi: 10.1128/AEM.01682-13.

36. Huijbers, P., Blaak, H., de Jong, M. et al. (2015). Role of the environment in the transmission of antimicrobial resistance to humans: a review. *Environmental Science & Technology* 49 (20): 11993–12004. doi: 10.1021/acs.est.5b02566.

37. Royal Veterinary College, University of London. 2016. A systematic review to assess the significance of the food chain in the context of antimicrobial resistance (AMR) with particular reference to pork and poultry meat, dairy products, seafood and fresh produce on retail sale in the UK [Online]. Available from: https://www.food.gov.uk/sites/default/files/amr-systematic-review-final-report-2016.pdf Dec. 29, 2017.

38. Watkinson, A., Micalizzi, G., Graham, G. et al. (2007). Antibiotic-resistant *Escherichia coli* in wastewaters, surface waters, and oysters from an urban riverine system. *Applied and Environmental Microbiology* 73 (17): 5667–5670. doi: 10.1128/AEM.00763-07.

39. Li, B., Yang, Y., Ma, L.P. et al. (2015). Metagenomic and network analysis reveal wide distribution and co-occurrence of environmental antibiotic resistance genes. *ISME Journal* 9 (11): 2490–2502. Presented in Zhang, T., Antibiotics and Resistance Genes in Wastewater Treatment Plants. AMR Control website. 2016 [Online]. Available from: http://resistancecontrol.info/2016/amr-in-food-water-and-the-environment/antibiotics-and-resistance-genes-in wastewater-treatment-plants/. Accessed December 29, 2017.

40. Zhang T. 2016. Antibiotics and Resistance Genes in Wastewater Treatment Plants. AMR Control website [Online]. Available from: http://resistancecontrol.info/2016/amr-in-food-water-and-the-environment/antibiotics-and-resistance-genes-in-wastewater-treatment-plants/. Accessed December 29, 2017.

41. Ye Z, Weinberg H., Meyer M. Occurrence of antibiotics in drinking water. 2004 [Online]. Available from: https://www.researchgate.net/publication/228581426_Occurrence_of_Antibiotics_in_Drinking_Water. Accessed December 29, 2017.

42. Martinez, J. (2009). The role of natural environments in the evolution of resistance traits in pathogenic bacteria. *Proceedings of the Biological Sciences/The Royal Society* 276: 2521–2530. doi: 10.1098/rspb.2009.0320.

43. Stoll, C., Sidhu, J.P., Tiehm, A., and Toze, S. (2012). Prevalence of clinically relevant antibiotic resistance genes in surface

water samples collected from Germany and Australia. *Environmental Science and Technology* 46 (17): 9716–9726. doi: 10.1021/es302020s.

44. Riquelme M, Metch J. Garner E, Vikesland P, Pruden, A. 2017. Global Survey of Antibiotic Resistance Genes in Wastewater Treatment Plants. Advancing Health Communities Through Environmental Engineering and Science – Research and Education Conference, Ann Arbor, MI [Online]. Available from: https://aeesp2017.com/program_list.php?sid=S4#a345. Accessed December 29, 2017.
45. Li, B. and Zhang, T. (2013). Removal mechanisms and kinetics of trace tetracycline by two types of activated sludge treating freshwater sewage and saline sewage. *Environmental Science and Pollution Research* 20 (5): 3024–3033. doi: 10.1007/s11356-012-1213-5.
46. Berendonk, T.U., Manaia, C.M., Merlin, C. et al. (2015). Tackling antibiotic resistance: the environmental framework. *Nature Reviews Microbiology* 13 (5): 310–317. doi: 10.1038/nrmicro3439.
47. Chen, Q., An, X., Li, H. et al. (2016). Long-term field application of sewage sludge increases the abundance of antibiotic resistance genes in soil. *Environment International* 92–93: 1–10. doi: 10.1016/j.envint.2016.03. 026.
48. Harris, S., Cormican, M., and Cummins, E. (2010). Antimicrobial residues and antimicrobial-resistant bacteria: impact on the microbial environment and risk to human health—a review. *Human and Ecological Risk Assessment* 18 (4): 767–809. doi: 10.1080/10807039.2012.688702.
49. Ikuma, K. 2017. Fate and Transport of Wastewater-Derived Antibiotic Resistance Genes. Advancing Health Communities Through Environmental Engineering and Science – Research and Education Conference, Ann Arbor, MI. Available from: https://aeesp2017.com/program_list.php?sid=S4#a327.
50. LaPara, T., Burch, T., McNamara, P. et al. (2011). Tertiary-treated municipal wastewater is a significant point source of antibiotic resistance genes into Duluth-Superior Harbor. *Environmental Science and Technology* 45: 9543–9549. doi: 10.1021/es202775r.
51. Zhou, L.J., Ying, G.G., Liu, S. et al. (2013). Occurrence and fate of eleven classes of antibiotics in two typical wastewater treatment plants in South China. *Science of the Total Environment* 452–453 (2013): 365–376. doi: 10.1016/j.scitotenv.2013.03.010.
52. Priyanka, U. (2014). Bacterial dynamics in antibiotic wastewater treatment – a review. *International Journal of Occupational Health and Safety, Fire and Environment – Allied Science* 2 (1) [Online]. Available at: https://www.researchgate.net/publication/278023524_Bacterial_Dynamics_in_Antibiotic_Wastewater_Treatment_-_A_Review): Accessed December 30, 2017.
53. Zhang, T. and Bing, L. (2011). Occurrence, transformation, and fate of antibiotics in municipal wastewater treatment plants. *Critical Reviews in Environmental Science and Technology* 41 ((11): 951–998. doi: 10.1080/10643380903392692.
54. Bouki, C., Venieri, D., and Diamadopoulos, E. (2013). Detection and fate of antibiotic resistant bacteria in wastewater treatment plants: a review. *Ecotoxicology and Environmental Safety* 91: 1–9. doi: 10.1016/j.ecoenv.2013.01.016.
55. Harris, S., Cormican, M., and Cummins, E. (2012). The effect of conventional wastewater treatment on the levels of antimicrobial-resistant bacteria in effluent: a meta-analysis of current studies. *Environmental Geochemistry and Health* 34 (6): 749–762. doi: 10.1007/s10653-012-9493-8.
56. Michael, I., Rizzo, L., McArdell, C.S. et al. (2013). Urban wastewater treatment plants as hotspots for the release of antibiotics in the environment: a review. *Water Research* 47: 957–995. doi: 10.1016/j.watres.2012.11.027.
57. Schwartz, T., Kohnen, W., Jansen, B., and Obst, U. (2003). Detection of antibiotic-resistant bacteria and their resistance genes in wastewater, surface water, and drinking water biofilms. *FEMS Microbiology Ecology* 43: 325–335. doi: 10.1111/j.1574-6941.2003.tb01073.x.
58. Young, S., Juhl, A., and O'Mullan, G.D. (2013). Antibiotic-resistant bacteria in the Hudson River estuary linked to wet weather sewage contamination. *Journal of Water and Health* 11: 297–310. doi: 10.2166/wh.2013.131.
59. Proia, L., von Shiller, D., Sánchez-Melsió, A. et al. (2016). Occurrence and persistence of antibiotic resistance genes in river biofilms after wastewater inputs in small rivers. *Environmental Pollution* 210: 121–128. doi: 10.1016/j.envpol.2015.11.035.
60. Rodriguez-Mozaz, S., Chamorro, S., Marti, E. et al. (2015). Occurrence of antibiotics and antibiotic resistance genes in hospital and urban wastewaters and their impact on the receiving river. *Water Research* 69: 234–242. doi: 10.1016/j.watres.2014.11.021.
61. Chu B, Petrovich M, Chaudhary A, Murphy B, Poretsky R., Wells G. 2017. Metagenomic analysis unveils how wastewater treatment effluent influences receiving water bodies in terms of antibiotic resistance genes, mobile genetic elements, and microbial community structure. Advancing Health Communities Through Environmental Engineering and Science – Research and Education Conference, Ann Arbor, MI [Online]. Available from: https://aeesp2017.com/program_list.php?sid=S4#a380. Accessed December 30, 2017.
62. Morris D, Harris S, Morris C, Commins E., M Cormican. 2016. Hospital Effluent: Impact on the Microbial Environment and Risk to Human Health. Irish Environmental Protection Agency Research Report #162 [Online]. Available from: http://www.epa.ie/pubs/reports/research/health/researchreport162.html. Accessed December 30, 2017.
63. Tan, D. and Shuaib, D. (2015). Research highlights- antibiotic resistance genes: from wastewater into the environment. *Environmental Science: Water Research and Technology* 1: 264. doi: 10.1039/c5ew90011e.
64. Miller, J., Novak, J., Knocke, W., and Pruden, A. (2016). Survival of antibiotic resistant bacteria and horizontal gene transfer control antibiotic resistance gene content in anaerobic digesters. *Frontiers in Microbiology* 7: 263. doi: 10.3389/fmicb.2016.00263.
65. Yuan, Q.B., Guo, M.T., and Yang, J. (2015). Fate of antibiotic resistant bacteria and genes during wastewater chlorination:

implication for antibiotic resistance control. *PLoS One* 10 (3): e119403. doi: 10.1371/journal.pone.0119403.

66. Clara, M., Strenn, B., Ausserleitner, M., and Kreuziner, N. (2004). Comparison of the behavior of selected micropollutants in a membrane bioreactor and a conventional wastewater treatment plant. *Water Science and Technology* 50: 29–36.
67. Leclercq, R., Oberle, K., Galopin, S. et al. (2013). Changes in enterococcal populations and related antibiotic resistance along a medical center–wastewater treatment plant–river continuum. *Applied and Environmental Microbiology* 79 (7): 2428–2434. doi: 10.1128/AEM.03586-12.
68. Ju F, Beck K, Yin X, McArdell C, Singer H, Johnson D, Zhang T., Buergmann H. 2017. Wastewater Treatment Plant Resistomes Are Structured by Microbial Composition and Actively Transcribed at All Process Stages. Abstract O24. 4th International Symposium on the Environmental Dimension of Antibiotic Resistance. Lansing, MI [Online]. Available from: http://www.antibiotic-resistance.de/. Accessed December 29, 2017.
69. Kreuzinger, N., Clara, M., Strenn, B., and Kroiss, H. (2004). Relevance of the sludge retention time (SRT) as design criteria for wastewater treatment plants for the removal of endocrine disruptors and pharmaceuticals from wastewater. *Water Science and Technology* 50 (5): 149–156.
70. Li, B. and Zhang, T. (2010). Biodegradation and adsorption of antibiotics in the activated sludge process. *Environmental Science and Technology* 44 (9): 3468–3473. doi: 10.1021/es903490h.
71. Petrovich M, Chu B, Wright D, Griffin J, Elfeki M, Poretsky R, Murphy B and Wells G. 2017. Metagenomic Analysis of Composition, Retention, and Mobilization of Antibiotic Resistance and Production Genes in Biofilm and Suspended Growth Wastewater Treatment Bioreactors. Advancing Health Communities Through Environmental Engineering and Science – Research and Education Conference. Ann Arbor, MI [Online]. Available from: https://aeesp2017.com/program_list.php?sid=S4#a380. Accessed December 29, 2017.
72. An X, Su J, Li B, Ouyang, W, Zhao Y, Chen Q, Cui L, Chen H, Gillings M, Zhang T and Shu Y. 2017. Tracking Antibiotic Resistance Genes During Wastewater Treatment. Abstract P106. 4th International Symposium on the Environmental Dimension of Antibiotic Resistance. Lansing, MI [Online]. Available from: http://www.antibiotic-resistance.de/. Accessed December 29, 2017.
73. Baygi A, Stadler L and Smith A. 2017. Comparative Evaluation of Antibiotic Resistance in Full-Scale Activated Sludge Systems and Bench-Scale Anaerobic Membrane Bioreactor. Advancing Health Communities Through Environmental Engineering and Science – Research and Education Conference. Ann Arbor, MI [Online]. Available from: https://aeesp2017.com/program_list.php?sid=S4. Accessed December 29, 2017.
74. Jäger T, Hembach N, Alexander J, Dötsch A., Schwartz T. 2017. Impact of UV/Ozone Treatment on Antibiotic Resistant Bacteria as Part of the Wastewater Effluent Microbiome. Abstract P136. 4th International Symposium on the Environmental Dimension of Antibiotic Resistance. Lansing, MI [Online]. Available from: http://www.antibiotic-resistance.de/. December 29, 2017.
75. Chen, J., Wei, X.D., Liu, Y.S. et al. (2016). Removal of antibiotics and antibiotic resistance genes from domestic sewage by constructed wetlands: optimization of wetland substrates and hydraulic loading. *Science of the Total Environment* 565: 240–248. doi: 10.1016/j.scitotenv.2016.04.176.
76. Zhang, S., Song, H.L., Yang, X.L. et al. (2016). Fate of tetracycline and sulfamethoxazole and their corresponding resistance genes in microbial fuel cell coupled constructed wetlands. *RSC Advances* 6 (98): 95999–96005. doi: 10.1039/C6RA20509G.
77. Global Health Security Agenda. 2017 [Online]. Available from: https://www.ghsagenda.org/. Accessed: December 29, 2017.
78. UN General Assembly. Resolution adopted by the General Assembly on 5 October 2016: Political Declaration of the High-Level Meeting of the General Assembly on Antimicrobial Resistance [Online]. Available from: http://www.who.int/antimicrobial-resistance/interagency-coordination-group/UNGA-AMR-RES-71-3-N1631065.pdf?ua=1. Accessed December 29, 2017.
79. World Health Organization. 2015. Global Action Plan on Antimicrobial Resistance [Online]. Available from: http://www.wpro.who.int/entity/drug_resistance/resources/global_action_plan_eng.pdf. Accessed December 29, 2017.
80. Draft Global Action Plan on Antimicrobial Resistance: Related Initiatives and Activities. World Health Organization [Online]. Available from: http://www.who.int/drugresistance/global_action_plan/ongoing_activities/en/. Accessed December 29, 2017.
81. P. Jorgensen. 2016. Antibiotic Resistance is the Next Great Global Challenge – We Must Act Now. World Economic Forum [Online]. Available from: www.weforum.org/agenda/2016/09/antimicrobial-resistance-is-the-next-global-commons-issue. Accessed December 29, 2017.
82. Robinson, T., Bu, D., Carrique-Mas, J. et al. (2016). Antibiotic resistance is the quintessential one health issue. *Transactions of the Royal Society of Tropical Medicine and Hygiene* 110 (7): 377–380. doi: 10.1093/trstmh/trw048.
83. G7 Ise-Shima Vision for Global Health. 2016 [Online]. Available from: http://www.mofa.go.jp/files/000160273.pdf. Accessed December 29, 2017.
84. World Health Organization. 2014. Briefing Note. Antimicrobial Resistance: An Emerging Water, Sanitation and Hygiene Issue. WHO/FWC/WSH/14.07. 2014 [Online]. Available from: http://www.who.int/water_sanitation_health/emerging/AMR_briefing_note.pdf. Accessed December 29, 2017.
85. Jimenez M. The Water Sector's Answer to Antibiotic Resistance. International Water Association [Online]. Available from: http://www.iwa-network.org/the-water-sectors-answer-to-antibiotic-resistance/. Accessed December 29, 2017.

15

GLOBAL SUBSTITUTION OF MERCURY-BASED MEDICAL DEVICES IN THE HEALTH SECTOR

Anitha Nimmagadda[1], Ivorie Stanley[1], Joshua Karliner[2], and Peter Orris[1]

[1] *Occupational and Environmental Medicine, University of Illinois Hospital and Health Sciences System, Chicago, IL, USA*

[2] *Health Care Without Harm, San Francisco, USA*

15.1 INTRODUCTION

Mercury, a heavy metal with neurotoxic effects, is liquid at room temperature and has long been used in medical devices, most commonly thermometers and sphygmomanometers (instruments for measuring blood pressure). This led to a paradox in which healthcare institutions and professionals, whose mission is to heal and promote health, have contributed to the significant global environmental health problem of mercury contamination through the use of their healthcare instruments.

In response, in the first decade of this century, the health sector in the United States largely phased out mercury thermometers. Most US hospitals no longer purchase mercury sphygmomanometers, and as of 2008, more than one-quarter of the US population lived in states that have banned or severely restricted their sale and/or use. Similarly, the European Union banned mercury thermometers in 2007 and extended the ban to other mercury-containing healthcare devices in 2014. The ministries of health in Argentina, Chile, Costa Rica, Nicaragua, the Philippines, Nepal, Sri Lanka, and Mongolia have all committed to mercury-free healthcare. Healthcare systems in Brazil, India, China, South Africa, and Mexico are also exploring mercury-free healthcare devices. This transition in many countries has been accelerated significantly by the recently negotiated Minamata Convention on Mercury, yet the health sector in much of the rest of the world is still using mercury-based medical devices.

In addition to eliminating this important source of mercury contamination, these efforts to substitute mercury-based medical devices educate and activate healthcare leaders around the world. Some have become advocates not only for cleaning up healthcare's mercury pollution but also for more comprehensive control of this global environmental contaminant, at the local, national, and global levels.

15.2 THE PROBLEM: MERCURY IN THE ENVIRONMENT

Mercury is a naturally occurring element found in the Earth's crust. At ambient temperature and pressure, it exists as a silvery liquid. Mercury has a higher vapor pressure than most heavy metals and thus is more volatile. When mercury vaporizes into the air, it may stay in the atmosphere for up to a year and can be transported and deposited globally, and then binds other elements, forming organic and inorganic compounds.

The total amount of mercury circulating in the Earth's atmosphere has increased significantly since the Industrial Revolution. Horowitz et al. recently reported that besides the 215 gigagrams (Gg) of mercury previously thought to be added to the atmosphere due to anthropogenic emissions since the advent of the industrial era, another 540 Gg of mercury has entered the environment since 1850 from other anthropogenic sources previously not considered [1].

Water and Sanitation-Related Diseases and the Changing Environment: Challenges, Interventions, and Preventive Measures, Second Edition. Edited by Janine M. H. Selendy.

Companion website: https://www.wiley.com/legacy/wileychi/selendy/

Current major sources of anthropogenic mercury emissions include artisanal small-scale gold mining and fossil fuel combustion [2]. Human exposure to mercury can result from a variety of sources, including, but not limited to, consumption of fish rich in methyl mercury and due to spills or leaks of the metallic element itself.

Mercury causes a variety of significant adverse impacts on human health and the global environment. If high levels of metallic mercury vapor are inhaled, it can quickly cause severe lung damage including pneumonitis and pulmonary edema. Although skin absorption is 50-fold less than lung, toxic levels have been reported due to handling of the liquid metal. This exposure is especially significant if the epithelial barrier has been broken due to cuts or abrasions. Target organs other than the lungs include the kidneys, the nervous system, and the gastrointestinal (GI) tract [3]. Anecdotal reports from hospitals utilizing mercury thermometers report breakage ranging from several to several hundred a month. These reports are paralleled by those noting leakages of mercury-containing sphygmomanometers as well, with the potential for significantly larger amounts released.

Yet of even more concern is the potential for developmental neurotoxicity produced by low-dose methylmercury exposure through food. Elemental mercury accumulates in lake, river, stream, and ocean sediments, where it is transformed into methylmercury, which then accumulates in fish tissue. This contamination of fish stock is ubiquitously present in oceans and lakes throughout the world, concentrating several hundred thousand times as it moves up the aquatic food chain [4].

Methylmercury is of special concern for fetuses, infants, and children because it impairs neurological development. When a woman eats seafood that contains mercury, it accumulates in her body, requiring months to excrete. If she becomes pregnant within this time, methylmercury easily crosses the placental barrier, where it can affect the fetus's central nervous system and cause adverse neurodevelopmental outcomes. Impacts on cognitive thinking, memory, attention, language, and fine motor and visual spatial skills have been documented in children with exposure in utero to levels of methylmercury found in some women of childbearing age in the United States [5, 6].

In 2005, Trasande et al., using national blood mercury prevalence data from the US Centers for Disease Control, estimated that in this century between 316 588 and 637 233 US children each year have core blood mercury levels >5.8 μg/l, a level associated with loss of IQ. They estimated that lost productivity due to this blood mercury level amounts to $8.7 billion annually (ranging from $2.2 to 43.8 billion) [7].

The United Nations Environment Programme (UNEP) and World Health Organization have identified the adverse effects of mercury pollution as a serious global environmental and health problem. The UNEP Governing Council has targeted reducing methylmercury accumulation in the global environment as a major global priority [8].

15.3 MERCURY IN HEALTHCARE

Waste mercury from thermometers and blood pressure devices add to those natural processes, such as volcanic emissions, spreading mercury on the Earth's surface where it is available for methylation and incorporation in the food chain. Additionally, these medical devices break and leak with regularity, exposing healthcare workers and patients to the acute effects of the inhalation of the metal itself. In view of this, as part of a global initiative to reduce the use and spread of mercury in all aspects of society, healthcare providers and institutions have begun to replace mercury-based medical devices with affordable, accurate, and safer alternatives.

The healthcare sector is far from the greatest source of organomercury compounds in the environment. Rather, coal-fired power plant emissions and mercury cell chloralkali plants, along with artisanal gold mining and battery disposal, are all far more significant polluters. However, the healthcare sector does play an important role as a source of global emissions, and can serve as a leader in efforts to phase out mercury in daily commerce.

Mercury can be found in many healthcare devices and is present in fluorescent bulbs as well as dental amalgam. Mercury is also found in many chemicals and measurement devices used in healthcare laboratories [9]. Medical waste incinerators, as well as municipal waste incinerators, emit mercury into the atmosphere when they burn wastes that contain mercury. According to the US Environmental Protection Agency (EPA), in 1996, prior to the mercury phase-out in US healthcare, medical waste incinerators were the fourth largest source of mercury emissions to the environment. Hospitals were also known to contribute 4–5% of the total wastewater mercury load. In addition, mercury fever thermometers alone contributed about 15 metric tons of mercury to solid waste landfills annually [10].

While no comprehensive figures are available, anecdotal evidence suggests that most mercury spills are not properly cleaned, nor is the waste segregated and managed properly. Rather, it is often incinerated, flushed down the drain, or sent via solid waste to a landfill, especially in poor countries or regions.

A recent study from the faculty of public health in Mahidol University in Bangkok estimated that approximately 800 kg of mercury annually were emitted from thermometers and sphygmomanometers in Thailand alone between 1960 and 2010 [11].

Until Buenos Aires began to switch over to alternatives in 2006, the city government ran 33 hospitals and more than 38 clinics for which it was purchasing 40 000 new thermometers a year, ultimately emitting in excess of 40 kg of mercury into the local hospital environment and into the global ecosystem every year [12].

In Mexico City, the 250-bed "Federico Gomez" Children's Hospital is a medical service, teaching, and research hospital affiliated with the National Autonomous University of Mexico. This national referral children's hospital documented a thermometer breakage rate of 385 per month, or well over 4000 per year. The mercury spilled was routinely collected along with other hospital waste and released into the environment. The total number of estimated broken thermometers in this children's hospital between 2002 and early 2007 is nearly 22 000, the equivalent of 22 kg of mercury [13].

In India, while far fewer thermometers are employed in many hospitals [14], a study of two New Delhi hospitals involved in phasing out the use of mercury conducted by Toxics Link, found "substantial presence of mercury in ambient air of both the hospitals." These levels, which ranged in differing areas from 1.12 μg/m to 3.78 μg/m, were all higher than US EPA and ATSDR Action Levels for indoor air [15]. Yet a survey study published in 2015 of healthcare professionals in India documented continued overall poor knowledge of awareness about mercury toxicity and the proper handling and disposal of mercury-containing items in healthcare [16]

Similarly, in Sri Lanka, a cross-sectional study of nurses at a tertiary care children's hospital performed in 2012 demonstrated that of 347 nurses using mercury thermometers, 67.1% had experienced breakages, and of 405 nurses using mercury sphygmomanometers, 20% had experienced mercury spillages in the three months prior to the study (see Table 15.1). In addition, the study reported that of the surveyed nurses, 36% had poor knowledge and 7.2% had very poor knowledge about mercury and clean-up procedures [17].

Besides the frequent breakage and spillage, mercury release, and contamination from needed regular sphygmomanometer calibration, is a common problem throughout the world. Louis Havinga, a former manager of Health Technology Services for the KwaZulu Natal Province Department of Health in South Africa described his experience that "... technicians were exposed to mercury when they repaired mercury column sphygmomanometers." To avoid this exposure, special precautions and equipment were needed when "working with mercury products like a dedicated fume/vapour extraction unit within the maintenance department. The mercury is extracted from the device and placed in a special marked container. The container must be able to be sealed and should remain inside the fume/vapour extraction unit. Once the container is full, the container must be disposed of in a well-documented and controlled manner by making use of a recognized hazardous waste disposal company. This process is very costly" [18].

TABLE 15.1 Distribution of study population with respect to number of thermometer breakages experienced in the 3 months prior to the study [17].

Number of broken thermometers	Frequency	Percentage
None	114	32.9
<10	150	43.2
10–20	62	17.9
21–30	16	4.6
31–50	0	0
More than 50	5	1.4
Total	347	100

While sphygmomanometers break less frequently than thermometers, the spillage and evaporation is significant and therefore problematic from an environmental health perspective.

Dental amalgam is another significant source of mercury in the healthcare sector. Approximately 340 tons of mercury are used in the dental industry annually and about 20–30% of this amount is thought to become solid waste [19]. Dental amalgam is approximately 50% mercury by composition and amalgam continuously off-gasses mercury. Mutter, writing for the European Commission's Scientific Committee in the *Journal of Occupational Medicine and Toxicology* in 2011, reported that autopsy studies have shown that dental amalgam constitutes the main source of the human total body mercury burden and that "many deceased individuals with dental amalgam have toxic levels of mercury in their brain and kidney tissues." Since the half-life of mercury in the brain varies from years to decades, toxic levels of mercury can accumulate in the brain over time from dental amalgams [20]. WHO has found that mercury-free alternatives to dental amalgam are available and are already being used by dental professionals in many nations [21], yet as mercury dental amalgam is still widely used in many nations, a complete immediate ban would be a significant public health challenge [22]. Recognizing this situation, the Minamata Convention on Mercury, which went into force in August of 2017, commits its signatory countries to phase down the use of dental amalgam taking into account the domestic circumstances and relevant international guidance including two or more practical measures, which can include promoting the use of cost-effective and clinically effective mercury-free alternatives for dental restoration [23].

15.3.1 Accuracy and Efficacy of Mercury-Free Alternatives

The phase-out of mercury containing medical devices cannot be achieved without ensuring the accuracy and efficacy of the substitutes. Multiple studies have evaluated mercury thermometers and sphygmomanometers and their mercury-free counterparts for efficacy, accuracy, and cost-effectiveness. A study comparing digital thermometers to mercury thermometers in a population of ambulatory

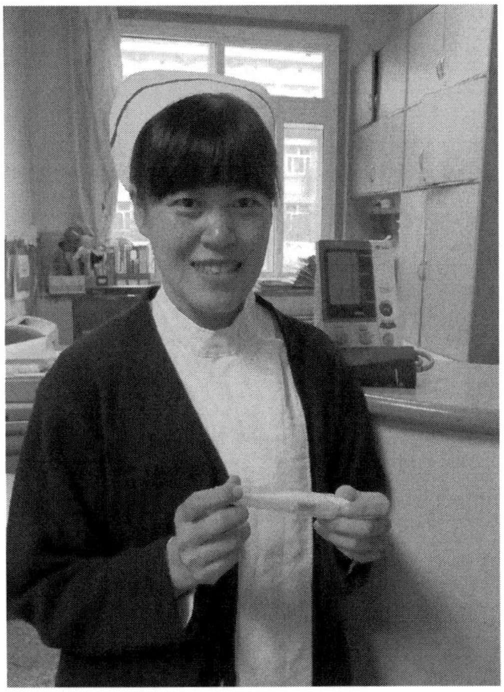

FIGURE 15.1 Beijing nurse with mercury-free Thermometer. Photo: Josh Karliner/Health Care Without Harm.

patients in Malaysia found digital thermometers to be reliable, accurate, safe, and cost-effective [24]. Mercury sphygmomanometers have not been used in Sweden since the late 1990s. A review of the Swedish experience with mercury-free sphygmomanometers concluded that the use of mercury-free sphygmomanometers did not result in any missed diagnoses from inaccurate measurements. Although mercury-free sphygmomanometers require more frequent calibration, they are still practical tools [25]. A 2003 American study found that alternatives to mercury thermometers and sphygmomanometers were readily available in the US and were reliable and cost-efficient [26]. A 2009 validation study found that aneroid sphygmomanometers were a viable replacement for mercury sphygmomanometers with no clinically significant difference in blood pressure measurements between the two devices [27] (Figure 15.1).

15.3.2 Mercury-Containing Medical Device Toxicity Outside the Healthcare Sector

In addition to spillage in a healthcare or home setting, the production of mercury thermometers and sphygmomanometers themselves pose serious occupational health and safety hazards to factory workers, as well as local and global pollution problems. The best-documented case in this regard is that of Kodaikanal, India.

In 1983, Cheeseborough Ponds Corporation relocated a factory from New York in the United States to the lakeside tourist destination of Kodaikanal in southern India. The factory produced mercury thermometers primarily for the US, European, South American, and Australian markets. In 1998, Hindustan Lever, a subsidiary of the Anglo-Dutch corporation Unilever, bought the factory and ran it until it was closed down in 2001.

Data from when the plant was operational report levels of airborne mercury well above internationally recognized safety limits. Not surprisingly, a series of serious mercury-related occupational health problems have been reported by former workers at the plant, including fatigue, headaches, nausea, skin complaints, respiratory dysfunction, kidney dysfunction, and more. Once it was closed, the plant and its surrounding areas, including the local lake, were found to be contaminated with high levels of mercury. Meanwhile, off-site, mercury-laden waste was sold to other businesses, including 7.4 tons to a local scrap dealer, thereby spreading the pollution [28]. Ultimately the company was compelled to ship back about 285 tons of mercury waste to the original suppliers of mercury in the US. More than an estimated 5000 tons of contaminated soil remains at the factory site, while off-site contamination in some places remains serious as well. This example indicates that a great deal of mercury employed in the thermometer production process may never make it to the hospital, but rather ends up as hazardous local and global pollution.

15.4 RESPONSES

15.4.1 The United States

In 1998, the NGO Health Care Without Harm began working with the US EPA, American Hospital Association, the American Nurses Association, and healthcare partners to reduce and eliminate mercury use in healthcare in the United States.

As a result, we have witnessed the progressive phase-out of the use of mercury-based medical devices from the US healthcare community both through voluntary initiatives and legislative mandates.

Over the course of 10 years, with support from Health Care Without Harm, Practice Greenhealth, and environmental health advocates across the country, the US healthcare sector has made significant progress in addressing mercury in healthcare. Noteworthy results include:

- All of the top pharmacy chains in the nation have stopped selling mercury thermometers, making it next to impossible to purchase a mercury thermometer in the United States [29].
- Mercury thermometer bans have been passed in 28 states and bans or severe restrictions on sphygmomanometers cover one-third of the US population.

- More than 1200 hospitals have signed a pledge to eliminate the use of mercury through Hospitals for a Healthy Environment and more than 400 have become virtually mercury-free.
- Most large Group Purchasing Organizations have taken mercury sphygmomanometers "off contract," including Consorta (which buys for 480 hospitals), Premier, and Novation (which buy for 3100 hospitals).

Over 97% of 554 hospitals surveyed by the American Hospital Association are aware of and have taken steps to address the mercury issue. These steps include the following:

Mercury in Clinical Devices

- Over 80% have completely eliminated mercury thermometers from their facilities and 18.7% have replaced some or most with a plan in place for eliminating the remainder.
- Over 73% have completely eliminated mercury sphygmomanometers, with 25% having replaced some or most with a plan in place for eliminating the remainder.
- About 75% have completely eliminated other clinical items (cantor tubes, bougies, etc.) with about 10% having replaced some or most with a plan in place for eliminating the remainder.

Mercury in Facilities

- Over 72% have inventoried all devices and labeled them as mercury-containing, where appropriate.
- About 75% are recycling fluorescent bulbs.

Other Environmental Improvements Made in the Healthcare Sector

- 80% report that they have a waste reduction policy.
- 90% have a regulated medical waste minimization program [30].

15.4.2 The European Union

The European Union has taken the lead in establishing a mercury free healthcare sector. In 2014, the European Union banned the sale and use of thermometers, sphygmomanometers, and other mercury-containing measuring devices with extremely limited exceptions.

Dental amalgam is the largest consumer use of mercury in the European Union. In November 2017, the European Union approved the Berlin declaration to end amalgam use in Europe by July 1, 2022. This action prohibits the use of dental amalgam in the European Union by a future date, which cannot be later than July 1, 2022. The European Union also voted to prohibit the use of dental amalgam in children, pregnant women, and breastfeeding women starting July 2018. In adopting the Berlin declaration, the European Commission noted that mercury-free alternatives to dental amalgam were readily available. Sweden and Norway already exclusively use mercury-free dental fillings [31].

In May 2017, the European Union ratified the Minamata Convention, furthering its commitment to decreasing global mercury usage. Prior to its approval of the Minamata Convention, the European Union had already established an export ban on mercury and several mercury compounds, prohibited the use of mercury in multiple products including batteries, accumulators, and electronic and electrical equipment, and established criteria for the management of metallic mercury waste.

15.4.3 Developing Countries

As it becomes clear to healthcare leaders in developing countries that accurate, cost-effective alternatives are available, more and more hospitals, healthcare systems, and entire nations are beginning to make the switch. The World Health Organization (WHO) policy on mercury in healthcare has provided a framework for this transition.

In 2005, the World Health Organization advised, in its eloquently crafted policy paper on the topic, a global transition of the healthcare sector toward the use of mercury-free care alternatives. Specifically, the paper projected a three-step approach:

Short Term. Develop and implement plans to reduce the use of mercury equipment and replace it with mercury-free alternatives. Address, clean-up, storage, and disposal.

Medium Term. Increase efforts to reduce the use of unnecessary mercury equipment in hospitals.

Long Term. Support a ban of mercury-containing devices and promote alternatives (WHO Mercury in Health Care Policy).

The United Nations Environment Programme's Global Mercury Partnership is a voluntary, multistakeholder partnership initiated in 2005 and designed to take action to protect human health and the environment from the release of mercury and its compounds [32].

In December 2008 the World Health Organization and Health Care Without Harm launched a Global Initiative to achieve virtual elimination of mercury-based thermometers and sphygmomanometers over the next decade and their substitution with accurate, economically viable alternatives.

Working under this framework, or in parallel to it, a number of developing countries are taking significant strides toward mercury elimination in the healthcare sector. The following are some examples:

Argentina. In February 2009, Argentina's Minister of Health issued a declaration banning the purchase of new mercury-based medical devices by the healthcare sector.

India. In May 2010, the Central Government issued guidelines to phase out mercury-containing equipment from all Government-run Hospitals. In Delhi, all government-run hospitals have stopped purchasing new mercury-based medical devices.

Mongolia. The Ministry of Health issued a Joint Order to ban acquisition of mercury-containing thermometers, sphygmomanometers, and dental amalgam starting January 15, 2011.

Philippines. In 2008, the government issued an order for the phase-out of mercury-based medical devices.

South Africa. The Province of Kwa Zulu Natal has issued a directive banning the purchase of mercury thermometers and sphygmomanometers [33].

15.4.4 The Minamata Convention on Mercury [23, 34, 35]

The Minamata Convention on Mercury is a United Nations treaty designed to address anthropogenic emissions and releases of mercury and mercury compounds worldwide, in an effort to reduce the risks to human health and the environment. This Convention was a result of UNEP efforts to bring the science of mercury poisoning to policy implementation.

- In 2001, UNEP was invited by its Governing Council to undertake a global assessment of mercury and its compounds, to include its chemistry and health effects, sources, long-range transport, as well as prevention and control technologies relating to mercury.
- The Governing Council considered this assessment in 2003, finding sufficient evidence of significant global adverse impacts from mercury release to warrant further international action.
- In February 2009, the Governing Council of UNEP decided to develop a global, legally binding instrument on mercury. The work to prepare this instrument was undertaken by an intergovernmental negotiating committee (INC) that met in five sessions from June 2010 until January 2013.
- The fifth session of the INC agreed to the final draft text of the Minamata Convention on Mercury, which was adopted and opened for signature for one year at a Diplomatic Conference (Conference of Plenipotentiaries), held in Kumamoto, Japan, from October 10 to 11, 2013.
- The Convention is made of 35 Articles and 5 Annexes with provisions relating to the entire life cycle of mercury.
- The treaty addresses controls and reductions across a range of products, processes, and industries where mercury is used, released, or emitted. This includes the export and import of mercury, the direct mining of mercury, its safe storage and its safe disposal. Convention implementation also includes identification of at-risk populations and advancing medical care, including improved training of healthcare professionals in identifying and treating mercury-related health effects.
- The Minamata Convention entered into force on August 16, 2017, the 90th day after the date of deposit of the 50th instrument of ratification, acceptance, approval, or accession.
- As of January 2018, 85 countries including the United States, the European Union, and China have ratified the Convention.

The Minamata Convention is named after the Japanese city of Minamata, which experienced severe mercury poisoning after industrial wastewater from a chemical factory was discharged directly into Minamata Bay for decades, beginning in the 1930s. The wastewater from the acetaldehyde plant contained methylmercury, which bioaccumulated in fish and shellfish in the bay. Local people who consumed contaminated seafood from Minamata Bay became severely ill; many were severely disabled and the high-level methylmercury exposure led to severe neurologic disorders in their offspring [35, 36].

15.4.5 The Importance of China

China, which signed the Minamata Convention in 2013, is the world's largest producer and consumer of mercury. According to the US Geological Survey, in 2016 China produced 4000 tons of mercury, approximately 89% of the world's production of mercury that year [37]. There are currently approximately 900 000 medical institutions in China serving a population of over one billion. The implementation of the Minamata Convention by China is necessarily critical to the global reduction of mercury emissions. The Convention requires China to stop the export of mercury thermometers and sphygmomanometers by 2020. However, China has applied under the convention for a 5-year extension for production of mercury-containing thermometers and sphygmomanometers for domestic market use, arguing that the internal market is too large to complete phase-out of mercury-containing medical devices by 2020 as per the Minamata Convention.

15.5 CONCLUSION

While significant progress has been made in the last two decades toward shifting global consumption of mercury-based medical devices toward safer, accurate, and affordable alternatives, much work remains to be done.

Challenges remain. With regards to the Minamata Convention, China is committed to end the export of mercury after the year 2020. This will pose significant problems for countries who have not made provisions to replace mercury-containing medical devices in their healthcare sector. The United Nations Development Programme (UNDP), in conjunction with the World Health Organization and Health Care Without Harm, is working with developing countries to phase out the use of mercury-containing medical devices and replace them with safe and cost-effective alternatives. Greater efforts in this vein are needed. A sudden, critical shortage of medical measuring devices may result in many developing countries if steps are not taken to anticipate the need for mercury-free substitutes in a timely fashion.

Disposal of mercury waste remains a significant issue. Finding a solution at the national, regional, and global levels to manage mercury waste, not just from healthcare but also from other important sources is imperative. The Global Medical Waste Project implemented by the UNDP in conjunction with the WHO and Health Care Without Harm was an international effort conducted between 2008 and 2015 that was designed to help countries reduce the emission of unintentional pollutants such as mercury from medical waste incineration and medical device breakage. The program succeeded in assisting 30 hospitals in 7 nations to become mercury free. Hospitals were taught best practices for mercury management including spill clean-up, mercury medical device phase-out and the introduction of appropriate substitutes, and training of healthcare staff. The program also worked with countries to develop a national plan to phase out mercury-containing medical devices and to appropriately evaluate and assess suitability of replacement devices. While these results are encouraging, further work needs to be done in this area so that all healthcare institutions in all countries adopt similar measures.

The education of not just the public, but of healthcare workers themselves, regarding the toxicity of mercury and the availability and efficacy of non-mercury medical devices, is critical. Knowledge, awareness and attitudes regarding mercury-containing medical devices and their alternatives varies significantly among healthcare workers, contributing not only to occupational health hazards but to possible resistance to adoption of mercury-free alternatives in the healthcare sector.

The entry into force of the Minamata Convention in August 2017 was a significant step forward in the pursuit of mercury-free healthcare. However, mercury is a global problem and will require the cooperation of all nations, not merely a portion. It is critical that nations that have already taken the lead in eliminating mercury-based devices from their healthcare sector continue to encourage, guide, and assist other nations in pursuing and implementing similar policies.

REFERENCES

1. Horowitz, H., Jacob, D.J., Amos, H.M. et al. (2014). Historical mercury releases from commercial products: global environmental implications. *Environmental Science and Technology* 48 (17): 10242–10250.
2. Sundseth, K., Pacyna, J.M., Pacyna, E.G. et al. (2017). Global sources and pathways of mercury in the context of human health. *International Journal of Environmental Research and Public Health* 14 (1): 105.
3. Tokar, E.J., Boyd, W.A., Freedman, J.H., and Waalkes, aP. Toxic effects of metals. In: *Casarett & Doull's Essentials of Toxicology*, 3e (ed. C.D. Klaassen and J.B. Watkins III). New York, NY: McGraw-Hill.
4. Centers for Disease Control and Prevention. 2017. Mercury Fact Sheet. 2017. Available from: https://www.cdc.gov/biomonitoring/Mercury_FactSheet.html.
5. United States Environmental Protection Agency. Health Effects of Exposures to Mercury. Available from: https://www.epa.gov/mercury/health-effects-exposures-mercury.
6. Mahaffey, K.R. (1998). Methylmercury Exposure and Neurotoxicity. *Journal of the American Medical Association* 280: 737–738.
7. Trasande, L., Landrigan, P.J., and Schechter, C. (2005). Public health and economic consequences of methyl mercury toxicity to the developing brain. *Environmental Health Perspectives* 113: 590–596.
8. United Nations Environment Programme (UNEP). Global Mercury Assessment. UNEP Chemicals, Geneva, 2002. Available from: http://web.unep.org/chemicalsandwaste/what-we-do/technology-and-metals/mercury/global-mercury-assessment.
9. Health Care Without Harm (HCWH). 2001. Instruments, Products, and Laboratory Chemicals Used in Hospitals That May Contain Mercury. Pub 2–03 in Going Green: A Resource Kit for Pollution Prevention in Health Care. Available from: https://noharm-uscanada.org/sites/default/files/documents-files/1063/Hosp_Products_Containing_Mercury.pdf.
10. United States Environmental Protection Agency (1996). Mercury Study Report to Congress. *Science Advisory Board Review Draft* 2: 4–19. ES-3.
11. Wongsoonthornchai, M., Scheidegger, R., Kwonpongsagoon, S., and Bader, H. (2017). Analysis of time-dependent mercury flows through the use of thermometers and sphygmomanometers in Thailand. *International Journal of GEOMATE* 12: 65–70.
12. Presentation by Dr Dra. Adriana Grebnicoff, Ministerio de Salud del GCBA, Coordinacion de Salud Ambiental, Primera Conferencia Latinoamericana sobre la Eliminacion del Mercurio en el Cuidado de la Salud, Buenos Aires, August 5, 2006.
13. HCWH/CATAA. Progress Report to the Commission on Environmental Cooperation: Partnership Project to Reduce Use, Discharges, and Emissions of Toxic Substances in the Health Care Sector in Mexico, With an Emphasis on Mercury Reductions, Mexico City/Buenos Aires, July 2007.

14. A 2004 study revealed that, on average, 70 thermometer breakages occur per month in a typical 300–500 bed hospital. Lurking Menace: Mercury in the Health Care Setting, Toxics Link, New Delhi, June 2004.
15. Pastore, P., Singh, R., and Jain, N. (2007). *Mercury in Hospital Indoor Air: Staff and Patients at Risk*. New Delhi: Toxics Link.
16. Halder, N. et al. (2015). Awareness assessment of harmful effects of mercury in a health care setup in India: a survey-based study. *Toxicology and Industrial Health* 31 (12): 1144–1151.
17. Senanayake, S.J. and Gunawardena, N.S. (2016). Knowledge, attitudes and practices regarding handling mercury containing medical devices among nurses in a tertiary care paediatric hospital in Sri Lanka. *Work* 55: 311–319.
18. Personal communication between Rico Euripidou, GroundWork South Africa and Louis Havinga, Manager, Health Technology Services, KwaZulu Natal Department of Health, Durban, June 20, 2007.
19. UNEP Global Mercury Assessment, 2013. Available from: http://web.unep.org/chemicalsandwaste/what-we-do/technology-and-metals/mercury/global-mercury-assessment.
20. Mutter, J. (2011). Is dental amalgam safe for humans? The opinion of the scientific committee of the European Commission. *Journal of Occupational Medicine and Toxicology* 6: 2.
21. World Health Organization. Future Use of Materials for Dental Restoration. Available from: http://www.who.int/oral_health/publications/dental_material_2011.pdf
22. World Health Organization. Media Centre. Mercury and Health. Available from:http://www.who.int/mediacentre/factsheets/fs361/en/
23. UNEP. 2013. Minamata Convention on Mercury. Available from: http://www.mercuryconvention.org/Portals/11/documents/Booklets/Minamata%20Convention%20on%20Mercury_booklet_English.pdf.
24. Fadzil, F.M., Choon, D., and Arumugam, K. (2010). A comparative study on the accuracy of noninvasive thermometers. *Australian Family Physician* 39 (4): 237–239.
25. KEMI (2005). *Mercury Free Blood Pressure Equipment: Experiences in the Swedish Health Care Sector*. Stockholm: Swedish Chemicals Inspectorate Available from: https://noharm-global.org/sites/default/files/documents-files/834/KEMI_Mercury-free_Equipment.pdf.
26. An Investigation of Alternatives to Mercury Containing Products. Prepared for Maine Department of Environmental Protection at Lowell Center for Sustainable Production, University of Massachusetts Lowell. 2003. Available from: http://www.sustainableproduction.org/downloads/An%20Investigation%20Hg.pdf.
27. Ma, Y. et al. (2009). Evaluating the accuracy of an aneroid sphygmomanometer in a clinical trial setting. *American Journal of Hypertension* 22 (3): 263–266.
28. Bridgen K, Stringer R. Atmospheric Dispersal of Mercury from the Hindustan Lever Limited Thermometer Factory, Kodaikanal, Tamil Nadu, India, Using Lichen as a Barometer. Greenpeace Research Laboratories, Exeter, November 2003; Karunasagar, D., et al. Studies of mercury pollution in a lake due to a thermometer factory situated in a tourist resort. Environ Pollut 2006; 143: 153–158.
29. Personal communication with Ken Scott Walden, Asset Management, 2005.
30. Making Medicine Mercury Free: A 2005 Report on the Status of Virtual Mercury Elimination in the Health Care Sector. Hospitals for a Healthy Environment, Arlington, 2005. Available from: https://noharm-uscanada.org/sites/default/files/documents-files/839/Making_Med_Mercury_Free_HCWH.pdf.
31. Berlin Declaration to End Amalgam Use in Europe by 1 July 2022. Available from: http://mb.cision.com/Public/15841/2406577/9b904fc67b4ddd7e.pdf.
32. UNEP. Global Mercury Partnership. Available from: http://web.unep.org/chemicalsandwaste/global-mercury-partnership.
33. HCWH. Q & A: Phasing Out Mercury Thermometers and Blood Pressure Devices. Available from: https://noharm-europe.org/sites/default/files/documents-files/2480/2013-01-09%20HCWH%20Mercury%20Factsheet%20for%20UN%20negotiations.pdf.
34. United Nations. Chapter XXVII – Environment, 17. Minamata Convention on Mercury. UN Treaty Collection. Available from: https://treaties.un.org/Pages/ViewDetails.aspx?src=TREATY&mtdsg_no=XXVII-17&chapter=27&clang=_en.
35. United States Environmental Protection Agency. Minamata Convention on Mercury. Available from: https://www.epa.gov/international-cooperation/minamata-convention-mercury.
36. Ekino, S., Susa, M., Ninomiya, T. et al. (2007). Minamata disease revisited: an update on the acute and chronic manifestations of methyl mercury poisoning. *Journal of the Neurological Sciences* 262 (1–2): 131–144.
37. Mineral Commodity Summaries 2017. US Geological Survey. Available from: https://minerals.usgs.gov/minerals/pubs/commodity/mercury/mcs-2017-mercu.pdf.

SECTION IV

WATER TREATMENT AND SAFE STORAGE

16

HOUSEHOLD WATER TREATMENT AND SAFE STORAGE IN LOW-INCOME COUNTRIES

Thomas F. Clasen

Department of Environmental Health, Rollins School of Public Health, Emory University, Atlanta, GA, USA

16.1 INTRODUCTION

The World Health Organization (WHO) estimates that improving water, sanitation, and hygiene could prevent at least 9.1% of the global burden of disease and 6.3% of all deaths [1]. Diarrhea represents a significant share of this burden, resulting in an estimated 4 billion cases and 1.4 million deaths each year of children under 5 years, or 19% of all such deaths in developing countries (GBD, or Global Burden of Disease [2]). Increasing evidence suggests that repeated environmental insults from enteropathogens are often subclinical and can also cause systemic inflammation and enteropathy associated with malnutrition, stunting, reduced resistance to infection, and impaired cognitive development [3].

As part of its Sustainable Development Goals (SDGs), the UN expressed its commitment by 2030 to "achieve universal and equitable access to safe and affordable drinking water for all" [4]. The SDGs were developed after the Millennium Development Goals (MDGs) expired in 2015. Unlike the MDGs, which assessed drinking water using only service level proxies, the SDGs include targeted goals and indicators to evaluate access to safe drinking water. In the past decade, some progress has been made in access to drinking water. In 2012, the WHO announced that the world had met the MDG of halving the proportion of people without access to safe drinking water [5]. However, as of 2015, an estimated 663 million people worldwide lacked access to improved drinking water sources [2] and even improved water supplies, such as protected wells and communal standposts, often fail to deliver safe drinking water in settings with poor sanitation due to infusion of fecal contamination [6]. Moreover, water that is microbiologically safe at the source or other point of distribution is subject to frequent and extensive fecal contamination during collection, transport, and storage in the home [7]. Thus, the health benefits of safe drinking water will remain elusive for vast populations for years to come.

Providing safe, reliable, piped-in water to every household is an essential goal, yielding optimal health gains while contributing to the SDG targets for adequate sanitation, poverty reduction, nutrition, childhood survival, school attendance, gender equality, and environmental sustainability. While committed strongly to this goal, and to incremental improvements in water supplies wherever possible, the WHO and others have called for targeted, interim approaches that will accelerate the health gains associated with safe drinking water for those whose water supplies are unsafe [8]. While careful not to encourage diversion of resources away from connected taps, public health officials have called for other approaches that will provide some of the health benefits of safe drinking water while progress is being made in improving infrastructure [9,10].

One such alternative is household water treatment and safe storage (HWTS) [11]. In many settings, both rural and urban, populations have access to sufficient quantities of water, but that water is unsafe for consumption as a result of microbial or chemical contamination. This is increasingly true even for piped-in water, since supplies are rarely provided on a 24–7 basis, forcing householders to store more water in the home in ways subject to recontamination and

Water and Sanitation-Related Diseases and the Changing Environment: Challenges, Interventions, and Preventive Measures, Second Edition. Edited by Janine M. H. Selendy.

Companion website: https://www.wiley.com/legacy/wileychi/selendy/

leading to microbial infiltration of poorly maintained systems. Effective treatment at the household level – often using the same basic approaches of filtration, disinfection, and assisted sedimentation, as characterize conventional water treatment – can remove, kill, or inactivate most microbial pathogens [12–15]. Moreover, by focusing at the point of use rather than the point of delivery, treating water at the household level minimizes the risk of recontamination that even improved water supplies can present [7].

Although HWTS is not new, its potential as a focused public health intervention strategy is just emerging. For centuries, householders have used a variety of methods for improving the appearance and taste of drinking-water, including filtering it through porous rock, sand, and other media or using natural coagulants and flocculants to reduce suspended solids. Even before germ theory was well established, successive generations were taught to boil water, expose it to the sun, or store it in metal containers with biocidal properties, all in an effort to make it safer to drink. In 2008, the WHO acknowledged the potential contribution of HWTS in its *Guidelines for Drinking-Water Quality* (GDWQ): "HWT technology has the potential to have rapid and significant positive health impacts in situations where piped water systems are not possible and where people rely on source water that may be contaminated, or where stored water becomes contaminated because of unhygienic handling during transport or in the home" [16]. The WHO also recognized the contribution that HWTS can make among people living with HIV/AIDS, both in providing safe drinking water for those on drug therapy and for the preparation of replacement feeds for mothers who are HIV+ and choose not to breastfeed in order to prevent transmission of the virus in breastmilk [17]. In 2009, UNICEF announced a seven-point strategy for the treatment and prevention of diarrhea among children that includes HWTS, calling for the adoption of "household water treatment and safe storage systems, such as chlorination and filtration, in both development and emergency situations to support reductions in the number of diarrhoea cases" [18].

Despite this support for HWTS as a targeted intervention for preventing diarrhea and other waterborne diseases among vulnerable populations, the promise of the intervention has yet to be fully realized. This is due in part to suboptimal methods for treating water at the household level, but also to the challenge of reaching the target population with affordable HWTS solutions and securing their correct, consistent, and sustained use [19].

The objective of this chapter is to review the evidence around HWTS as a health intervention in low-income settings. The chapter begins by reviewing some of the leading methods for treating water at the household level and the research on their microbiological performance. It then summarizes the evidence concerning the effectiveness and cost-effectiveness of HWTS to prevent diarrheal disease. The chapter ends by noting the major challenges in scaling-up HWTS by achieving coverage and sustained uptake among populations that might benefit most from the intervention, as well as commentary on the threat climate change may pose to the sustainability of such interventions.

16.2 HWTS METHODS

In 2000, the WHO commissioned a comprehensive study to review HWTS methods and practices. The review identified 37 different options for treating and safely storing water at the household level and assessed the available evidence on their microbiological effectiveness, health impact, acceptability, affordability, sustainability, and scalability [9]. In 2014 the WHO developed a methodology for assessing the microbiological performance (http://www.who.int/water_sanitation_health/water-quality/household/scheme-household-water-treatment/en/) and in 2016 completed the testing of certain technologies that were submitted for testing under the methodology [20]. This section summarizes those options for which there is significant research on microbiological effectiveness and health impact. For additional details regarding these and other HWTS interventions, readers are referred to the websites on the intervention maintained by the WHO (http://www.who.int/household_water/en/) and the US Centers for Disease Control and Prevention (CDC) (http://www.cdc.gov/safewater), which both contain other useful links.

16.2.1 Boiling

Boiling or heating with fuel is perhaps the oldest means of effectively disinfecting water at the household level. It is certainly the most common, with an estimated 600 million people in low- and middle-income countries reporting that they usually boil their water before drinking it [21]. If practiced correctly, boiling is also one of the most effective, killing or deactivating all classes of waterborne pathogens, including bacterial spores and protozoan cysts that have shown resistance to chemical disinfection and viruses that are too small to be mechanically removed by microfiltration. Moreover, while chemical disinfectants and filters are challenged by turbidity and certain dissolved constituents, boiling can be used effectively across a wide range of waters. While some authorities recommend boiling water for 10, 20, and even 25 minutes, the WHO GDWQ simply recommend bringing water to a rolling boil of one minute as an indication that a disinfection temperature has been achieved [22].

For decades, governments, NGOs, and others have promoted boiling, both in developing countries where water is routinely of uncertain microbial quality and in developed countries when conventional water treatment fails or water supplies are interrupted due to disasters or other emergencies

[23]. Among householders who report that they "always" or "almost always" treat their water by boiling, the practice has been associated with reductions of fecal bacteria of 99% in Vietnam [24], 97% in India [25], and 86.2% in Guatemala [21]. In China, the use of electric kettles for boiling drinking water is associated with substantial reductions in fecal contamination [26]. In rural Kenya, pasteurization of water using a simple wax indicator to show householders when water reached 70 °C increased the number of households whose drinking water was free of coliforms from 10.7 to 43.1% and significantly reduced the incidence of severe diarrhea compared to a control group [27]. Given this evidence of microbiological effectiveness and the widespread use of the practice, Clasen et al. [25] have argued that boiling is the benchmark against which all other HWTS methods should be measured.

Nevertheless, boiling water presents certain disadvantages that may limit its scalability as a means of routinely treating drinking water. First, the evidence does show that as actually practiced in the home, boiling and storing water often does not always yield microbiologically safe drinking water. Among 137 households in Pakistan who reported boiling as their only method for treating water, only 24 (17.5%) of samples from stored water were free of fecal coliforms [28]. In hygiene-challenged emergency settings, householders who reported boiling their drinking water were no more likely than others to have water that was free of fecal contamination [29]. Outside the emergency context, boiling has been shown to be more effective, though a quarter of water samples in field studies in Vietnam and India still contained medium or high levels of fecal contamination [25].

Second, more than 40% of the world's population relies chiefly on wood, charcoal, and other biomass for their energy supplies [30]. In sub-Saharan Africa, Southeast Asia, and the Western Pacific regions, the figures are 77, 74, and 74%, respectively. The procurement of these fuels represents a substantial commitment of time and energy, primarily by women and girls, and may detract from other productive and potentially health-promoting activities [31]. An alternative means of treating water that does not require the use of such fuels may reduce the time spent collecting the same. Third, because many of those who must treat their own water rely on high-emission biofuels and frequently cook indoors on low-efficiency stoves in poorly ventilated rooms, boiling can aggravate what is often already poor indoor air quality [32]. The use of these fuels is also environmentally unsustainable in many countries and contributes to greenhouse gases. Fourth, boiling water at home has also been associated with higher levels of burn accidents, especially among young children [33]. Finally, the long-term cost of boiling is greater than some alternatives [25] and may be the principal reason why the practice is not even more widespread. Research on the affordability of boiling in a village in Bangladesh found that families in the lowest income quartile would have had to spend 22% of their yearly income on fuel; even those in the highest income bracket would have spent 10% [23]. For a typical family in the lowest income quartiles, boiling of drinking water would require an 11% increase in household budget.

16.2.2 Chlorination

Chlorination is the most widely practiced means of treating water at the community level; it has been used continuously in Europe and North America for more than 100 years [8]. Combined in most cases with some level of filtration to remove solids and reduce chlorine demand, community-based chlorination of drinking water has been credited with nearly half the total mortality reduction, three-quarters of the infant mortality reduction, and nearly two-thirds of the child mortality reduction in major cities in the early twentieth century, the period that recorded the largest gains in childhood health [34].

Chlorination is also used increasingly at the household level. An estimated 5.6% of households in middle- and low-income countries, or 67 million people, report that they usually treat their water with bleach (sodium hypochlorite) or another source of chlorine at home before drinking it [21]. Tablets formed from dichloroisocyanurate (NaDCC), a leading emergency treatment of drinking water, and novel systems for on-site generation of oxidants such as chlorine dioxide, also have a role in household water treatment. At doses of a few mg/l and contact time of about 30 minutes, free chlorine inactivates more than 99.99% of enteric pathogens, the notable exception being *Cryptosporidium*, which the Global Enteric Multicenter Study (GEMS) found to be highly prevalent in young children with moderate-to-severe diarrhea [35].

The impact of chlorine interventions in improving water quality and reducing diarrheal diseases in field studies has been documented. In a systematic review of 21 intervention studies of point-of-use water treatment with chlorine, Arnold and Colford [36] reported that the intervention significantly reduced the risk of stored water contamination with *Escherichia coli* (pooled relative risk: 0.20; 95%CI: 0.13–0.30). They also reported that the intervention reduced the risk of child diarrhea (pooled relative risk: 0.71; 95%CI: 0.58–0.87). A major finding of their review, however, was an attenuation of the intervention's reduction of child diarrhea in longer trials. The authors noted that this could be attributable to seasonal variations in microbiological performance or to user fatigue that may be more acute for interventions such as chlorine that adversely impact water aesthetics (taste and odor). The diminished effect of chlorine and other disinfection interventions over time has been contrasted with filter interventions, leading some investigators to conclude that this absence of sustained effect, after adjusting for possible reporting bias, raises questions about the suitability of scaling up chlorine interventions at the household level until further efforts to secure long-term uptake can be shown [37].

The Safe Water System (SWS) is a programmatic chlorination intervention developed by the US Centers for Disease Control and Prevention in response to a cholera outbreak in Latin America. It combines bottles of dilute sodium hypochlorite with safe storage and behavior change techniques (http://www.cdc.gov/safewater). The microbiological efficacy of the SWS has been demonstrated [14,15] and the effectiveness of the intervention in reducing diarrheal disease has been reported in a variety of settings [38–43].

One chlorine alternative to sodium hypochlorite is sodium dichloroisocyanurate (NaDCC), also known as sodium dichloro-s-triazinetrione or sodium troclosene. For more than 30 years, an effervescent tablet version of NaDCC (Aquatabs® sold by Ireland-based Medentech, Ltd) has been used for the emergency treatment of water. Its widespread use in non-emergency household water treatment applications began after review of the chlorinated isocyanurates by the United States Environmental Protection Agency (USEPA), WHO, and the Food and Agriculture Organization of the United Nations for the routine treatment of drinking water. NaDCC may also offer certain other advantages over sodium hypochlorite in terms of safety, shelf life, up-front cost, and convenience, although these have not yet been shown to impact coverage or uptake [44]. A blinded field trial has shown the product to be effective in treating drinking-water [45]. However, a blinded health impact trial has in Ghana and India reported no impact from the intervention on diarrhea [46,47]. In Ghana, the authors suggested that this might have been attributable to low levels of fecal contamination in the water or to the use of a control (improved storage vessel) that may have also been protective [47]. In India, uptake of the intervention was low and inconsistent [46].

Despite impressive gains in coverage, the adoption and long-term use of chlorine-based HWTS options has been challenging. Follow-up studies on populations exposed to the intervention on a programmatic basis have reported mixed but often low levels of adoption despite extensive programmatic campaigns by implementation partners [48]. As discussed in Section 16.5, achieving correct, consistent use of the intervention by the target population is a significant challenge for all HWTS options. In an effort to address the deficiencies with respect to chlorine, investigators have begun experimenting with the placement of chlorine dispensers at protected springs and other water sources [49]. After collecting a vessel of water, users turn a valve to deliver a measured dose of a stock solution of sodium hypochlorite directly into their water, starting the 30 minutes contact time even before they reach home. There is some initial evidence that these chlorine dispensers have higher levels of consistent uptake than chlorine products that are purchased for household use [49]. Research is underway to explore this alternative in greater scale and to investigate strategies for ensuring that the dispensers are filled when necessary. As most of the cost of the SWS and similar sodium hypochlorite solutions is in the packaging, bulk chlorine dispensers offer an alternative that is significantly lower in cost per liter or cost per household. Whether there is a way to recover the cost of treatment is not yet clear.

16.2.3 Solar Disinfection

Solar disinfection has been subject to rigorous efficacy testing, both in the laboratory and under field conditions, and evaluated for effectiveness in preventing diarrheal disease. While such testing has included permanently mounted panels and other configurations [50], the "Sodis" bottle system, which simply involves filling 1–2 l plastic PET bottles with water and exposing them to the sun for 6 or more hours, has been particularly well documented. Testing in the laboratory has demonstrated the microbiological efficacy of the method against a variety of pathogens [51,52].

The Sodis method has been shown to be effective in reducing diarrhea [53–55] and cholera [56]. However, in one of the largest and most rigorous (though not yet published) studies of the Sodis method, Mausezahl et al. [57] found no statistically significant reduction in diarrhea among children <5 in 22 villages in Bolivia (RR 0.81; 95%CI: 0.59–1.12). The cluster-randomized trial followed 725 children from 425 households for one year after a local NGO implemented the intervention to half the villages and allowed the other half to follow their existing water management practices. The investigation was designed to assess the effectiveness of the intervention as actually deployed by a program implementer; accordingly, the local NGO conducted a standardized interactive Sodis promotion and hygiene education campaign. The investigators found, however, that mean compliance with the intervention was only 32.1% and ascribed the insignificant result to poor uptake.

Unfortunately, none of these health impact studies included assessments of the microbiological performance of the intervention. Accordingly, they do not shed additional light on the contribution of Sodis in improving water quality. To date, only Rainey and Harding [58] have reported on the microbiological performance of the Sodis method in a field study. This was an efficacy study in 40 households in Nepal. The investigators reported that Sodis reduced the level of fecal contamination under household conditions, with 9 of 10 household water samples free of fecal bacteria despite high levels of contamination in source water. However, the study was small, follow-up limited (three times over five months) and the intervention was delivered in a research rather than programmatic context. The investigators also reported that only 10% of study households actually adopted the intervention.

16.2.4 Filtration

Unlike boiling, chlorination and solar disinfection, household filtration encompasses a variety of different processes and products for improving the microbiological quality of drinking water. Mechanical filtration involves the physical removal of suspended solids (including microbes) from water by employing a porous media whose pore size is smaller than the target contaminant. Common media include cloth, sand, porous rock, and unglazed ceramics. Advanced membranes for microfiltration, ultrafiltration, nanofiltration, and reverse osmosis are also used, but must be specially configured for gravity–pressure applications common in low-income settings. Adsorption involves the retention of contaminants within the medium itself, much like a sponge retains water. Activated carbon, either in block or granulated form, is the most common adsorption media for treating water, but is rarely used in low-income settings because of the inability to determine when its adsorption capacity has been exhausted.

Slow-sand filters, which remove suspended solids and microbes by means of a slime layer (*schmutzdecke*) that develops within the top few centimeters of sand, are capable of removing 99% or more of enteric pathogens if properly constructed, operated, and maintained [59]. A simpler but more advanced version, known as the "bio-sand" filter, was specifically designed for intermittent use and is more suitable for household applications. It has been tested both in the laboratory and the field [60,61] and is being deployed widely in development settings by the Centre for Alternative Water and Sanitation Technologies (www.cawst.org) and others [19].

Ceramic filters have been used for treating drinking water for more than a century [8]. Higher quality ceramic filters treated with bacteriostatic silver have been shown to be effective in the lab at reducing waterborne protozoa by more than 99.9% and bacteria by more than 99.9999%; to date, however, they have not met targets for reducing waterborne viruses, though viral sorption and inactivation can be enhanced by special processing [62]. Commercial ceramic filters are typically formed into hollow cylindrical "candles," which are mounted into the top of a two-compartment vessel; locally fabricated filters usually consist of a ceramic pot that serves as the upper chamber and is designed to fit directly into a lower plastic vessel. Pathogens are removed as contaminated water passes through the ceramic in the top compartment to the lower holding compartment. Because the filtered water can only be accessed from this lower compartment by a tap or spigot, it is protected from another significant risk–recontamination prior to consumption. Their potential usefulness as a public health intervention has been reported in development and emergency settings [63,64]. The improving quality of locally fabricated silver-coated ceramics is particularly promising as a sustainable and low-cost alternative [64].

FIGURE 16.1 Vestergaard–Frandsen LifeStraw Filter 2.0. (*See insert for color representation of this figure.*)

In recent years, several types of commercial filters have been promoted for application at the household level in low-income settings. While some of these, such as those that rely on reverse osmosis or ultraviolet technology, require water pressure or electric power or both, other systems that use mechanical filtration and/or disinfection are intended for use in settings like ceramic filters that rely simply on gravity. Few of these filters have been tested rigorously in the laboratory or shown to meet WHO testing standards. One device that has been tested under the WHO methodology is the LifeStraw Family Filter 2.0 (Figure 16.1). This device has also been assessed in a field study in Rwanda [65].

16.2.5 Combination Flocculation and Disinfection

Sachets combining flocculation and disinfection agents were developed in South Africa more than two decades ago. The original products, still sold under the Watermaker™ and Chlor-floc™ brands, employ alum to reduce turbidity and chlorine-resistant protozoan cysts and dichloroisocyanurate to inactivate bacteria and viruses. In 2002, Procter & Gamble Company began field testing its own flocculation/disinfectant sachets. Marketed under the PUR® brand, the product uses ferric sulfate as the flocculant and calcium hypochlorite as the disinfectant, and was designed to address perceived deficiencies in other combination products. Users open the sachet, pour the contents into 10 L of water, stir it repeatedly for several minutes until the floc settles out in the bottom of the vessel, pour the supernatant through a clean cloth into another vessel, and then allow it to stand for 30 minutes.

Few HWTS technologies have been tested as extensively as PUR sachets, both in the laboratory and the field. Laboratory tests demonstrated that the product is highly efficacious, not only against bacteria (>99.99999% reduction), virus (>99.99%), and cysts (>99.95%), but also in reducing levels of arsenic, a significant chemical health hazard in many South Asian water supplies [14]. Aside from boiling and certain commercial filters that have not yet been deployed widely among low-income populations (e.g., Hindustan Unilever "Pureit®" and the Vestergaard-Frandsen "LifeStraw Family®"), PUR sachets are the only household water treatment option designed for use in low-income settings that would appear to satisfy the requirements of a microbiological water purifier established under the EPA Protocol and NSF P-231 [15]. Significantly, the sachets have also shown efficacy against arsenic, an important waterborne risk in many Asian countries [15,66]

A series of rigorous field trials have been undertaken, mainly to assess the efficacy of the intervention in reducing diarrheal disease. Some of these studies also assessed the microbiological performance of the intervention [38,41,43,67,68].

16.2.6 Safe Storage

Households without access to piped-in water are normally required to store water in the home. Research has consistently found that even water that is safe at the point of collection is subject to frequent and extensive microbial contamination during collection, transport, and storage in the home, usually as a result of poor hand hygiene and the practices of drawing drinking water by dipping into a large-mouth vessel [7]. The microbiological quality of water from a contaminated surface or other sources will be aggravated as a result of poor storage practices. While there is some debate over the pathogenicity of microbes circulating in the household (domestic domain) as opposed to those outside the home (public domain) [69], rigorous field trials have reported substantial reductions in diarrheal disease from the distribution and use of improved vessels for storing water [70]. Some of the key factors influencing the impact of storage vessels and conditions on household water quality are: (i) portability and ease of use, based on capacity, size, shape, weight, presence of handles; (ii) durability, weight, and other properties related to resistance and longevity; (iii) presence of a coverable (preferably screw-cap) opening for filling and cleaning access but small enough to reduce the potential for introducing contaminants by contaminated hands, dipping utensils and other vehicles (e.g. airborne dust), vectors, or other sources; (iv) ability to withdraw water in a sanitary manner, such as via a tap, spigot, spout, or other narrow orifice; and (v) presence and accessibility of documentation describing how to properly use the container for water treatment and sanitary storage [8].

16.3 EFFECTIVENESS AND COST-EFFECTIVENESS OF HWTS TO PREVENT DIARRHEA

16.3.1 Effectiveness

Two decades ago, Esrey and colleagues reviewed previous studies on the impact of environmental interventions on diarrhea, and found improvements in water quality to be considerably less effective than those aimed at water quantity, water availability, and sanitation [71,72]. Ubiquitously cited in both professional journals and practical guides, the reviews have led to the dominant paradigm respecting water supply and sanitation interventions: that to achieve broad health impact, greater attention should be given to safe excreta disposal and proper use of water for personal and domestic hygiene rather than to drinking water quality. The corollary has become equally established: that interventions aimed solely at improving drinking water quality would have relatively little impact in reducing diarrheal disease.

As a result of an increasing research on HWTS, however, the dominant paradigm has been refined. Esrey's conclusions that water-quality improvements could reduce diarrheal disease by 15–17% were based exclusively on studies involving interventions at the point of distribution, such as protected wells and springs. Thus, it did not capture the potential additional health gains that could be achieved by ensuring water quality to the point of use by treating it at the household level or by preventing recontamination after collection, transport, and storage in the home. When subsequent reviews subgrouped studies on water quality to separate conventional interventions at the source from point-of-use interventions at the household level, they found only HWTS to be protective against diarrhea [19,36,73,74].

In the most recent systematic review, Clasen et al. [19] came to the same conclusion. Significantly, conventional source-based interventions such as protected wells and springs provided no statistically significant protective effect against diarrhea. Notably, however, there were no trials of piped-in water connections to the home. Household-based interventions, however, were generally effective against diarrhea. On average, distributing water disinfection products for use at the household level may reduce diarrhea by around one-quarter (home chlorination products: RR 0.77; 95% CI: 0.65–0.91; 14 trials, 30 746 participants; flocculation and disinfection sachets: RR 0.69; 95% CI: 0.58–0.82, 4 trials, 11 788 participants). Point-of-use filtration systems probably reduce diarrhea by around a half (RR 0.48; 95% CI: 0.38–0.59, 18 trials, 15 582 participants). Important reductions in diarrhea episodes were shown with ceramic filters, biosand systems, and LifeStraw® filters (Ceramic: RR 0.39; 95% CI: 0.28–0.53; 8 trials, 5763 participants; Biosand: RR 0.47; 95% CI: 0.39–0.57; 4 trials, 5504 participants, LifeStraw: RR 0.69; 95% CI: 0.51–0.93; 3 trials, 3259 participants). Solar water disinfection (SODIS) by distribution of plastic bottles with instructions to leave filled bottles in direct sunlight for at least six hours before drinking probably reduces diarrhea by around a third (RR 0.62; 95% CI: 0.42–0.94; 4 trials, 3460 participants). In subgroup analyses, larger effects were seen in trials with higher adherence and trials that provided a safe storage container. In most cases, the reduction in diarrhea shown in the studies was evident in settings with improved and unimproved water sources and sanitation.

There is increasing evidence, however, that the results from HWTS interventions may be exaggerated due to the placebo effect and reporting bias. Clasen and colleagues observed that while more than 45 studies of HWTS interventions reported a protective effect, only 1 of the 9 studies that attempted to blind the intervention with a placebo found the effect to be statistically significant [19]. This and other evidence of the lack of successful strategies for achieving adoption of HWTS interventions at scale has led some investigators to conclude that efforts to scaleup the intervention are premature [75]. Other investigators have noted, however, that if the estimates of effect of HWTS interventions are discounted by the exaggerated effect associated with open trials of reported outcomes (like self-reported diarrhea), the interventions are still protective, particularly for those HWTS methods whose sustainability can be maintained [37,76]. The blinded studies may have also been impacted by low compliance and relatively low exposure [19].

There is a need for additional assessments of HWTS interventions using placebos and objective outcomes (such as anthropometrics) in order to determine the actual protective effect of HWTS interventions to prevent diarrhea. Large-scale multifactorial trials of WASH interventions including chlorine-based HWTS that include both reported diarrhea and stunting as primary outcomes have recently been completed in Kenya and Bangladesh, though the results have not yet been reported [77]. Even so, the size of the effect, if any, is likely to depend largely on the prevailing conditions. The major factor influencing the effect is more likely to be the extent to which water is the dominant pathway for transmission of diarrheagenic agents. Thus, any given trial, blinded or open, is unlikely to deliver an estimate of effect that is fully generalizable, since the underlying effect will depend on transmission dynamics that vary according to settings, seasons, and other factors. This is consistent with the large range of effect sizes and heterogeneity reported in the meta-analyses that accompanied each of these systematic reviews.

16.3.2 Cost-Effectiveness

While HWTS offers the potential for superior health gains, the economic advantages over conventional improvements in water supplies are also compelling. The cost of implementing water quality interventions varies, from a low of US$0.63 per person per year (solar disinfection) and US$0.66 (chlorination) to US$3.03 (ceramic filters) and US$4.95 (combined flocculation/disinfection); this compares to an average US$1.88 per person per year for installing and maintaining wells, boreholes, and communal tap stands in Africa [78]. The combination of lower cost and higher effectiveness renders household-based chlorination the most cost-effective of water quality interventions to prevent diarrhea, with a cost-effectiveness ratio in Africa of US$53 per disability-adjusted life year (DALY) averted, compared to US$123 for conventional source-based interventions. When health cost savings are included in the analysis, implementing low-cost HWTS interventions actually results in net savings to the public sector [78]. A WHO-sponsored analysis also concluded that household-based chlorination was among the most cost-beneficial of the various options for pursuing international targets for drinking water, yielding high returns on every dollar invested, mainly from lower healthcare costs but also increased productivity and value of school attendance [79]. Finally, there is considerable evidence that the target population is willing and able to pay for some or all of the cost of household-based water treatment products [80], leveraging public sector and donor funding and allowing it to be more focused on the base of the economic pyramid.

In summary, (i) the up-front cost of providing low-cost household water treatment is about half that of conventional source-based interventions; (ii) most or all of that cost can be borne directly by the beneficiary, not the public sector; and (iii) the public sector will nevertheless recover more than the full cost of implementation from reduced health costs for disease treatment. On the other hand, HWTS does not improve access to water or the quantity of water available, both of which are usually higher priorities for low-income populations and can deliver other important gains in productivity and health. Moreover, there is evidence that HWTS provides little benefit in settings without sufficient quantities of water [19]. Accordingly, the intervention is only appropriate in situations where basic levels of access and quantity have been met.

16.4 OPTIMIZING THE POTENTIAL OF HWTS: ACHIEVING TARGETED COVERAGE AND UPTAKE AT SCALE

Despite the long history of HWTS, the practice is still at a nascent stage in terms of public health interventions. There is considerable research to support the microbiological effectiveness of certain approaches and a growing body of research about its health impact on diarrheal disease. In order to realize the full potential of the intervention, however, it is essential that effective and affordable HWTS options reach the most vulnerable populations at scale (coverage) and that those populations use the intervention correctly and consistently over the long term (uptake). To date, there is mixed evidence on both of these fronts.

16.5 SCALING UP COVERAGE AMONG TARGET POPULATIONS

Using data from household surveys from 67 middle- and lower-income countries, Rosa and Clasen [21] estimate that 33% of households – or more than 1.1 billion people – report treating their water at the household level before drinking it. However, more than 20% of this figure is using practices such as filtering their water through a cloth or letting it stand and settle, which are unlikely to render the water safe. Most of the balance consists of those who reported boiling, evidence of which has shown to be practiced less frequently and more inconsistently than the Joint Monitoring Program for Water, Sanitation and Hygiene (JMP) household surveys reflect [81]. Moreover, the practice of boiling is still largely regional and probably does not correspond with risk. Rural populations, for example, are much less likely to have piped water supplies or other improved water sources; in 2006, rural dwellers represented 84% of the population worldwide using unimproved sources of drinking water [82]. However, in all geographical regions, rural populations are less likely to be boiling their water [21]. With only 58% of its population having access to improved water sources, Sub-Saharan Africa is one of the regions most likely to rely on unsafe water supplies. Nevertheless, only 4.9% of the population in 17 Africa region countries report boiling their water before drinking it. This compares to 20.4% in Latin America and the Carribean, 21.2% in South East Asia, 34.8% in Central Europe, and 68.3% in the Western Pacific region. Like other HWTS methods, householders in the richest quintile are more than twice as likely to practice boiling than those of the poorest quintile [21]. There is also evidence that children under five years, who are most vulnerable to diarrheal diseases, may not be benefiting from the HWTS, even if it may be practiced by other members of the household [21].

A report commissioned by the WHO used data from manufacturers and program implementers to estimate coverage of HWTS in developing countries [76]. In 2007, the combined efforts of the above-described HWTS products – exclusive of boiling and emergency applications – produced approximately 15.5 billion liters of treated water. This represents an average annual growth of 25.5% over 2005 and 2006 levels of 9.9 billion and 12.2 billion liters treated, respectively. Manufacturers and program implementers reported 18.8 million users of these HWTS products as of the end of 2007, compared with 12.0 million in 2005 and 15.5 million in 2005 – an average annual growth rate of 25.1%. It is noted that these figures do not include boiling, which Rosa and Clasen [21] estimated to be practiced by nearly 600 million people. They also do not include emergency use of HWTS products – an important role for point-of-use water treatment, but one that does not contribute to the overall levels of coverage under routine conditions. The largest single contribution to the figures supplied by the manufacturers and program implementers was socially marketed sodium hypochlorite, representing approximately 57% of the 18.8 million users in 2007. While social marketing is expressly targeted to low-income populations, there is evidence, however, that these initiatives also result in inequitable uptake that favors urban and higher-income households [83], much like HWTS practices generally.

These figures are impressive, especially given that most of these HWTS programs have been under way for less than 10 (and some less than 5) years. However, given 884 million people relying on unimproved water supplies and many more whose water is not consistently safe for drinking, the results to date for HWTS methods other than boiling provide a perspective that is more sobering. Using the figures from the WHO report, these HWTS methods will not cover even 100 million users for many years, if ever. Except for the SWS in Zambia and, arguably, Madagascar, and some promising ceramic filter coverage in Cambodia, no HWTS method other than boiling can be said to have achieved scale in coverage [76].

Coverage must also focus on the most vulnerable populations, especially children under five years, persons living with HIV/AIDS who have compromised immune systems, and people at the base of the economic pyramid who are most likely to rely on untreated water supplies and are least likely to be practicing effective HWT practices or to have access to effective healthcare. Commercial and quasi-commercial strategies such as social marketing have sought to target some of these markets, using strategies such as microfinance linked to women's self-help groups and bicycle vendors to reach the "last mile" of the distribution chain (http://www.path.org/projects/safe_water.php; http://www.psi.org/our-work/healthy-lives/diarrheal-disease). However, it is likely that the public sector and donors will need to play a leading role in securing coverage among these target populations [76].

16.5.1 Ensuring Uptake (Correct, Consistent Use)

Coverage is an important metric for measuring progress toward scaling up. Like most health interventions, HWTS promoters must actually reach the target population with safe, effective, appropriate, and affordable solutions. This can be a daunting challenge, even for an intervention such as a vaccine. However, unlike vaccines and certain other interventions, water treatment in the home requires householders to embrace and routinely use the intervention in order to provide protection; even occasional consumption of untreated water may neutralize the potential health benefits of the intervention [84]. Thus, the real potential for the benefits of

HWTS depends not only on the extent to which it can be made available to the target population but also on the extent to which it is adopted by that population and used correctly and consistently (uptake) on a sustained basis.

Most of the evidence to date on uptake of the HWTS comes from research-driven efficacy trials. These typically involve intensive campaigns at the outset of the study to ensure compliance with the intervention, an effort that cannot always be duplicated programmatically. Because these studies typically involve continuous water sampling and health surveillance, the study population is also visited frequently, which may itself increase compliance (Hawthorne effect) in ways that will not characterize the programmatic delivery of the intervention by governments, NGOs, and entrepreneurs. Even so, these efficacy studies report a wide range of compliance. A review by Arnold and Colford [36] reported that the percentage of households with detectible chlorine in their water supplies ranged from 36 to 100%. There is some evidence of higher uptake of chlorine from dispensers placed at water collection points [49]. There is also some evidence of longer-term uptake of ceramic and biosand filters [76]. Procter & Gamble ultimately withdrew its flocculant/disinfectant product from the market when penetration did not meet its expectations, reaching about 15% in the Philippines and only 5% in Guatemala [85]. Even among participants in a prior intervention study, an aggressive six-month marketing campaign did not increase adoption beyond such 5%, leading investigators to conclude that demonstrated effectiveness of the intervention was not enough to increase uptake [86].

Recent assessments of programs by NGOs to introduce HWT among vulnerable populations demonstrate the challenges of achieving uptake and the corresponding lack of any measurable health impact. Arnold et al. [87] assessed the uptake and health impact of a three-year program by NGOs in Guatemala to promote HWT (boiling, Sodis, and chlorination) and handwashing with soap (HWWS). The six-month study used propensity scoring to match and compare 600 households in 30 villages (15 intervention/15 control). They found no statistically meaningful difference in adoption of intervention and control households for HWT (9 versus 3%) or HWWS. Consistent with the low sustained behavior adoption, the investigators found no difference between intervention and control villages in child diarrhea, respiratory infections, or growth. Mausezahl and colleagues used a cluster-randomized controlled trial design in 22 rural communities in Bolivia to evaluate the effect of Sodis in reducing diarrhea among children under the age of five. A local NGO conducted a standardized interactive Sodis-promotion campaign in 11 villages targeting households, communities, and primary schools. Mothers completed a daily child health diary for one year. Despite this extensive promotion campaign, investigators found only 32% compliance with the intervention and no strong evidence for a substantive reduction in diarrhea among children (RR 0.81; 95% CI: 0.59–1.12). Boisson et al. [46] assessed the effectiveness of a programmatically delivered chlorine tablet intervention in Orissa, India. They also found the intervention to be ineffective in preventing diarrhea (LP 0.95; 95% CI: 0.79–1.13). They noted, however, that although coverage in the intervention was high, actual use (as confirmed by the presence of chlorine) was low and inconsistent. Poor, inconsistent, and exaggerated use of HWTS has also been found among households that claim to practice household water treatment in response to JMP questions on national surveys [88,89].

Finally, as noted above, there is evidence that the health impact of some HWTS interventions diminishes over time [36,37,74]. Sustained uptake of healthful interventions involving behavior change is a common problem; recent evidence suggests that adoption of oral rehydration solution (ORS) is falling from levels achieved a generation ago [90]. There are also issues concerning the sustainability of common HWTS options themselves, due to affordability and the need to ensure continued supplies of consumables and replacement filters [91]. However, there is evidence that HWWS has been able to achieve sustained adoption in some settings [74]. Promoters of HWTS have begun to follow successful hygiene promotion strategies, such as delivery in schools where children become change agents for influencing behaviors at home [92]. Although there is increasing research on behavior change and communication in the context of environmental interventions such as HWTS [93], the lessons learned have rarely been incorporated effectively into HWTS campaigns. HWTS may actually benefit by being delivered synergistically with hygiene promotion both to increase sustained adoption and to reduce costs [19].

16.6 IMPACT OF CLIMATE CHANGE

There is increasing evidence that climate change will alter environmental conditions, putting accessibility to safe drinking water at risk. Climate change is expected to increase precipitation extremes and raise the temperature, threatening drinking water quality [94]. Elevated temperatures can increase the viral count of pathogens in water and can cause greater persistence of the expression of bacterial virulence genes [95,96]. A study by Konkel and colleagues found *Shigella* is triggered at 37 °C, when the bacterium can be repressed at 30 °C [95]. In a 2016 systematic review, Levy and colleagues reported heavy rainfall and flooding can directly change the pathways of pathogens, potentially leading to contamination of surface and ground waters [97]. The review also noted several studies that recorded an increase in diarrhea or disease outbreaks after dry periods, indicating a relationship between heavy rainfall conditions and the buildup of pathogens in the environment. Heavy rainfalls and floods can indirectly impact water quality by damaging water infrastructure and displacing populations to areas without adequate water systems [97]. A WHO/DFID Vision 2030 report documented that droughts, which are expected to become more frequent and widespread with climate change, may hinder household water accessibility by heightening the risk of water scarcity [98]. Typically droughts are associated with periods of inactivation and immobilization of fecal pathogens. However, research has proven that droughts may threaten water quality by magnifying the risk of encroachment into piped water systems and increasing the proportion of wastewater in surface water [84,99,100].

Climate change has also been associated with the elevation of sea levels. The rising of sea levels presents a risk of saline intrusion into household water sources. Saline contamination could become a major challenge, as most current household water treatments, except for reverse osmosis and distillation technologies, do not have the capability of addressing salinity [101]. The lessons from current research suggests that the effects of climate change will threaten the quantity and quality of accessible drinking water, as well as potentially hinder the effectiveness and sustainability of most HWTS methods. HWTS may need to adapt additional practices and technologies to ensure access to safe drinking water is resilient to the effects of climate change.

16.7 CONCLUSION

Drinking water that is free from fecal contamination is not a sufficient condition to human health, but it is a necessary one that is far from the reach of billions of people worldwide. Combined with improved sanitation and hygiene, HWTS can prevent disease and save lives. Importantly, HWTS provides a tool for the most vulnerable populations to take charge of their own water security while they patiently wait for the pipe to finally reach them.

Additional work is necessary to develop affordable HWTS solutions that operate effectively and sustainably in the challenging environments facing low-income populations and that will be embraced and used correctly and consistently. Research is also required to identify and target the populations that could benefit most from HWTS and to develop delivery strategies that can be implemented at scale.

ACKNOWLEDGMENT

The author acknowledges with gratitude the assistance of Julia Krauss in reviewing and commenting on this chapter.

REFERENCES

1. Prüss-Üstün, A., Bos, R., Gore, F., and Bartram, J. (2008). *Safe Water, Better Health: Costs, Benefits and Sustainability of Interventions to Protect and Promote Health*. Geneva: World Health Organization.
2. WHO/UNICEF (2015). *Progress on Sanitation and Drinking Water: 2015 Update and MDG Assessment*. Geneva: UNICEF and World Health Organization.
3. Guerrant, R.L., DeBoer, M.D., Moore, S.R. et al. (2013). The impoverished gut – a triple burden of diarrhoea, stunting and chronic disease. *Nat. Rev. Gastroenterol. Hepatol.* 10 (4): 220–229.
4. United Nations (2015). United Nations Sustainable Development Goals. General Assembly Res. 66/288 (10 September 2014).
5. WHO/UNICEF (2012). Progress on drinking Water and Sanitation: 2012 update. United States of America: WHO/UNICEF Joint Monitoring Programme for Water Supply and Sanitation.
6. RADWQ (2008). Summary of Results of Rapid Drinking Water Quality Analysis. Geneva: World Health Organization.
7. Wright, J., Gundry, S., and Conroy, R. (2003). Household drinking water in developing countries: a systematic review of microbiological contamination between source and point-of-use. *Trop. Med. Int. Health* 9 (1): 106–117.
8. Sobsey, M.D. (2002). *Managing Water in the Home: Accelerated Health Gains from Improved Water Supply*. Geneva: World Health Organization).
9. Mintz, E., Bartram, J., Lochery, P., and Wegelin, M. (2001). Not just a drop in the bucket: expanding access to point-of-use water treatment systems. *Am. J. Public Health* 91 (10): 1565–1570. [This paper summarizes salient technologies for treating water at the household level and their health impact.].
10. WHO (2005). *World Health Report 2005*. Geneva: World Health Organization.
11. WHO (2007). *Combating Waterborne Disease at the Household Level*. Geneva: World Health Organization.
12. Luby, S., Agboatwalla, M., Razz, A., and Sobel, J. (2001, 2001). A low-cost intervention for cleaner drinking water in Karachi, Pakistan. *Int. J. Infect. Dis.* 5 (3): 144–150.
13. Quick, R., Venczel, L., Gonzalez, O. et al. (1996). Narrow-mouthed water storage vessels and in situ chlorination in a Bolivian community: a simple method to improve drinking water quality. *Am. J. Trop. Med. Hyg.* 54: 511–516.
14. Rangel, J.M., Lopez, B., Mejia, M.A. et al. (2003). A novel technology to improve drinking water quality: a microbiological evaluation of in-home flocculation and chlorination in rural Guatemala. *J. Water Health* 1: 15–22.
15. Souter, P.F., Cruickshank, G.D., Tankerville, M.Z. et al. (2003). Evaluation of a new water treatment for point-of-use household applications to remove micro-organisms and arsenic from drinking water. *J. Water Health* 1 (2): 73–84.
16. WHO (2008). *Guidelines for Drinking- Water Quality, Vol. 1. Recommendations*. Second Addendum to Third Edition. Geneva: World Health Organization.
17. WHO (2008). *Essential Prevention and Care Interventions for Adults and Adolescents Living with HIV in Resource-Limited Settings*. Geneva: World Health Organization.
18. UNICEF (2009). *Diarrhea: Why Children Are Still Dying and What Can Be Done*. New York: United Nations Children's Fund.
19. Clasen, T.F., Alexander, K.T., Sinclair, D. et al. (2015). Interventions to improve water quality for preventing diarrhoea. *Cochrane Database Syst. Rev.* (10): CD004794.

20. WHO 2016. Results of Round I of the WHO International Scheme to Evaluate Household Water Treatment Technologies. Geneva: World Health Organization.

21. Rosa, G., Miller, L., and Clasen, T. (2010). Microbiological effectiveness of disinfecting water by boiling in rural Guatemala. *Am. J. Trop. Med. Hyg.* 82 (2): 289–300.

22. WHO (2004). *Drinking-Water Quality Standards*, 3e. Geneva: World Health Organization.

23. Gilman, R.H. and Skillicorn, P. (1985). Boiling of drinking-water: can a fuel-scarce community afford it? *Bull. World Health Organ.* 63: 157–163.

24. Clasen, T., Do Hoang, T., Boisson, S., and Shippin, O. (2008). Microbiological effectiveness and cost of boiling to disinfect water in rural Vietnam. *Environ. Sci. Tech.* 42 (12): 4255–4260. doi: 10.1021/es7024802.

25. Clasen, T., McLaughlin, C., Nayaar, N. et al. (2009). Microbiological effectiveness and cost of disinfecting water by boiling in semi-urban India. *Am. J. Trop. Med. Hyg.* 79 (3): 407–413.

26. Cohen, A., Tao, Y., Luo, Q. et al. (2015). Microbiological evaluation of household drinking water treatment in rural China shows benefits of electric kettles: a cross-sectional study. *PLoS One* 10 (9): e0138451.

27. Iijima, Y., Karama, M., Oundo, J.O., and Honda, T. (2001). Prevention of bactierial diarrea by pasteurization of drinking water in Kenya. *Microbiol. Immunol.* 45 (6): 413–416.

28. Luby, S.P., Syed, A.H., Atiullah, N. et al. (1999). Limited effectiveness of home drinking water purification efforts in Karachi, Pakistan. *Int. J. Infect. Dis.* 4: 3–7.

29. Gupta, S.K., Suantio, A., Gray, A. et al. (2007). Factors associated with *E. coli* contamination of household drinking water among tsunami and earthquake survivors, Indonesia. *Am. J. Trop. Med. Hyg.* 76: 1158–1162.

30. Bonjour, S., Adair-Rohani, H., Wolf, J. et al. (2013). Solid fuel use for household cooking: country and regional estimates for 1980–2010. *Environ. Health Perspect.* 121 (7): 784–790.

31. Biran, A., Abbot, J., and Mace, R. (2004). Families and firewood: the costs and benefits of children in firewood collection and use in two rural communities in sub-Saharan Africa. *Hum. Ecol.* 32 (1): 1–25.

32. Rehfuess, E., Mehta, S., and Pruss-Ustun, A. (2006). Assessing household solid fuel use: multiple implications for the Millennium Development Goals. *Environ. Health Perspect.* 114 (3): 373–378.

33. Houangbevi, A. (1981). Burns in children in Africa (experience of the Burns Department at the Lome University Hospital Center). *Ann. Anesthesiol. Fr.* 22 (4): 346–350. (in French).

34. Cutler, D. and Miller, G. (2005). The role of public health improvements in heath advances: the twentieth-century United States. *Demography* 42: 1–22.

35. Sow, S.O., Muhsen, K., Nasrin, D. et al. (2016). The burden of *Cryptosporidium* diarrheal disease among children < 24 months of age in moderate/high mortality regions of sub-Saharan Africa and South Asia, utilizing data from the global enteric multicenter study (GEMS). *PLoS Negl. Trop. Dis.* 10 (5): e0004729.

36. Arnold, B. and Colford, J. (2007). Treating water with chlorine at point-of-use to improve water quality and reduce child diarrhea in developing countries: a systematic review and meta-analysis. *Am. J. Trop. Med. Hyg.* 76 (2): 354–364.

37. Hunter, P. (2009). Household water treatment in developing countries: comparing different intervention types using meta-regression. *Environ. Sci. Technol.* 43 (23): 8991–8997.

38. Luby, S.P., Agboatwalla, M., Hoekstra, R.M. et al. (2004). Delayed effectiveness of home-based interventions in reducing childhood diarrhea, Karachi, Pakistan. *Am. J. Trop. Med. Hyg.* 71 (4): 420–427.

39. Quick, R.E., Venczel, L.V., Mintz, E.D. et al. (1999). Diarrhoea prevention in Bolivia through point-of-use water treatment and safe storage: a promising new strategy. *Epidemiol. Infect.* 122 (1): 83–90.

40. Quick, R.E., Kimura, A., Thevos, A. et al. (2002). Diarrhoea prevention through household-level water disinfection and safe storage in Zambia. *Am. J. Trop. Med. Hyg.* 66 (5): 584–589.

41. Reller, M.E., Mendoza, C.E., Lopez, M.B. et al. (2002). A randomized controlled trial of household-based flocculant-disinfectant drinking water treatment for diarrhoea prevention in rural Guatemala. *Am. J. Trop. Med. Hyg.* 69: 411–419.

42. Semenza, J.C., Roberts, L., Henderson, A. et al. (1998). Water distribution system and diarrhoeal disease transmission: a case study in Uzbeckistan. *Am J Trop Med Hyg* 59 (6): 941–946.

43. Chiller, T.M., Mendoza, C.E., Lopez, M.B. et al. (2004). Reducing diarrhea in Guatemalan children: a randomized controlled trial of a flocculant-disinfectant for drinking water. *Bull. WHO* 84 (1): 28–35.

44. Clasen, T. and Edmondson, P. (2006). Sodium dichloroisocyanurate (NaDCC) tablets as an alternative to sodium hypochlorite for the routine treatment of drinking water at the household level. *Int. J. Hyg. Environ. Health* 209: 173–181.

45. Clasen, T., Saeed, T., Boisson, S. et al. (2007). Household-based chlorination of drinking water using sodium dichloroisocyanurate (NaDCC) tablets: a randomized, controlled trial to assess microbiological effectiveness in Bangladesh. *Am. J. Trop. Med. Hyg.* 76 (1): 187–192.

46. Boisson, S., Stevenson, M., Shapiro, L. et al. (2013). Effect of household-based drinking water chlorination on diarrhoea among children under five in Orissa, India: a double-blind randomised placebo-controlled trial. *PLoS Med.* 10 (8): e1001497.

47. Jain, S., Sahanoon, O.K., Blanton, E. et al. (2010). Sodium dichloroisocyanurate tablets for routine treatment of household drinking water in periurban Ghana: a randomized controlled trial. *Am. J. Trop. Med. Hyg.* 82 (1): 16–22.

48. Clasen, T. (2015). Household water treatment and safe storage to prevent diarrheal disease in developing countries. *Curr. Environ. Health Rep.* 2 (1): 69–74. doi: 10.1007/s40572-014-0033-9.

49. Kremer M, Miguel E, Mullainathan S, Null C, Zwane A (2008). Trickle Down: Chlorine Dispensers and Household Water Treatment. Workshop on Scaling Up the Distribution of Water Treatment Technologies, Harvard University, Cambridge, MA, December 12, 2008.

50. Kang, G., Roy, S., and Balraj, V. (2006). Appropriate technology for rural India – solar decontamination of water for

emergency settings and small communities. *Trans. R. Soc. Trop. Med. Hyg.* 100 (9): 863–866.

51. McGuigan, K.G., Méndez-Hermida, F., Castro-Hermida, J.A. et al. (2006). Batch solar disinfection (SODIS) inactivates oocysts of *Cryptosporidium parvum* and cysts of *Giardia muris* in drinking water. *J. Appl. Microbiol.* 101 (2): 453–463.
52. Wegelin, M., Canonica, S., Mechsner, K. et al. (1994). Solar water disinfection: scope of the process and analysis of radiation experiments. *J. Water SRT-Aqua.* 43 (4): 154.
53. Conroy, R.M., Elmore-Meegan, M., Joyce, T. et al. (1996). Solar disinfection of drinking water and diarrhoea in Maasai children: a controlled field trial. *Lancet* 348 (9043): 1695–1697.
54. Conroy, R.M., Meegan, M.E., Joyce, T. et al. (1999). Solar disinfection of water reduce diarrhoeal disease: an update. *Arch. Dis. Child.* 81 (4): 337–338.
55. Rose, A., Roy, S., Abraham, V. et al. (2006). Solar disinfection of water for diarrhoeal prevention in southern India. *Arch. Dis. Child.* 91 (2): 139–141.
56. Conroy, R.M., Meegan, M.E., Joyce, T. et al. (2001). Solar disinfection of drinking water protects against cholera in children under 6 years of age. *Arch. Dis. Child.* 85 (4): 293–295.
57. Mausezahl, D., Christen, A., Duran Pacheco, G. et al. (2009). Solar drinking water disinfection (SODIS) to reduce childhood Diarrhoea in rural Bolivia: a cluster-randomized, controlled trial. *PLoS Med.* 6 (8): e1000125. doi: 10.1371/journal.pmed.1000125.
58. Rainey, R.C. and Harding, A.K. (2005). Acceptability of solar disinfection of drinking water treatment in Kathmandu Valley, Nepal. *Int. J. Environ. Health Res.* 15 (5): 361–372.
59. Hijnen, W.A.M., Visser, A., Schijven, J.F. et al. (2004). Elimination of viruses, bacteria and protozoan oocysts by slow sand filtration. *Water Sci. Technol.* 50 (1): 147–154.
60. Stauber, C.E., Elliott, M.A., Koksai, F. et al. (2006). Characterisation of the biosand filter for *E. coli* reductions from household drinking water under controlled laboratory and field use conditions. *Water Sci. Technol.* 54: 1–7.
61. Tiwari, S.S., Schmidt, W.P., Darby, J. et al. (2009). Intermittent slow sand filtration for preventing diarrhoea among children in Kenyan households using unimproved water sources: randomized controlled trial. *Tropical Med. Int. Health* 14 (11): 1374–1382.
62. Brown, J. and Sobsey, M.D. (2009). Ceramic media amended with metal oxide for the capture of viruses in drinking water. *Environ. Technol.* 30 (4): 379–391.
63. Clasen, T., Brown, J., Suntura, O. et al. (2004). Reducing diarrhoea through household-based ceramic filtration of drinking water: a randomized, controlled trial in Bolivia. *Am. J. Trop. Med. Hyg.* 70 (6): 651–657.
64. Brown, J., Sobsey, M.D., and Proum, S. (2007). *Use of Ceramic Water Filters in Cambodia*. Washington, DC: World Bank, Water, Sanitation and Hygiene.
65. Kirby, M.A., Nagel, C.L., Rosa, G. et al. (2017). Use, microbiological effectiveness and health impact of a household water filter intervention in rural Rwanda-A matched cohort study. *Int. J. Hyg. Environ. Health* 220 (6): 1020–1029.
66. Norton, D.M., Rahman, M., Shane, A.L. et al. (2009). Flocculant-disinfectant point-of-use water treatment for reducing arsenic exposure in rural Bangladesh. *Int. J. Environ. Health Res.* 19 (1): 17–29.
67. Crump, J.A., Otieno, P.O., Slutsker, L. et al. (2005). Household based treatment of drinking water with flocculant-disinfectant for preventing diarrhea in areas with turbid source water in rural western Kenya: cluster randomized controlled trial. *BMJ* 331 (7515): 478–484.
68. Doocy, S. and Burhnam, G. (2006). Point-of-use water treatment and diarrhoea reduction in the emergency context: an effectiveness trial in Liberia. *Tropical Med. Int. Health* 11 (10): 1542–1552.
69. Cairncross, S., Blumenthal, U., Kolsky, P. et al. (1996). The public and domestic domains in the transmission of disease. *Tropical Med. Int. Health* 1 (1): 27–34.
70. Roberts, L., Chartier, Y., Chartier, O. et al. (2001). Keeping clean water clean in a Malawi refugee camp: a randomized intervention trial. *Bull. World Health Organ.* 79: 280–287.
71. Esrey, S.A., Feachem, R.G., and Hughes, J.M. (1985). Interventions for the control of diarrhoeal diseases among young children: improving water supplies and excreta disposal facilities. *Bull. WHO* 64: 776–782.
72. Esrey, S.A., Potash, J.B., Roberts, L., and Shiff, C. (1991). Effects of improved water supply and sanitation on ascariasis, diarrhoea, dracunculiasis, hookworm infection, schistosomiasis, and tracoma. *Bull WHO* 69: 609–621.
73. Fewtrell, L., Kaufmann, R., Kay, D. et al. (2005). Water, sanitation, and hygiene interventions to reduce diarrhoea in less developed countries: a systematic review and meta-analysis. *Lancet Infect. Dis.* 5: 42–52.
74. Waddington, H., Snilstveit, W.H., and Fewtrell, L. (2009). *Water, Sanitation and Hygiene Interventions to Combat Childhood Diarrhoea in Developing Countries*. Delhi: International Initiative for Impact Evaluation.
75. Schmidt, W.P. and Cairncross, S. (2009). Household water treatment in poor populations: is there enough evidence for scaling up now? *Environ. Sci. Technol.* 43 (4): 986–992.
76. Clasen, T. (2009). *Scaling up Household Water Treatment among Low-Income Populations*. Geneva: World Health Organization.
77. Arnold, B., Null, C., Luby, S. et al. (2013). Cluster-randomized controlled trials of individual and combined water, sanitation, hygiene, and nutritional interventions in rural Bangladesh and Kenya: the WASH benefits design and rationale. *BMJ Open* 3 (8): e003476.
78. Clasen, T., Haller, L., Walker, D. et al. (2007). Cost-effectiveness analysis of water quality interventions for preventing diarrhoeal disease in developing countries. *J. Water Health* 5 (4): 599–608.
79. Hutton, G. and Haller, L. (2004). *Evaluation of the Costs and Benefits of Water and Sanitation Improvements at the Global Level*. Geneva: World Health Organization.
80. Ashraf N, Berry J, Shapiro J (2006). Can Higher Prices Stimulate Product Use? Evidence from a Field Experiment on Clorin in Zambia. Presentation at the Berkeley Water Center

Workshop on water sanitation and hygiene, Berkeley, California, November 6, 2006.

81. Makutsa, P., Nzaku, K., Ogutu, P. et al. (2001). Challenges in implementing a point-of-use water quality intervention in rural Kenya. *Am. J. Public Health* 91 (10): 1571–1573.

82. WHO/UNICEF (2008). *Progress in Drinking-Water and Sanitation: Special Focus on Sanitation*. Geneva: World Health Organization and United Nations Children's.

83. Rheingans R, Dreibelbis R (2007). Disparities in Sûr'Eau Use and Awareness: Results from the 2006 PSI TraC Survey (Preliminary Results).

84. Hunter, P.R., Zmirou-Navier, D., and Hartemann, P. (2009). Estimating the impact on health of poor reliability of drinking water interventions in developing countries. *Sci. Total Environ.* 407 (8): 2621–2624.

85. Hanson, M. and Powell, K. (2006). *Procter & Gamble and Population Services International (PSI): Social Marketing for Safe Water*. Paris: Institut Européen d'Administration des Affaires.

86. Luby, S.P., Mendoza, C., Keswick, B.H. et al. (2008). Difficulties in bringing point-of-use water treatment to scale in rural Guatemala. *Am. J. Trop. Med. Hyg.* 78 (3): 382–387.

87. Arnold, B., Arana, B., Mausezahl, D. et al. (2009). Evaluation of a pre-existing, 3-year household water treatment and handwashing. *Int. J. Epidemiol.* 2009: 38 (6): 1651–38 (6): 1661.

88. Rosa, G., Huaylinos, M.L., Gil, A. et al. (2014). Assessing the consistency and microbiological effectiveness of household water treatment practices by urban and rural populations claiming to treat their water at home: a case study in Peru. *PLoS One* 9 (12): e114997.

89. Rosa, G., Kelly, P., and Clasen, T. (2016). Consistency of use and effectiveness of household water treatment practices among urban and rural populations claiming to treat their drinking water at home: a case study in Zambia. *Am J Trop Med Hyg* 94 (2): 445–455.

90. Sreeramareddy, C.T., Low, Y.P., and Forsberg, B.C. (2017). Slow progress in diarrhea case management in low and middle income countries: evidence from cross-sectional national surveys, 1985–2012. *BMC Pediatr.* 17 (1): 83.

91. Sobsey, M.D., Stauber, C.E., Casanova, L.M. et al. (2008). Point of use household drinking water filtration: a practical, effective solution for providing sustained access to safe drinking water in the developing world. *Environ. Sci. Technol.* 42 (12): 4261–4267.

92. O'Reilly, C.E., Freeman, M.C., Ravani, M. et al. (2008). The impact of a school-based safe water and hygiene programme on knowledge and practices of student and their parents: Nyanza Province, western Kenya, 2006. *Epidemiol. Infect.* 136 (1): 80–91.

93. Figueroa, M.E. and Kincaid, D.L. (2009). *Social, Cultural and Behavioral Correlates of Household Water Treatment and Storage*. Geneva: World Health Organization.

94. Intergovernmental Panel on Climate Change (IPCC) (2014). *Climate Change 2014: Impacts, Adaptation, and Vulnerability. Part A: Global and Sectoral Aspects. Contribution of Working Group II to the Fifth Assessment Report of the Intergovernmental Panel on Climate Change*. Cambridge, UK and New York, NY: Cambridge University Press, p. 1132.

95. Konkel, M.E. and Tilly, K. (2000). Temperature-regulated expression of bacterial virulence genes. *Microbes Infect.* 2 (2): 157–166.

96. Shapiro, R.S. and Cowen, L.E. (2012). Thermal control of microbial development and virulence: molecular mechanisms of microbial temperature sensing. *MBio* 3 (5): e00238-12.

97. Levy, K., Woster, A.P., Goldstein, R.S., and Carlton, E.J. (2016). Untangling the impacts of climate change on waterborne diseases: a systematic review of relationships between diarrheal diseases and temperature, rainfall, flooding, and drought. *Environ. Sci. Technol.* 50 (10): 4905–4922.

98. WHO/Department for International Development (DFID) (2009). Summary and Policy Implications Vision 2030: The Resilience of Water Supply and Sanitation in the Face of Climate Change.

99. Besner, M., Prevost, M., and Regli, S. (2011, 2011). Assessing the public health' risk of microbial intrusion events in distribution systems: conceptual model, available data, and challenges. *Water Res.* 45: 961–979.

100. Moors, E., Singh, T., Siderius, C. et al. (2013). Climate change and waterborne diarrhoea in northern India: impacts and adaptation strategies. *Sci. Total Environ.* 468–469 (Suppl): S139–S151.

101. Greenlee, L.F., Lawler, D.F., Freeman, B.D. et al. (2009). Reverse osmosis desalination: water sources, technology, and today's challenges. *Water Res.* 43 (9): 2317–2348.

SECTION V

CLIMATE CHANGE AND HUMAN HEALTH

17

CHANGING GEOGRAPHIC DISTRIBUTION OF DISEASE VECTORS

Mary E. Wilson

School of Medicine, University of California, San Francisco, CA, USA and Harvard T.H. Chan School of Public Health, Boston, MA, USA

17.1 INTRODUCTION

Arthropod vectors of human disease are responsive to environmental conditions. The survival, distribution, abundance, and behavior of vectors, such as mosquitoes and ticks, are intimately linked to the landscape, the local ecology, and the physicochemical environment. The built environment can also affect vector distribution and behavior, sometimes providing protected niches where vectors can survive. Human activities, especially via travel and trade, move vectors around the world, providing abundant opportunities for vectors to become established in new geographic areas [1]. Ticks and perhaps other vectors can be carried long distances by migratory birds.

This chapter will examine how vectors change in geographic distribution and the factors that influence whether the presence of a vector will be associated with transmission of human disease and, if so, the intensity of transmission [2]. The most important vectors of human disease globally, mosquitoes, are inextricably linked to water, which is essential for their propagation. Other vectors, such as ticks, sandflies, other flies, mites, lice, and fleas, can also transmit human diseases, but the primary focus in this chapter will be on mosquitoes, with a few examples given of other vectors. Vector-borne pathogens span many types of pathogenic organisms, with the most important being the malaria parasite, viruses (e.g., dengue, West Nile, chikungunya, Japanese encephalitis, Zika, yellow fever viruses), bacteria (e.g., Lyme disease spirochetes, rickettsiae) [3], and helminths (e.g., filarial worms, onchocerciasis) and non-malaria protozoa (e.g., those causing Chagas disease, African trypanosomiasis) (see Table 17.1). Although this chapter will focus on vector-borne infections of humans, it is important to note that vector-borne infections also affect animals and plants [4]. Infections in these populations can have profound implications for food security and access to materials derived from plants and animals. Disruption of food sources can affect socioeconomic and political stability of countries and regions. Among emerging infections, vector-borne infections are relatively common [5]. In an analysis of emerging infectious diseases (EID) in events between 1940 and 2004, Jones and colleagues found that vector-borne diseases were responsible for 22.8% of EID events and 28.8% in the last decade [6]. The recent global spread of chikungunya and Zika virus infections and the ongoing active spread of yellow fever virus in outbreaks in Africa and South America (especially in Brazil) show that vector-borne infections continue to create challenges for the global health community [7–12].

17.2 HOW VECTORS MOVE

17.2.1 Mosquitoes

17.2.1.1 Aedes aegypti Although some movement of vectors occurs through natural movement, such as flying or passive carriage on winds or with water, extensive transport of vectors is a result of human activity [13]. The flight distance of mosquitoes can be measured in meters and

Water and Sanitation-Related Diseases and the Changing Environment: Challenges, Interventions, and Preventive Measures, Second Edition. Edited by Janine M. H. Selendy.

Companion website: https://www.wiley.com/legacy/wileychi/selendy/

TABLE 17.1 Most common human vector-borne infections by vector.

Vector	Pathogens/infections
MOSQUITO	
Aedes	Dengue, yellow fever, Zika, chikungunya viruses; filarial pathogens
Anopheles	Malaria (*Plasmodium falciparum*, *P. vivax*, *P. malariae*, *P. ovale*, *P. knowlesi*); filarial pathogens
Culex	West Nile virus, Japanese encephalitis, filarial pathogens
FLIES	
Sandflies	Leishmaniasis, Mayaro virus, sandfly fever virus
Tsetse flies	African trypanosomiasis
Blackflies (Simulium)	Onchocerciasis
TICKS	
Argasidae (soft ticks)	Borrelia (relapsing fever)
Ixodidae (hard ticks)	*Borrelia burgdorferi* (Lyme), rickettsiae, ehrlichiae, babesia
BUGS (Hemiptera)	
Reduviidae (genus Triatoma)	Chagas disease (*Trypanosoma cruzi*)

infrequently has been demonstrated to be as long as 1–2 km [14, 15]. Adult mosquitoes, eggs, and immature forms, in contrast, can easily be carried around the world by air, sea, or land. In past centuries, mosquitoes were transported in ships and introduced into new areas, where they could become established if ecoclimatic conditions were appropriate [16]. *Aedes aegypti*, the most important vector for dengue viruses, is thought to have originated in West Africa but was introduced into the Americas, probably aboard slave ships in the fifteenth through the seventeenth centuries. Although the voyage from Africa to the Americas could take 4–6 weeks, the water storage casks on the ships were suitable breeding sites, and blood meals were available from the humans on board. Outbreaks of yellow fever occurred on board ships. From the coastal areas where *Ae. aegypti* was introduced and became established, it subsequently spread into the interior. Its presence allowed the introduction and establishment of the yellow fever virus, which caused massive outbreaks in the Americas as far north as Boston [16]. The yellow fever virus also became established in non-human primates in the Americas, where the virus is maintained today and can emerge from a sylvatic cycle. Although an effective yellow fever vaccine has been available since the 1930s, human cases still occur when unvaccinated humans come into contact with mosquitoes carrying the virus. Mosquitoes can also transmit the yellow fever virus from one human to another in urban settings. Today an estimated 200 000 cases of yellow fever and 29 000–60 000 deaths occur worldwide each year, the majority in Africa [17]. The number of human cases has increased in the last two decades and there are currently active outbreaks in Africa and South America, where there is risk of establishment of urban transmission cycles given the wide distribution of *Ae. aegypti* (and to a lesser extent *Ae. albopictus*) in cities on both continents [11]. Many of these urban populations lie within regions where yellow fever vaccine is not routinely administered, so large susceptible populations exist. Maps from Kraemer show the distribution of *Ae. aegypti* and *Ae. albopictus* [18]. See Figures 17.1 and 17.2.

To control spread in the face of inadequate supplies of yellow fever vaccine an approach used in some African countries in 2017 and currently being started in Brazil is the use of fractional doses of yellow fever vaccine (a full dose is divided among four or five recipients). Studies and models suggest that this will provide excellent short-term protection but not necessarily the lifetime of protection conferred by a full dose of the vaccine [10, 12, 19].

Because of its importance in the transmission of the deadly yellow fever virus in the Western Hemisphere, *Ae. aegypti*, an introduced species, was targeted by the Pan American Health Organization (PAHO), which adopted a resolution in 1947 calling for the hemisphere-wide eradication of *Ae. aegypti* [20]. Twenty-two countries in the Americas were successful (Figure 17.3, a map of the Americas showing distribution of *Ae. aegypti* during three different years) and were declared free of *Ae. aegypti*.

The largest country, Brazil, eradicated the mosquito by 1958, but the mosquito reinfested the country in 1967 [20]. Attempts at re-eradication followed and were initially successful, but the country was reinfested in 1976 and has been infested since then. Today all countries of Latin America are infested with *Ae. aegypti* with the exception of mainland Chile. *Ae. aegypti* is also present in the parts of southern US and in expanding foci in the western US (see Figures 17.1 and 17.2). *Ae. aegypti* is now also widely distributed widely in Asia and is also found in parts of Australia [16]. It was once established in large areas of Europe.

Ae. aegypti is extremely well adapted to the contemporary urban environment. It enters houses and prefers humans (to animals) as a source of blood meals [21]. It is often found in dark places, such as closets and under beds, and may rest among clothing. It lays its eggs just above the water line. These dessication-resistant (resist drying) eggs can survive several months until rain reappears. The female tends to lay a small number of eggs at multiple sites, so if one site is destroyed, others may survive. The female mosquito can take multiple blood meals, feeding on several different human hosts during one gonotrophic cycle, thereby increasing the probability of viral transmission. Thus, one mosquito can infect multiple humans. The mosquito is small; many humans who are bitten do not recall having been

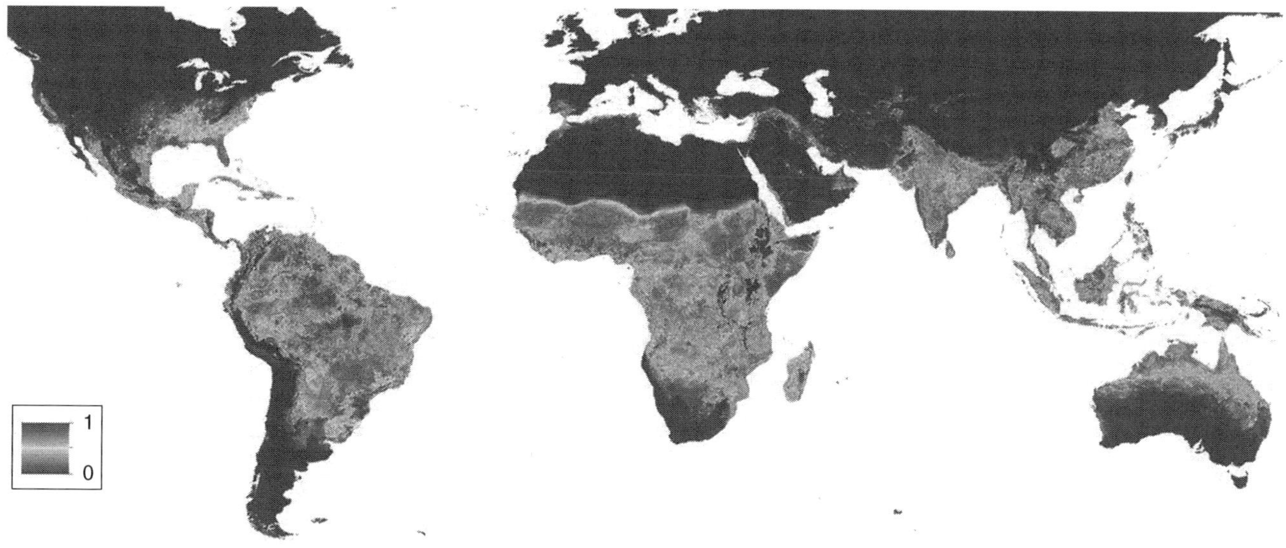

FIGURE 17.1 Distribution *Ae. aegypti*. (*Source*: Reproduced from [18]. (*See insert for color representation of this figure.*)

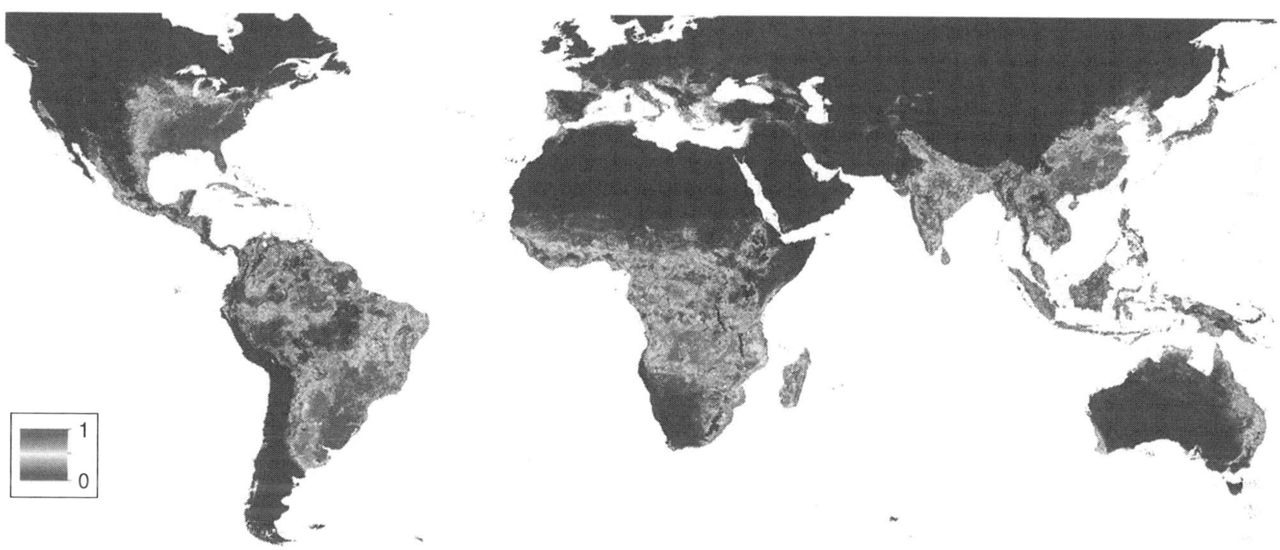

FIGURE 17.2 Distribution *Ae. albopictus* (globally) and areas with rapidly expanding range [18]. MAPS Figures 17.1 and 17.2: This is an open-access article free of all copyright and may be freely reproduced, distributed, transmitted, modified, built upon, or otherwise used by anyone for any lawful purpose. The work is made available under the Creative Commons CC0 public domain dedication. (*See insert for color representation of this figure.*)

bitten. It has now been shown that the same mosquito can be co-infected with more than one virus (e.g., dengue, chikungunya, and Zika) and can transmit more than one virus in a single bite [22, 23]. In a longitudinal study in Nicaragua [24] where dengue, chikungunya, and Zika were co-circulating, investigators assessed individuals with suspected arbovirus infection using a multiplex PCR test for diagnosis. Among 346 individuals with suspected arbovirus infection, 263 were infected with one virus and 71 tested positive for infection with two or all three. Overall 20% of patients were simultaneously infected with two or three arboviruses.

Another feature of the mosquito biology is vertical transmission of the virus from the female mosquito to eggs, another way to maintain virus in a geographic area. This means that the virus can persist in the eggs and the mosquito that emerges does not have to feed on a viremic host to be able to transmit virus. Vertical transmission has now been demonstrated for three important arboviruses – dengue, chikungunya, and Zika viruses [25–27].

The mosquito usually stays within 100 m of human habitats and breeds in artificial containers holding water, such as discarded plastic cups, flowerpots, buckets, and discarded tires [14]. For many residents of urban areas in the tropics and subtropics piped water is unavailable or supplies are erratic, leading them to store water in containers in the home. In an urban area in Rio de Janeiro state, Brazil, investigators studied the relative contribution of different containers and sites to the population of *Ae. aegypti* mosquitoes [28]. They inspected 747 containers in 300 dwellings for the presence of *Ae. aegypti* larvae and pupae. The containers with the highest percentage positive for larvae were tires, jugs, barrels, cisterns, and water tanks – primarily containers that are necessary because of inadequate or unreliable piped water supplies. Overall they found that 90.2% of the larvae and 89.9% of pupae were in containers related to the water supply or trash. Water storage containers also hold a large volume of water and hence can support larger populations of larvae and pupae than small volume containers (Figure 17.4: see the photo of Salvador and water tanks in slum areas, by Felix Lam).

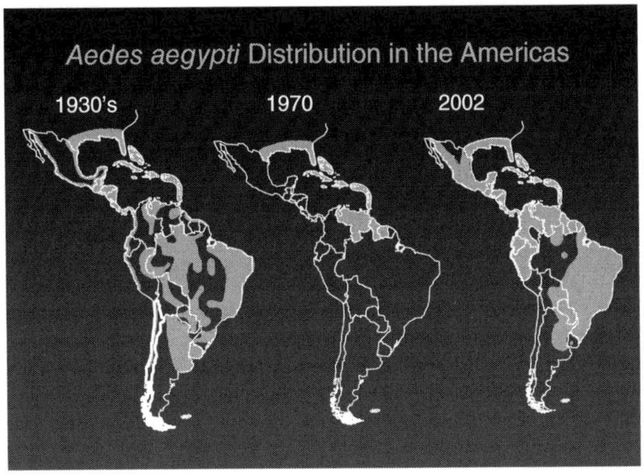

FIGURE 17.3 Map of the Americas showing distribution of *Ae. aegypti* during three different years.

Although programs to cover the water storage tanks have been started in many areas, many tanks remain uncovered or covers have deteriorated over time, allowing tanks to again become infested, creating a productive habitat. The habit of *Ae. aegypti* of distributing its eggs among several oviposition sites, noted above, makes control of these mosquito populations especially difficult [29].

In an analysis of the projected impact of climate change on the range of *Ae. aegypti* in Australia, where drought is already a serious problem, investigators concluded that changes in water storage practices (shifting to storage in and around the home) likely would have a greater effect on mosquito populations and risk for vector-borne infections, such as dengue, than climate change would [30, 31]. Many other investigators have assessed or tried to model the potential impact of climate change on the distribution and activity of *Ae. aegypti* [32–35].

With the availability of better and longer-term data, more recent studies have modeled the receptivity of areas to transmission under various scenarios of climate change. The general conclusion is that warming, an increase in extreme events, and other changes associated with climate change are likely to have several consequences for vector-borne infections, including increasing the geographic range vulnerable to transmission, lengthening the months of transmission, and increasing the intensity of transmission. The latter two

FIGURE 17.4 Photo of Salvador and water tanks in Pau da Lima slum areas, by Felix Lam (from First Edition).

changes can occur in the absence of a change in the geographic distribution of the arthropod vectors [36–39]. Increasing population growth in tropical and subtropical areas will also increase the burden of these diseases because a larger population will inhabit at-risk areas. Global population growth today is occurring predominantly in urban, middle- and lower-income countries, and in low-latitude areas, especially favorable for the transmission of the trio of *Aedes*-transmitted infections (dengue, chikungunya, Zika). These mosquito vectors thrive in an urban environment – and the world today is increasingly urbanized [40, 41]. All of these factors point to a potential increased burden from these vector-borne infections.

17.2.1.2 Aedes albopictus *Aedes albopictus*, the so-called Asian tiger mosquito, continues to invade new areas; in some regions it has displaced populations of *Ae. aegypti* [13]. It is less discriminating than *Ae. aegypti* in its selection of hosts for blood meals and will feed on many vertebrates, including birds and reptiles, though it prefers mammals and will feed on humans if they are available. In a study of *Ae. albopictus* in the Chinese Territory of Macao (situated in the Pearl River delta) for example, humans were the most common host, accounting for 44% of blood meals in the mosquitoes examined [42]. *Ae. albopictus* can breed in artificial containers, including used tires, but also breeds in tree holes, other plants, including the decorative bromeliads, and other sites in the natural environment. It is more abundant in areas with vegetation, and is found in suburban and rural areas as well as in cities.

Although it originated in Asia, *Ae. albopictus* has been transported to most continents and has become established widely in the Americas, Europe, and Africa. It was identified in the US in Memphis, Tennessee, in 1983 [43]. When it was found near Houston, Texas, in 1985, local surveillance studies determined that it was already widespread and common in that area [44]. A survey of 12 states in 1985 found that it was already present in 84% of the 57 counties surveyed. It has subsequently spread broadly in the continental US [45]. It is able to survive in colder environments than *Ae. aegypti* and has been found as far as northern US. It has become established in Mexico and all countries of Central and South America with the exception of continental Chile. It is present in at least 16 European countries with Italy being the most heavily infested (see Figure 17.5a and b). It was first reported to be present in Italy in 1990. In many European countries, the presence of *Ae. albopictus* was first documented after 2000. The mosquito is also established in parts of Africa (at least Nigeria, Gabon, Equatorial Guinea, and Cameroon) [46].

Much of the global dispersal of *Ae. albopictus* can be attributed to international trade in used tires, which provide an ideal habitat for mosquito eggs or immature forms. One examination of 22 000 used tires arriving in Brazil from Japan found that 25% of them held water [47]. Investigators found *Ae. aegypti* and three other species of mosquitoes that were not native to that area. Tires are typically shipped in large containers stacked on transport ships. Millions are shipped each year with some ships large enough to carry more than 14 500 units. The containers may be sent directly from the dockside to the destination, making any meaningful inspection virtually impossible. The characteristics of the *Ae. albopictus* mosquitoes found in the US suggested that they had originated in northern Asia, perhaps Japan or South Korea [44]. At least one of the infestations in Italy is thought to have originated in used tires shipped to Italy from Atlanta, USA. Interstate traffic of used tires is also thought to be important in the movement of the mosquito within countries.

An analysis of international ship traffic movements coupled with climatic information showed that the areas of vector dispersal were remarkably predictable [48]. Tatem and colleagues found that the volume of shipping traffic was more than twice as high between climatically similar ports where *Ae. albopictus* had become established in comparison with those where it had not yet invaded.

Among the reasons for the intense concern about the expanding distribution of *Ae. albopictus* is that this mosquito is competent to transmit a number of viruses that are pathogenic for humans. In the wild, *Ae. albopictus* mosquitoes infected with Eastern equine encephalitis and LaCrosse viruses have been found. It has been the primary vector for several of the outbreaks of chikungunya fever, a viral infection that has caused explosive outbreaks in parts of Africa, the Indian Ocean islands, in many parts of Asia, and even caused a summertime outbreak in northern Italy in 2007 [49–52]. A mutation has been identified in the envelope protein gene of the chikungunya virus that allows it to replicate more efficiently in *Ae. albopictus* mosquitoes. The presence of the mutation is associated with increased midgut infectivity of the mosquito and more rapid viral dissemination into the mosquito salivary glands [53]. In one outbreak, the mutated strain was absent initially but then became the predominant strain, accounting for >90% of viral sequences, suggesting a fitness advantage of this virus [54].

Ae. albopictus has also been the vector for several outbreaks of dengue, including ones in Macao [42] and Hawaii [55]. In the laboratory, it can be infected with a number of other arboviruses that can cause severe disease in humans, including the yellow fever virus. Because it can survive in forests as well as in urban areas it has the potential to serve as a bridge vector, bringing viruses from non-human primates, and perhaps other animals, into urban areas [56, 57]. Concern continues that yellow fever virus could be introduced into urban areas by humans infected and viremic with the yellow fever virus and thus able to spark an urban epidemic of yellow fever in unvaccinated populations in areas heavily infested with *Ae. aegypti* [58]. Although

(a)

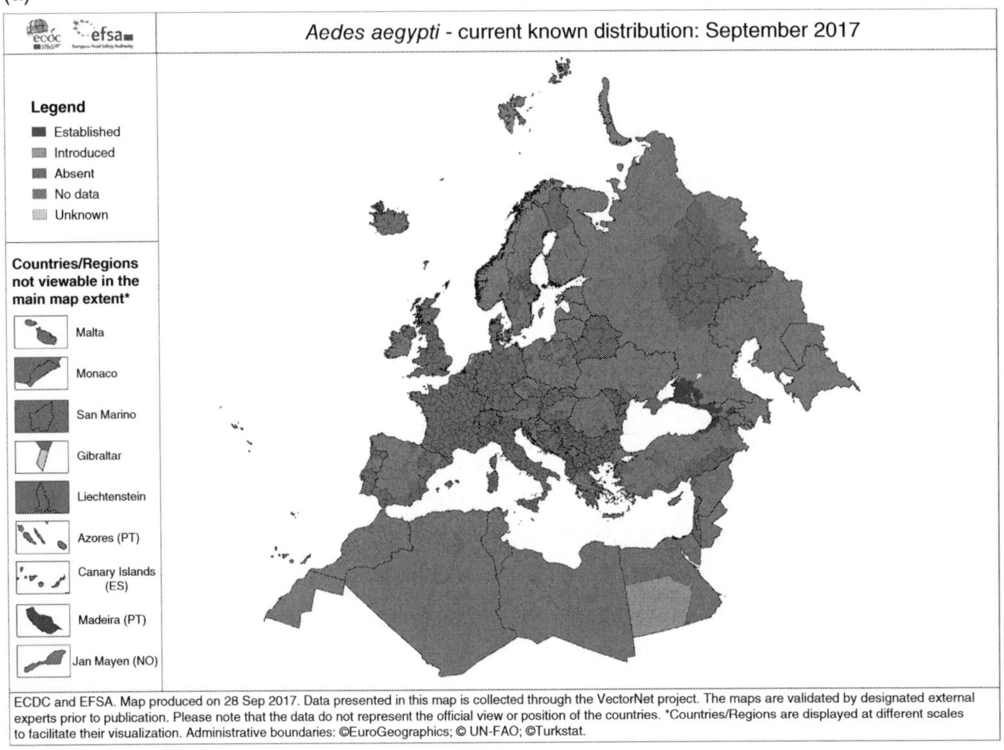

(b)

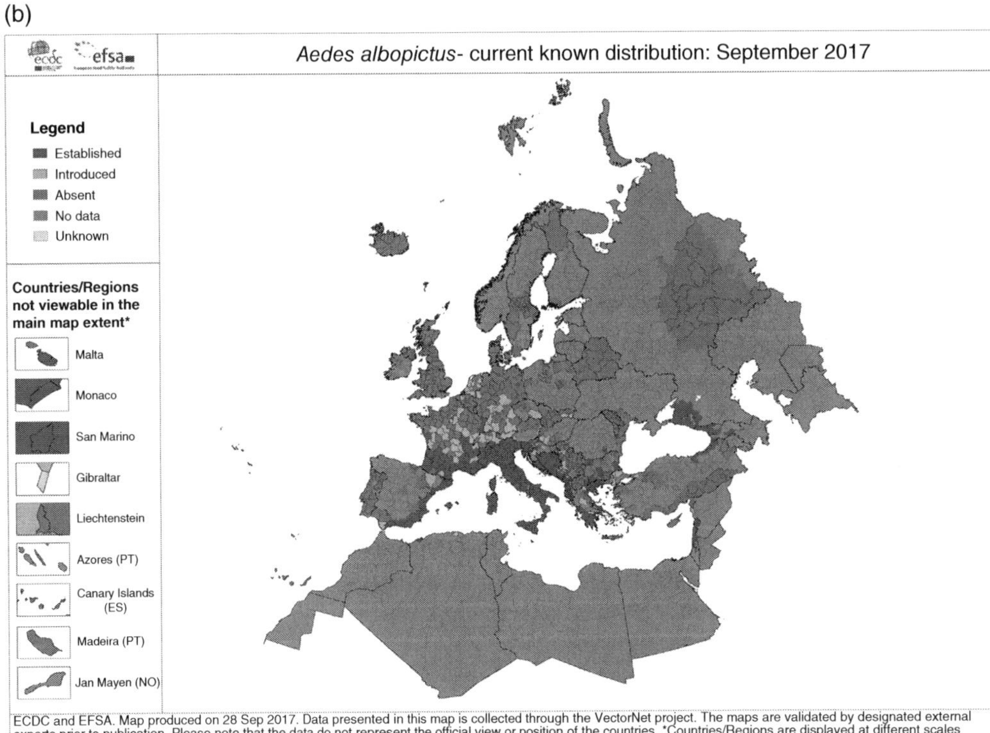

FIGURE 17.5 Maps: current distribution of *Ae. aegypti* (a) and *Ae. albopictus* (b) in Europe from ECDC. Available from: http://ecdc.europa.eu/en/activities/pages/programme_on_emerging_and_vector-borne_diseases_maps.aspx. (*See insert for color representation of this figure.*)

Ae. albopictus mosquitoes naturally infected with yellow fever virus have not yet been identified, this vector could enter forested areas inhabited by primates, which can be infected with yellow fever virus in some tropical areas. Areas at highest risk would be those lying in the tropics or subtropics [46, 58, 59].

Laboratory studies of *Ae. aegypti* and *Ae. albopictus* strains of mosquitoes from Florida (USA) have shown that both are susceptible to a chikungunya virus from recent outbreaks and capable of transmitting the virus [60]. In the summer of 2009, a cluster of dengue fever cases occurred in the Florida Keys [61]. Imported cases of chikungunya fever occurred sporadically in the US for many years in travelers who had visited areas with active transmission. In 2013 for the first time local transmission was documented in the Western Hemisphere, in the Caribbean island of Saint Martin [62]. Transmission has since spread rapidly throughout the Caribbean and Central and South America, with high attack rates in some areas, especially on some of the Caribbean Islands (e.g., 90.4% in Suriname). More than 2.9 million cases have been suspected or confirmed (as of 2016) and 296 deaths are attributed to chikungunya. Transmission continues. As in other areas, outbreaks have left a legacy of individuals with chronic, sometimes disabling, joint pain. Although limited transmission has occurred in Florida and Texas, the brunt of transmission has been borne by countries that are warmer and have fewer curbs on the spread of vector-borne infections (e.g., screens, air conditioning, effective vector control programs, other).

Although mosquitoes are changing in distribution, it is human travelers who introduce the viruses into new areas. If competent mosquito vectors are present and climatic conditions are appropriate, local transmission can occur [63]. In temperate areas, local transmission is possible only during the hot months of the year. The movement of dengue, chikungunya, and Zika viruses globally is almost exclusively attributable to the movement of people [64–66].

In 2004 Zika virus, an arbovirus first identified in monkeys in Uganda in 1947, caused explosive outbreaks on the Pacific Islands [7, 8]. From there it was carried to South America and has now spread widely. Although consequences of Zika infection are still being fully defined, microcephaly – now understood to be part of a broader constellation of findings manifested by the congenital Zika syndrome (CZS) – first became apparent in northeastern Brazil in 2015. Testing of stored blood specimens showed serologic evidence of Zika infection in Rio de Janeiro as early as 2013 [67]. It is likely that the virus circulated silently for months to years before large numbers were infected and consequences of infection became visible. In this instance, as with chikungunya virus, competent mosquito vectors were already in place throughout the Americas to allow extensive transmission once the virus entered. This virus has left a legacy of infants with CZS as well as other neurologic consequences, such as Guillain Barre syndrome. Before the recent outbreaks Zika virus was thought to cause only mild, self-limited illness.

Zika has also changed the thinking about the transmission and movement of a vector-borne infection as it is now understood that this virus can persist for weeks to months in semen [68] and can be sexually transmitted. This requires revising models that predict the spread both in numbers and geography given that this vector-borne infection can spread without a mosquito vector [69–71]. Zika virus, because of its capacity to be spread sexually, provides the unusual example of a primarily vector-borne infection that can extend its geographic reach because of an alternative route of spread. Its long-lasting impact can be felt far from endemic areas of transmission when a pregnant traveler acquires infection and returns to a non-endemic area and gives birth to an infant with CZS. Men who carry the Zika virus in semen can transmit infection months after acute infection, even if the initial infection caused no symptoms and was not recognized.

17.2.1.3 Anopheles gambiae and Brazil An historical example is relevant in demonstrating the importance of the specific vector in risk of transmission of an infection. In 1930, in Natal, Brazil an entomologist saw, to his surprise, the mosquito *Anopheles gambiae*. Furthermore, he counted approximately 2000 larvae [72]. This was unexpected because this mosquito was an African mosquito, not known to inhabit South America. Because dikes had been built along the local river, the saltwater flats had been transformed into freshwater hay fields that could support the breeding of mosquitoes. It was later postulated that rapid French boats (avisos) that made mail runs between Dakar, Senegal and Natal, Brazil were the most likely source. The boats were able to cover the 3300 km in <100 hours and anchored within 1 km of the site where the larvae were found. The possibility of introduction via an aircraft could not be excluded. From this small colony that *An. gambiae* established in the marsh near the coast, the mosquitoes spread along the coastal region and moved 200 miles inland into the Jaguaribe River valley. Factors that favored its dissemination from Natal were the confluence of traffic linking the port to other areas. Natal was an ocean port and received steamships that traveled along the coast as well as international steamships. Small sailing vessels that traveled to fishing villages stopped in Natal, and the city served as the terminus of two railroad lines and was the center of a truck transportation network. Malaria was already present in this region of Brazil, so the parasite was present, but the local mosquitoes were not efficient vectors for the malaria parasite. *An. gambiae*, in contrast, entered houses, bred in open pools outside of the forest, and was highly efficient in transmitting malaria parasites. In 1938 and 1939, almost a decade after the mosquito had first been identified in this region, outbreaks of malaria exploded, causing more than 100 000 cases and about 20 000 deaths.

At this point a massive and intense campaign was launched to eradicate this invading vector. With a staff of >3000 persons and at a cost of > $2 million (1940 dollars), the mosquito was eradicated from Brazil. The last Brazilian-bred *An. gambiae* was collected in November 1940 [72]. This eradication effort was completed without DDT (dichlorodiphenyltrichlorine), which was not yet available. Today, increasing resistance of vectors to insecticides complicates control efforts [73].

Tatem and colleagues also looked at *An. gambiae*, which, in contrast to *Ae. albopictus*, has rarely spread from its geographic origin in Africa. They postulated that the low volume of sea traffic from African ports to climatically similar destinations may partly explain this. The increase in the volume of air traffic from Africa that is currently underway could increase the risk of introduction of this mosquito into climatically similar areas.

17.2.1.3.1 Air Transport of Vectors Mosquitoes and other insects can survive airplane flights and for this reason disinfection (fumigation of the airplane) is sometimes carried out [74]. After the disastrous consequences of the establishment of *An. gambiae* in Natal in the 1930, the Brazilian government required inspections of aircraft arriving from Africa. In one 9-month period in 1941–1942, *An. gambiae* was found seven times on aircraft. Inspectors also found 132 mosquitoes and two live tsetse flies (vector for African trypanosomiasis, also known as sleeping sickness) [75]. The US Public Health service published results of insects found in aircraft during a 13-year period. They reported finding 20 000 insects, including 92 species of mosquitoes, 51 of which were not known to live in mainland US, Hawaii, or Puerto Rico [76]. Mosquitoes have been found in the airplane cabins, in luggage storage compartments, in storage areas, and in airfreight containers. It is clear that mosquitoes infected with malaria parasites can complete a long international flight and survive long enough to infect individuals in areas that are not endemic for malaria, such as Europe. As of August 1999, 89 cases of airport malaria had been reported, the overwhelming majority from Europe (France, Belgium, and UK had the highest numbers), reflecting the large volume of non-stop flights from Africa [77]. In 1994 during a three-week period when six cases of airport malaria occurred near Charles de Gaulle Airport, an estimated 2000–5000 anopheline mosquitoes were imported in the 250–300 aircraft that arrived from malaria-endemic areas of Africa [78]. Mosquitoes can even travel, rarely, in suitcases (baggage malaria). At least two cases related to suitcase transport have been reported [75].

Although mosquitoes can survive air travel, most arrivals by air travel do not lead to their establishment in the new region. On aircraft, typically only a few adults arrive, in contrast to a boat with tires and water containers, in which large numbers of immature forms can be imported. In many instances, the ecoclimatic conditions in the new environment are not favorable for survival and spread.

A recent study assessed the probability of *Plasmodium falciparum* or dengue being introduced by infected mosquitoes on aircraft [79]. They found the probability of introduction was low because there are few mosquitoes per aircraft, infection prevalence among mosquitoes is low, and mosquitoes have high mortality (do not have long survival). They estimated that introduction of one of these pathogens was 1000 times more likely via an infected human than a mosquito for *P. falciparum* and 200 times more likely for dengue virus. The authors conclude that disinfection has a limited role in preventing the spread of vector-borne disease (VBD).

As West Nile virus spread throughout continental US, Kilpatrick and colleagues assessed the pathways by which West Nile virus could potentially reach Hawaii, an event that could be devastating for the local bird populations, as well as causing human disease and disrupting tourism to the islands. They concluded that the virus could reach Hawaii from North America in humans, wind-transported mosquitoes, human-transported mosquitoes, human-transported birds or other vertebrates, and migratory birds. After doing a quantitative estimate for each, they concluded that the highest risk was posed by mosquitoes transported by airplanes and human-transported birds that are exempt from quarantine [80].

17.2.1.3.2 JE Encephalitis Transmission of the mosquito-transmitted Japanese encephalitis (JE) virus has expanded into new territories in Asia, in addition to Australia, as described below. JE virus is transmitted primarily by the mosquito *Culex tritaeniorhynchus*, which often breeds in irrigated rice fields. Water birds serve as reservoir hosts, but pigs are amplifying hosts; other animals, as well as humans, can be infected. In contrast to dengue, movement of infected humans is unimportant to JE epidemiology because the level of viremia in humans is usually low grade and of short duration, even though infection can be fatal or leave those infected with severe neurologic impairment. Migrating birds may help to disperse the virus. Although an effective vaccine is available, it is not used routinely in the poorer countries in the region. Reasons for the expansion of JE human infections in the regions affected in Asia include a marked increase in pig rearing and expansion of irrigated rice farming in size and intensity. Because of population growth and the changes in rice irrigation and pig rearing, the number of people living in high-risk areas has increased. In Myanmar, for example, between 1990 and 2005, the rice paddy area increased 47% and pork production increased almost fourfold [81]. The largest number of people living in close proximity to rice irrigation live in Bangladesh (more than 116 million), a country where JE vaccine is not routinely used [81]. The epidemiology of JE could shift in the future if climate changes lead to alterations in bird migratory patterns or agricultural practices. Overall, the number of cases may decrease in the future because of large-scale vaccination programs in China, but JE human infections are likely to continue and potentially worsen in countries that do not use the vaccine.

17.2.1.3.3 Shift in Distribution Related to Climate Change Because vectors are responsive to ecoclimatic conditions, changes in temperature and rainfall and increased occurrence of extreme events could influence geographic range, abundance, and activity of some vectors of human infections [82]. Cases of tick-borne encephalitis, for example, have occurred at higher altitudes as well as latitudes in Europe [83]. This is thought to be attributable, at least in part, to rising temperatures. The main tick vector, *Ixodes ricinus* in Europe, has extended its range north and westwards [84]. For many vector-borne infections it is difficult to determine the relative contribution of climate change when many other ecological and environmental factors and human activities are changing simultaneously [32]. Warmer or drier conditions in some areas can potentially disrupt or reduce transmission of some infections. Under warmer conditions, insects may be able to overwinter in areas where they currently cannot [85]. See also the earlier section.

17.2.1.3.4 Dispersal Without Human Assistance Although mosquitoes are weak flyers, unusual weather patterns may carry mosquitoes long distances. Japanese encephalitis, a mosquito-borne viral infection, first appeared in Australia in 1995 in the Torres Strait. Investigators used a computer simulation to assess whether weather conditions before the 1995 and 1998 JE outbreaks generated winds that were sufficient to have carried mosquitoes from Papua New Guinea (PNG) to the Torres Strait and Cape York of Australia [86]. They found that during both years low-pressure systems west of the Cape York Peninsula provided conditions that could have carried mosquitoes to Australia. For this to have occurred, large numbers of JE-infected mosquitoes must have been present in PNG and the mosquito species involved must be able to reach high altitudes. The authors note that female *Culex annulirostris* mosquitoes, the implicated species, have been collected at altitudes of 310 m. Transport of infected vectors (Culicoides midges) by wind has been thought to be responsible for incursions of bluetongue virus, an animal pathogen, into northern Australia from Indonesia [87].

In contrast, some species of black flies, for example, the *Simulium damnosum* – the vector for onchocerciasis – have been reported to travel 150–200 km without human assistance. Movement may be aided by wind with the passage of the InterTropical Convergence Zone [88].

17.2.1.3.5 Role of Migratory Birds Migrating birds that carry ticks may introduce these potential vectors and the pathogens they carry into new areas. In Sweden, Bjoersdorff and colleagues examined migratory passerine birds for ticks at a stopover site on the east coast of Oland in 1996 [89]. Of the 3054 birds of 56 species examined, *I. ricinus* ticks were collected from 2.4% of birds of 18 species. Ehrlichia DNA was identified in 8% of the 112 nymph-stage ticks. Based on the estimated 100 million birds that migrate through Sweden annually, the authors projected that almost 600 000 Ehrlichia-infected ticks were imported into Sweden that year. The birds that they examined arrived from continental Europe, North Africa, and Sub-Saharan Africa. The tick species found is also the main vector for *Borrelia burgdorferi* (the cause of Lyme disease) and tick-borne encephalitis virus in Europe.

Migrating birds can also play a role in transporting human pathogens into areas where competent vectors already exist [90]. In the fall of 1998, West Nile virus was isolated from migrating white storks (*Ciconia ciconia*) that had landed in a town in southern Israel [91]. The flock of 1200 had migrated from southern Europe, where West Nile virus had been causing outbreaks since 1996. See also the earlier discussion about JE virus infections.

17.2.1.3.6 Increase in Intensity of Transmission or Severity of Disease Without Change in Vector Distribution Even if a vector species and distribution remain unchanged, intensity of transmission and/or severity of disease can change under certain circumstances. Mutations in the virus that permit more efficient replication and dissemination in a mosquito vector has been documented for chikungunya virus and *Aedes albopictus* [92]. Other studies have looked at whether this might play a role with Zika virus [93].

Mutations in the Zika virus have been postulated to help to explain the dramatic appearance of CZS. Yuan et al. have identified a mutation in the Zika virus that confers increased neurovirulence and may contribute to the CZS [94, 95].

Transmission of malaria parasites that have become resistant to chemotherapeutic agents can cause an increase in human morbidity and mortality even without a change in the species or abundance of the mosquito vector.

Elegant work by Duong and colleagues has demonstrated that pre-symptomatic and asymptomatic individuals with dengue are more infectious to mosquitoes than are those with symptomatic dengue. This underscores the need to consider the role of asymptomatically infected individuals in spread of dengue in the community [96].

These examples are a reminder of the complexity of the vector–pathogen–host–ecosystem interaction and multiple influences on it.

17.2.1.3.7 Vector Without Disease The presence of a disease vector in a region does not mean that a vector-borne infection will occur. Although multiple species of mosquitoes that were competent to transmit West Nile (WN) virus infested the Americas, the disease, West Nile fever, did not exist (except for occasional instances of imported cases) until the virus was introduced into the eastern US in 1999. It then rapidly spread across North America. By the end of 2007, more than 11 000 cases of neurologic infection from WN virus and >1000 deaths had been reported in the US [97]. Although the means of introduction were never documented, it was most likely imported in an infected mosquito or bird [97]. Genetic sequences of the WN isolates found in New York were similar to those in isolates from Israel, suggesting the likely origin was the Middle East or Eastern Europe [98, 99]. Over a period of several years, WN virus also spread into some parts of the Caribbean and Central and South America. Already in place were abundant potential vectors, amplifying hosts (birds), and favorable environmental conditions to allow the virus to flourish and expand its distribution. Because birds, the most important reservoir host, could fly and some migrated, the virus spread easily. It is now endemic in parts of the Americas and is expected to continue to cause cases of human infection. Although human infections have a distinct seasonal pattern, being more common in the warmer months of the year, the virus can survive the winter in hibernating (diapause) adult female mosquitoes, probably in some birds, and perhaps also in rodents [97].

As already noted, mosquitoes that are competent to transmit dengue and yellow fever viruses – as well as JE virus and malaria parasites – infest parts of the United States (as well as parts of Europe and many other regions).

Populations can be naturally resistant or can be made resistant to some vector-borne infections. One common approach is the use of vaccines. Available vaccines, for example, are highly effective in protecting against JE and yellow fever infections, but they are often not administered to those populations at highest risk for infection.

17.2.1.3.8 Presence of Vectors and Pathogens in the Absence of Recognized Disease Vectors and pathogens can be present in an area where human disease is not recognized. In a recent study investigators collected triatomine bugs, the vector for Chagas disease (American trypanosomiasis), in and around homes in the Tucson, Arizona (southwestern US) areas [100]. They examined the triatomines with polymerase chain reaction (PCR) for the presence of *Trypanosoma cruzi*. They found that 42% of the 164 insects collected were infected. The authors speculated that the absence of reports of human disease in this area might be attributed to good housing conditions (absence of crevices where insects can reside), availability of animal hosts for blood meals, misdiagnosis of human cases, and characteristics of the vector or the local parasite associated with reduced infectivity or virulence.

Studies of urban populations along the US (Texas)–Mexico border on either side of the Rio Grande River found that residents of Mexico were more likely to have serologic evidence of dengue infection than were residents in Texas. Although competent vectors (*Ae. aegypti*) infested urban areas on both sides, air conditioning was more common in Texas and was associated with a reduced likelihood of being seropositive for dengue infection [101]. Housing conditions, such as the presence of screens and availability of running water (so that water is not stored in containers in the household) could influence exposure to some vector-borne infections.

17.2.2 Tools

Many new tools have been useful in tracing the origin of vectors and in anticipating where they might become established. Analysis of genetic characteristics of vectors, isozyme analyses, examination of DNA sequences in the mitochondrial ND5 subunit of NADH dehydrogenase, and other approaches have been used to determine the relatedness and origin of vectors. Surveillance networks, such as GeoSentinel, and organizations, such as WHO, CDC, ECDC, and ProMED (Program for Monitoring Emerging Diseases) can help to track the spread of vector-borne infections. Other tools via the Internet provide access to essential data, such as the Disease Vector Database, a global, free, public, Web-accessible resource with data on the geographical distribution of vectors of infectious diseases and reservoirs [102]. Groups have also convened, drawing on many kinds of expertise including disease modeling, climate sciences, and entomology, to develop risk maps, so that public health experts can better prepare for the potential introduction and spread of vectorborne infections. The European Centers for Disease Control (ECDC) regularly produces risk maps for *Ae. albopictus*, *Ae. aegypti*, and other invasive mosquitoes that have been documented in Europe and also has maps with distribution of other known vectors, such as ticks and sandflies (see Figures 17.1 and 17.2).

Recent papers review new and old vector control strategies [103, 104] and new studies provide more detailed analysis of the impact of *Wolbachia* on *Aedes* mosquitoes [105].

17.3 SUMMARY

Global changes today, including urbanization and population growth in low latitude areas, will lead to continued changes in the geographic changes in vectors and the diseases that they spread among humans, animals, and plants. In an analysis of emerging infectious diseases (EID) events between 1940 and 2004, Jones and colleagues found that vector-borne diseases were responsible for 22.8% of EID events and 28.8% in the last decade [6].

Among the most potent forces changing the distribution of vectors are travel and trade, though changes in the natural environment through human activities (e.g., deforestation), alterations in the built environment (e.g., dams, urbanization), climate change, and human demography and activities all contribute as well. A theme of water is a common element throughout – too little, too much, wrong place.

REFERENCES

1. Wilson, M.E. (2010). Global travel and emerging infections. In: Infectious Disease Movement in a Borderless World. Microbial Threats Forum, Institute of Medicine. Workshop Summary, 90–104, 126–129. Washington, DC: The National Academies Press.
2. Wilson, M.E. (2017). Geography of infectious diseases. In: Infectious Diseases, 4e (ed. J. Cohen, W.G. Powderly and S.M. Opal), 938–947. Elsevier.
3. Walker, D.H., Paddock, C.D., and Dumler, J.S. (2008). Emerging and reemerging tick-transmitted rickettsial and ehrlichial infections. *Med. Clin. N. Am.* 92 (6): 1345–1362.
4. Bram, R.A., George, J.E., Reichard, F.E., and Tabachnick, W.J. (2002). Threat of foreign arthropod-borne pathogens to livestock in the United States. *J. Med. Entomol.* 39 (3): 405–416.
5. Gubler, D.J. (1998). Resurgent vector-borne diseases as a global health problem. *Emerg. Infect. Dis.* 4: 442–450.
6. Jones, K.E., Patel, N.G., Levy, M.A. et al. (2008). Global trends in emerging infectious diseases. *Nature* 451: 990–994.
7. Baud, D., Gubler, D.J., Schaub, B. et al. (2017). An update on Zika virus infection. *Lancet* 390: 2099–2109.
8. Gubler, D.J., Vasilakis, N., and Musso, D. (2017). History and emergence of Zika virus. *J. Infect. Dis.* 216 (Suppl. 10): S860–S870.
9. Weaver, S.C. and Lecuit, M. (2015). Chikungunya virus and the global spread of a mosquito-borne disease. *N. Engl. J. Med.* 372: 1231–1239.
10. WHO (2016). Fractional dose yellow fever vaccine as a dose-sparing option for outbreak response. Geneva: World Health Organization.
11. Wilder-Smith, A.L. and Monath, T.P. (2016). Responding to the threat of urban yellow fever outbreaks. *Lancet Infect. Dis.* Epub 2016 Dec 23.
12. Wu, J.T., Peak, C.M., Leung, G.M., and Lipstich, M. (2016). Fractional dosing of yellow fever vaccine to extend supply: a modelling study. *Lancet* 388 (10062): 2904–2911. Published online Nov 9. Available from: http://dx.doi.org/10.1016/S0140-6736(16)31838-4.
13. Lounibos, L.P. (2002). Invasions by insect vectors of human disease. *Annu. Rev. Entomol.* 47: 233–266.
14. Reiter, P. (2007). Oviposition, dispersal, and survival in *Aedes aegypti*: implications for the efficacy of control strategies. *Vector-borne Zoonotic Dis.* 7 (2): 261–273.
15. Shannon, R. and Davis, N. (1930). The flight of *Stegomyia aegypti* (L.). *Am. J. Trop. Med. Hyg.* 10: 145–150.
16. Reiter P. Yellow Fever and Dengue: A Threat to Europe? Eurosurveillance. Published on March 11, 2010a. Available from: www. eurosurveillance.org.
17. World Health Organization. Yellow Fever Fact Sheet. May 2016.
18. Kraemer, M.G.U., Sinka, M.E., Duda, K.A. et al. (2015). The global distribution of the arbovirus vectors *Aedes aegypti* and *Aedes albopictus*. *eLife* 4: e08347. doi: 10.7554/eLife.08347.
19. Roukens, A.H., Vossen, A.C., Bredenbeek, P.J. et al. (2008). Intradermally administered yellow fever vaccine at reduced dose induces a protective immune response: a randomized controlled non-inferiority trial. *PLoS One* 3: e1993.
20. PAHO A Timeline for Dengue in the Americas to December 31, 2000 and Noted First Occurrences. Prepared by J. Schneider and D. Droll, June, 2001. Available from: www.paho.org.
21. Suwonkerd, W., Mongkalangoon, P., Parbaripai, A. et al. (2006). The effect of host type on movement patterns of *Aedes aegypti* (Diptera: Culicidae) into and out of experimental huts in Thailand. *J. Vector Ecol.* 31 (2): 311–318.
22. Furuya-Kanamori, L., Liang, S., Milinovich, G. et al. (2016). Co-distribution and co-infection of chikungunya and dengue viruses. *BMC Infect. Dis.* 16: 84.
23. Goertz, G.P., Vogels, C.B.F., Geertsema, C. et al. (2017). Mosquito co-infection with Zika and chikungunya virus allows simultaneous transmission without affecting vector competence of *Aedes aegypti*. *PLoS NTD* 11 (6): e0005654.
24. Waggoner, J.J., Gresh, L., Vargas, M.J. et al. (2016). Viremia and clinical presentation in Nicaraguan patients infected with Zika virus, chikungunya virus, and dengue virus. *Clin. Infect. Dis.* 63: 1584–1590.
25. Agarwal, A., Dash, P.K., Singh, A.K. et al. (2014). Evidence of experimental vertical transmission of emerging novel ECSA genotype of chikungunya virus in *Aedes aegypti*. *PLoS NTD* 8 (7): e2990. doi: 10.1371/journal.pntd.0002990.
26. Ciota, A.T., Bialosuknia, S.M., Ehrbar, D.J., and Kramer, L.D. (2017). Vertical transmission of Zika virus by *Aedes aegypti* and *Ae. albopictus* mosquitoes. *Emerg. Infect. Dis.* 23: 880–882.
27. Thangamani, S., Huang, J., Hart, C.E. et al. (2016). Vertical transmission of Zika virus in *Aedes aegypti* mosquitoes. *Am. J. Trop. Med. Hyg.* 95 (3): 1169–1173.
28. Medronho, R.A., Macrini, L., Novellion, D.M. et al. (2009). *Aedes aegypti* immature forms distribution according to type of breeding site. *Am. J. Trop. Med. Hyg.* 80 (3): 401–404.
29. Reiter, P., Cmador, M.A., Anderson, R.A., and Clark, G.G. (1995). Short report: dispersal of *Aedes aegypti* in an urban area after blood feeding as demonstrated by rubidium marked eggs. *Am. J. Trop. Med. Hyg.* 52 (2): 177–179.
30. Beebe, J.W., Cooper, R.D., Mottram, P., and Sweeney, A.W. (2009). Australia's dengue risk driven by human adaptation to climate change. *PLoS NTD* 3 (5): e429.
31. Kearney, M., Porter, W.P., Williams, C. et al. (2009). Integrating biophysical models and evolutionary theory to predict climatic impacts on species' ranges: the dengue mosquito *Aedes aegypti* in Australia. *Funct. Ecol.* 23: 528–538.
32. Gubler, D.J., Reiter, P., Ebi, K.L. et al. (2001). Climate variability and change in the United States: potential impacts on vector- and rodent-borne diseases. *Environ. Health Perspect.* 109 (Suppl. 2): 223–233.

33. Hales, S., deWet, N., Maindonald, J., and Woodward, A. (2002). Potential effect of population and climate changes on global distribution of dengue fever: an empirical model. *Lancet* 360: 830–834.
34. Jetten, T.H. and Focks, D.A. (1997). Potential changes in the distribution of dengue transmission under climate warming. *Am J Trop Med Hyg* 57 (3): 285–297.
35. Johansson, M.A., Dominica, F., and Glass, G.E. (2009). Local and global effects of climate on dengue transmission in Puerto Rico. *PLoS NTD* 3 (2): e382.
36. Boyce, R., Reyes, R., Matte, M. et al. (2016). Severe flooding and malaria transmission in the Western Ugandan Highlands: implications for disease control in an era of global climate change. *J. Infect. Dis.* 214 (9): 1403–1410.
37. Caminade, C., Turner, J., Metelmann, S. et al. (2017). Global risk model for vector-borne transmission of Zika virus reveals the role of El Nino 2015. *PNAS* 114: 119–124.
38. Rochlin, I., Ninivaggi, D.V., Hutchinson, J.L., and Farajollahi, A. (2013). Climate change and range expansion of the Asian titer mosquito (*Aedes albopictus*) in Northeastern USA: implications for public health practitioners. *PLoS One* 8: e60874.
39. Struchiner, C.J., Rocklov, J., Wilder-Smith, A., and Massad, E. (2015). Increasing dengue incidence in Singapore over the past 40 years: population growth, climate and mobility. *PLoS One* 10: e0136286.
40. Gubler, D.J. (2011). Dengue, urbanization and globalization: the unholy trinity of the 21st century. *Trop. Med. Health* 39: 3–11.
41. Weaver, S.C. (2013). Urbanization and geographic expansion of zoonotic arboviral diseases: mechanisms and potential strategies for prevention. *Trends Microbiol.* 21: 360.
42. Almeida, A.P.G., Baptista, S.S.S.G., Sousa, C.A.G.C.C. et al. (2005). Bio-ecology and vectorial capacity of *Aedes albopictus* (Diptera: Culicidae) in Macao, China, in relation to dengue virus transmission. *J. Med. Entomol.* 42 (3): 419–428.
43. Reiter, P. and Darsie, R. (1984). *Aedes albopictus* in Memphis, Tennessee (USA): an achievement of modern transportation? *Mosquito News* 44: 396–399.
44. Hawley, W.A., Reiter, P., Copeland, R.S. et al. (1987). *Aedes albopictus* in North America: probable introduction in used tires from northern Asia. *Science* 236: 1114–1116.
45. Moore, D.G. and Mitchell, C.J. (1997). *Aedes albopictus* in the United States: ten-year presence and pubic health implications. *Emerg. Infect. Dis.* 3 (3): 329–334.
46. Reiter, P. (2010b). A Mollusc on the Leg of a Beetle: Human Activities and the Global Dispersal of Vectors and Vector-Borne pathogens. In: In: Infectious Disease Movement in a Borderless World. Microbial Threats Forum, Institute of Medicine. Workshop Summary, 150–165, 175–178. Washington, DC: The National Academies Press.
47. Craven, R.B., Eliason, D.A., Francy, D.B. et al. (1988). Importation of *Aedes albopictus* and other exotic mosquito species into the United States in used tires from Asia. *J. Am. Mosq. Control Assoc.* 4 (2): 138–142.
48. Tatem, A.J., Hay, S.I., and Rogers, D.J. (2006). Global traffic and disease vector dispersal. *Proc. Natl. Acad. Sci.* 103 (16): 6242–6247.
49. Charrel, R.N., de Lamballiere, X., and Raoult, D. (2007). Chikungunya outbreaks – the globalization of vector-borne diseases. *N. Engl. J. Med.* 356 (8): 769–771.
50. Reiter, P., Fontenille, D., and Paupy, C. (2003a). *Aedes albopictus* as an epidemic vector of chikungunya virus: another emerging problem? *Lancet Infect. Dis.* 3: 463.
51. Rezza, G., Nicolietti, L., Angelini, R. et al. (2007). Infection with chikungunya virus in Italy: an outbreak in a temperate region. *Lancet* 370: 1840–1846.
52. Simon, F., Savini, H., and Parola, P. (2008). Chikungunya: a paradigm of emergence and globalization of vector-borne diseases. *Med. Clin. North Am.* 92 (6): 1323–1434.
53. Tsetsarkin, K.A., Vanlandingham, D.L., McGee, C.E., and Higgs, S. (2007). A single mutation in chikungunya virus affects vector specificity and epidemic potential. *PLoS Pathog.* 3 (12): e201.
54. Schuffenecker, I., Iteman, I., Michault, A. et al. (2006). Genome microevolution of chikungunya viruses causing the Indian Ocean outbreak. *PLoS Med.* 3: e263.
55. Effler, P.V.L., Pang, L., Kitsutani, P. et al. (2005). Hawaii Dengue Outbreak Investigation Team. Dengue fever, Hawaii, 2001–2002. *Emerg. Infect. Dis.* 11 (5): 742–749.
56. Lourenco-de-Oliveira, R., Castro, M.G., Braks, M.A.H., and Lounibos, L.P. (2004). The invasion of urban forest by dengue vectors in Rio de Janeiro. *J. Vector Ecol.* 29 (1): 94–100.
57. Maciel-de-Freitas, R., Neto, R.B., Goncalves, J.M. et al. (2006). Movement of dengue vectors between the human modified environment and an urban forest in Rio de Janeiro. *J. Med. Entomol.* 43 (6): 1112–1120.
58. Monath, T. (1999). Facing up to the re-emergence of urban yellow fever. *Lancet* 353 (9164): 1541.
59. Massad, E., Burattini, M.N., Coutinho, F.A.B., and Lopez, L.F. (2003). Dengue and the risk of urban yellow fever reintroduction in Sao Paulo State, Brazil. *Rev. Saude Publica* 37 (4): 477–484.
60. Reiskind, M.H., Pesko, K., Westbrook, C.J., and Mores, C.N. (2008). Susceptibility of Florida mosquitoes to infection with chikungunya virus. *Am. J. Trop. Med. Hyg.* 78: 422.
61. Centers for Disease Control and Prevention (2010). Locally acquired dengue – Key West, Florida 2009-2010. *MMWR* 59 (19): 577–581.
62. Yactayo, S., Staples, J.E., Millot, V. et al. (2016). Epidemiology of chikungunya in the Americas. *J. Infect. Dis.* 214 (S 5): S441–S445.
63. Charrel, R.N., de Lamballerie, X., and Raoult, D. (2008). Seasonality of mosquitoes and chikungunya in Italy. *Lancet* 8: 5–6.
64. Harrington, L.C., Scott, T.W., Lerdthusnee, K. et al. (2005). Dispersal of the dengue vector *Aedes aegypti* within and between rural communities. *Am. J. Trop. Med. Hyg.* 72 (2): 209–220.
65. Stoddard, S.T., Morrison, A.C., Vazquez-Prokopic, G.M. et al. (2009). The role of human movement in the transmission of vector-borne pathogens. *PLoS NTD* 3 (7): e481.

66. Wilder-Smith, A. and Gubler, D.J. (2008). Geographic expansion of dengue: the impact of international travel. *Med. Clin. North Am.* 92: 1377–1390.
67. Passos, S.R.L., MABd, S., Cerbino-Neto, J. et al. (2017). Detection of Zika vírus in April 2013 patient samples, Rio de Janeiro, Brazil. *Emerg. Infect. Dis.* 23 (12): 2120–2121.
68. Paz-Bailey, G., Rosenberg, E.S., Doyle, K. et al. (2017). Persistence of Zika virus in body fluids – preliminary report. *N. Engl. J. Med.* doi: 10.1056/NEJMoa1613108.
69. Blohm, G.M., Lednicky, J.A., Marquez, M. et al. (2018). Evidence for mother-to-child transmission of Zika virus through breast milk. *Clin. Infect. Dis.* 66 (7): 1120–1122.
70. Gregory, C.J., Oduyebo, T., Brault, A.C. et al. (2017). Modes of transmission of Zika virus. *J. Infect. Dis.* 216 (Suppl. 10): S875–S883.
71. Moreira, J., Peixoto, T.M., Siqueira, A.M., and Lamas, C.C. (2017). Sexually acquired ZIka virus: a systematic review. *Clin. Microbiol. Infect.* 23: 296–305.
72. Soper, F.L. and Wilson, D.B. (1943). Anopheles gambiae in Brazil, 1930–1940, 262. New York: The Rockefeller Foundation.
73. Hemingway, J. and Ranson, H. (2000). Insecticide resistance in insect vectors of human disease. *Annu. Rev. Entomol.* 45: 371–391.
74. Russell, R.C. (1987). Survival of insects in the wheel bays of a Boeing 747B aircraft on flights between tropical and temperate airports. *Bull. WHO* 65: 659–662.
75. Gratz, N.G., Steffen, R., and Cocksedge, W. (2000). Why aircraft disinsection? *Bull. WHO* 78: 995–1004.
76. Hughes, J.H. (1961). Mosquito interceptions and related problems in aerial traffic arriving in the United States. *Mosquito News* 21: 93–100.
77. Isaaccson, M. (1989). Airport malaria: a review. *Bull. World Health Organ.* 67: 737–743.
78. Giacomini, T., Mouchet, J., Mathieu, P., and Petithory, J.C. (1995). Study on six cases of malaria contracted near Roissy-Charles de Gaulle in 1994. Preventive measures necessary in airports. *Bul.l de l'Acad. Natl. Med. (France)* 179 (2): 335–353. (in French).
79. Mier-y-Teran-Romero, L., Tatem, A.J., and Johansson, M.A. (2017). Mosquitoes on a plane: disinsection will not stop the spread of vector- borne pathogens, a simulation study. *PLoS Negl. Trop. Dis.* 11 (7): e0005683. doi: 10.1371/journal.pntd.0005683.
80. Kilpatrick, A.M., Gluzberg, Y., Burgett, J., and Daszak, P. (2004). Quantitative risk assessment of the pathways by which West Nile virus could reach Hawaii. *EcoHealth* 1: 205–209.
81. Erlanger, T.E., Weiss, S., Keiser, J. et al. (2009). Past, present, and future of Japanese encephalitis. *Emerg. Infect. Dis.* 15 (1): 1–7.
82. Dohm, D.J., O'Guinn, M.L., and Turell, M.J. (2002). Effect of environmental temperature on the ability of Culex pipiens (Diptera: Culicidae) to transmit West Nile virus. *J. Med. Entomol.* 39 (1): 221–225.
83. Lukan, M., Bullova, E., and Petko, B. (2010). Climate warming and tickborne encephalitis, Slovakia. *Emerg. Infect. Dis.* 16 (3): 524–526.
84. Mansfield, K.L., Johnson, N., Phipps, L.P. et al. (2009). Tick-borne encephalitis virus – a review of an emerging zoonosis. *J. Gen. Virol.* 90: 1781–1794.
85. Bale, J.S. and Hayward, S.A.L. (2010). Insect overwintering in a changing climate. *J. Exp. Biol.* 213: 980–994.
86. Richie, S.A. and Rochester, W. (2001). Wind-blown mosquitoes and introduction of Japanese encephalitis into Australia. *Emerg. Infect. Dis.* 7 (5): 900–903.
87. Melville, L.F., Prichard, L.I., Hunt, N.T. et al. (1997). Genotypic evidence of incursions of new strains of bluetongue viruses in the Northern Territory. *Arbovirus Res. Aust.* 7: 181–186.
88. Garms, R. and Walsh, J.F. (1987). The migration and dispersal of black flies: *Simulium damnosum* S.L., the main vector of human onchocerciasis. In: Black Flies: Ecology, Population Management, and Annotated World List (ed. K.C. Kim and R.W. (e.) Merritt), 201–214. University Park, PA, Pennsylvania University Press.
89. Bjoersdorff, A., Bergstrm, S., Massung, R.F. et al. (2001). Ehrlichia-infected ticks on migrating birds. *Emerg. Infect. Dis.* 7 (5): 877–879.
90. Rapole, J.H., Derrickson, S.R., and Hubalek, Z. (2000). Migratory birds and spread of West Nile virus in the Western Hemisphere. *Emerg. Infect. Dis.* 6 (4): 319–328.
91. Malkinson, M., Banet, C., Weisman, Y. et al. (2001). Intercontinental transmission of West Nile virus by migrating white storks. *Emerg. Infect. Dis.* 7 (3): 540.
92. Tsetsarkin, K.A. and Weaver, S.C. (2011). Sequential adaptive mutations enhance efficient vector switching by chikungunya virus and its epidemic emergence. *PLoS Pathog.* 7: e1002412.
93. Liu, Y., Liu, I., Du, S. et al. (2017). Evolutionary enhancement of Zika virus infectivity in *Aedes aegypti* mosquitoes. *Nature* 545: 482–486.
94. Screatib, G. and Mongkolsapaya, J. (2017). Evolution of neurovirulent Zika virus. *Science* 358: 863–864.
95. Yuan, L., Huang, S.Y., Liu, Z.Y. et al. (2017). A single mutation in the prM protein of Zika virus contributes to fetal microcephaly. *Science* 358 (6365): 933–936.
96. Duong, V., Lambrechts, L., Paul, R.E. et al. (2015). Asymptomatic humans transmit dengue virus to mosquitoes. *PNAS* 112 (47): 14688–14693.
97. Peterson, L. and Hayes, E. (2008). West Nile virus in the Americas. *Med. Clin. North Am.* 92 (6): 1307–1322.
98. Deubel, V., Gubler, D.J., Layton, M., and Malkinson, M. (2001). West Nile virus: a newly emergent disease. *Emerg. Infect. Dis.* 7 (3): 536.
99. Gubler, D.J. (2007). The continued spread of West Nile virus in the western hemisphere. *Clin. Infect. Dis.* 45 (8): 1039–1046.
100. Reisenman, C.E., Lawrence, G., Guerenstein, P.G. et al. (2010). Infection of kissing bugs with *Trypanosoma cruzi*, Tucson, Arizona, USA. *Emerg. Infect. Dis.* 16 (3): 400–405.

101. Reiter, P., Lathrop, S., Bunning, M. et al. (2003b). Texas lifestyle limits transmission of dengue virus. *Emerg. Infect. Dis.* 9: 86–89.

102. Moffett, A., Strutz, S., Guda, N. et al. (2009). A global public database of disease vector and reservoir distributions. *PLoS NTD* 3 (3): e378.

103. Achee, N.L., Gould, F., Perkins, T.A. et al. (2015). A critical assessment of vector control for dengue prevention. *PLoS Negl. Trop. Dis.* 9 (5): e0003655. doi: 10.1371/journal.pntd.0003655.

104. Bowman, L.R., Donegan, S., and McCall, P.J. (2016). Is dengue control deficient in its effectiveness or evidence? Systematic review in meta-analysis. *PLoS NTD* 10 (3): e0004551.

105. Carrington, C.T., BCN, T., NTH, L. et al. (2018). Field- and clinically derived estimates of Wolbachia-mediated blocking of dengue virus transmission potential in *Aedes aegypti* mosquitoes. *PNAS* 115 (2): 361–366.

18

REASSESSING MULTIPLE-INTERVENTION MALARIA CONTROL PROGRAMS OF THE PAST: LESSONS FOR THE DESIGN OF CONTEMPORARY INTERVENTIONS

BURTON H. SINGER[1] AND MARCIA C. CASTRO[2]

[1] *Emerging Pathogens Institute, University of Florida, Gainesville, FL, USA*

[2] *Department of Global Health and Population, Harvard School of Public Health, Boston, MA, USA*

18.1 INTRODUCTION

Malaria is no novelty in the historical medical and public health literature. Sumerians, Assyrians, Chinese, Indians, Egyptians, Greeks, and Romans left a series of records of deadly fevers [1–3].[1] One of the best descriptions of malaria was provided by Hippocrates (about 460 BC – about 377 BC), considered the Father of Medicine, who observed a relationship between stagnant waters and enlarged spleens on the nearby population, and highlighted the importance of the season of the year and place of residence for one's health [1–3, 5, 6]. Although naïve in their formulation, hypotheses about relations between malaria and the environment (particularly stagnant waters) led to the adoption of engineering works aimed at reducing marshes and swamps. Probably one of the earliest records of such initiatives dates back to 490–430 BC and refers to the work of Empedocles of Agrigentum, a citizen of Sicily, who altered the course of rivers to avoid the formation of stagnant and bad smelling water in the city of Selinus [7]. This was followed by similar initiatives over the centuries, including the construction of surface drainage, removal of swamps, and drainage of coastal and wetland areas [8]. All these initiatives were based on the miasmatic theory of malaria transmission[2] and continued after the etiology and life cycle of the disease were elucidated. Countries such as Italy, Israel, and the United States (Tennessee Valley Authority) conquered malaria through massive drainage and land reclamation by hydraulic work [9–11].

Gradually accumulating evidence about malaria over centuries was augmented dramatically by discoveries from the late nineteenth century to the present day. They revealed the nature of the disease, its form of transmission, the characteristics of the vector, the life cycle of the parasite, a diversity of strategies for prevention and control, alternatives for treatment, and the genome sequencing of malaria vectors and parasites [12–19]. Yet, apart from all the accumulated knowledge, there were an estimated 216 million malaria cases worldwide in 2016, the vast majority (90%) in Africa. The disease accounted for an estimated 445 000 deaths in 2016, most of which were recorded in Africa [20]. Nevertheless, the case number and the deaths represent nearly 50% reductions over the prior eight-year period. Children younger than five years are often at higher risk, and Africa accounts for 9 out of every 10 child deaths due to malaria [20].

[1] The earliest genetic evidence of *Plasmodium falciparum* malaria was established after the examination of 16 mummies from the 18th dynasty (c. 1550–1295 BC). The malaria parasite was found in four of the mummies, including king Tutankhamun [4].

[2] Malaria was believed to be caused by breathing in the foul vapors emanating from stagnant and fetid waters. Indeed, the word malaria (of Roman origin) means bad air (*mala aria*).

Water and Sanitation-Related Diseases and the Changing Environment: Challenges, Interventions, and Preventive Measures, Second Edition. Edited by Janine M. H. Selendy.

Companion website: https://www.wiley.com/legacy/wileychi/selendy/

Geographically, malaria is currently restricted to some tropical areas in Central and South America, Africa, and Central and Southeast Asia [21, 22]. This is a very different picture than that observed in the eighteenth century, when malaria was present in all continents [1, 23], a pattern that continued well through the mid-twentieth century [23, 24]. The changing spatial distribution of malaria is a result of organized and successful efforts to reduce the burden of malaria. The nature, focus, and coverage of these efforts changed over time [1, 6, 25–30]. Funding for malaria control in the public sector has increased fourfold in the past two decades, and strategies focus on prevention (42%), treatment (31%), health systems strengthening (14%), and program support (13%) [31]. A recent Global Fund report estimates that in 12 African countries (Democratic Republic of the Congo, Ethiopia, Ghana, Kenya, Mauritania, Mozambique, Nigeria, Rwanda, Senegal, Sierra Leone, United Republic of Tanzania, and Zambia), approximately 384 000 child lives were saved between 2000 and 2009 through scaling-up the use of insecticide-treated mosquito nets (ITNs) and intermittent preventive treatment for pregnant women (IPTp) [31].

In 2007, a call for malaria elimination by 2015 and long-term eradication was made during the Gates Malaria Forum [32, 33]. More recently, targets have moved to the year 2020 and beyond, as the subtlety and complexity of achieving elimination has become ever more apparent. Major emphasis is currently placed on a substantial research agenda targeted at the development of new tools [34]. As of 2007, recommended interventions to achieve this goal included long-lasting insecticidal nets (LLINs), indoor residual spraying (IRS), and IPTp; adoption of other vector control strategies, however, were awaiting what would be viewed as "sound scientific evidence" for their efficacy [35]. As a result of the above emphases, environmental management for malaria control has not been a priority, despite its long history of success in reducing malaria transmission [36] and its inclusion in the Chinese National Malaria Control Program from 1955 to the present time [37, 38]. Of particular interest is the use of intermittent irrigation strategies for control of rice-field malaria in China [37].

Current challenges for malaria elimination are immense and could be exacerbated by the consequences of climatic change (e.g., favorable conditions for vector and parasite development in areas previously inhospitable for malaria; increased occurrence of extreme climatic events, such as flooding and drought; and climate-induced population displacements), and by the construction of large-scale infrastructure and development projects – some in responses to climatic changes (e.g., dams, water transfer schemes, irrigation, and sea walls) [39]. All these factors involve significant environmental modification, commonly associated with increases in malaria transmission, which could be mitigated through environmental management strategies for vector control [40]. Despite our emphasis on malaria, there is a more wide-ranging strategy focused on integrated vector control across multiple diseases that share a common geography, e.g., malaria, dengue, and lymphatic filariasis [41]. Such strategies could be highly cost-effective and deserve serious consideration looking ahead.

In this paper we review a variety of multiple-intervention malaria control programs where vector control strategies played a decisive role. We present examples of integrated malaria control programs in which the choice of control strategies, as well as any needed fine-tuning and modification of the program components over time, were based on regular surveillance and performance evaluation. We selected five case studies for consideration: Federated Malay States, Northern Rhodesia, Panama Canal Zone, Borneo, and China. These examples have a particular focus on the relationships between local environmental modification and local hydrology.

We organize the discussion in five sections. We start (Section 18.2) with a brief summary of the cycle of malaria transmission, types of vectors and parasites, and common symptoms of the disease. We also discuss a systemic view of the disease based on a multidisciplinary framework of analysis. In Section 18.3 we review the types of control interventions currently available. In Section 18.4 we discuss the components and structural organization and management of past successful malaria control interventions, illustrated by five case studies. In Section 18.5 we discuss the need to revisit the successful interventions (as illustrated by the case studies) in order to set up dynamic decision systems for current control programs. Finally, we conclude with some remarks on lessons learned.

18.2 MALARIA: THE DISEASE

Human malaria is a disease of the blood caused by parasites of the genus *Plasmodium*. Except for the zoonotic form, *Plasmodium knowlesi*, transmission involves two hosts: humans and mosquitoes. Humans are the intermediate host where the asexual phase of the parasite takes place and a sexual phase in which gametocytes, which are infective to mosquitoes, are produced. Hybridization of parasites occurs in the mosquito. The life cycle of the disease has three basic stages. The first is called the human liver stage and starts when an infected mosquito feeds on a human. *Plasmodium* cells in the form of sporozoites enter the human bloodstream, reaching and infecting the liver, where they mature and multiply asexually, producing merozoites. The second, named human blood stage, starts when merozoites infect red blood cells, and additional asexual reproduction occurs until male and female gametocytes are formed. Following emergence from the liver stage, the first clinical manifestations of malaria can be detected, and the disease can be diagnosed by a blood test. The final stage starts when a mosquito bites an infected human, ingesting blood that contains gametocytes. Inside the mosquito's stomach, gametocytes are transformed into gametes. When fertilized, gametes generate zygotes, which develop into sporozoites. The cycle of the disease is perpetuated when this infected mosquito bites another human, introducing the sporozoites into the human bloodstream [30].

Five types of parasites are responsible for human malaria: *P. falciparum*, *P. vivax*, *P. malariae*, *P. knowlesi,* and *P. ovale*. The first, *P. falciparum*, is the most severe and lethal parasite, responsible for the majority of deaths attributed to malaria; *P. ovale* is the least prevalent parasite. Infections with the monkey parasite *P. knowlesi* (until recently commonly misdiagnosed as *P. malariae*) have been reported across Malaysian Borneo and Peninsular Malaysia, Vietnam, and Thailand [42–46]. Mixed infections with two or more parasites also occur with considerable frequency and augment the challenge of elimination [47–49].

Common symptoms of malaria include intermittent fevers and chills, usually accompanied by headache, myalgias, arthralgias, weakness, vomiting, nausea, abdominal pain, anorexia, and diarrhea. Other clinical manifestations include splenomegaly (enlarged spleen), anemia, thrombocytopenia, hypoglycemia, pulmonary or renal dysfunction, and neurological changes. *P. falciparum* cases, if not well treated, can progress to cerebral malaria, acute renal and liver failure, icterus, severe anemia and coagulation defects, pulmonary and cerebral edema, adult respiratory distress syndrome, coma, and death. Complications of *P. vivax* cases can result in splenomegaly, while those of *P. malariae* include nephritic syndrome [30, 42, 50].

Female mosquitoes of the genus *Anopheles* are the vectors responsible for the transmission of malaria. There are approximately 400 species of *Anopheles* mosquitoes, but only 60 or so are able to carry malaria, and roughly half of them are important for malaria transmission. The female *Anopheles* is able to release eggs every 2 to 4 days, and it takes between 7 and 20 days for the egg to turn into an adult mosquito, depending on temperature and humidity conditions [51, 52]. Each species may have different characteristics regarding biting pattern (unimodal versus bimodal biting hours, and indoors versus outdoors), feeding preferences (humans versus animal blood), flight range, and choice of breeding sites [5]. This coarse-grained picture of *Anopheles* mosquitoes is in the process of being dramatically refined using contemporary molecular technology. For example, the discovery that the single mosquito taxon called *Anopheles gambiae*, first described in 1902, was actually the name for an entire complex of closely related but morphologically indistinguishable sibling species led to the use of whole-genome reference assemblies to understand the evolution by introgression of this vector species [53].

One episode of malaria does not offer immunity to the individual, and a person can be infected as many times as he or she is bitten by infected mosquitoes. However, exposure to malaria over several years can provide acquired immunity, attenuating the most severe symptoms of the disease and resulting in asymptomatic infections [5, 43, 54, 55]. It has been shown that immunity to the most severe effects of malaria occurs after one or two prior infections [56]. Acquired immunity is lost when exposure to a malaria infection ceases (e.g., migration to a malaria-free area). Two genetic mutations confer some protection against a malaria infection. The first, red blood cell Duffy antigen negativity is widespread in Africa and renders protection against *P. vivax* without producing any negative consequences for the carrier [57, 58]. However, a few cases of *P. vivax* malaria infection among Duffy-negative individuals have been reported [59–62]. The second is called sickle cell hemoglobin and offers some protection against *P. falciparum*. Children may inherit a sickle cell gene from one parent and a normal gene from the other, in which case there is protection against the most severe consequences of *P. falciparum*, especially against cerebral malaria [55]. The negative consequence of this mutation, however, is that children who inherit the sickle cell gene from both parents develop sickle cell anemia and exhibit significant morbidity and mortality [63].

18.2.1 A Systemic View of Malaria

Understanding of the relationships in the triangle human–vector–parasite demands a multidisciplinary approach, which combines epidemiology with environmental sciences, demography, anthropology, entomology, economics, politics, geography, hydrology, and molecular biology [64]. Figure 18.1 presents a systemic view of malaria, where the triangle human–vector–parasite exists within (and interacts with) a local environment [65, 66]. The human vertex of the triangle includes variables that characterize individuals (e.g., their susceptibility to infection, their knowledge and behavior, and ways through which they can transform the local environment) and the context in which individuals interact (e.g., political and social). The vector vertex includes characteristics that are crucial for choosing the ideal package of interventions to launch an integrated vector management program [67, 68] and for measuring local exposure to malaria infection. The parasite vertex refers to issues that determine the effectiveness of drug-based intervention and that indicate the levels of disease severity. Finally, the local environment includes both the natural environment and the built habitat. The three vertices are not static, but rather interact among each other and with the local environment, producing unique local profiles of malaria risk.

18.3 MALARIA CONTROL

The discoveries of Charles-Louis-Alphonse Laveran (1845–1922) in 1880 (who saw the malaria parasite under a microscope), of Ronald Ross (1857–1932) in 1898 (who discovered that malaria was transmitted by female *Anopheles* mosquitoes), of Carlos Chagas (1879–1934) in 1905 (who proposed the concept of malaria as a household infection), of Malcolm Watson (1873–1955) and N.H. Swellengrebel (1885–1970) in

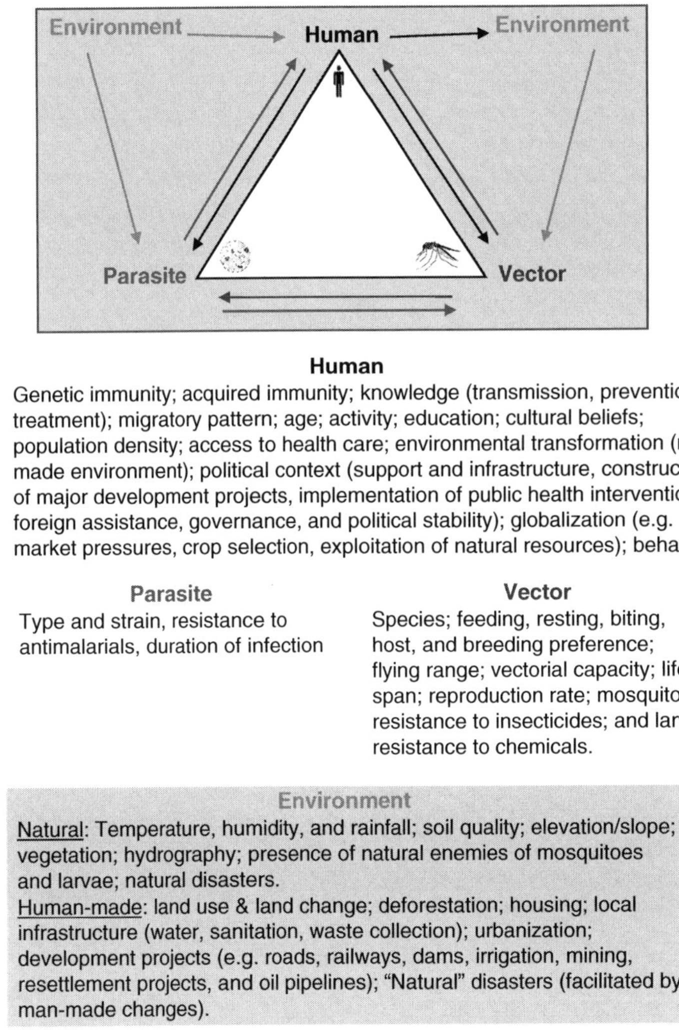

FIGURE 18.1 Systemic view of malaria.

the 1910s (who found that only some species of *Anopheles* were able to transmit malaria, coining the term "species sanitation"), and of Paul H. Müller (1899–1965) in 1948 (who discovered the insecticidal properties of dichlorodiphenyltrichlorine (DDT)) brought about significant changes in the strategies utilized for malaria control [3, 28, 69]. Moreover, the experiences from past control efforts and the knowledge accumulated over the years indicate that no single package of interventions can be adopted in all malaria endemic areas. Instead, efforts need to address local idiosyncrasies in malaria transmission resulting from the dynamic interaction between humans, vector, parasite, and the environment (Figure 18.1) [70].

18.3.1 Types of Control Interventions

Traditional measures for malaria control, which focus on the mechanisms of disease transmission and on the local environment, can be grouped according to their main objectives [42], as shown in Figure 18.2.

Regarding antimalarials, until the 1930s quinine was the only available effective drug against malaria [71]. Since then, a handful of drugs have been produced, but widespread drug resistance developed at a faster pace than new drug development [71]. At present, the World Health Organization (WHO) recommends the use of artemisinin-based combination therapy (ACT)[3] as the first-line drug treatment [72]. However, artemisinin-resistant *P. falciparum* malaria has been reported in the Cambodia–Thailand border [73], which raises much concern since there is no new drug available to replace artemisinin.

[3] Four ACT combinations are recommended: artemether plus lumefantrine (AL), artesunate plus amodiaquine (AS+AQ), artesunate plus mefloquine (AS+MQ), and artesunate plus sulfadoxine-pyrimethamine (AS+SP). The choice depends on the level of drug resistance of the partner medicine in the combination. In addition, dihydroartemisinin plus piperaquine (DHA+PPQ) has been recommended for the treatment of uncomplicated *P. falciparum* malaria [72].

Objective of the intervention	Type of the intervention
1. Reduce vector-human contact	Site selection, improvement of house conditions (e.g., closing eaves), house screening, mosquito nets, indoor residual spraying (IRS), repellents, and protective clothing
2. Reduce breeding habitats	Environmental management: (i) environmental modifications - irrigation systems, filling, sanitation, intermittent drying of rice fields and drainage (ii) environmental manipulations - water-level fluctuations, regulation and improvement of natural watercourses, intermittent irrigation, sluicing, salinity, vegetation clearing, and aforestation/deforestation
3. Reduce *Anopheles* larvae	Chemical (e.g., oil) and biological larviciding (e.g., *Bacillus thuringiensis* and *B. sphericus*), predators
4. Reduce adult *Anopheles*	IRS, insecticide treated nets (ITNs), long lasting insecticide treated nets (LLITNs), genetic control*
5. Eliminate parasite in humans	Antimalarial drugs (for both prophylactic and therapeutic purposes), intermittent preventive treatment for malaria in infants (IPT_i), intermittent preventive treatment of malaria during pregnancy (IPTp), vaccine*

FIGURE 18.2 Types of control interventions. * Indicates that the intervention in under research, and therefore not yet available.

Control efforts may also address other issues highlighted in Figure 18.1, such as individual behavior and knowledge about malaria transmission and prevention. Community-based approaches and sensitization campaigns are ideal strategies to address these issues, but their incorporation in a malaria control program is still limited [74]. A notable exception is the Chinese national malaria control program from 1955 to the present [75].

18.4 MALARIA AND HYDROLOGY

Next, we present five case studies that illustrate successful integrated malaria control efforts, constrained by local ecological and cultural characteristics. Here success means that malaria was substantially suppressed relative to what it was in the absence of control measures. Only one of these examples, the Chinese national program, resemble current National Malaria Control Programs implemented in malaria-endemic countries. Four of the programs are essentially corporate initiatives where the primary objective is the maintenance of a healthy workforce. Nevertheless, lessons from these experiences have much to contribute to the ongoing discourse about malaria elimination/eradication.

18.4.1 Case Study 1 – Malaria Control in the Federated Malay States (FMS)

The control program began in 1901 with initiation of planning and implementation in the town of Klang and the nearby seaport, Port Swettenham [26]. In qualitative terms, the baseline conditions were: (i) a large percentage – by far the majority – of hospital patients at Klang had malaria; (ii) the town of Klang and the people along the entire coastline were suffering from a major outbreak of malaria; (iii) there was a proposal to eliminate the town of Jugra – 20 miles south of Klang – because of high malaria rates; and (iv) only a small fraction of the sick ever came to a hospital. It was anticipated that if curative treatment was the response to the malaria problem, the hospital system, however extended, could not accommodate the demand. Furthermore, focusing on treatment would provide no benefit to the overall community. Every well, ditch, and swamp teemed with *Anopheles* larvae.

18.4.1.1 Control at Klang Town Based on the above assessment, Malcolm Watson, who had been appointed Surgeon General and who was responsible for mitigation of the malaria problem, made the decision that prevention should be attempted. Malaria was to be studied outside the hospital ward. He assembled baseline data having two components: (i) a survey conducted to ascertain *where* the documented malaria cases came from and the case rates (measured by spleen rates) at selected communities; and (ii) a survey of breeding sites followed by preparation of a map indicating their location.

With this information at hand, an initial intervention plan for malaria control had to be delineated. A combination of preventive and curative tools was available, including the extant hospital system for treatment. The options that, in principle, could be utilized were: quinine (for treatment and prophylaxis, whether in hospital or outside), bednets, house screening, spreading of petroleum on swamps, continuous sweeping out and elimination of individual collections of water, and provision of permanent drains and filling of water bodies. There were also a priori constraints on implementation: (i) it was impossible to force use of quinine dispensed to Chinese workers and it was difficult to get consistent use of it from Indian workers; (ii) it was impossible to enforce use of bednets, however persuasive the advertising about their benefits, as night visitation for surveillance would be required; (iii) compulsory screening of all houses was impossible because of financial limitations; (iv) use of petroleum on swamps was not practical due to cost and temporary efficacy; (v) the vast area at Klang prevented sweeping out or dealing with individual collections of water on any continuous, ongoing basis; and (vi) it would have been necessary to constantly supervise coolies on measures that would suppress mosquitoes only as long as they were applied.

Thus, measures such as quinine distribution, house screening, and bednets could be anticipated to be used sporadically and with low population coverage. The limited experience prior to 1901 supported this contention. This led to a proposed trial policy and intervention; namely, any expenditure should be on works of a permanent nature (e.g., draining and filling of standing water bodies). The decision was made and implemented: Klang town should be drained.

Surveillance of malaria cases (spleen rates) and mortality was carried out from the time when the drain construction was initiated. Within three years malaria ceased to be a problem of any practical consequence at Klang. In particular, it no longer inhibited work associated with production and shipping of rubber to Port Swettenham and out of the FMS to international markets. Ongoing maintenance of the drains was the only follow-up at Klang. There were no a priori targets of spleen rates or mortality rates that were designated as representing "success" of the program. There was also no control, or comparison, community designated as part of the evaluation of the drainage and filling intervention at Klang. Comparisons with pre-intervention levels – i.e., historical controls – were the basis for evaluation. Achieved spleen rates below 5% were pronounced to be acceptable – a decline from anywhere between 30 and 80+% depending on the specific locality. Total mortality counts dropped from above 250 per year to 50 per year by 1903, with these levels sustained and/or improved upon through 1921. The cost of installation of the drainage system and filling of swamps, covering 300 acres, was US$30 000 in 1901 dollars.

18.4.1.2 Control at Port Swettenham The opening of the port on 15 September 1901 was followed by an abrupt explosion of malaria transmission, which prevented shipping by both boat and railway. A proposal was made to close the port by the end of November 1901 until it could be made safe. Although the port was not formally closed, a package of antimalarial interventions was immediately put in place. These were drainage, oiling of all pools of water, and universal offering of quinine. By 1904, malaria ceased to be a problem at the port. This continued to be the case until 1906 when the first signs of recrudescence appeared. Monitoring of drains revealed that some malaria cases coincided with the placing of cement inverts into the road drains at a level that blocked the flow of water into all lateral drains that had previously been open to them. This led to standing water that was the site of new *Anopheles* larval development. The situation was made worse by the extension of a railroad embankment. With the rapid flare-up of malaria cases, the government authorized new expenditures to repair and enhance the drainage system. This was immediately followed by a sharp reduction in the malaria case rate.

The ongoing surveillance and adaptive tuning of interventions exemplified by the Port Swettenham drain modifications was a central feature of the FMS program as it continued to develop in urban areas and rural rubber estates. Of particular interest was the fact that not only the malaria death rate but also the all-cause death rate dropped precipitously at Port Swettenham following the initial introduction of a drainage system. The high all-cause mortality rate prior to the interventions was a consequence of many malaria asymptomatic cases becoming vulnerable to other diseases (e.g., dysentery, diarrhea). Eliminating malaria resulted in reduced case rates and mortality from the other causes. The cost of drainage and filling of swamps, covering 100 acres, was US$50 000 in 1901 dollars.

18.4.1.3 Control at Flat Land Coastal Rubber Estates The baseline information consisted of knowledge that *An. umbrosis* was the primary malaria vector and that it bred in stagnant jungle pools. Spleen rate and mortality rate were high and the constraints on enforcing bednet usage and proper use of quinine were the same as listed above for Klang. The following interventions were initiated: (i) drainage work, consisting of 37 miles of drains, 20 ft wide, which allowed rubber planters to drain 24 000 acres; (ii) drainage of stagnant jungle pools at estate sites, and cultivation was designed to eliminate pools locally; and (iii) surveillance of children via blood samples and of the general population via spleen rates. Death rates were also monitored. Annual surveillance of malaria cases was designed to cover 500 square miles. Regular monitoring of breeding sites at estates was initiated.

Initial findings indicated that there was no malaria among individuals with houses and work sites located at more than 0.5 miles from breeding sites and that death rates were correlated with spleen rates. Recommended actions based on one to two years of monitoring and interventions were: (i) to remove houses to distances exceeding 0.5 miles from jungle pools or (ii) to fell jungle and drain pools for at least 0.5 miles from extant houses and work areas. Implementation of these two options at flat land estates resulted in a sharp decrease in spleen and mortality rates. Subsequently, new flat land estates were opened and designed based on these two recommendations, and malaria was not a problem. However, at one of the new estates, the surveillance revealed a sudden fresh outbreak of malaria cases. Close checking of these cases revealed that the coolies had been working near some new lines on the estate that were close to a point where the estate boundaries bent inward toward the rubber plants. Jungle clearance and pool drainage had not taken place outside the bent-in boundary, thereby putting the coolies less than 0.5 miles from *An. umbrosis* breeding sites. Jungle clearance and shifting of the boundaries for coolie lines solved the problem. This particular step in the program of flat land estate malaria control represents an instance of adaptive tuning of interventions over time.

18.4.1.4 Control at Hilly Estates In 1909, 99% of hillside rubber estates had high malaria rates. There had been no improvement since the estates opened several years earlier. The only water was from crystal springs and brooks. There was a very high hospital admission rate and high mortality rates. Further, quinine did not stop or significantly reduce transmission, although it did serve to reduce hospital admission rates and the mortality rate. The principal vector was *An. maculatus*. Its larvae can survive even strong currents and streams filled with grass, as long as the water is fresh. Indeed, a preliminary experiment at the Seafield estate showed that thoroughness of quinine dispensing and weeding of the edges of streams had not been effective at sharply reducing malaria rates.

With this background information and the experience from low land estates at hand, a new intervention experiment was implemented at the Seafield estate. The coolie population was placed – housing and working – in selected "sanitary areas" and all water less than 0.375 miles from them was to be carried off by underground pipes. The idea was to create a mosquito-free circle on hillsides, with drainage up to 0.375 miles from the coolie line. The smaller drainage zone than that used at the lowland estates (0.375 miles rather than 0.5 miles) was motivated by the fact that *An. maculatus* was a much more delicate vector than *An. umbrosis* and thus would be likely to have a shorter dispersal range. This intervention led to a short-term (within two years) reduction in malaria cases, but the reduction was decidedly inadequate for maintaining a healthy workforce. The drainage was then extended to 0.5 miles, but it needed larger pipes to carry away rainy season water. The result, although costly, was a major reduction in malaria cases.

The above experimental experience led to the following general specifications for malaria control at hilly estates: (i) rebuild coolie lines on flat land at 0.5 miles from the hills; (ii) if a low-lying swamp is present, open it and move coolie lines proximal to it (recall that *An. maculatus* requires very clean water and swamps are not hospitable to their larvae); and (iii) if there are hills behind and flatland in front of coolie lines, pipe drain hills where *An. maculatus* is breeding. Following this protocol, there was a decline in the sickness rate for workers from 13.5% in November 1911 to 2.2% in December 1912. The spleen rate declined in the same period from 91 to 14%.

18.4.2 Case Study 2 – Malaria Control at Copper Mines in Northern Rhodesia, 1928–1950

Economic exploitation of copper in the former Northern Rhodesia (now Zambia) began in 1909, following the completion of a railway into the copper belt. Until the 1920s, Northern Rhodesia's copper production (entirely from outcrop copper deposits) was insignificant compared to the neighboring Belgian Congo. However, starting in 1923 there was a sharp increase in the world demand for copper and an accompanying rise in copper prices on the world market. Four copper mining companies were started with the initiation of major production during the years 1929–1936.

A primary prior obstacle to copper production was difficulty in recruiting an adequate labor force, as a result of intense malaria transmission at the mining sites [27]. Health records from the medical department of the Roan Antelope mine, the first of the four mines in the area, indicate that when Europeans arrived there were 105 malaria attacks per 1000 people in a single month. As with most data from health clinics, this was an underestimate of the extent of the malaria problem. Indeed, parasite rates and spleen rates among children in villages proximal to the Roan Antelope mine ranged between 36 and 45% [76]. *An. gambiae* and *An. funestus* were the vectors responsible for transmission at the mining communities.

Members of the mining boards quickly realized that sustaining a healthy labor force and recruiting new workers would require effective disease control measures. An initial malaria control effort consisting of house screening, use of mosquito nets, and administration of quinine for prophylaxis and treatment was not sufficiently effective in reducing illness rates and the malaria incidence rates remained high. There is clear documentation that this unsuccessful attempt at malaria control prior to 1929 resulted in migrant workers abandoning the Roan Antelope site. The initial interventions were supplemented by an environmental management program, initiated in 1929, and consisting of vegetation clearance, swamp drainage, river boundary modification (to enhance flowing water), and regular application of oil to open water bodies. This package of interventions at the Roan Antelope mine was adaptively tuned and maintained, with the addition of DDT in 1946, for two decades. Comparable packages of interventions were implemented at Mufulira, Nkana-Kitwe, and Nchanga mines between 1930 and 1936, and maintained until 1950. The baseline annual malaria incidence rate at the Roan Antelope mine was 514 per 1000. During the first three years of the full intervention program, the annual incidence rate was reduced to 135 per 1000 and maintained below this level until 1946, when DDT was introduced. This led to a further reduction in annual incidence to 21 per 1000. The net result of the malaria control initiative was a dramatic increase in population at the mining sites from an estimated 10 964 persons in 1930 to 140 368 by 1949. The workforce at the mining sites grew by a factor of 5 during this period. To ensure steady revenue flows, the management boards of the Northern Rhodesian mines closely monitored the work shifts lost due to malaria and regularly tuned the package of control measures to minimize this outcome measure. Ultimately, the copper production also increased dramatically, and Northern Rhodesia, which contributed about 0.3% of annual worldwide copper extraction between 1925 and 1930, became the third largest producer of copper ore in the world by 1940, with a global share just exceeding 11% [77].

It is important to observe that the FMS and Northern Rhodesian malaria control programs were carried out under corporate sponsorship. Their objectives were to ensure a healthy workforce. They did virtually nothing to improve public health on a broader scale than communities proximal to the work sites. The large up-front expenses needed to install drainage systems and the use of packages of interventions for the full programs were not at issue due to the large financial returns. Discourse about cost-effectiveness of single interventions such as bednets, insecticidal spraying, and drugs that pervades the contemporary tropical public health arena was absent from the colonial corporate environment of the FMS and Northern Rhodesian programs. In addition, stable management of the programs facilitated their successful operation for 20+ years. There is no counterpart to this in the public health arena, with the exception of the Chinese national malaria control program that we discuss in Section 18.4.5.

18.4.3 Case Study 3 – Malaria Control in the Panama Canal Zone

The control of malaria was a necessity for the construction of the Panama Canal. The initial groundwork for malaria control in Panama was an important effort directed by Surgeon Major William Gorgas (1854–1920), initially focused on yellow fever, during the American occupation of Havana, Cuba. Here the US Army imposed a strategy of screening houses and extensive drainage to reduce, if not eliminate, breeding sites of mosquitoes. Yellow fever was eradicated from Havana, and as a by-product malaria transmission was also greatly reduced.

The Isthmus of Panama was (in the early 1900s), and still is, an ideal environment for many species of mosquitoes. High temperatures persist throughout the year. The rainy season lasts for nine months and the interior of the Isthmus is tropical jungle. Antimalarial work was primarily carried out in rural areas along 47 miles of railroad track between Panama and Colon. Approximately 80 000 people lived within one-half mile of the railroad, occupying 30 villages and camps. It was estimated that roughly 17% of the population in Colon experienced at least one malarial attack each week.

Starting in 1904, a multiple-intervention malaria control program was initiated. It consisted of seven interventions: (i) drainage; (ii) brush and grass cutting; (iii) oiling; (iv) larviciding; (v) quinine distribution; (vi) house screening; and (vii) killing adult mosquitoes. Each of these interventions was monitored, evaluated analogous to the performance-based ratings used in the FMS and Northern Rhodesian programs, and adaptively modified/tuned to facilitate reduction in the *Anopheles* mosquitoes and, correlatively, reduction in malaria incidence.

The drainage component consisted of elimination of all pools of water within 200 yards of village boundaries and within 100 yards of individual houses. Subsoil drainage was preferred, followed by concrete ditches and open ditches. Paid inspectors were responsible for maintaining drains free of obstructions. All bush and grass was kept at a height of less than 1 ft over the same distances as used for pool drainage. When drainage was impossible, oil was added along edges of ponds and swamps to kill mosquito larvae. Since there were no commercially available insecticides in the early 1900s, a mixture of carbolic acid, resin, and caustic soda was used as a larvicide. Finally, the killing of adult mosquitoes was carried out by feeding collectors who gathered adult mosquitoes that remained in houses and tents during the day. This was an era long before DDT and other modern adulticides.

Because of the control program, the death rate due to malaria for canal construction employees dropped from 11.6 per 1000 in November 1906 to 1.23 per 1000 in December 1909. Malaria-specific deaths in the general population proximal to the railroad fell from 16.2 per 1000 in July 1906 to 2.6 per 1000 in December 1909. The percentage of employees hospitalized for malaria dropped steadily from 9.6% in December 1905 to 1.6% in 1909. Although malaria was never eliminated from the Canal Zone, it was controlled at a sufficiently low level that construction could be completed.

This brief summary of the Panama Canal project does not begin to do justice to the complex day-to-day decision making, performance-based rating evaluations, and intervention adjustments that had to be made throughout the program. These details, from various perspectives, have been extensively documented by the program managers [77–81]. However, a particularly useful comparative analysis was assembled by Watson [14], where he contrasts his own FMS program with the Panama Canal efforts, and provides unparalleled insight – valuable to the present day – about the nuances of adaptively working toward an overall effective malaria control program.

18.4.4 Case Study 4 – Malaria Control in Borneo

Beginning with his arrival in Borneo in 1938, John McArthur (1901–1996) initiated surveys to determine the distribution of malaria in areas proximal to where transmission was already known to occur. His overall goal was to establish a program of reduction, and even eradication if possible, of malaria from the area [82]. *An. maculatus*, a well-known malaria vector in other parts of Asia, was present in abundance in villages on open sunlit plains, but there was no human malaria at these localities. Surprisingly, villages on hillsides adjacent to dense jungle, where there was no *An. maculatus*, were found to be highly malarious, with spleen rates above 80%.

Extensive experience with *An. maculatus* in Malaya [26], where transmission was promoted whenever jungle was disturbed, was automatically translated into expensive efforts with oiling, drainage, and other antimosquito measures in Borneo by the local malaria service. These efforts, however, were totally ineffective in Borneo. Here, following a remarkable piece of detective work by McArthur [83], *An. leucosphyrus* was incriminated as the primary malaria vector in Borneo. *An. leucosphyrus* is a purely jungle breeding mosquito, operating only in seepages under jungle shade and very difficult to detect. With this knowledge in hand, it was easily controlled by localized jungle clearance around the breeding places.

An important experiment was conducted on the Tambunan plain in North Borneo from 1949 to 1952 to determine whether the admission of sunlight to seepages, followed where possible by land cultivation, was able to eradicate *An. leucosphyrus* and malaria cases from jungle villages. An area of approximately 10 miles of jungle-covered hills, where spleen rates had been maintained at above 80% for at least 14 years, was selected as the experimental site. Jungle was cleared from approximately 200 seepages along 32 miles of waterway. This was followed by a 95% reduction in vector breeding that was maintained for three years with only minimal maintenance. Spleen rates declined during the first year from 86 to 56%. The decline continued to 53% and then 45% by the third year. There was some transmission remaining, as demonstrated by parasites in the blood of infants. This was attributed to infected mosquitoes from surrounding uncontrolled villages with very high malaria rates. Through the first two years of the experimental period, the spleen rates in a control area remained unchanged at nearly 100%. In the third year, the village chief in the control sites decided to imitate the program from experimental villages. This led to a precipitous drop in spleen rates, as had occurred in the first year at the experimental sites. While the experimental protocol (originally intended to run several years) was destroyed by this action, the drop in spleen rates actually added further evidence to support the control strategy for *An. leucosphyrus*, as well as to demonstrate that community-based jungle clearance was feasible.

In a further follow-up, McArthur suggested the following as a continuation to the Tambunan study: "The clearing in the experimental area lasted much longer than was at first expected, and such clearing can be maintained permanently if the people cultivate, or graze their cattle in, the moist fertile areas thus exposed. Newer methods of clearing by spraying with selective herbicides – one spraying of which is claimed to be effective for five years – may prove an economical means of malaria control which is worthy of further study" [82, p. 656].

Following his 1950s journal articles [82–85], the publication of a book [86], and a presentation to the World Health Organization [87], McArthur's work – and its potential – regrettably seems to have vanished from the contemporary international discourse on malaria control. This is unfortunate in light of the recent documentation of transmission of *P. knowlesi* from monkeys to humans via *An. leucosphyrus* in Borneo [44]. *P. knowlesi* could easily have been misclassified as *P. malariae* by McArthur, using the microscope diagnostic technology of the 1940s and 1950s. However, contemporary polymerase chain reaction (PCR) analyses [44] document *P. knowlesi* infections in humans in Borneo, thereby clarifying that five species of plasmodia, rather than the long-established four species, infect humans and that one of them is a zoonosis with *An. leucosphyrus* as the primary transmitter.

18.4.5 Case Study 5 – Malaria Control in China

China has had the longest running successful public health initiative focused on malaria of any country in the world. Starting gradually in 1950, systematically organized with a National Malaria Control Program in 1955 [38, 75], and continuing to the present day, this adaptive, multiple-intervention, locally tuned effort at malaria suppression warrants in-depth examination and much more attention in contemporary discourse than it is receiving. Using 1949 as a starting point for baseline statistics, there were more than 30 million malaria cases in the country, and the mortality rate was approximately 1% per annum. Malaria was epidemic in 1829

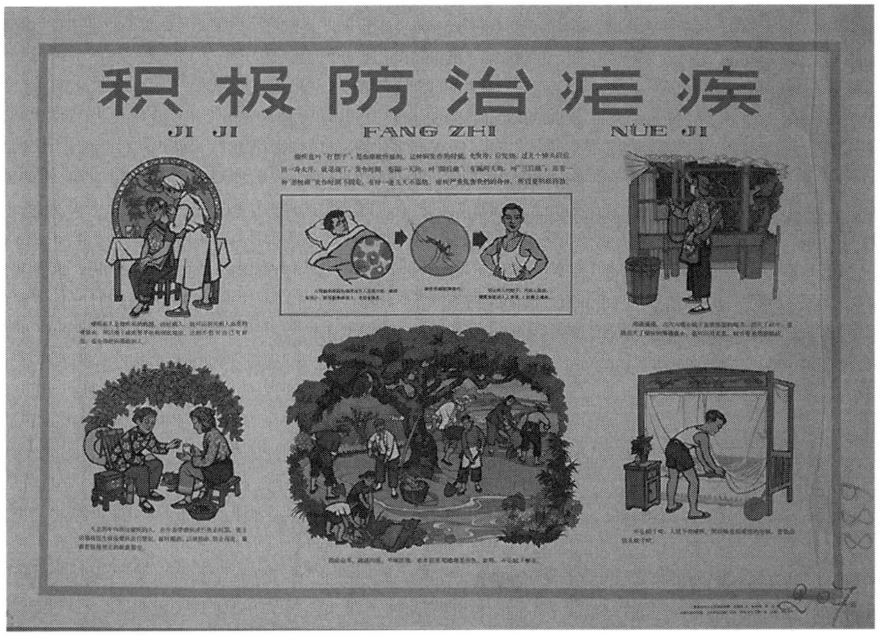

FIGURE 18.3 Poster from the Chinese National Malaria Control Program. *Source*: Available at the US National Library of Medicine, National Institutes of Health website: http://www.nlm.nih.gov/exhibition/chineseantimalaria/gallery.html. Accessed March 2010. **Note**. The poster emphasizes malaria prevention and timely treatment. The title is *Actively Prevent and Treat Malaria*. Text was written by Wang Liancheng and the artwork was performed by Zhao Shuqi and Mo Gong. The images have the aesthetic features of Chinese New Year Pictures and paper cutting, two popular media of Chinese folk art. The poster was published by Science Popularization Press and printed by People's Education Printing Factory, May 1963.

counties/cities, accounting for 70–80% of all counties in the country. Further, malaria represented 61.8% of the total recorded cases of acute infectious diseases in China in 1949.

When the national control program was initiated in 1955, it relied on primary healthcare networks as an organizational base and made extensive use of community participation to respond to local needs. An intensive educational program was put in place that featured advertising of integrated sets of interventions, giving balanced emphasis to both prevention and curative medicine. This balance has persisted to the present day. Indeed, successful suppression of malaria in the diverse Chinese eco-systems owes much to this holistic philosophy. Figure 18.3 shows one of the many posters that were used for education at the local level as part of the broad-based Chinese antimalarial campaign. It emphasizes community work toward the elimination of breeding habitats and the use of preventive measures.

By the year 2000, there were 1.202 billion people in 2787 counties/cities living in areas where malaria incidence was less than 0.1 per 1000 (including original malaria-free areas as well as endemic areas now free from the disease). There were 30.62 million people in 72 counties/cities with an incidence of 0.1–1.0 per 1000, and 3.94 million in 17 counties/cities with an incidence of 1.1–10 per 1000. There was no county or city in the entire country with an incidence above 10 per 1000. Figure 18.4 shows the national level trends in incidence and mortality rates from 1955 to 1998 [38]. This coarse level of description masks considerable variation from community to community. It also masks the many examples of recrudescence of malaria that were detected in diverse communities, but followed by adjustment and/or intensification of interventions leading to reduced transmission and malaria case rates [75, 88–90].

The guiding framework for malaria control in China is the adaptive tuning of multiple interventions guided by performance-based ratings carried out over time. However, an additional form of monitoring needs to be included here, namely, assessment of drug resistance. Not surprisingly, this phenomenon was, and continues to be, a challenging feature of antimalarial drug distribution in China. However, a research program focused on drug resistance and the development of new drugs was initiated in response to this problem [75, 90] and plays an important role in the current version of the national program.

18.5 DISCUSSION

Critiques of vector control strategies suggesting inadequate "scientific evidence" for their performance indicate the need to reformulate Watson's intervention strategies in contemporary statistical terms. This exercise has the potential to contribute to the ongoing efforts of national malaria control programs implemented in malaria-endemic countries. Here,

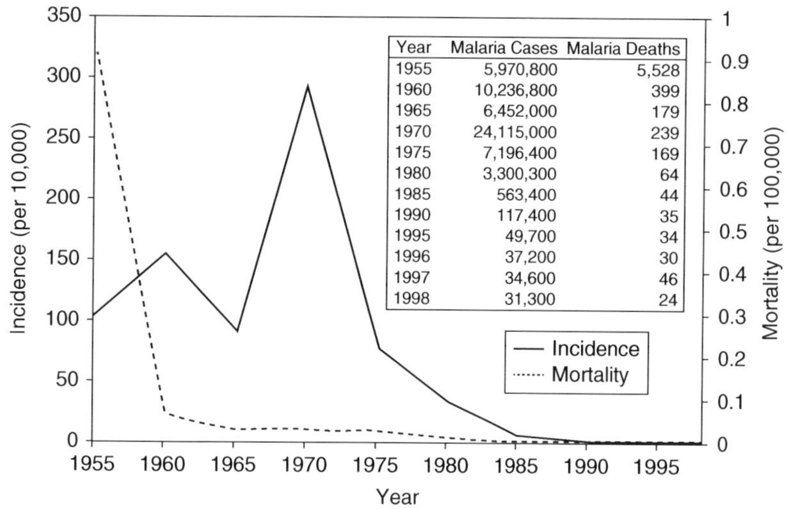

FIGURE 18.4 Trends in Malaria incidence and mortality in China, 1955–1998. *Source*: Tang et al. [38].

we offer a few suggestions and an outline, without attempting to fully develop the model.

We begin by focusing on a single outcome measure (e.g. spleen rate or mortality rate) and view it as functionally dependent on the intensity of application of a set of K interventions (e.g., drugs, larvacides, drainage, bednets, etc.), ecological conditions, and calendar time. Thus, if $p(t)$ is the outcome measure, we can write it formally as

$$p(t) = f(x_1, x_2, \ldots, x_K; \mathbf{y}, t), \tag{18.1}$$

where x_i is a measure of the intensity of application of the ith intervention (e.g., percentage of the population using bednets, the percentage of the potential larval habitats with no *Anopheles* larvae, etc.), and $\mathbf{y}$ is a vector of ecological conditions. The independent time variation can represent seasonal effects such as rainfall. The x_i values and $\mathbf{y}$ can also vary with time. Thus, Equation (18.1) represents a response surface representing the consequences of application of any one, or a combination, of K interventions. The objective of a control program is to reduce and maintain $p(t)$ at a low level for all times t beyond some initial start-up period. If the target is eradication, then obviously we want $p(t)=0$ for all t after some prescribed intervention period.

Operating with a fixed set of K interventions, we assume that $\partial f/\partial x_i < 0$ for all i. This simply means that we anticipate that each of the interventions will reduce p, even if only by a very small amount. Changes in one or more x-variables can influence ecological conditions (one or more components of $\mathbf{y}$) – and it is not necessarily the case that changes in components of $\mathbf{y}$ lead to a decrease in p. Indeed, a striking example of initiation of interventions leading to deleterious ecological conditions that increase an outcome, p, is the sharp rise in malaria mortality at Klang town in the Federated Malay States between 1901 and 1902, consequential to the initiation of drain construction. This activity simultaneously created new *An. umbrosis* breeding sites and brought in a fresh cadre of exposed workers.

Within the context of malaria control, we do not necessarily know all of the x variables that should be included in (18.1). For example, if vaccines or the utilization of transgenic mosquitoes were to become viable forms of intervention, they could be included in a future specification of (18.1). In addition, we may not know the region $R(x_1, \ldots, x_K)$ that leads to the lowest level of p consistent with possible constraints on application of some of the interventions. Third, we do not know the functional form of f, and observations on the x variables and p are subject to moderate, and sometimes large, errors. This is precisely the set of conditions that are the setting for Evolutionary Operation (EVOP) put forth in the chemical engineering and statistics literature in the 1960s [91].

In terms of Watson's FMS program, we begin with baseline values of p, $\mathbf{y}$, and some of the x variables. Then one or more additional interventions were applied and we can examine the extent to which p is reduced. Further, associated with each of the interventions is a separate measure for rating its performance on each round of application. These performance-based ratings serve as a guide for which interventions to intensify at the next round of testing. An example of a table of performance-based ratings tuned to malaria control is shown in Table 18.1 [92]. The values in the table are qualitatively consistent with judgments in Watson's decision-making. However, the central point is that such a rating system was utilized in the FMS program, and with greater detail in Watson's malaria control programs at copper mines in Northern Rhodesia [27], as discussed in Section 18.4.4. Chemical sprays, intervention type 8 in Table 18.1, were not utilized in the FMS programs. However, they did appear with DDT in 1946, and this was a late-stage part of the Rhodesian copper mine programs.

TABLE 18.1 Performance-based ratings of malaria control programs according to potential problems associated with different types of interventions.

Possible problem with intervention	Type of intervention	Rating Index[a] categories for program outcome			Factors associated with poor performance
		Good	Fair	Poor	
Non-compliance	1. Antimalarial drugs	a>80	20–80	<20	Inadequate community education; inadequate training of medical staff
Non-compliance	2. Bednets	b>60	30–60	<30	Inadequate community education
Non-compliance	3. House screening	b>85	30–85	<30	Inadequate education and training
Inadequate coverage	4. Education	c>85	50–85	<50	Poor contact strategy
Inadequate coverage	5. Drainage	d<10	10–70	>70	Improper drainage; lack of trained field staff
Inadequate coverage	6. Vegetation clearance	d<10	10–70	>70	Previously unsuspected breeding sites; incomplete cutting
Inadequate coverage	7. Larvicides	d<10	10–70	>70	Low oil supply; insufficient effort by staff
Non-compliance	8. Chemical sprays	b>70	30–70	<30	Inadequate education; poor relationship between program staff and household heads

Source: Adapted from Singer [92].

[a] The Rating Index is defined as:

- a = percent of compliance among those self-selecting for treatment.
- b = percent of households in compliance with intervention.
- c = percent of those with symptoms volunteering for treatment.
- d = maximum value of the ratio (mean mosquito catch in current year/mean mosquito catch in baseline year*100) observed over surveillance sites.

In EVOP, as formulated by Box and Draper [91], decisions about new interventions to put in place are based on factorial experiments that involve two or more interventions. This allows the investigator to determine more points on the response surface represented in (18.1) than is possible when experimenting with one or a combination of interventions applied in a single run, as in the malaria control context. In the malaria context there is also no randomization of experimental protocols across communities, as identification of ecologically matched control communities is a virtual impossibility. Interestingly, this is also an issue in the chemical process control context: "If runs can be randomized without too much difficulty, then this should be done. Our experience has been, however, that a random sequence is difficult to organize under actual production conditions and for the 2^2 design with added center points" [91, p. 75].

Decisions about new intervention protocols in the malaria context are also based on regular surveillance with updating of the performance-based ratings chart. The analog of this is the information board in the EVOP set-up [91, p. 13]. Beyond the routine assessments in performance-based ratings – and information boards – there is also an inevitable component of subjective judgment in deciding on new runs of interventions. Watson's explication of his own judgments in the FMS program and in the copper mine programs [14, 26, 27] provides considerable insight about the limits of automating decision processes in malaria control. This has an exact counterpart in chemical process control, as described in Box and Draper [91]. Yet, Watson's programs ran successfully for periods of 20 years, prior to its formalization by Box and Draper [91].

An interesting open question in the malaria context concerns the extent to which you can have an "Expert Computer System" that could be widely employed for ongoing decision-making in the adaptive tuning and ongoing operation of a multiple intervention control program. An integral part of such a system would inevitably be a suite of malaria transmission models, which currently exist in great abundance. A disconcerting aspect of the vast malaria modeling literature – more than a century old – is that there is to-date no example of a transmission model that has been employed to guide a control program in real time. The analogue of real-time modeling for the West African Onchocerciasis Program (OCP) has yet to make an appearance in the context of malaria control [93]. The point of raising these issues is that if effective malaria control programs are to be dependent on a manager with Malcolm Watson's remarkable acuity and insight, then this is a very tall order for the tropical disease control community. Answers here lie in the future and will require a great deal of local experimentation.

18.6 CONCLUSION

An understanding of the mechanism of malaria transmission has been at hand since the late nineteenth century. Combining this information with a large literature on *Anopheles* vector ecology and bionomics, multiple-intervention malaria control programs were successfully implemented more than a century ago. There has also been dramatic progress in our understanding of the biology of vectors and parasites, much of it deriving from molecular-level studies, as well as considerable advances in pharmacological and larvicide/adulticide interventions. These more recent scientific advances have been accompanied by a decline in use and emphasis on vector control strategies that played such a central role during the first 60 years of the twentieth century.

With an eye toward restoring a balance between preventive and curative strategies as components of malaria control programs, each finely tuned to local ecological conditions, we have presented five case examples of successful integrative programs from the past. One example, the Chinese National Malaria Control Program is a public-sector initiative that continues to operate to the present day. Three of the examples derive from the British colonial experience in Asia and Africa, and one of them – malaria control to facilitate construction of the Panama Canal – is an unusual exception in terms of money invested and intensity of effort expended. Nevertheless, all of these programs had adaptive tuning of a multiplicity of interventions, together with ongoing performance-based ratings of the component tools, and measurement of outcomes as a basic mode of operation that led to the successful suppression of malaria.

As a preliminary response to the contemporary criticism of vector control strategies as lacking a base of "scientific evidence," we have outlined a framework (building on EVOP concepts) for providing rigorous evaluation of the multiple-intervention programs of the past, as well as contemporary initiatives. A full development of EVOP tuned to malaria control would go beyond the scope of the present paper. However, our brief outline does at least open a new avenue for providing what has recently been regarded as the missing base of "scientific evidence" for a diverse set of vector control methods, integrated with drugs, bednets, IRS, and IPTp. Expanding on this outline and linking it with cost-effectiveness analyses could be an important step toward bringing back the many positive features of integrative malaria control programs of the past, and redressing them in modern clothes.

Finally, a few remarks are in order regarding the contemporary malaria/elimination/eradication initiatives. First, the examples described in this paper, augmented by some contemporary corporate sponsored control programs – e.g., Newmont Mining at Batu Hijau, Indonesia [94] – indicate that sustained suppression of transmission in complex ecological settings is possible without achieving elimination. Second, getting from low transmission to elimination is exceedingly difficult, expensive, and demanding of considerable further research [34]. While being thoroughly supportive of the scientific research associated with malaria, in all of its aspects, we raise the question of whether a global elimination program warrants the cost and effort needed to succeed – and maintain this condition – in settings where many other diseases – e.g., dengue, HIV, schistosomiasis, Zika virus, re-emerging yellow fever, tuberculosis – and severe treatment problems – e.g., antibiotic resistance – must compete for resources and attention of the global health community. In this regard, the contemporary tuberculosis situation is especially vexing [95].

REFERENCES

1. Bruce-Chwatt, L.J. (1988). History of malaria from prehistory to eradication. In: Malaria: Principles and Practice of Malariology, vol. 1 (ed. W.H. Wernsdorfer and I. McGregor), 1–59. Edinburgh and New York: Churchill Livingstone.
2. Bruce-Chwatt, L.J. and Zulueta, J. (1980). The Rise and Fall of Malaria in Europe: A Historic-Epidemiological Study. English Language Ed. Oxford, 240. New York: Oxford University Press.
3. Desowitz, R.S. (1991). The Malaria Capers: More Tales of Parasites and People, Research and Reality, 288. New York: W.W. Norton.
4. Hawass, Z., Gad, Y.Z., Ismail, S. et al. (2010). Ancestry and pathology in king Tutankhamun's family. *Journal of the American Medical Association.* 303 (7): 638–647.
5. Bailey, N.T.J. (1982). The Biomathematics of Malaria, 210. London: C. Griffin.
6. Boyd, M.F. (ed.) (1949). Historical review. In: Malariology: A Comprehensive Survey of all Aspects of this Group of Diseases from a Global Standpoint, vol. 1, 3–25. Philadelphia: W.B. Saunders.
7. Slater, L.B. (2009). War and Disease: Biomedical Research on Malaria in the Twentieth Century, 249. New Brunswick, NJ: Rutgers University Press.
8. Celli, A. and Celli-Fraentzel, A. (1933). The History of Malaria in the Roman Campagna from Ancient Times, 226. London: Bale & Danielsson.
9. Carter, E.D. (2007). Development narratives and the uses of ecology: malaria control in Northwest Argentina, 1890–1940. *Journal of Historical Geography* 33: 619–650.
10. Ross, R. (1910). The Prevention of Malaria, 669. New York: E.P. Dutton & Company.
11. Snowden, F.M. (2006). The Conquest of Malaria: Italy, 1900–1962, 296. New Haven: Yale University Press.
12. Kitron, U. and Spielman, A. (1989). Suppression of transmission of malaria through source reduction: antianopheline measures applied in Israel, the United States, and Italy. *Reviews of Infectious Diseases* 11 (3): 391–406. Epub 1989/05/01. PubMed PMID: 2665000.
13. Warrell, D.A. and Gilles, H.M. (2002). Essential Malariology, 4e, 348. London and New York: Arnold.
14. Watson, M.S. (1915). Rural Sanitation in the Tropics. Being Notes and Observations in the Malay Archipelago, Panama and Other Lands, 320. London: J. Murray.
15. Wernsdorfer, W.H. and McGregor, I. (1988). Malaria: Principles and Practice of Malariology. Edinburgh, 1818. Edinburgh and New York: Churchill Livingstone.
16. Gardner, M.J. et al. (2002). Genome sequence of the human malaria parasite *Plasmodium falciparum*. *Nature* 419: 498–511.

17. Holt, R.A. et al. (2002). The genome sequence of the malaria mosquito *Anopheles gambiae*. *Science* 298: 129–149.
18. Carlton, J.M., Adams, J.H., Silva, J.C. et al. (2008). Comparative genomics of the neglected human malaria parasite *Plasmodium vivax*. *Nature* 455 (7214): 757–763. Epub 2008/10/10. doi: 10.1038/nature07327 PubMed PMID: 18843361; PubMed Central PMCID: PMCPMC2651158.
19. Tene Fossog, B., Ayala, D., Acevedo, P. et al. (2015). Habitat segregation and ecological character displacement in cryptic African malaria mosquitoes. *Evolutionary Applications* 8 (4): 326–345. Epub 2015/05/01. doi: 10.1111/eva.12242 PubMed PMID: 25926878; PubMed Central PMCID: PMCPMC4408144.
20. World Health Organization (2017). World Malaria Report 2016. Geneva: World Health Organization.
21. Malaria Atlas Project [Internet]. Available from: https://map.ox.ac.uk/. Accessed 20 January 2018). 2017.
22. Howes, R.E., Battle, K.E., Mendis, K.N. et al. (2016). Global epidemiology of *Plasmodium vivax*. *The American Journal of Tropical Medicine and Hygiene* 95 (Suppl. 6): 15–34. doi: 10.4269/ajtmh.16-0141 PubMed PMID: PMC5198891.
23. Hay, S.I., Guerra, C.A., Tatem, A.J. et al. (2004). The global distribution and population at risk of malaria: past, present, and future. *Lancet Infectious Diseases* 4: 327–336.
24. Bruce-Chwatt, L.J. (1985). Essential Malariology, 2e, 452. New York: John Wiley & Sons Inc.
25. Konradsen, F., van der Hoek, W., Amerasinghe, F.P. et al. (2004). Engineering and malaria control: learning from the past 100 years. *Acta Tropica* 89 (2): 99–108. Epub 2004/01/21. PubMed PMID: 14732233.
26. Watson, M. (1921). The Prevention of Malaria in the Federated Malay States: A Record of Twenty Years' Progress, 2e, 381. London: J. Murray.
27. Watson, S.M. (1953). African Highway: The Battle for Health in Central Africa, 1e, 294. London: J. Murray.
28. Bradley, D.J. (1994). Watson, Swellengrebel and species sanitation: environmental and ecological aspects. *Parassitologia* 36 (1–2): 137–147.
29. Sallares, R. (2002). Malaria and Rome: A History of Malaria in Ancient Italy, 341. Oxford: Oxford University Press.
30. Phillips, M.A., Burrows, J.N., Manyando, C. et al. (2017). Malaria. *Nature Reviews Disease Primers* 3: 17050. Epub 2017/08/05. doi: 10.1038/nrdp.2017.50 PubMed PMID: 28770814.
31. UNICEF, World Health Organization, PATH (2010). Malaria Funding and Resource Utilization: The First Decade of Roll Back Malaria, 98. Geneva: World Health Organization.
32. Roberts, L. and Enserink, M. (2007). Did they really say . . . eradication? *Science* 318: 1544–1545.
33. Tanner, M. and de Savigny, D. (2008). Malaria eradication back on the table. *Bulletin of the World Health Organization* 86 (2): 82–83.
34. malERA – An updated research agenda for malaria elimination and eradication. PLoS Collections. Available from: http://collectionsplosorg/malera-refresh. Accessed 19 January 2018). 2017.
35. Roll Back Malaria Partnership (2008). The Global Malaria Action Plan: for a Malaria-Free World, 274. Geneva: World Health Organization.
36. Keiser, J., Singer, B.H., and Utzinger, J. (2005). Reducing the burden of malaria in different eco-epidemiological settings with environmental management: a systematic review. *The Lancet Infectious Diseases* 5 (11): 695–708.
37. Baolin L. Environmental management for the control of rice-field-breeding mosquitoes in China. In: Institute IRR (ed.), Vector-Borne Disease Control in Humans through Rice Agroecosystem Management: Proceedings of the Workshop on Research and Training Needs in the Field of Integrated Vector-Borne Disease Control in Riceland Agroecosystems of Developing Countries, 9–14 March 1987. Manila: International Rice Research Institute in collaboration with the WHO/FAO/UNEP Panel of Experts on Environmental Management for Vector Control; 1988. pp. 111–121.
38. Tang, L.H., Qian, H.L., and Xu, S.H. (1991). Malaria and its control in the People's Republic of China. *Southeast Asian Journal of Tropical Medicine and Public Health* 22 (4): 467–476.
39. Castro, M.C. (2017). Malaria transmission and prospects for malaria eradication: the role of the environment. *Cold Spring Harbor Perspectives in Medicine* 7 (10). Epub 2017/05/12): doi: 10.1101/cshperspect.a025601 PubMed PMID: 28490534.
40. WHO (1982). Manual on Environmental Management for Mosquito Control: With Special Emphasis on Malaria Vectors, 283. Geneva: World Health Organization.
41. Golding, N., Wilson, A.L., Moyes, C.L. et al. (2015). Integrating vector control across diseases. *BMC Medicine* 13 (1): 249. doi: 10.1186/s12916-015-0491-4.
42. Bruce-Chwatt, L.J. (1987). Malaria and its control: present situation and future prospects. *Annual Review of Public Health.* 8: 75–110.
43. Institute of Medicine (US). Malaria: Obstacles and Opportunities. A Report of the Committee for the Study on Malaria Prevention and Control: Status Review and Alternative Strategies, Division of International Health, Institute of Medicine. Oaks SC, Mitchell VS, Pearson GW, Carpenter CJ (eds). Washington, DC: National Academy Press; 1991. 309 pp.
44. Cox-Singh, J., Davis, T., Lee, K. et al. (2008). *Plasmodium knowlesi* malaria in humans is widely distributed and potentially life threatening. *Clinical Infectious Diseases* 46 (2): 165–171.
45. Van den Eede, P., Vythilingam, I., Ngo, D.T. et al. (2010). *Plasmodium knowlesi* malaria in Vietnam: some clarifications. *Malaria Journal* 9 (20. Epub 2010/01/20): doi: 10.1186/1475-2875-9-20 PubMed PMID: 20082717; PubMed Central PMCID: PMCPMC2817702.
46. Jongwutiwes, S., Putaporntip, C., Iwasaki, T. et al. (2004). Naturally acquired *Plasmodium knowlesi* malaria in human, Thailand. *Emerging Infectious Diseases* 10 (12): 2211–2213.
47. Lover, A.A. and Coker, R.J. (2015). Do mixed infections matter? Assessing virulence of mixed-clone infections in experimental human and murine malaria. *Infection, Genetics and Evolution: Journal of Molecular Epidemiology and Evolutionary Genetics in Infectious Diseases* 36: 82–91.

Epub 2015/09/04. doi: 10.1016/j.meegid.2015.08.041 PubMed PMID: 26334940.

48. Ginouves, M., Veron, V., Musset, L. et al. (2015). Frequency and distribution of mixed *Plasmodium falciparum-vivax* infections in French Guiana between 2000 and 2008. *Malaria Journal* 14: 446. Epub 2015/11/12. doi: 10.1186/s12936-015-0971-1 PubMed PMID: 26555553; PubMed Central PMCID: PMCPMC4641342.
49. Marchand, R.P., Culleton, R., Maeno, Y. et al. Co-infections of *Plasmodium knowlesi*, *P. falciparum*, and *P. vivax* among humans and *Anopheles dirus* mosquitoes, southern Vietnam. *Emerging Infectious Diseases* 2011, 17 (7): 1232–1239. doi: 10.3201/eid1707.101551 PubMed PMID: PMC3381379.
50. Benenson, A.S. (1990). Control of Communicable Diseases in Man: An Official Report of the American Public Health Association, 15e, 532. Washington DC: American Public Health Association.
51. Consoli, R.A.G.B. and Oliveira, R.L. (1998). Principais Mosquitos de Importância Sanitária no Brasil, 228. Rio de Janeiro: Editora Fiocruz.
52. Deane, L.M., Causey, O.R., and Deane, M.P. (1946). Studies on Brazilian Anophelines from the Northeast and Amazon Regions, 50. Baltimore: The Johns Hopkins Press.
53. Fontaine, M.C., Pease, J.B., Steele, A. et al. (2015). Extensive introgression in a malaria vector species complex revealed by phylogenomics. *Science (New York, NY)* 347 (6217): 1258524. doi: 10.1126/science.1258524 PubMed PMID: PMC4380269.
54. McGregor, I.A. and Wilson, R.J.M. (1988). Specific immunity: acquired in man. In: Malaria: Principles and Practice of Malariology, vol, vol. 1 (ed. W.H. Wernsdorfer and I. McGregor), 559–619. Edinburgh and New York: Churchill Livingstone.
55. Molineaux, L. (1988). The epidemiology of human malaria as an explanation of its distribution, including some implications for its control. In: Malaria: Principles and Practice of Malariology, vol., vol. 1 (ed. W.H. Wernsdorfer and I. McGregor), 913–998. Edinburgh and New York: Churchill Livingstone.
56. Gupta, S., Snow, R.W., Donnelly, C.A. et al. (1999). Immunity to non-cerebral severe malaria is acquired after one or two infections. *Nature Medicine* 5 (343): 340.
57. Webb, J.L.A. (2009). Humanity's Burden: A Global History of Malaria, 236. Cambridge and New York: Cambridge University Press.
58. Cavalli-Sforza, L.L., Menozzi, P., and Piazza, A. (1994). The History and Geography of Human Genes, 518. Princeton, NJ: Princeton University Press.
59. Ménard D, Barnadas C, Bouchier C, Henry-Halldin C, Gray LR, Ratsimbasoa A, et al. *Plasmodium vivax* Clinical Malaria is Commonly Observed in Duffy-Negative Malagasy People. Proceedings of the National Academy of Sciences. 2010; Published online before print March 15, 2010. doi: 10.1073/pnas.0912496107.
60. Ryan, J.R., Stoute, J.A., Amon, J. et al. (2006). Evidence for transmission of *Plasmodium vivax* among a duffy antigen negative population in western Kenya. *American Journal of Tropical Medicine and Hygiene* 75 (4): 575–581.
61. Cavasini, C.E., Mattos, L.C., Couto, A.A. et al. (2007). *Plasmodium vivax* infection among Duffy antigen-negative individuals from the Brazilian Amazon region: an exception? *Transactions of the Royal Society of Tropical Medicine and Hygiene* 101 (10): 1042–1044.
62. Cavasini, C.E., LCd, M., Couto, A.A.D.A. et al. (2007). Duffy blood group gene polymorphisms among malaria vivax patients in four areas of the Brazilian Amazon region. *Malaria Journal* 6 (167): doi: 10.1186/1475-2875-6-167.
63. Ashley-Koch, A., Yang, Q., and Olney, R.S. (2000). Sickle hemoglobin (*Hb S*) allele and sickle cell disease: a HuGE review. *American Journal of Epidemiology* 151 (9): 839–845.
64. Castro, M.C., Monte-Mór, R.L., Sawyer, D.O., and Singer, B.H. (2006). Malaria risk on the Amazon frontier. *Proceedings of the National Academy of Sciences* 103 (7): 2452–2457.
65. Ault, S.K. (1989). Effect of demographic patterns, social, structure, and human behavior on malaria. In: Demography and Vector-Borne Diseases (ed. Service MW), 283–301. Boca Raton, FL: CRC Press.
66. Ault, S.K. (1989). Effect of malaria on demographic patterns, social structure, and human behavior. In: Demography and Vector-Borne Diseases (ed. Service MW), 271–282. Boca Raton, FL: CRC Press.
67. Beier, J.C., Keating, J., Githure, J.I. et al. (2008). Integrated vector management for malaria control. *Malaria Journal* 7 (Suppl. 1): S4.
68. WHO (2004). Global Strategic Framework for Integrated Vector Management, 15. Geneva: World Health Organization, WHO/CDS/CPE/PVC.
69. Cruz, O. (1910). Prophylaxis of malaria in central and southern Brazil. In: The Prevention of Malaria (ed. R. Ross), 390–399. New York: E.P. Dutton & Company.
70. Carter, R., Mendis, K.N., and Roberts, D. (2000). Spatial targeting of interventions against malaria. *Bulletin of the World Health Organization* 78 (12): 1401–1411.
71. Peters, W. (1987). Chemotherapy and Drug Resistance in Malaria, 2e, 1085. London and Orlando: Academic Press.
72. World Health Organization (2010). Guidelines for the Treatment of Malaria, 2e, 194. Geneva: World Health Organization.
73. Dondorp, A.M., Yeung, S., White, L. et al. (2010). Artemisinin resistance: current status and scenarios for containment. *Nature Reviews Microbiology* 8: 272–280. doi: 10.1038/nrmicro2331.
74. Bjørndal, A. (2009). Improving social policy and practice: knowledge matters. *Lancet* 373: 1829–1831.
75. Yin, J.H., Zhou, S.S., Xia, Z.G. et al. (2014). Historical patterns of malaria transmission in China. *Advances in Parasitology* 86: 1–19. Epub 2014/12/06. doi: 10.1016/b978-0-12-800869-0.00001-9 PubMed PMID: 25476879.
76. Utzinger, J., Tozan, Y., and Singer, B.H. (2001). Efficacy and cost-effectiveness of environmental management for malaria control. *Tropical Medicine and International Health* 6 (9): 677–687. PubMed PMID: ISI:000171126300003.
77. Haskin, F.J. (1913). The Panama Canal, 386. Garden City, NY: Doubleday Page & Company.

78. Darling, S.T. (1910). Factors in the transmission and prevention of malaria in the Canal Zone. *Annals of Tropical Medicine and Parasitology* 4: 179–223.
79. Darling, S.T. (1910). Studies in Relation to Malaria, 2e, 42. Mount Hope: Isthmian Canal Commission, Laboratory of the Board of Health. Quartermaster's Department.
80. Gorgas, W.C. (1910). Malaria prevention on the isthmus of Panama. In: The Prevention of Malaria (ed. R. Ross), 346–352. New York: E.P. Dutton & Company.
81. Le Prince, J.A. (1910). Anti-malarial work on the isthmus of Panama. In: The Prevention of Malaria (ed. R. Ross), 353–368. New York: E.P. Dutton & Company.
82. McArthur, J. (1953). Malaria control in Borneo. *The Lancet* 262 (6787): 655–656.
83. McArthur, J. (1947). The transmission of malaria in Borneo. *Transactions of the Royal Society of Tropical Medicine and Hygiene* 40 (5): 537–558.
84. McArthur, J. (1951). The importance of *anopheles leucosphyrus*. *Transactions of the Royal Society of Tropical Medicine and Hygiene* 44 (6): 683–694.
85. McArthur, J. (1954). The control of malaria in Borneo (an account of the Tambunan experiment). *Transactions of the Royal Society of Tropical Medicine and Hygiene* 48 (3): 234–241.
86. McArthur, J. (1948). Malaria in Borneo: An Account of the Work of the Malaria Research Department, North Borneo, 1939–42, 250. Labuan: Malaria Research Headquarters.
87. McArthur J. Malaria Control by Methods other than Insecticides. Malaria Carried by *Anopheles leucosphyrus*. The World Health Organization. WHO/MAL/84. Available from: whqlibdoc.who.int/malaria/WHO_MAL_84.pdf. Accessed March 2010, 1953.
88. Sleigh, A.C., Lin, X.-L., Jackson, S., and Shang, L. (1998). Resurgence of vivax malaria in Henan Province, China. *Bulletin of the World Health Organization* 76: 265–270.
89. Jackson, S., Sleigh, A.C., and Lin, X.-L. (2002). Cost of malaria control in China: Henan's consolidation programme from community and government perspectives. *Bulletin of the World Health Organization.* 80 (8): 653–659.
90. Yang, H., Yang, Y., Yang, P. et al. (2008). Monitoring *Plasmodium falciparum* chloroquine resistance in Yunnan Province, China, 1981–2006. *Acta Tropica* 108 (1): 44–49.
91. Box, G.E.P. and Draper, N.R. (1969). Evolutionary Operation: A Statistical Method for Process Improvement, 237. New York: John Wiley & Sons, Inc.
92. Singer, B. (1989). Social science and the improvement of tropical disease control programs. *Annals of the New York Academy of Sciences* 569 (1): 275–287. doi: 10.1111/j.1749-6632.1989.tb27375.x.
93. McKenzie, F.E. and Samba, E.M. (2004). The role of mathematical modeling in evidence-based malaria control. *The American Journal of Tropical Medicine and Hygiene* 71 (Suppl. 2): 94–96. Epub 2004/08/28. PubMed PMID: 15331824; PubMed Central PMCID: PMCPMC2518450.
94. Castro, M.C., Krieger, G.R., Balge, M.Z. et al. (2016). Examples of coupled human and environmental systems from the extractive industry and hydropower sector interfaces. *Proceedings of the National Academy of Sciences* 113 (51): 14528–14535. doi: 10.1073/pnas.1605678113.
95. Bloom, B.R. (2018). A neglected epidemic. *New England Journal of Medicine* 378 (3): 291–293. doi: 10.1056/NEJMe1714609 PubMed PMID: 29342389.

19

ECOSYSTEM HEALTH AS THE BASIS FOR HUMAN HEALTH

TOM BARKER[1,2] AND JANE FISHER[1,3]

[1] *Graduate School of the Environment, Centre for Alternative Technology, Machynlleth, Wales, UK*
[2] *Department of Geography & Planning, University of Liverpool, Liverpool, UK*
[3] *Learning and Teaching Institute, University of Chester, Chester, UK*

19.1 INTRODUCTION. THE CONTEXT OF FRESHWATER ECOSYSTEMS AND BIODIVERSITY

The function of fresh water in society is important for poverty, equity, disease prevention, production of food and aesthetic cultural and emotional fulfillment. The central role played by freshwater ecosystems in a sustainable global system, and their doubtful future in the face of environmental change, cannot be comprehended without a reasonable understanding of the roles of ecology and biodiversity, and how they underpin healthy populations in urban as well as rural environments.

Ecosystems work at multiple interconnected scales and, as scientific understanding improves, are increasingly recognized as more than the commonly held perception of distinct habitats such as may be perceived of a wooded area or a pond. Tansley's early [1] concept of an ecosystem inferred a localized interconnected system of relations between species and the abiotic environment, but there is a new emphasis today on connectivity in order for ecosystems to flourish. Wetlands and lakes in particular are dependent on the broader landscape and cannot rightly be regarded as independent ecosystems. As human society has emerged from wild nature, its growth has increasingly been an incursion into nature's space. When considered from this viewpoint, it is perhaps easier to appreciate that industrioeconomic expansion always has negative consequences for natural systems. A lake is a good example because it is apparently self-contained, yet it can be sustained only with constant inputs of resources including water, energy, and organic materials, as well as removal of wastes, and it is essential for many organisms that visit it for just part of their life-cycle or at different times of the year. Fragmentation of the broader habitat leads to extinctions and loss of functioning [2].

The role of biodiversity in ecosystems is complex. The term biodiversity includes diversity of genetic traits within and between species as well as numbers of species, numbers of individuals, their assemblages and ecological functions and their geographical distributions. The term is much more than a description of life; it is life in all its aspects and is the reason and method of the existence and persistence of all life on Earth.

The tree of life is vast and recent studies have added new leaves to it. It is thought that there are 8.7 million species on Earth [3] with less than a quarter described so far, though this has to be matched by the increasing rate of extinctions of eukaryotic organisms, caused primarily by human activities, that currently are estimated to run at 1000 times natural background rates [4] and are set to increase by a factor of 5 or 10 in coming decades [4, 5]. As well as these millions of "leaves", the tree of life also has roots, the microbial processors that make higher life possible, and these are far greater in numbers of species. There is a mind-boggling number of species on Earth, and most of them are microorganisms. They all do different things, though perhaps some do similar things in different places. Recently refined genetic techniques based on identification of the last universal common

Water and Sanitation-Related Diseases and the Changing Environment: Challenges, Interventions, and Preventive Measures, Second Edition. Edited by Janine M. H. Selendy.

Companion website: https://www.wiley.com/legacy/wileychi/selendy/

ancestor (LUCA and variations of it) have led to a reassessment of the evolutionary tree from the accepted three major domains of Bacteria, Archaea (bacteria-like), and Eukarya (cells having a bound nucleus, including animals, plants, and fungi) to one of just two domains, Bacteria and Arkarya, the latter containing both the Archaea and the eukaryotes [6, 7]. The Eukarya are placed within the new domain because it seems they arose from the Archaea, with the help of bacteria and their ready ability to swap genetic information [8]. The point is that the new tree of life (Figure 19.1) immensely favors microbes, with eukaryotes a more recent side branch, and Opisthokonts, i.e., animals and fungi, together in an offshoot of that. With this knowledge, we might question some assumptions that nature exists for the benefit of human society.

The microbial community extends downwards for some meters, processing leached materials and materials entering via subsoil routes such as seepage and aquifers. There is evidence that microbial diversity strongly influences aboveground biodiversity and functioning locally and globally [9]. Though many microbial communities exist independently of larger forms, there are many well-known examples of interactions; e.g., mineralizing soil communities provide nutrients to plants, use their resources, and decompose their wastes, including conversion of plant and animal matter passed through animals and via intricate fungal associations (e.g., [10]), and microbes attack plants and animals as pathogens. They have often unseen roles in industry, such as the removal of oil from seepages and spills, which is carried out by microbes [11] and which can be fed with nutrients to hasten recovery, famously for the Exxon Valdez environmental disaster [12], which spilled crude oil into the sea.

There is a direct, complex relationship between biodiversity and ecosystem functioning, and the implications for global functioning are only now being assessed. It boils down to transactions in the ecological use of resources including assimilation of wastes. These affect local ecologies by changing habitats and life cycles of species, which have a range of impacts on the well-being and survival of people. The growing problems of broad environmental change mean we can no longer ignore the global dimension of ecosystems. The Earth ecosystem is responsive to its component parts.

An appreciation of environmental sustainability, e.g., as redefined by Morelli [13] ("meeting the resource and services needs of current and future generations without compromising the health of the ecosystems that provide them") can only truly be had upon realization of where human society and economy sits within nature's universal set. James Lovelock's Gaia hypothesis [14], today advanced through the modern science of "Earth systems science" (see, e.g., [15]), showed us that an equable and breathable atmosphere is the result of the activities of living organisms. In the chemical equilibrium that exists on other planets and would be expected to exist here on Earth in the absence of life (e.g., mostly CO_2), life as we know it would be impossible, but uptake of certain elements and compounds and deposition of others holds the Earth's atmosphere out of chemical equilibrium and suitable for life. In this way, Lovelock showed that species are not, as hitherto understood, merely eking out an existence where the chemical substrate is favorable but, through an often complex series of negative feedbacks, themselves bring about environmental change, such as CO_2 removal, sedimentary rock creation, cloud formation, and weathering (see, e.g., [10, 16–18]). The result is that the presence of life makes the planet suitable for life, including us. It seems that we tamper with this at our peril.

19.2 HOW DO ECOSYSTEMS WORK?

Energy enters an ecosystem directly from the sun, where it is used in photosynthesis to make chemical energy in the form of sugars. It also arrives in already fixed (chemical) forms in the shape of materials. These might be animals and their products, air-borne particles and chemicals including CO_2, water as rain and snow, surface and subsurface flow and all the chemicals and materials gathered and carried with it. Leaves, fruits and insects, droppings, and dead animals fall from trees into water bodies, where they are consumed by fish and invertebrates or colonized by fungi, which itself forms food for water invertebrates. These then hatch and the adult forms supply nutrients to terrestrial species such as birds and bats [19]. The cycle perpetuates as energy and materials are recycled constantly, and constitutes an important resource for both the aquatic and terrestrial ecosystems. Eventually, energy and materials are replaced as, transformed, they leave the ecosystem by air, water, and leaching through the soil, or they might just sit there for millennia, e.g. as peat.

19.2.1 The Importance of Connectivity

Although some species live their entire lives in one area, there is a continual recruitment and loss of species and organisms that spend part of their time or life cycle in an ecosystem and then move on. In a wetland, for example, invertebrates may breed and live as larvae in open water but leave the area as adults, e.g., malaria mosquitoes. Predators will enter and leave as they roam different parts of their territory for prey, and the flow of materials from outside can revitalize an ecosystem. Life cycles of certain species will be tuned to regular occurrences as resource availability dictates the behavior of animals and the production of the ecosystem. A very satisfying example is provided by the last contributions of salmon (*Oncorhynchus* spp.) to Alaskan forests courtesy of the brown bear (*Ursus arctos*) and other species [20]. Salmon migrating upstream to breed have spent many

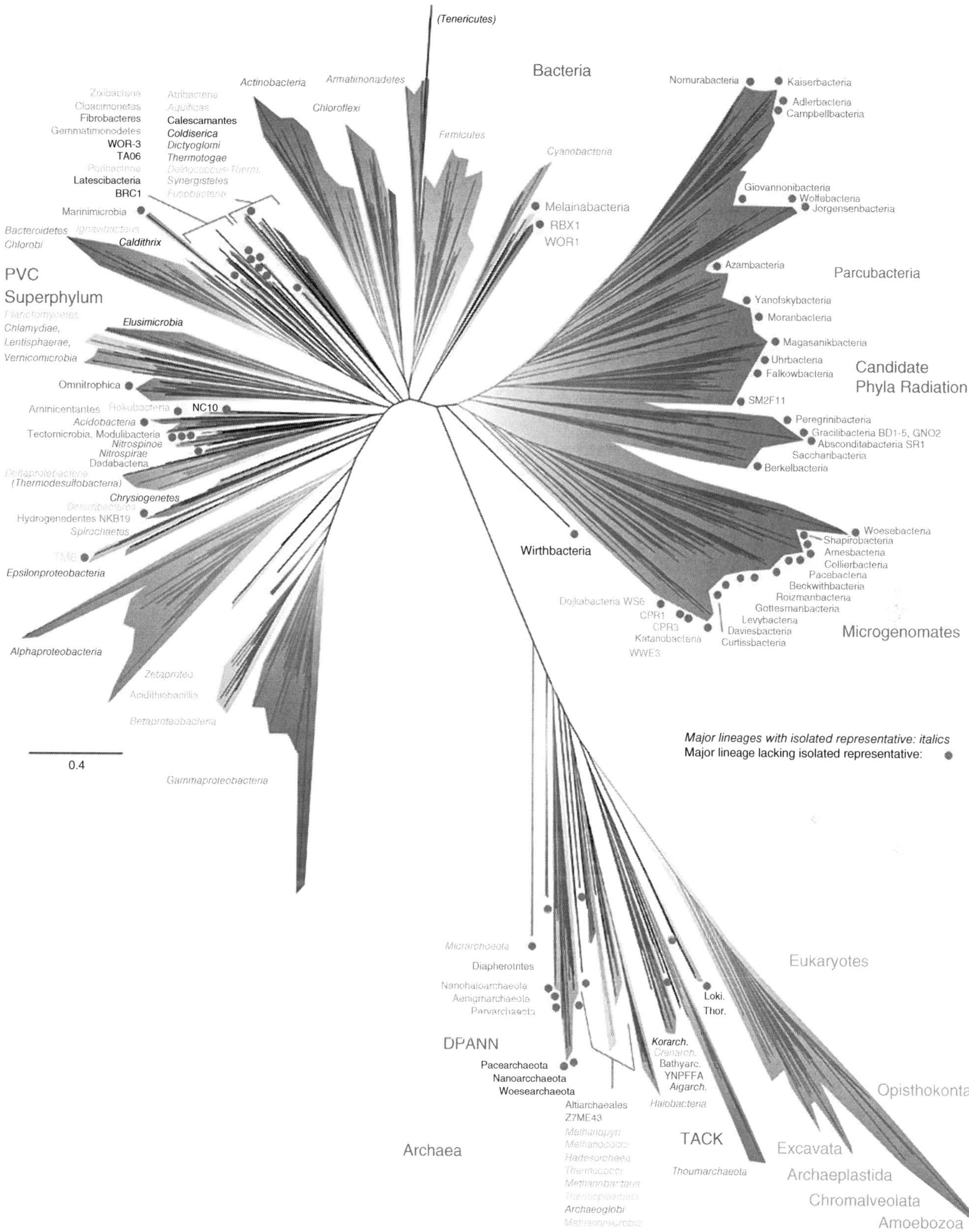

FIGURE 19.1 The new tree of life, based on 1000 newly sequenced genomes, places, all of the plants (the Archaeplastida, bottom right corner), animals and fungi (the Opisthokonta, bottom right corner) on an evolutionary edge, having emerged from the microbes that support our continued existence on Earth. For explanation of names and relationships, see the *source*: Hug et al. [8]. Available from: https://www.nature.com/articles/nmicrobiol201648#s6. Licensed under CC BY 4.0. (*See insert for color representation of this figure.*)

years in the sea growing to their breeding size, and represent a large input of nutrients from the sea to the stream ecosystems, where the fish spawn and die, and to woodland areas nearby [21]. The main vector of these nutrients to the woodland soil is the brown bear. They and other predators catch the fish on their way to or at the breeding sites and scavenge the bodies that remain after spawning. Bears take their catches to land to eat and also excrete as they later wander in the woodland. There are several processes involved once the salmon carcasses are left lying. In streams they become entangled in woody debris dropped by trees and left under such unmanaged conditions, and their nutrients are used by fungi to colonize low-nutrient plant matter such as leaves and wood. These fungi render the leaves palatable and thus a nutritious meal for invertebrates as the nutrients enter the food web. On land, the carcasses, and the feces of bears are consumed by birds, mammals, and invertebrates, and colonized by bacteria that mineralize the nutrients within them. These seep into the soil whereupon they are taken up by the roots of plants. The plants in such areas can therefore draw on increased fertility and are thus more productive than those further from the stream. Studies of nitrogen isotopes have confirmed that around 25% of nitrogen and phosphorus in riparian plants of these areas originates in the sea [21, 22], which means that these nutrients come from the salmon. All of these processes in turn supply nutrients and invertebrate food for the salmon hatchlings (parr) once they have ventured from the egg nutrient reserves, and for other species downstream. Thus it all adds up to a mass movement of nutrients from the sea to the upper reaches of rivers in Alaskan forests, where they promote growth of trees, mammals, birds, invertebrates, and, not least, the next generation of salmon.

There is another side to the coin in that organisms always exploit any niche they can develop, and the relationships between migrating salmon and bears provide opportunities for parasites to complete their life cycles as they swap hosts. In one such infection, and there are several equivalent pathways using different parasite and host species, a freshwater copepod, the handsome *Cyclops strenuus*, is the first host of the tapeworm parasite *Diphyllobothrium nihonkaiense*, which subsequently provides food for migrating salmon (*Oncorhynchus* spp.). The salmon act as the second intermediate host before themselves being eaten by their final host, the brown bear (*U. arctos*). However, bears are not the only consumers of salmon, and with the increasing popularity of sushi in the affluent world, a re-emergence of diphyllobothriasis in humans has been reported in Japan, Russia, and Alaska particularly, and also in Europe via imported salmon [23].

Intricate cycles thus supply living organisms with the mineral resources they need to survive, and living organisms, each acting according to their own short-term needs, have evolved in concert to exploit them. For that to continue, and thus for the species to survive, we humans as managers of the biosphere must ensure that this connectivity is maintained and not fragmented by conversion of ecosystems for some industrial purpose. Similar sorts of linkages between ecosystems can be found wherever semi-natural connected ecosystems exist, and their loss represents a major blow to biodiversity protection [24].

Without free movement of organisms across a broad interconnected ecosystem, isolated populations in ecosystem fragments are vulnerable to disease, disaster, or predation, and a narrowing of the genetic pool. Large predators in particular require large roaming distances to feed and mate, and are important links in connecting nutrient flows between ecosystem patches and controlling grazer behavior, factors that might have devastating effects on vegetated areas if left unchecked. As Colinvaux [25] pointed out so eloquently, predators are both maintained and limited by their prey. Moss [26] likened the policy of protecting isolated areas of a particular habitat as wildlife reserves to a situation of a physician who treats someone's organs individually but ignores the fact that they are connected by the flow of nutrients, wastes, and chemicals in the blood. The prognosis for that person is gloomy. So too is the maintenance of ecological integrity in habitats that are separated from others. In zoos, breeding programs use careful control of genetic information in attempts to protect endangered species [27], and increasingly are becoming involved in the protection of their habitats, reflecting that it is not merely because the species have been eradicated by hunting but also by habitat pressures. The loss of connectivity of habitat has resulted in isolated populations that lead to genetic damage by inbreeding of individuals forced into permanent close proximity.

Is it perhaps partly because ecosystems are considered to be self-contained functioning units that they are mistakenly assumed by policy-makers to be able to survive in isolation from the broader landscape? They are not. Conceivably, it is the tendency of people, who evolved after all in small communities, to view all systems whether ecological or urban as self-contained, the natural state of institutions and populations being small ([28], pp. 52–54). This viewpoint might explain the generally held assumption of people living and working in cities that the environment is out there, other, and not relevant to city life, despite the fact that the ecological footprint [29], i.e. the area needed to provide resources and assimilate wastes, of a city might be a 1000 times bigger than the area of the city itself [30].

Seen as independent units, ecosystems and their species may be regarded by society as safe if they are secured as reserves protected against development and agriculture, but, more recently, the essential requirement for links, particularly to external flows of species and their genes, has led to calls for extensive connections that reflect the ranges of species, including top predators [31]. Without these links, the species that characterize the ecosystems, their ecological functions, and the services to humans supported by them are jeopardized. We will return to this topic later.

19.2.2 Resource Use

Organisms generally follow the path of least resistance according to their evolutionary history. If one encounters a resource, it will attempt to exploit it maximally using the route that costs it least energy. If one encounters a competitor, it will respond in terms of advantage gained minus risk weighed against the potential evolutionary gain (e.g., [32]); mates can be a mixture of these things. In other words, it wants to consume nutrients, reproduce, and protect itself from being eaten or injured. Through this lens, the complex interactions of an ecosystem can be glimpsed.

A newly collected resource, e.g. wind-blown dust accumulating in a crevice, combined with water retained from previous rain and perhaps a dead leaf is very quickly likely to be colonized by fungi and bacteria. Arriving invertebrates such as insects or worms would find the fungally processed plant matter palatable and consume it along with some of the microbes themselves. Macroscopic predators and parasites such as centipedes and nematodes would soon follow, and the seed of a plant would find the moist, mineralized nutrients that were produced by microbes from plant matter and animal frass (dung) enough to support growth. Here, in this sheltered crevice can be found the essential elements of an ecosystem, their interconnections supported by processes mediated by organisms: a supply of substrate, dead organic matter as a resource for microbes, water, microbial decomposers, animal consumers of soils and microbes, predators of invertebrates, mineralized nutrients, and a plant or two to gather light and add biomass to the wind-blown total – in short, the beginnings of an ecosystem. Each component organism that arrives makes use of what it finds. If it cannot use a material that is available there, whether that is microbial, plant, or animal, it moves on or dies. As the embryonic ecosystem grows and develops, the steadily increasing range and volume of available resources makes it ever more likely that the next organism to arrive will be able to survive there. Thus the ecosystem will grow until it is hindered by some physical constraint such as drought, or the limit of the area where water and dust can accumulate.

As a depiction of such a rudimentary ecosystem that was found between rocks, Figure 19.2 is necessarily simplistic because many species undoubtedly present are not indicated, and relationships are complicated by the fact that species do not sit neatly within narrow definitions such as herbivore, predator, and parasite but are omnivorous [33]. Nevertheless, it gives a clue to how species fit together into a productive system though each only seeks to maximize its chances to live and reproduce. One wonders if Adam Smith was impressed by a similar observation.

It is usual for resources to be available in patchy distribution spatially or temporally, or in gradients of concentration as a result of physical and biological factors. For example,

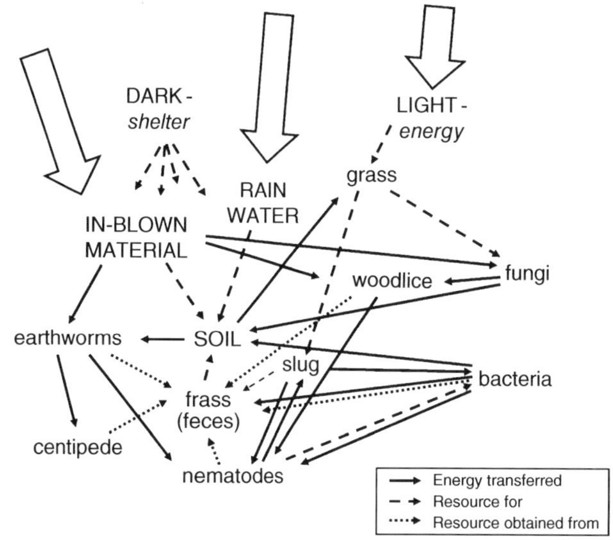

FIGURE 19.2 A very simplified drawing of the rudimentary beginnings of an ecosystem found among rocks. It shows the main organisms detected and the key relationships. Ecosystem processes include colonization by fungi and bacteria, remineralization of nutrients, consumption and excretion, and support for primary production. Ecosystem services are nutrient cycling, primary production, soil formation and waste removal. Solid lines show energy transferred, dashed lines show physical resources and dotted lines the contributions to soil formation.

water might arrive via rainfall or overland flow and some of it might be retained in a hollow, some will seep into the ground, some will be taken up by plants, and some will evaporate. This creates gradients from dry through various levels of dampness to wet. It might have plant-available nutrients associated with it. If this is a semi-permanent situation, an examination of the ground will reveal microbial, invertebrate, vertebrate, and plant species that are characteristic of the degree of wetness or fertility in that place. A water-loving plant such as *Caltha palustris* L. will not survive in drier areas and yarrows (*Achillea* sp.) will not tolerate wet or waterlogged conditions; many species lie in between these more extreme conditions and some, e.g. mosquitoes, can exploit temporary pools to complete their life cycle. Species grow where the environmental conditions suit their physiology. Animals reside there according to their own physiological needs, though larger animals, being usually unconstrained to a particular site, may wander in and out, or use just one part of the gradient for their life cycle.

Seasons and the population dynamics of species create temporal heterogeneity, while topography and gradients in abiotic variables lead to spatial heterogeneity [34]. Encroachment, whether natural or anthropogenic adds or removes complexity temporally or permanently.

19.2.3 Species in Their Habitat – The Niche

Comparisons of habitats from across the globe have revealed that there are several factors that have a bearing on the numbers of species able to evolve and survive in a particular place. A species exists within a niche, which is the net resource space necessary for it to survive and breed. It is further shaped by predators and competitors, which variously restrict an individual's ability to use the resources it needs. Thus a species with tolerance ranges for, say, water availability, temperature, salinity, availability of food or nutrients (and many other things) will also be constrained by the overlapping niches of predators and parasites that might kill it and competing species that will try to take the same resources. The "realized niche" of a species, i.e. the space in which it actually manages to survive and reproduce (e.g. [35]), emerges from the attempts to balance of all these often competing factors.

Opportunities for niches, and factors that constrain species richness are influenced by the size of an ecosystem and the time it has been in existence and also how favorable the environment is in terms of stability, because these enable species to specialize and thus exist in smaller niches, e.g. there are far more species in old than young lakes. The ancient Lake Victoria contains a diverse range of haplochromid cichlids that specialize in many different foods and species, whereas younger Holocene lakes left by the retreat of ice are comparatively depauperate in species numbers. Where human interference alters the evolved balance of species, widespread knock-on effects can disrupt an entire ecosystem, causing many extinctions [36]. Deforestation, agriculture, and in particular the introduction of the Nile perch (*Lates niloticus*) into Lake Victoria in the 1950s and 1960s brought ecosystem changes that led to the extinction of many of the native fish species, including many haplochromine cichlids. This removed subsistence livelihoods of local people in favor of the industrial-scale Nile perch and tilapia fishery and, combined with overfishing and eutrophication, eventually brought about near ecological collapse [37, 38]. The industrial catch of Nile perch, Nile tilapia (*Oreochromis niloticus*), and *dagaa*, a pelagic cyprinid (*Rastrineobola argentea*), is approximately one million t a^{-1}, while typical catches for local fishing people are reduced [39].

More productive habitats tend to provide more resources and permit more specialization and thus speciation, as competition for resources leads to intricate specialization between species that live in the same habitat. The physical, structural heterogeneity of the habitat also provides more opportunities for species to specialize, again enabling the evolution of more species to cram into an area. Predator niches can inhibit the prevalence of some abundant species, which might act to the benefit of others. Two useful papers that discuss these concepts are by Ricklefs [40, 41].

While many species in their niches in biodiverse environments are dependent on fairly narrow ranges of resource values in order to survive and breed, other species are ubiquitous. Rats (*Rattus norvegicus* and *R. rattus*) and pigeons seem to be able to survive almost anywhere; they are the opposite of specialists. The feral pigeon or rock dove (*Columba livia* and various subspecies), for example, has radiated from its native Eurasia and North Africa to exist in many countries around the world, having been introduced by pigeon fanciers. Its requirements are broad and varied and it has replaced many species that were ousted as their environment was changed by human activities, an occurrence that perhaps should lead us to expect ubiquitous pest species to increase proportionally as more specialized species become extinct due to environmental change.

19.2.4 Resilience to Perturbation

Resilience is the name given to the capacity of an ecosystem to withstand and recover from external pressures and shocks that change the physical and community structure. If a system recovers enough to restore its former community of species, it can be said to be resilient, but resilience is therefore dependent not only on the capacity of the ecosystem to survive relatively intact but also on the type, degree, and duration of disturbance. Ecosystems that normally undergo change, e.g. due to seasonal weather, fire, or migrations, are likely to have an evolved capacity for resilience to those factors, and they might have comparatively fewer species, which are adapted to frequent change. Other ecosystems, in more environmentally stable places might be less able to withstand an external shock without loss of species. An ecosystem damaged by a fire that wipes out a swathe of woodland might recover if sufficient soil moisture is available and the area of remaining woodland is large enough to seed the burnt area, but if the destruction is complete, the land might revert to grassland. If the destruction comprises razing of the woodland, perhaps a tropical forest, and replacing it with a palm plantation for growing oil for cakes and sweets to be sold in Western shops, the change is likely to be permanent.

Resilience to an external perturbation depends on numerous factors, but is underpinned by the species present and their diversity of genetic traits. Losses of certain keystone species can have devastating and widespread effects on resilience [42], resulting in changes to the character of an ecosystem.

19.2.5 Ecosystem Change

In a habitat modified by humans, the ecosystem structure always suffers and species are always put under strain and lost. This is because human societies, whether gatherer-hunter or industrial, impose conditions on ecological communities, placing some species under more strain (they might go locally extinct) while making life easier for others. Such a disturbed ecosystem, provided it survives, will eventually readjust to a new structure, and if it has not lost too many species, might still perform approximately the same ecological functions, though its resilience can be expected to be reduced [43]. The ability of ecosystems to perform functions will be addressed in more detail below. Habitats that are modified by humans may not appear immediately to be damaged and their resilience can enable ecological functioning to persist in the face of environmental pressures until a point of no return is reached.

The concept of alternative states in ecosystems was elaborated for shallow lakes [44, 45] and reveals how ecosystems and perhaps dynamic systems generally can suddenly change dramatically and in a non-linear (not readily reversible) fashion, i.e. a tipping point, if a sufficiently powerful perturbation pushes one into a new "basin of attraction" [46]. In nature, there are winners and losers, and frequently the change is to a very degraded state, e.g. from a clear water lake characterized by lots of plants and animals to a murky pea soup of a lake with toxic blue-green algae and loss of structure (Figure 19.3). This happens because negative feedbacks that keep a system within certain boundaries are weakened, allowing other species to dominate (Figure 19.4). When they do, the system can flip into a new ecosystem condition that resists a reverse flip (back to its former state) according to the strength of the newly established negative feedback mechanisms that support the new quasi-stable state, completely changing the character of the system. This has since been shown in different types of dynamic system such as in trees and savannah in Africa [49], enteric microbes [50], desertification [51] and, potentially, for the global climate system and the global economy ([52, 53, pp. 386–411, 54], among others).

The crucial thing is that it cannot be estimated with any degree of certainty at all where a system currently lies on the axis (curve A in Figure 19.4), where a tipping point might be, or how close to it the system is at any particular time, so a precautionary approach is needed in order to avoid sudden system-wide and irreversible change. Precautionary approaches for environmental reasons are not the sort of things governments prioritizing short-term economic matters generally do unless they are relatively cheap, as seen in delay after delay in climate change legislation and incentives ([55, 56]; though also see [57]), so academic and popular efforts to persuade legislators have seen a steady increase [58–63].

Signs of an impending switch in lakes have been seen in a slowing of recovery time (resilience) in perturbed ecosystems compared with intact ecosystems [51] and in the degree of variance, i.e. increased ecosystem state fluctuations around the mean [64, 65], probably associated with loss of resilience, which can be identified in the data of functioning in the period preceding a flip to a new state.

Are we seeing similarly increased variance in our weather systems? Increased incidence of floods, storms, droughts, and *El Nino* [66] might be signs of greater changes

FIGURE 19.3 The rapid transition in a shallow lake from clear water dominated by one (a) or many species (b) of plant, supported mutually by various aquatic animals, to a turbid state (c) where large plants and the animals that depend on and support them are excluded in favor of microscopic algae and cyanobacteria. The turbid water shades the sediments so large plants cannot root and the loss of plants removes the refuge for zooplankton (from fish predators), so they can no longer sweep the water clear of phytoplankton. Shown are communities in water taken from Hickling Broad, Norfolk, UK, for a study of alternative stable states [47, 48]. (a) Shows clear water with *Chara intermedia*, (b) shows clear water with *Myriophyllum spicatum* L, *Ceratophyllum demersum* L, *Enteromorpha intenstinalis* L, *Potamogeton pectinatus* L, *Lemna trisulca* L, and *Elodea canadensis* Michx, (c) shows the turbid, green, water of the degraded state, empty of plants, and dominated by species of microscopic algae and cyanobacteria. (*See insert for color representation of this figure.*)

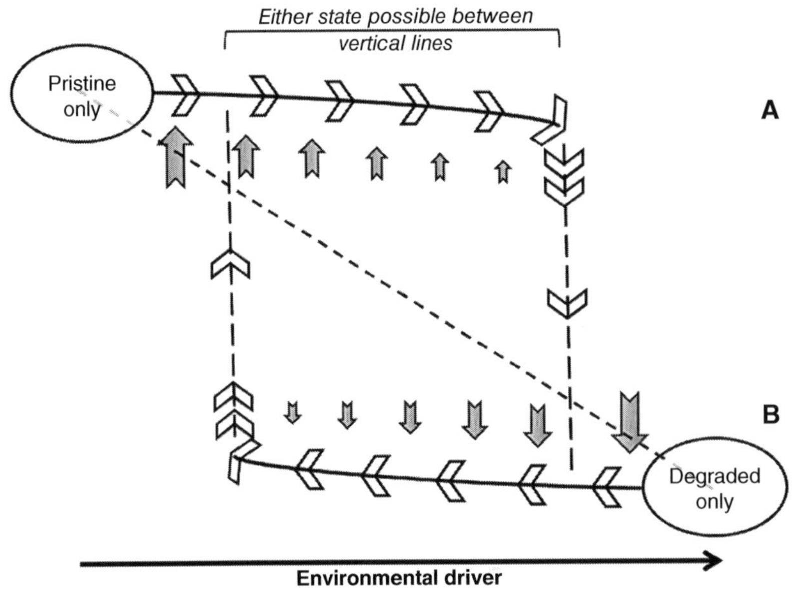

FIGURE 19.4 Developed to describe non-linear regime shift in shallow lakes (Figure 19.3) and thought to be common to many dynamic systems, the environmental driver is an increasing perturbation, e.g. the rate of natural resource use by human industrial society, or an increasing nutrient loading in shallow lakes. When the driver is very weak (left-hand end), the system sits in a pristine state (upper ellipse on curve: A); when it is very strong (right-hand end), it can only reside in a very degraded state (lower ellipse on curve: B). In between these two extremes (vertical dashed lines) either state is possible depending on the effectiveness of stabilizing mechanisms, which maintain the current state until the strength of the driver forces it to flip into a new state. In shallow lakes, these mechanisms include the presence or absence of plants and grazers of algae, and the shading potential (color) of the water. The diagonal dashed line represents the expected state of the lake (in a linear system) as the driver increases in the absence of stabilizing buffers, e.g. as if the color of the water changed from clear to green by the increasing addition of a green dye. Chevron arrows indicate the shift of the actual system state with the driver. Vertical arrows indicate the strength of the stabilizing mechanisms, which are weakened according to the strength of the driver. A = good quality system; B = poor quality system. *Source*: Based on Scheffer et al. [46] and Moss [26].

to come. What will those changes be and how will they affect communities and populations around the world? Time will tell, but one thing is sure. The time for delay tactics is over, and governments around the world need to grasp the nettle of climate change mitigation and adaptation, biodiversity loss, and the intimately linked issue of global inequity very soon. Another thing that is highly likely is that drastic change will be imposed on us by environmental forces if we do not act first to mitigate the drivers of environmental change [67]. This is a bit like a person heading for Type Two diabetes. Although the reasons for reducing calorific intake may have been clear and strong for decades, only once the symptoms and diagnosis of diabetes are confirmed is the patient forced to adopt a diet likely to mitigate the effects of the disease, by which time it is too late to avoid it. The same appears true for the global system: we will have to live within the limits of the planet eventually. Far better to take the foot off the pedal that is driving environmental change while it is still under our control than be battered by the winds of global ecological change after a global systemic flip. The Earth system is vastly more powerful than human industrial systems.

19.3 THE ROLES, IMPORTANCE, AND VALUE OF ECOSYSTEMS AND BIODIVERSITY

19.3.1 Valuing Ecosystems

Ecosystems and the biological functions comprising them, though important in themselves, also keep us alive and happy. In recent years people have tried to quantify what is provided by nature to human society. They do this because these values are usually ignored when decisions are made that have impacts on the environment. Economic means of expressing nature's values are regarded as an effective way to bring conservation of nature to the attention of politicians, economists, and corporations whose interests are primarily in increasing profitability and revenue, and who understand the value of money but not the value of life support systems (*sensu* [68]) originating from biogeochemical cycles, themselves driven by the interplay of organisms and their environment. As a result of the "unvalued" nature of these systems they are lost in the pursuit of business objectives.

The problem is a failure to recognize the value inherent in nature, or by the mistaken impression that there is sufficient nature for it not to matter if a habitat is converted for commercial purposes which, after all, increases GDP. Species and biodiversity, unless they are marketable products such as fish, timber, and tourism, are regarded as having no direct value, and their preservation is thus not a good economic reason to deny development. Other reasons and values are not economically important, and can therefore be disregarded. There is an urgent need to address this incorrect attitude to reflect the true value of nature.

It is hard to pin down what that value is. There is no doubt that many people experience a less than tangible feeling of connectedness when in contact with semi-wild nature, and it can have a strong effect on attitudes. People speak of feeling happier when an intimate relationship with nature is achieved, and the bonding inherent in a close association with nature is strong even though it is hard to define or link empirically to physiology. There is a growing body of evidence for mental health benefits from contact with nature (e.g., [69–72]) and some for physical health benefits, e.g. blood pressure and general perceived health [73, 74] derived from contact with nature. People report "feelings of rejuvenation and renewal" when in association with natural environments [75], but these feelings cannot yet be directly traced in a functional sense. In a review, Russell et al. [76] selected four "channels" of human–ecosystem interaction in an attempt to quantify the ways in which people benefit from contact with nature. These were *knowing* (about an ecosystem), *perceiving* (one from afar), *interacting* (with nature in a sensory way), and *living within* a natural setting. These were used in an attempt to discover how experiences with nature can fulfil basic human needs (*sensu* [77]). They found that, although both mental and physical health were positively affected by contact with nature, as well as being essential for the lives of the world's indigenous groups, evidence is scant for direct causal links with positive effects. Though the authors confirmed links, some stronger than others, they came no nearer to defining how nature comes to be precious to many people in an emotional way.

The strikingly high popularity of natural history TV programs suggests that there is a large cohort of people who enjoy learning about nature, whether or not they report feelings similar to others. A solution to environmental destruction, which is now overwhelming natural systems, might be found somewhere in the empowerment of that cohort to express the values they place on ecosystems over and above those that they place on consumer goods with built-in obsolescence. The continuous production and consumption of those consumer goods is a major cause of environmental destruction and degradation.

Studies in valuing nature have predominantly concentrated on quantification of the direct values of products (such as timber, in which market price is used) or regulatory functions such as control of water volume (e.g. in preventing flooding downstream, in which avoided costs of damage and repair are estimated). Approaches to economic valuation of personal preferences and other less tangible aspects of ecosystems include estimates of economic value based on stated preferences of people in terms of "willingness to pay" or "willingness to accept" to experience or forego a particular place, scene, or quality of life (e.g. [78]). Such economic valuations are useful for judging the importance people attribute to their experiences of nature but are always inadequate as estimates of value [79, 80]. They are also a long way from becoming part of our normal currency and of becoming a means of comparing the value of goods and services from the natural and anthropogenic world.

19.3.2 How Ecosystem Processes Provide Functions and Services

Ecological *processes* are things species do, such as grow. They do this by taking up nutrients from the external environment and using them to build tissues, perhaps using photosynthesis. Thus the *functions* flowing from these processes are increase of biomass and uptake of nutrients. These functions might provide *services* to human society in the form of food or fiber (biomass) and removing elements and compounds that otherwise might build up to harmful concentrations in the environment.

Individual organisms and species complete different functions as they struggle to survive and reproduce. Apart from growing, animals (also some fungi and bacteria) consume vegetation or hunt, build nests and protective structures, migrate, and die. In doing so they redistribute nutrients, remove pests, pollinate, hydrate the landscape, and contribute to the life cycles of other species, performing as they go various services that are useful to human society. The list of "ecosystem services," as they have become known, provided by species groups living naturally in the wild is very long, and the benefits they provide to human society are vital to life. These services include provision of oxygen to keep us alive, biomass that we can eat, mineralization of organic matter to provide fertility, formation of soil, storage of carbon, pollination of crops, removal of pathogens, assimilation of wastes, and provision of clean water. Nature also provides people with emotional support that helps give a sense of identity, place, and contact with the natural world, and is a huge educational and scientific resource. A handy categorization of the benefits provided to human society by ecosystem services, divided into regulating (e.g. climate, flooding), provisioning (e.g. food, timber), cultural (e.g. recreation, personal), and supporting (e.g. pollination, soil formation) services was provided by the Millennium Ecosystem Assessment (Figure 19.5).

The processes that lead to provision of ecosystem services are carried out by organisms as they interact with their environment. The more organisms and species present, the more processes will be carried out, and the more secure the

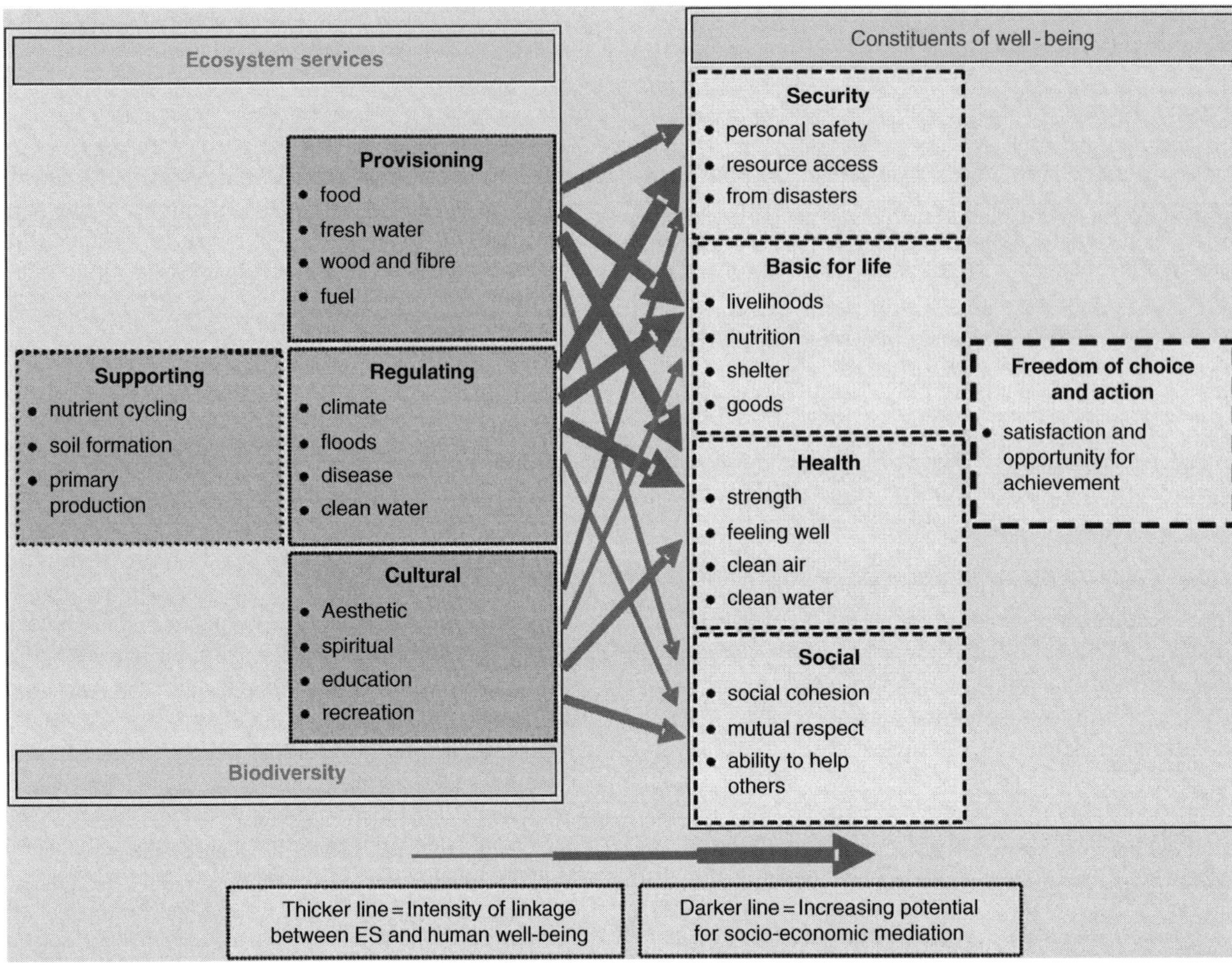

FIGURE 19.5 The ecosystem services (ES) classification system of the Millennium Ecosystem Assessment. The processes and functions of ecosystems provide the services categorized on the left panel and lead to the benefits shown on the right. *Source*: Adapted from MA [81].

performance of the functions that result from those processes. It is not quite as simple as it sounds because the organisms that perform different functions are arranged into what are called functional groups. For example, water plants are a functional group in that they all remove nutrients (chiefly nitrogen and phosphorus) from the water column and store them in their tissues. This particular function is nutrient removal, which is important for reducing the likelihood of toxic algal blooms [82], preventing the local extinction of nutrient-sensitive plants [47], and for direct economic benefits such as reducing the water cleaning costs faced by water companies.

Numerous different species might perform the same function and the exact species doing so is irrelevant from the perspective of the water manager seeking that outcome, even though other differences will be present in the biology of particular species. This had led to the generally accepted concept of "redundancy" [83], wherein species able to perform the same task are regarded as surplus in a functional sense. In other words, once sufficient species exist for the function to be performed, others are considered redundant. This rather anthropocentric argument was challenged conceptually by Naeem [84] and recent studies (e.g. [85]) have reinforced the idea that "redundant" species are not redundant at all. They might perhaps be better called "contingent," since they all have separate roles and can be regarded as insurance against the disappearance of other species. The role in ecosystem resilience of comparatively rare species [86, 87] is testament to the inadequacy of the term "redundancy" and the importance of preserving biodiversity for its "existence value" as well as the continued provision of ecosystem services for their contributions to human well-being.

Figure 19.6 provides a simplified illustration of how biodiversity and ecological functioning are linked. It is easy to understand how higher biodiversity might lead to a greater range of functions and thus to increased provision of ecosystem services than would be expected in a comparatively species-poor ecosystem. In an ecosystem subject to high concentrations of toxins or ecological degradation due to invasions or conversion to industrial use, there might be

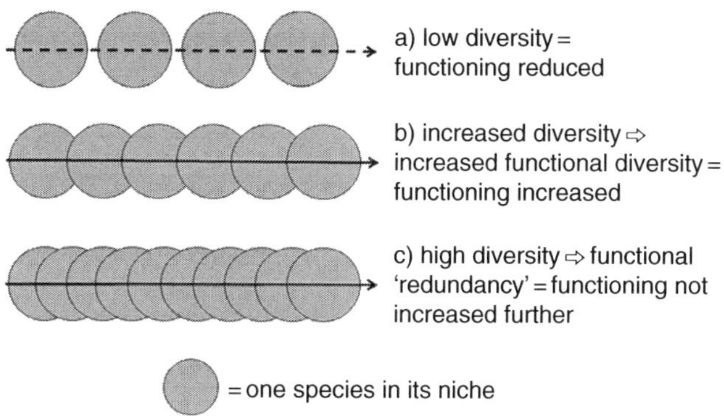

FIGURE 19.6 The links between species diversity and ecosystem functioning. Circles represent a species in its niche in an ecosystem. The number of species is related to ecosystem functioning. Different species constitute "functional groups" that together perform some function that leads to provision of an ecosystem service; e.g. they might pollinate plants. (a) Few species in a depauperate ecosystem might not be able to perform all of the potential services likely for that type of ecosystem. (b) Greater numbers of species (and individuals) are likely to be able to perform the full range of potential services. (c) Once the functional groups are saturated by species, the presence of more species will not increase service provision, leading to the erroneous assumption that these species are "redundant." We now know that these extra species ensure ecosystem functioning and provide continuity of service provision at times of disease or disturbance (see main text).

insufficient species present to fulfil all of the functional roles potentially operating in a similar but intact ecosystem.

We now know that high biodiversity itself maintains healthy ecosystems and that reduced biodiversity impairs ecosystem functioning and leads to reduced provision of ecosystem services in a degraded ecosystem [88]. In other words, we need very high biodiversity in order to maintain ecological systems on this planet, i.e. the systems that keep us alive and healthy.

19.3.3 The Role of Policy

The importance of high biodiversity, e.g. of "contingent" species, is not obvious to policy-makers and corporations, and may only become clear when the consequences of its loss are considered, particularly in economic terms. The primary outcome of climate change is to reduce the provision of ecosystem services. The industrioeconomic causes of climate change (e.g. [67]) also drive degradation of habitats across the globe. These are of primary importance to human society: sea level rise removes the terrestrial environment, threatening major world cities; weather changes such as in wind speeds, precipitation and temperature reduce habitability, reduce the capacity of land to provide food, and force migrations of people and animals, exposing them to new threats. All these things together cause extinctions of so many species that we are now in a "mass extinction" event [89] with the consequent expectation that more than 75% of all species will be lost [90].

Climate change eradicates very high numbers of species. Conversely, high biodiversity represents an insurance against many of the impacts of climate change, and we need to preserve it. The effects of changing climate are seen, for example, in increasing occurrences of extreme weather that can reduce crop yields. Many genetic traits, such as tolerance to drought and heat, and comparatively lower requirements for inputs of nutrients and irrigation can be found in wild relatives of crops [91]. Preserving the natural habitats and thus the genetic diversity of wild species can help us maintain provision of our core food crops. For example, in rice (*Oryza sativa*), 22 related wild species provide a valuable source of traits that can be cross-bred into cultivated varieties, providing protection against diseases such as blight and stunt, and insect pests such as planthoppers and stemborers [92, 93]. It will thus pay in economic as well as ethical terms to preserve this genetic diversity. Environmental change now under way will present increasing challenges to food production for a larger global population in a climate-changed world. This has given a new impetus to plant breeding [94], to investigating naturally occurring traits, and importantly to agroecological methods [95] of increasing food production that center on the maintenance of biodiversity through sustainable environmental management.

19.4 ECOSYSTEM SERVICES SUPPORT HUMAN EXISTENCE

Ecosystems in all biomes produce a vast resource for human existence [81] and the loss and degradation of ecosystems has severe consequences for human populations at the global scale, but especially for the rural poor in the short term. Table 19.1 provides an outline of the most obvious services separated into categories of provision, regulation, cultural, and habitat, as defined by The Economics of Ecosystems and Biodiversity [96].

One of the most cited examples of ecosystem services is pollination, which has become the focus of campaigns against neonicotinoid pesticides that are thought to be implicated in an inability of bees to find their way home, leading to colony collapse [99], although a recently seen parasite, *Nosema ceranae*, found in the guts of bees apparently also plays a part in bee memory [100]. The presence of wild bee species near to croplands supports populations of honeybees (*Apis mellifera* L.) and augments pollination significantly, reinforcing the point that maintenance of areas of wild nature is important for crop production [101]. A recent focus on pollination as an ecosystem service has usefully served to remind the wider public of the need for preservation of habitats, but in fact to concentrate on pollination alone would be

TABLE 19.1 Ecosystem service (ES) categories and examples, based on the format of TEEB [96].

ES listed in TEEB are from Balmford et al. [97] and [98]	Examples of uses and components
Provisioning services	
Provision of food	Supporting primary and secondary production of wild, arable, livestock, and marine foods, including pollination, soil ecology and wild relatives with valuable genes
Water provision	Rainfall, drinking supply, and irrigation through provision of precipitation
Water regulation	Flood prevention and drought alleviation through impoundment, retentivity of ecosystems, ice, and timing of flows
Water purification	Removal of pathogens and nutrients in soils and vegetation, often in preparation for potable supply
Fuels and fibers (includes timber, sugar, and oils)	Domestic and industrial uses including for construction, textile manufacture, fuel, and fuel manufacture from biomass crops, including cultivation of algae
Genetic resources	Increases the ability to resist new pests and diseases, particularly with increasing climate change, and enhances plant and animal breeding and sources of new crops
Medicinal and other biochemical resources	Sources of nutrients, pharmaceuticals, industrial and agricultural chemicals, including for "green chemistry"
Ornamental resources	Flowers, plants, plumage, essences removed for ornamental use, and exotic species in parks and cages
Regulating services	
Air and environmental quality in towns	Air cleaning, urban-heat-island mitigation, noise reduction, evaporative cooling, maintenance of health
Climate regulation	Global temperature control (greenhouse effect), water distribution, heat transference, interactions between soils, vegetation, oceans, and atmosphere
Moderation of extreme events	Protection from inland and marine flooding, storm surges and tsunamis, landslides, avalanches, high winds, and fires
Prevention of erosion	Water retention, soil stability from roots and protection by above-ground biomass
Maintenance of soil quality	Fertility and nutrient cycling including nitrogen fixation supported by microbes and organic materials
Pollination	Provision of fruit and vegetable crops and support for other species and soil ecology
Biological control	Predator and parasite control of pest and disease species. Plant and animal genetic defenses against pests and diseases
Habitat services	
Life-cycle support	Dispersal of nutrients and resources, and support in terms of food, breeding, nursery, and rest sites for stages in the lives of mobile species
Maintenance of genetic diversity	Evolution and flows of genetic materials on micro and macro scales to increase, diversify, and mix genes of all species
Cultural and amenity services	
Aesthetics	Organisms and settings of beauty and awe
Inspiration for culture, art and design	Source materials and inspiration for cultural arts, norms and icons, and for biomimetic design
Information for cognitive development	Resource for education and awareness of natural processes and laws that are common to all aspects of life
Spiritual development	Focus for sense of place, emotional, superstitious, and religious meaning and feelings of connection that enhance well-being
Opportunities for recreation and tourism	Sites and habitats for sports, nature hobbies, and tourism

The category of Habitat Services was included in TEEB [96] in place of the category "Supporting Services" (e.g., nutrient cycling, soil formation, and primary production, considered to be contained within ecological processes) used in MA [81] in recognition of the values of what are often termed "biological hotspots" and the general interconnectedness of ecosystem components, which together underpin genetic diversity.

to miss out on the much larger picture of the essential need for functioning ecosystems to support human survival.

It might seem glib to point out that everything we eat, drink, and breathe has been provided for us by the natural environment, but it is in fact true. The vastness of nature and inherent time lags of global processes mean that society can dump a lot of waste materials in the environment before the global scale consequences of them can be seen, even by scientists looking for them. Arrhenius predicted global warming from industrial activity in 1896 [102] and clear evidence for the effect has been available since the 1950s [103, 104]. Mounting evidence and scientific concern about anthropogenic changes in the climate (e.g. [105]) built up to impossible-to-ignore levels in the 1980s until eventually, the Earth Summit (UNFCCC; http://www.un.org/geninfo/bp/enviro.html/) was held in 1992. Even then, it was not until the publication, 14 years later, of a report into the economic consequences of climate change [106] that some governments started to act decisively.

Politics being what it is, further and continued delays to mitigation of the problem are thought now to have jeopardized the ability of the planet to maintain conditions suitable for the existence of human civilization. A study by the Stockholm Resilience Institute [107] concluded that the comparatively equable climate since the end of the last ice age, which has seen the development and expansion of human civilization, is threatened because of biophysical changes to the regulatory feedback mechanisms that hold the planet in comparative homeostasis. Rockström et al. [107] established the concept of "planetary boundaries," the breaching of which is considered likely to cause planet-wide instability sufficient to cause unacceptable and irreversible environmental change. The nine planetary boundaries (Figure 19.7), discussed further in Steffen et al. [108], include climate change, biosphere integrity, land-use change, biochemical flows, and acidification of the oceans.

The causes of the approach and transgression of different boundaries are to be found in economic and industrial expansion to provide high standards of living to increasing numbers of people on Earth. That is, while millions live in extreme poverty in poor environments, the rest of us are living considerably beyond what the Earth can sustain. Development that changes the structure of a community from subsistence agriculture to production of some commercial export crop, benefits only comparatively wealthy people who may work for the company and so be able to afford ostensibly equivalent services to those provided by nature. Poor people are left impoverished, with associated health impairments and susceptibilities [109]. While the primary bearers of the consequences of environmental degradation are the rural poor [81], especially in the short term, there is an urgent need to act decisively and fast [110, 111] in order for the global population, and the economy that acts in its name, to keep to within the defined planetary boundaries by reducing stocks of atmospheric carbon, enhancing ecosystems and biodiversity, and maintaining healthy populations.

Sustainability is essentially about living within our means because most environmental problems are caused either by taking from the environment at a rate greater than that provided by nature or by dumping into the environment at a rate greater than can be assimilated by nature. If society cannot comply, resources are depleted or waste materials such as CO_2, build up in the environment. Dense populations

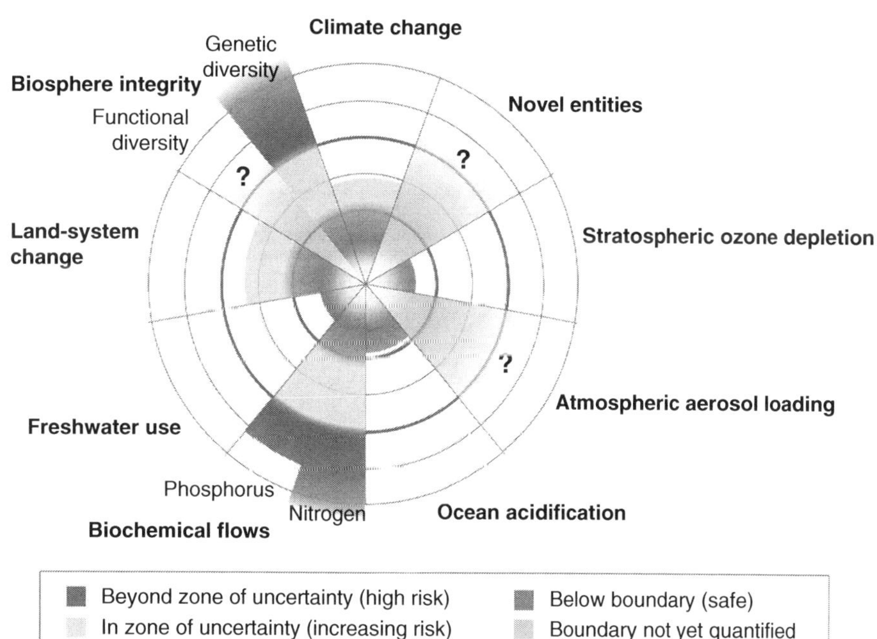

FIGURE 19.7 The planetary boundaries view of environmental degradation at the global scale. Theoretical boundaries have been quantified for seven of the nine identified resource- and waste-related variables (atmospheric CO_2 concentration, ozone depletion, ocean acidification, flows of phosphorus and nitrogen, freshwater use, land-system change, and biodiversity). Genetic diversity (through extinctions) and phosphorus and nitrogen use (from agriculture and other industry) have already transgressed limits determined as "safe" for humanity, and another two are close to transgression. Green (the central shaded area) = safe; yellow (the intermediate shaded areas) = uncertain; red (the outer shaded areas) = transgressed; grey (indicated by '?') = boundary not yet quantified. *Source*: Steffen et al. [108].

and the requirement of globalization to trade for reasons of "comparative advantage" lead to hot-spots of environmental degradation, generally where legislation is lax enough to permit environmental and social exploitation. The practice of 'comparative advantage' also leads to widespread transfer of species outside their natural ranges, where they disrupt local ecologies, causing extinctions that can reduce provision of ecosystem services. The effects of the globalized economy can be seen in ecosystem changes that affect local people through land-use change and displacement of populations, which may fuel conflict. Increasingly, equity is seen as a core component of global sustainability, and the equable treatment of all people is frequently at the root of ecological problems faced by communities [112].

19.5 HUMAN USES OF FRESHWATER ECOSYSTEMS

Societies in developing and developed countries use freshwater ecosystems and their functions for cultural and commercial uses including tourism and recreational purposes, agriculture, fishing, domestic water supply, industry, wastewater treatment and many other uses. Roggeri [113] listed the functions and their values provided by wetlands in the Amu Darya Delta, Uzbekistan. These include the regulatory and supporting functions of groundwater recharge, climate stabilization, a windbreak effect, organic matter recycling and toxicant storage, soil stabilization, maintenance of biodiversity, and provision of nursery and migratory habitats for fish, waterfowl, and mammals. Together, these functions provide support to living conditions; health in the form of quality of consumed products, especially water; subsistence production and support to subsistence for people external to the Delta; support to economic production and added value and employment for economic sectors, especially irrigated agriculture; fisheries and fish canneries; muskrat-based industry and other animal and agriculture products; and salt and dust control. Ecologically, regulatory functions include support to ecosystems beyond the wetlands, control of desertification in the Aral Sea, and support to biodiversity in cross-boundary and intergenerational ways. The wetlands also support human settlements, agriculture, animal husbandry, nature conservation, tourism, and recreation [113], and produce food, fodder, clothing, medicinal plants, and products. They also provide emotional support in the form of aesthetic living conditions, and are important for the cultural identity of the people of Karakalpakstan.

Equivalent lists could be assembled for rivers, from which a range of ecosystem services are derived, including culture, religion and inspiration, transport, power generation, industrial and domestic supply, fisheries, irrigation, and fertility. The presence of dams has restricted the functions of floodplain wetlands and migrating species in many rivers [114]. Floodplain wetlands are one of the most valuable environments in terms of protecting downstream areas from flooding [115], but are often difficult to justify to planners and policy-makers.

Agriculture has frequently been listed as a provider of ecosystem services (e.g. [116]) but is probably better considered as a major industrialized user of ecosystem services. Much modern large-scale agriculture, i.e. sponsored by global agribusiness, compacts and impoverishes soils, often leading to erosion. It displaces people, abstracts water in unsustainable quantities, pollutes water bodies, eradicates wild nature, and involves the use of heavy machinery and transport that consumes large amounts of fossil fuels [117]. It also leads to the dependence of rural people on expensive technologies that must be bought in from elsewhere. The problems associated with how to farm sensitively and provide healthy food for people, i.e. as opposed to just another commodity for corporate purposes, include the fairly recent concept of agroecology [118], which focuses on maintenance of soil ecology, recycling of materials, diets, empowerment of populations, and well-being of growers and consumers [119–122]. This type of approach, which retrieves some ancient practices of soil husbandry for modern use, can be a pointer to sensitive use of nature to protect biodiversity and ecosystem functioning alongside aspects of human well-being such as dignity, cultural identity, and provision of food, water, and waste management services. For wetlands, this is known as "wise use" (www.ramsar.org/about/the-wise-use-of-wetlands) and some parallel principles are provided for agroecology by Koohafkan et al. [123].

In terms of water cleansing, ecosystems, and in particular wetlands and forested areas, remove particulate matter and the nutrients and toxins bound to them, for example from soil in runoff water. Available nutrients are removed through uptake by plants and microorganisms and particulate matter is settled out in areas of still water. In an update to the well-known 1997 paper [124a] on valuation of ecosystem services, Costanza et al. [124b] valued waste treatment by wetlands at \$219 500 ha^{-1} a^{-1} at 2011 prices. For this as well as for reasons of biodiversity protection, large amounts of money are spent in Western economies in cleaning up polluted lakes through the reduction of nutrient loading and the reintroduction of water plants in attempts to restore biodiversity and hence provision of ecosystem services [125].

We need to protect natural systems before we can be confident that they will continue to provide benefits that we need for the continuation of society in the West and for the improvement of conditions for people in developing countries and regions. The global population is growing to an unknown extent and industrial expansion and urbanization is advancing. These factors appear to be in direct conflict with a desire for the restriction or contraction of human incursions into wild nature that is needed all over the world in every biome and every community. How can we square this circle?

19.6 THE IMPACTS OF DEGRADATION OF NATURE AND ECOSYSTEM FUNCTIONING ON HUMAN HEALTH

Schumacher [126] lamented that, "*Modern man does not experience himself as a part of nature but as an outside force destined to dominate and conquer it. He even talks of a battle with nature, forgetting that, if he won the battle, he would find himself on the losing side.*" Here, surely, is the crux of the problem. Human society has come to regard nature as "other." It is unruly, invasive and not readily controlled, so where it gets in the way of development it must be eradicated or subdued, and then replaced by human-made structures.

Technology can replace the services formerly provided by nature, but only some of them, and at a cost. Concrete infrastructure to protect a community from flooding, for example, can cost much more than protection of the ecosystems that would prevent flooding in the first place, as the United States Army Corps of Engineers showed clearly for the Charles River floodplains [127]. The costs of destruction of nature can far exceed the benefits of development. An example is provided by the clearance of mangroves for conversion to shrimp farms, which is widespread in tropical coastal areas, and typically results in loss of vital ecosystem services used by subsistence fishing communities. Mangroves are the source of their livelihood and security but are replaced with aquaculture jobs for a minority of inhabitants and profits for external investors. In a study of shrimp aquaculture in Sri Lanka, Gunawardena and Rowan [128] found that the costs to the community were between 6 and 11 times greater than the benefits from the new shrimp farm and that there were "disproportionately large impacts on traditional livelihoods and social welfare."

Mangroves constitute natural barriers to flooding and destruction from storm surges [129] as well as providing nursery habitats for fish, with the ecological integrity and area of mangroves tending to be proportional to their protective function [130]. Back in 2002, Balmford et al. [131] concluded that if economic costs from lost provision of eco system services were fully accounted, no new development of wild nature could be justified in the context of the global economy. This relationship is not clear to the lay person, who over time has witnessed both the conversion of nature to industrial use and vast improvements in human health [132]. The destruction of ecosystems has been used to fuel economic growth and has funded medical and scientific advances, but we have reached a tipping point where the natural ecosystems, or rather the losses of them, are coming back to bite us.

Globally, the benefits of maintaining healthy ecosystems easily outweigh the costs in lost development. The total economic value of global ecosystem services was calculated to be \$124.8 trillion a^{-1} in 2011 [124], compared with global GDP in 2011 of £75.2 trillion a^{-1}, which equaled just 60% of the (otherwise uncounted) revenue from ecosystems. The loss of ecosystems because of their destruction for agriculture and other industrial uses equated to an economic loss of \$4.3 trillion a^{-1} in 2011, with the costs of degradation, rather than total loss, of remaining ecosystems being unknown. Much of these losses are considered "non-tradable public benefits," in other words they cannot enter the market economy, and the costs of their loss fall mainly on poor communities and future generations [124]. These impacts are largely invisible to politicians and businesses who operate in the market economy on short timescales and they do not prevent the conversion of wild nature to enable the transfer of economic benefits from the many to the few [132].

The changing pattern of weather associated with climate change has meant droughts and floods [133], especially in tropical and subtropical areas [134]. In addition to climate change effects, irrigation and impoundment schemes for agriculture and hydroelectric power alter land uses and water availability for many people. Land-use change has removed the ability of people to support themselves using their local environmental resources. In developed nations, even where water is a scarce resource, it is still used with impunity by industry [135]. In developing nations, international corporations seek to control the vital, subsistence, water resources of local people for industrial purposes [136, 137]. The technological solutions to these problems are designed for rich societies, are often more expensive than the natural processes they replace, have negative externalities in the form of further environmental degradation, and feed the system that is rendering the global environment dysfunctional by increasing the GDP of developed nations. Are there viable alternatives to technological responses?

The catalogue of environmental destruction is large and ranges from local to global in scale, from removal of forest for crops [138] or mangroves for fish farms [139, 140], both of which displace people and their livelihoods, to very large-scale climatic changes induced by global warming [141, 142]. There are numerous examples of the environmental destruction of freshwater and wetland habitats and their impacts on human health and well-being. Pollution incidents can lead to severe, but relatively short-term threats to human health such as the devastating cyanide spills (from gold mines) into the River Tisza from Romania, flowing through Hungary in January and February 2000, which not only killed all of the fish but also more than 90% of all phytoplankters and zooplankters [143]. This incident caused the total collapse of the river ecosystem and removed the ability of local people to fish, for subsistence and commercially, or use the most polluted water for drinking or recreationally [144].

Habitat destruction may be a factor in increasing the risk of malaria in some cases. For instance, malarial risk along the Peruvian Amazon has been shown to increase with an increase in the presence of the Malarian vector, *Anopheles darlingi*, in wetlands where there had been a reduction in forest cover and ecological disturbance [145]. Conversely, the destruction of wetland habitats via drainage has led to a reduction in the incidents of malaria [146]. Other authors have linked extreme weather events and climate change to outbreaks of waterborne diseases (e.g. [147, 148]). Prüss-Ustün et al. [149] estimated that ecosystem changes were responsible for over 25% of all disease occurrences worldwide.

The relationship between ecosystem degradation and human health was reviewed by Myers et al. [131], who concluded that human health and propensity to disease are inextricably linked to human-driven ecosystem change.

Equally, some ecological functions can provide "disservices," a term given to negative characteristics of ecosystems, and perhaps a primary example is the infection of people by disease vectors such as *Simulium* spp., e.g. *S. guianense*, where onchocerciasis (river blindness) is a very high price to pay for the particulate filtering capabilities of blackfly larvae, which remove particulate matter from sub-Saharan African and Yemeni streams. This particular example contains some hope because medical technology and advances in methods look to eradicate the disease within decades [150].

The dynamic interactions between the diseases, their invertebrate and vertebrate hosts, local lifestyles, agriculture and livestock, and ecosystem type were studied by Leach et al. [151] across gradients of vegetation and precipitation in Kenya (Rift Valley Fever via *Aedes* mosquitoes and livestock), Zambia (trypanosomiasis via tsetse flies), Zimbabwe (trypanosomiasis via tsetse flies), Sierra Leone (Lassa fever via rodents), and Ghana (henipaviruses via bats). Occurrences such as heavy rain and standing water in pools, behind dams, and in irrigation systems are related to infection rates. The research revealed that changes in land use and behavior patterns bring poor people in particular into contact with areas of vegetation and irrigation systems that are linked with mosquito abundance, and this combines with poverty to increase Rift Valley Fever vulnerability.

In an interesting review of perceptions in colonial explorers of mangrove forests in the nineteenth century from sources such as expedition reports in English, Friess [152] found many references to ecosystem services and disservices. The explorers were there for colonial purposes and looking for export products, and their views of mangroves reflected their immediate experiences. References reviewed referred particularly to disease (a disservice reported in 35% of articles), provisioning of products, especially wood for building and fuel, and food; "gloominess/scariness" and threats from animals and people (other disservices); and observations of the importance of regulating services in the form of wave attenuation and erosion control. Cultural services for local populations were mentioned less often, mostly by anthropologists who perhaps were less focused on finding products to support an empire. More recently, Leach et al. [151] found that people visiting woodland–wildlife interfaces are more at risk from tsetse fly infection, while those staying near agriculture are less vulnerable. The ecosystem service benefits derived from woodland and wild species can thus be balanced in some circumstances by greater exposure to risk.

An environment that has been degraded following economic development is likely to have greater consequences for the poorer members of society than for wealthier people. Zoonotic diseases tend to align with climatic patterns of temperature and rainfall, and are shifting with climate change, which appears to be having an increasing influence on infection rates, e.g. through the incidence of sudden downpours associated with El Nino [153, 154]. Studies of disease transmission in Mauritania by Caminade et al. [155] found that incidences of heavy rain that occurred following a dry spell after the main rainy season were associated with disease outbreaks of Rift Valley Fever among farmers. This was consistent with the watering of animals at ponds during the rainless period. The sudden heavy rain coincided with the hatching of already infected (by the mother) *Aedes* mosquitoes that would normally be expected to hatch in the next rainy season.

On local scales, the interactions between human communities and ecosystems can be both positive and negative in the same areas. They vary between type of ecosystem, type of community group, and type of interaction, often representing trade-offs between the provision of benefits and occurrence of disease [151], and it is these interactions that underpin local health-based decisions.

Degradation of wetland habitats also has a more indirect impact on human health, such as via the release of greenhouses gases to atmosphere and the reduction or elimination of their function as a carbon sink. For example, Murdiyarso et al. [156] have shown that mangrove destruction contributes up to 30% of carbon emissions from land-use change in Indonesia. It is known that ponds, shallow lakes, and wetlands are extremely efficient at sequestering carbon, perhaps more so than woodland [157], and Gorham [158] has estimated that 20–25% of global soil carbon storage is in wetlands. If such wetlands become degraded then the release of carbon from their sediment stores will contribute further to climate change, bringing with it the myriad of consequential health risks to the global human population. In addition, it is known that warming stimulates the emission of the greenhouse gases CO_2 and CH_4 from wetlands, further driving warming [159, 160].

Less obvious but equally harmful are the changes involving species ranges, invasive species, local extinctions, pollution, and other ecological processes. For example, species whose survival depends on being particularly mobile are likely to suffer losses of links between functional groups such as between pollination, seed dispersal, and pest control, and environmental change will enable new links that facilitate the spread of invasive species that can be disease vectors and carriers of toxic substances [161]. This type of change degrades the ecological integrity that maintains functioning, leading to unpredictable changes. Environmental drivers such as climate change, increased runoff, pollution, water abstraction for irrigation, or loss of genetic diversity can change an ecosystem in fundamental and unexpected ways.

19.7 LINKING HUMAN WELL-BEING AND ENVIRONMENTAL SUSTAINABILITY

A key message of this chapter is that the industrioeconomic society, climate change, environmental loss and degradation, mass extinction, continued widespread poverty, landlessness, and inequity, as well as extreme wealth are all part of the same thing, and are all implicated in localized issues of health and well-being as much as they are global issues of critical concern. We have failed to live within the ecological limits of the planet or regions within it. Contrary views, such as that of Simon [162], however noble and humanist, are based on assumptions of substitutability, which works only up to the point where there is something to substitute with. Technological idealism like *Jurassic Park* notwithstanding, extinctions are of course forever, and the fact that we are in a mass extinction and that CO_2 is building up in the environment to damaging levels while its natural sinks' ability to absorb it are declining [163] is testament to real limits that cannot be argued away. Technological solutions such as carbon capture and storage and mirrors in space might have applications but can be expected to have serious limitations, will be much more expensive than natural methods of achieving the same ends, and will have other consequences. They also will not address parallel problems such as ocean acidification and biodiversity loss [164, 165]. We should have noted that when renewable resources started running out, it is an indication that we are consuming too fast for the environment to match us in production. Ecosystem degradation centers around biodiversity loss in terms of species and individuals, and is both a result of ecosystem degradation and a cause of loss of functioning.

The expansion of human activity inevitably causes biodiversity losses. The Millennium Ecosystem Assessment ([81], p. 15) made projections, starting from existing conditions and trends, based on its findings for two alternative paths of structural development: "increased globalization" and "increased regionalization." Each of these paths was given two plausible scenarios (an economic growth-driven future or an environmental management-driven future). None of the four projections resulted in environmental sustainability. All of them revealed continued or increased pressures leading to further biodiversity losses, and the two economic growth-driven scenarios pointed to greatly increased environmental harm and biodiversity losses.

It is clear that economic and industrial expansion and conversion of nature leads to net loss of nature, its integrity in terms of interrelationships between species, and the accompanying provision of services to human health and well-being. The Millennium Ecosystem Assessment [81] concludes that ecosystem destruction must be reversed in order to escape its trajectory of human and environmental disaster. The UN initiative, TEEB (www.teebweb.org) develops guidance for different sectors such as business and agriculture (see www.teebweb.org/our-publications/all-publications/) in order to help economic decision-makers understand methods of incorporating true values into decision-making.

Fortunately, a great deal of thought and effort has been invested in identifying potential solutions to current global problems and their implications, and finding solutions to improve human health by rehabilitating ecosystem health is central to them. Summaries have been published in the popular press (see, for example, [126, 166–169]) and the academic literature (e.g., [63, 170–178]). This is not new. An example from the mid-nineteenth century is the poor state of the River Thames, which was linked to cholera outbreaks in the city of London. In that instance, engineering works were carried out to channel sewage away from the river, resulting in local water quality improvements and an end to the cholera outbreaks [179]. Today, we are faced with similar problems repeated on a vast scale. Potential solutions will need to be applied across scales from the global to the local, but within an agreed framework of environmental sustainability.

Locally, it seems likely that the reintroduction of predatory organisms and the habitats they need for their life cycles reduces disease incidence by reducing population sizes of disease vectors, thus protecting human health [180]. For example, in Senegal, human infection rates of schistosomiasis were reduced after the reintroduction of the freshwater water prawn *Macrobrachium vollenhovenii*, which preys on the aquatic snails that act as hosts of the Schistosoma helminth parasite [181]. It is likely that other environmentally transmitted diseases that reside in organisms could be controlled via predation from a natural predator [181], which in turn can be supported by conservation or ecosystem restoration work. Ecosystem restoration of plants has also been shown to reduce disease. For instance in the USA, the removal of an invasive tree species reduced the incidence of Lyme disease [182].

Marine Protected Areas (MPAs) and artificial reefs have been found to improve biodiversity, fish catches, and recreational activities, thus bringing economic and well-being benefits [183], though MPAs have also been found to exclude local economic activity by forcing the fishing into other areas [184]. This illustrates the need for habitat restoration to consider ways to accommodate habitats and wise use, rather than to prohibit use across large areas.

Research to elaborate the ecology of disease organisms is still in its infancy and much is still unknown about the actual links between ecological disease vectors, ecosystem service restoration, and human health [181, 185]. Myers et al. [131] call for a more systematic study of the links between ecosystems and human health to ensure that policy-makers are informed about the health implications of environmental changes. This assumes that we can recognize what constitutes "sustainable" environmental change, which is unlikely given the complexity of ecosystems. The examples above illustrate that improvements to ecosystem health have brought about improvements to human health, and this link, once it is widely recognized and valued, will go some way to informing management decisions.

19.8 POLITICAL CHALLENGES IN ACHIEVING SUSTAINABILITY

Linking ecological degradation with climate change and the economy can be a conceptual leap that depends on the understanding of cause and effect in ecology. The world is vast and has an enormous capacity to assimilate wastes and provide resources for human society. When these resources are compromised on local scales, we see removal of ecosystem services for local people, resulting in increased poverty and disease as remaining resources are too few to support local communities. At a global scale, the buildup of materials added by human society, whether that be CFCs or CO_2, can be seen to directly change the Earth system services (i.e. global-scale ecosystem services) that maintain the homeostasis that is essential for modern human civilization to survive. In between the local and the global, we observe degradation of functioning at the biome level: the tropical and boreal forests, the Arctic and Antarctic ice sheets, and permafrost. At all scales, the degradation of these ecological systems has impacts on the biogeochemical processes and feedbacks that maintain relative global stasis (e.g., [107, 108]).

This represents a challenge to orthodox economics that relies on continual economic growth to fend off recession and all the misery that frequently results from it. How can we maintain healthy and fulfilling lifestyles, particularly in the developed world where most resources are consumed and discarded, without continuing to degrade the biosphere that keeps us alive? How can the people of developing countries achieve similar well-being without making the same mistakes as made by the developed countries? There is no place here for a meaningful discussion of the global inequities that maintain poor people in poverty, but the economic structures that perpetuate global and regional inequity are intimately connected with environmental management (see, for example, [186–189]).

In answer to the question, "what can we do?," the first thing must surely be to recognize the problem for what it is – a global threat to the survival of everything we hold dear. The second is to find ways of defining sustainability as the real bottom line in any economic or political decision.

A regional or global response to ecological degradation from local to international scales is likely to require local solutions embedded in a framework organized nationally and coordinated internationally. This is expected to involve the transformation of societal norms to incorporate the principles of sustainability in all aspects and at all levels of society [67, 190, 191]. This means not just taking economic steps to invest in preserving biodiversity and access to nature. It means fundamental transformations of economic, industrial, and social systems [58], with new forms of living, working, and playing across all sectors of society [192], especially in the developed world and in the practices and policy of transnational organizations.

The call for "transformational adaptation" as a defense against environmental change has come about as part of a final realization that we depend on nature for our health and well-being on the human scale, and economically at local, regional, and global levels. A certain amount of resistance to the idea is present and inevitable in the future because it threatens to undermine the national and corporate economic strategies of the world, but those very economies represent the major driver of global ecological degradation and even potential societal collapse, and the conclusions that lead to their necessary overhaul are irresistible [61, 193].

The declared purpose of economic growth is to support well-being [194], yet economic growth is seen to be the main driver of ecological destruction. It has been shown that people report higher subjective well-being (SWB) according to income only up to a point (Figure 19.8), after which wealth becomes increasingly irrelevant to well-being. This picture is complicated by expressions of SWB that correlate partly with income in the short term only, with other factors becoming more important in the medium and long term [196, 197].

Since economic growth is the key driver of environmental destruction and has direct benefits for its declared purpose while personal incomes are comparatively low (Figure 19.8), it makes sense that a sensible goal for sustainability should be the maximization of well-being for everybody while economic growth, as measured by GDP, is minimized. In recent decades, well-being as estimated by the Genuine Progress Indicator, appears to have fallen [198]. This is not a good reason to pursue unfettered economic growth that is steadily eroding the means of survival of humans and most other species. What to do about it is for others to argue, but we can no longer ignore the calls to find a new economics.

An emerging argument that might hold promise, appears logical, and is likely to be achievable is the promotion of increased local self-reliance for resources including food and the assimilation of society's wastes. This would represent a reversal of globalization and thus would place the political and economic forces of the global economy in opposition to environmental sustainability. Localization follows the principle of subsidiarity, which states that

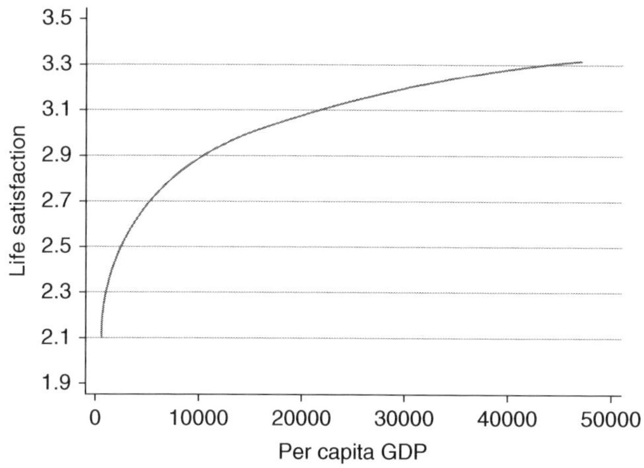

FIGURE 19.8 The relationship between reported life satisfaction and per capita GDP (US$ 2008) from 195 pooled observations for 89 countries. $R^2=0.452$. *Source*: Easterlin and Angelescu [195].

decisions should be made at the lowest practicable level [126, 199] and would be dependent on achieving international agreements to limit industrial expansion and pollution. The problem, according to TEEB, is that our current developed world economic model is based on the assumption that we may extract, consume, and produce without any awareness or concerns about limits or consequences for the environment or for others, and that the global economy rewards the accumulation of things and wealth rather than societal well-being and nature [200]. That is the enormity of the conceptual task before us. Technology can help but it cannot solve the problems on its own, as any reasonably careful study of the literature on sustainability will reveal (e.g. [190, 201, 202]). It will be solved by political and economic cooperation, for example, as elaborated by Nobel Prize winner Elinor Ostrom [203].

19.9 SUMMARY – FUTURE CHALLENGES

While the casual observer might be excused for ignoring the physical, chemical, and biological connections linking an ecosystem or a city with the wider world that sustains it, policy-makers and economists cannot legitimately claim that excuse, and only ineptitude to govern and create the rules for society can explain it. The scientific evidence is that intact ecosystems are able to perform functions that are essential for human existence, as well as for our health and well-being. Biodiversity is required to ensure the control of beneficial and harmful organisms, as well as pests, diseases, and parasites. In healthy freshwater or wetland ecosystems some parasites, pests, and pathogens of humans will inevitably exist. Humans, like all organisms, are part of nature and there will be infections. We have seen that freshwater and wetlands are essential for regulating our atmosphere, mitigating against climate change, and providing food and fuel, clean water, and flood reduction. We cannot therefore seek to eradicate wetlands in the hope of eradicating naturally occurring waterborne diseases. We must instead trust science to find cures and prevention. We do know though that many wetland diseases are more prolific in polluted waters, in waters where a predator is missing, invasive species are present, or where the environment is degraded or damaged. We require intact ecosystems, characterized by biological interactions between functional groups to ensure that such diseases, pests, and parasites do not become overly abundant.

However, we live in a world where world leaders typically are not scientists. When Moss [53] listed the backgrounds of world leaders, whether presidents, prime ministers, or dictators, 75% had emerged from the fields of economics, politics, administration, law, business, military, and engineering. One only out of nearly two hundred had any kind of ecological background (a Master's degree in maritime studies), and he was later deposed in a military coup [204].

There is a mismatch between what is understood scientifically and what we know to be the "known unknowns" in the fields of human health and environmental sustainability and what is understood and appreciated in the developed world. As discussed, this lack of understanding is pervasive throughout society from the "normal" citizen up to national leaders. We cannot expect ecosystems to be protected and restored to the benefit of human health if those societies that cause the most damage do not perceive that the two are linked. Stern et al. [205] noted that lack of understanding of environmental problems stems from its complex and varied causes, wide geographic area, and complex and wide-ranging solutions. Studies show that the public understanding of science has little improved for decades [206]. It is unrealistic for all of Westernized society to gain an education sufficient to understand these complex environmental problems, but developing a general appreciation of the life support given by the natural world is possible, as is a trust in scientists to advise, educate, and play their role in forming policy and economic decision-making. The integration of environmental scientific realities into global economic and policy-making, at all levels and all geographies, is a necessity for human health, well-being, and ultimately survival.

REFERENCES

1. Tansley, A.G. (1935). The use and abuse of vegetational concepts and terms. *Ecology* 16: 284–307.
2. Dudgeon, D., Arthington, A.H., Gessner, M.O. et al. (2005). Freshwater biodiversity: importance, threats, status and conservation challenges. *Biological Reviews* 81: 163–182. doi: 10.1017/S1464793105006950.
3. Mora, C., Tittensor, D.P., Adl, S. et al. (2011). How many species are there on Earth and in the ocean? *PLoS Biology* 9 (8): e1001127. doi: 10.1371/journal.pbio.1001127.

4. De Vos, J.M., Joppa, L.N., Gittleman, J.L. et al. (2015). Estimating the normal background rate of species extinction. *Conservation Biology* 29 (2): 452–462. doi: 10.1111/cobi.12380.
5. Pimm, S.L. and Raven, P. (2000). Extinction by numbers. *Nature* 403: 843–845.
6. Forterre, P. (2015). The universal tree of life: an update. *Frontiers in Microbiology* 6 ((Article 717)): doi: 10.3389/fmicb.2015.00717.
7. Williams, T.A., Foster, P.G., Cox, C.J., and Embley, M. (2013). An archaeal origin of eukaryotes supports only two primary domains of life. *Nature* 504: 231–236. doi: 10.1038/nature12779.
8. Hug, L.A., Baker, B.J., Anantharaman, K. et al. (2016). A new view of the tree of life. *Nature Microbiology* 1: 16048. doi: 10.1038/nmicrobiol.2016.48.
9. Bardgett, R.D. and van der Putten, W.H. (2014). Belowground biodiversity and ecosystem functioning. *Nature* 515: 505–511.
10. Landeweert, R., Hoffland, E., Finlay, R.D. et al. (2001). Linking plants to rocks: ectomycorrhizal fungi mobilize nutrients from minerals. *Trends in Ecology & Evolution* 16 (5): 248–254.
11. Head, I.M., Jones, D.M., and Röling, W.F.M. (2006). Marine microorganisms make a meal of oil. *Nature Reviews* 4: 173–182.
12. Bragg, J.R., Prince, R.C., Harner, E.J., and Atlas, R.M. (1994). Effectiveness of bioremediation for the Exxon Valdez oil spill. *Nature* 368 (6470): 413–418.
13. Morelli, J. (2011). Environmental sustainability: a definition for environmental professionals. *Journal of Environmental Sustainability* 1 (1): 2. doi: 10.14448/jes.01.0002.
14. Lovelock, J. (1979). *Gaia: A New Look at Life on Earth*. Oxford: Oxford University Press.
15. Wilkinson, D. (2006). *Fundamental Processes in Ecology, an Earth Systems Approach*. Oxford and New York: Oxford University Press.
16. Berner, R.A. (2003). The long-term carbon cycle, fossil fuels and atmospheric composition. *Nature* 426: 323–326.
17. Mannion, A.M. (2009). Reconstructing environmental changes over the last 3 million years. In: *Principles of Environmental Sciences* (ed. J.J. Boersema and L. Reijnders). Dordrecht: Springer. doi: 10.1007/978-1-4020-9158-2_3.
18. Ruddiman, W.F. (2001). *Earth's Climate. Past and Future*. San Francisco: W.H. Freeman and Company.
19. Richardson, J.S., Zhang, Y., and Marczak, L.B. (2010). Resource subsidies across the land–freshwater interface and responses in recipient communities. *River Research and Applications* 26 (1): 55–66.
20. Helfield, J.M. and Naiman, R.J. (2006). Keystone interactions: salmon and bear in riparian forests of Alaska. *Ecosystems* 9 (2): 167–180.
21. Ben-David, M., Hanley, T.A., and Schell, D.M. (1998). Fertilization of terrestrial vegetation by spawning Pacific salmon: the role of flooding and predator activity. *Oikos* 83: 47–55.
22. Hilderbrand, G.V., Hanley, T.A., Robbins, C.T., and Schwartz, C.C. (1999). Role of brown bears (*Ursus arctos*) in the flow of marine nitrogen into a terrestrial ecosystem. *Oecologia* 121: 546–550.
23. Arizono, N., Yamada, M., Nakamura-Uchiyama, F., and Ohnishi, K. (2009). Diphyllobothriasis associated with eating raw pacific salmon. *Emerging Infectious Diseases* 15 (6): 866.
24. Valiente-Banuet, A., Aizen, M.A., Alcántara, J.M. et al. (2015). Beyond species loss: the extinction of ecological interactions in a changing world. *Functional Ecology* 29 (3): 299–307.
25. Colinvaux, P.A. (1980). *Why Big Fierce Animals Are Rare*. London.: George Allen & Unwin.
26. Moss, B. (2014). Models from ecohydrology and hydrobiology can inform our human future. *Ecohydrology and Hydrobiology* 14: 21–32.
27. Willoughby, J.R., Fernandez, N.B., Lamb, M.C. et al. (2015). The impacts of inbreeding, drift and selection on genetic diversity in captive breeding populations. *Molecular Ecology* 24 (1): 98–110.
28. Kohr, L. (1957). *The Breakdown of Nations*. Totnes: Routledge and Kegan Paul, London. Republished in 2001 by Green Books.
29. Wackernagel, M. and Rees, W. (1998). *Our Ecological Footprint: Reducing Human Impact on the Earth*. New Society Publishers.
30. Rees, W.E. and Moore, J. (2013). Ecological footprints, fair earth-shares and urbanization. In: *Living within a Fair Share Ecological Footprint* (ed. R. Vale and B. Vale), 3–32. Abingdon: Earthscan.
31. Ripple, W.J., Estes, J.A., Beschta, R.L. et al. (2014). Status and ecological effects of the world's largest carnivores. *Science* 343 (6167): 1241484.
32. Parker, G.A. and Maynard Smith, J. (1990). Optimality theory in evolutionary biology. *Nature* 348: 27–33.
33. Tanabe, K. and Namba, T. (2005). Omnivory creates chaos in simple food web models. *Ecology* 86: 3411–3414. doi: 10.1890/05-0720.
34. Whittaker, R.H. (1975). *Communities and Ecosystems*. New York: Macmillan Publishing.
35. Austin, M.P. (1992). Modelling the environmental niche of plants: implications for plant community response to elevated CO_2 levels. *Australian Journal of Botany* 40 (5): 615–630.
36. Estes, J.A., Terborgh, J., Brashares, J.S. et al. (2011). Trophic downgrading of planet earth. *Science* 333 (6040): 301–306.
37. Cornelissen I.J. (2015). Eutrophication, Nile Perch and Food–Web Interactions in South-East Lake Victoria. PhD Thesis, Wageningen University.
38. van Rijssel, J.C., Hecky, R.E., Kishe-Machumu, M.A., and Witte, F. (2017). Changing ecology of Lake Victoria cichlids and their environment: evidence from C^{13} and N^{15} analyses. *Hydrobiologia* 791 (1): 175–191.
39. Kolding, J., Medard, M., Mkumbo, O., and van Zwieten, P. (2014). Status, trends and management of the Lake Victoria Fisheries, Chapter 4. In: *Inland Fisheries Evolution and Management. Case Studies from Four Continents*, Fisheries and Aquaculture Technical Paper 579 (ed. R.L. Welcomme, J. Valbo-Jorgensen and A.S. Halls). Rome: FAO.
40. Ricklefs, R.E. (2004). A comprehensive framework for global patterns in biodiversity. *Ecology Letters* 7: 1–15. doi: 10.1046/j.1461-0248.2003.00554.x.

41. Ricklefs, R.E. (2006). Evolutionary diversification and the origin of the diversity-environment relationship. *Ecology* 87: S3–S13. doi: 10.1890/0012-9658(2006)87[3:EDATOO]2.0.CO;2.
42. Fischer, J., Lindenmayer, D.B., and Manning, A.D. (2006). Biodiversity, ecosystem function, and resilience: ten guiding principles for commodity production landscapes. *Frontiers in Ecology and the Environment* 4 (2): 80–86.
43. Sundstrom, S.M., Allen, C.R., and Barichievy, C. (2012). Species, functional groups, and thresholds in ecological resilience. *Conservation Biology* 26: 305–314. doi: 10.1111/j.1523-1739.2011.01822.x.
44. Irvine, K., Moss, B., and Balls, H.R. (1989). The loss of submerged plants with eutrophication II. Relationships between fish and zooplankton in a set of experimental ponds, and conclusions. *Freshwater Biology* 22: 89–107.
45. Scheffer, M., Hosper, S.H., Meijer, M.-L. et al. (1993). Alternative equilibria in shallow lakes. *Trends in Ecology and Evolution* 8: 275–279.
46. Scheffer, M., Carpenter, S., Foley, J.A. et al. (2001). Catastrophic shifts in ecosystems. *Nature* 413: 591–596.
47. Barker, T., Hatton, K., O'Connor, M. et al. (2008a). Effects of nitrate load on submerged plant biomass and species richness: results of a mesocosm experiment. *Fundamental and Applied Limnology/Archiv für Hydrobiologie* 173 (2): 89–100.
48. Barker, T., Hatton, K., O'Connor, M. et al. (2008b). Control of ecosystem state in a shallow, brackish lake: implications for the conservation of stonewort communities. *Aquatic Conservation: Marine and Freshwater Ecosystems* 18 (3): 221–240. doi: 10.1002/aqc.819.
49. Staver, A.C., Archibald, S., and Levin, S. (2011). Tree cover in sub-Saharan Africa: rainfall and fire constrain forest and savanna as alternative stable states. *Ecology* 92 (5): 1063–1072.
50. Faust, K. and Raes, J. (2012). Microbial interactions: from networks to models. *Nature Reviews Microbiology* 10 (8): 538–550.
51. Dakos, V., Kéfi, S., Rietkerk, M. et al. (2011). Slowing down in spatially patterned ecosystems at the brink of collapse. *The American Naturalist* 177 (6): E153–E166.
52. Folke, C., Carpenter, S.R., Walker, B. et al. (2010). Resilience thinking: integrating resilience, adaptability and transformability. *Ecology and Society* 15 (4): 20.
53. Moss, B. (2012). *Liberation Ecology: The Reconciliation of Natural and Human Cultures*, Excellence in Ecology series No. 24. Oldendorf/Luhe: International Ecology Institute See p. 11, Figure 1.4.
54. Scheffer, M., Carpenter, S.R., Lenton, T.M. et al. (2012). Anticipating critical transitions. *Science* 338 (6105): 344–348.
55. Brulle, R.J. (2014). Institutionalizing delay: foundation funding and the creation of US climate change counter-movement organizations. *Climatic Change* 122 (4): 681–694.
56. Carter, N. and Clements, B. (2015). From "greenest government ever" to "get rid of all the green crap": David Cameron, the Conservatives and the environment. *British Politics* 10 (2): 204–225.
57. Sprinz, D. and Vaahtoranta, T. (1994). The interest-based explanation of international environmental policy. *International Organization* 48 (1): 77–105.
58. Eriksen, S., Aldunce, P., Bahinipati, C.S. et al. (2011). When not every response to climate change is a good one: identifying principles for sustainable adaptation. *Climate and Development* 3 (1): 7–20.
59. Feola, G. (2015). Societal transformation in response to global environmental change: a review of emerging concepts. *Ambio* 44 (5): 376–390.
60. IPCC (2014a). Summary for policymakers. In: *Climate Change 2014: Impacts, Adaptation, and Vulnerability. Part A: Global and Sectoral Aspects. Contribution of Working Group II to the Fifth Assessment Report of the Intergovernmental Panel on Climate Change* (ed. C.B. Field, V.R. Barros, D.J. Dokken, et al.), 1–32. Cambridge, UK and New York: CUP.
61. Lorek, S. and Spangenberg, J.H. (2014). Sustainable consumption within a sustainable economy – beyond green growth and green economies. *Journal of Cleaner Production* 63: 33–44.
62. O'Brien, K., Reams, J., Caspari, A. et al. (2013). You say you want a revolution? Transforming education and capacity building in response to global change. *Environmental Science and Policy* 28: 48–59.
63. Termeer, C.J.A.M., Dewulf, A., and Biesbroek, R. (2017). Transformational change: governance interventions for climate change adaptation from a continuous change perspective. *Journal of Environmental Planning and Management* 60 (4): 558–576. doi: 10.1080/09640568.2016.1168288.
64. Brock, W. and Carpenter, S. (2006). Variance as a leading indicator of regime shift in ecosystem services. *Ecology and Society* 11 (2).
65. Wang, R., Dearing, J.A., Langdon, P.G. et al. (2012). Flickering gives early warning signals of a critical transition to a eutrophic lake state. *Nature* 492 (7429): 419–422.
66. Cai, W., Borlace, S., Lengaigne, M. et al. (2014). Increasing frequency of extreme El Niño events due to greenhouse warming. *Nature Climate Change* 4 (2): 111–116.
67. Beddoe, R., Costanza, R., Farley, J. et al. (2009). Overcoming systemic roadblocks to sustainability: the evolutionary redesign of worldviews, institutions, and technologies. *PNAS* 106 (8): 2483–2489.
68. Odum, E.P. and Barrett, G.W. (1971). *Fundamentals of Ecology*. Philadelphia: W.B. Saunders & Co.
69. Bratman, G.N., Hamilton, P.J., and Daily, G.C. (2012). The impacts of nature experience on human cognitive function and mental health. *Annals of the New York Academy of Sciences* 1249: 118–136.
70. Maller, C., Townsend, M., Pryor, A. et al. (2006). Healthy nature healthy people:'contact with nature' as an upstream health promotion intervention for populations. *Health Promotion International* 21 (1): 45–54.
71. Mayer, S.F. and Frantz, C.M. (2004). The connectedness to nature scale: a measure of an individuals feeling in community with nature. *Journal of Environmental Psychology* 24: 503–515.

72. Nisbet, E.K., Zelenski, J.M., and Murphy, S.A. (2011). Happiness is in our nature: exploring nature relatedness as a contributor to subjective well-being. *Journal of Happiness Studies* 12 (2): 303–322.
73. Shanahan, D.F., Bush, R., Gaston, K.J. et al. (2016). Health benefits from nature experiences depend on dose. *Scientific Reports* 6 (28551).
74. Sugiyama, T., Leslie, E., Giles-Corti, B., and Owen, N. (2008). Associations of neighbourhood greenness with physical and mental health: do walking, social coherence and local social interaction explain the relationships? *Journal of Epidemiology and Community Health* 62 (5): e9.
75. Willis, C. (2015). A human needs approach to revealing nature's benefits for visitors to the coast. *Annual Review of Environment and Resources* 47: 422–428. doi: 10.1111/area.12206.
76. Russell, R., Guerry, A.D., Balvanera, P. et al. (2013). Humans and nature: how knowing and experiencing nature affect well-being. *Annual Review of Environment and Resources* 38 (1): 473–502.
77. Max-Neef, M.A. (1991). *Human Scale Development: Conception, Application and Further Reflections*. New York: Apex Press.
78. Bateman, I.J., Carson, R.T., Day, B. et al. (2002). *Economic Valuation with Stated Preference Techniques: A Manual*. Cheltenham: Edwards Elgar.
79. Kosoy, N. and Corbera, E. (2010). Payments for ecosystem services as commodity fetishism. *Ecological Economics* 69 (6): 1228–1236.
80. Spash, C.L. and Aslaksen, I. (2015). Re-establishing an ecological discourse in the policy debate over how to value ecosystems and biodiversity. *Journal of Environmental Management* 159: 245–253.
81. MA (2005). *Ecosystems and Human Well-being – Synthesis. Millennium Ecosystem Assessment*. Washington, DC: Island Press The four MA development scenarios can be found from page 15.
82. Smith, V.H. and Schindler, D.W. (2009). Eutrophication science: where do we go from here? *Trends in Ecology and Evolution* 24 (4): 201–207.
83. Walker, B.H. (1992). Biodiversity and ecological redundancy. *Conservation Biology* 6: 18–23.
84. Naeem, S. (1998). Species redundancy and ecosystem reliability. *Conservation Biology* 12 (1): 39–45.
85. Scheffer, M., Vergnon, R., van Nes, E.H. et al. (2015). The evolution of functionally redundant species; evidence from beetles. *PloS One* 10 (10): e0137974.
86. Flöder, S., Jaschinski, S., Wells, G., and Burns, C.W. (2010). Dominance and compensatory growth in phytoplankton communities under salinity stress. *Journal of Experimental Marine Biology and Ecology* 395 (1): 223–231.
87. Mouillot, D., Bellwood, D.R., Baraloto, C. et al. (2013). Rare species support vulnerable functions in high-diversity ecosystems. *PLoS Biology* 11 (5): e1001569. doi: 10.1371/journal.pbio.1001569.
88. Hooper, D.U., Adair, E.C., Cardinale, B.J. et al. (2012). A global synthesis reveals biodiversity loss as a major driver of ecosystem change. *Nature* 486 (7401): 105–108.
89. Ceballos, G., Ehrlich, P.R., Barnosky, A.D. et al. (2015). Accelerated modern human–induced species losses: entering the sixth mass extinction. *Science Advances* 1 (5): e1400253. doi: 10.1126/sciadv.1400253.
90. Barnosky, A.D., Matzke, N., Tomiya, S. et al. (2011). Has the Earth's sixth mass extinction already arrived? *Nature* 471 (7336): 51–57.
91. Warschefsky, E., Penmetsa, R.V., Cook, D.R., and von Wettberg, E.J. (2014). Back to the wilds: tapping evolutionary adaptations for resilient crops through systematic hybridization with crop wild relatives. *American Journal of Botany* 101 (10): 1791–1800.
92. Brar, D.S. and Khush, G.S. (1997). Alien introgression in rice. *Plant Molecular Biology* 35 (1): 35–47.
93. Khush, G.S. and Brar, D.S. (2017). Alien introgression in rice. *Nucleus* 60 (3): 251–261. doi: 10.1007/s13237-017-0222-7.
94. Tester, M. and Langridge, P. (2010). Breeding technologies to increase crop production in a changing world. *Science* 327 (5967): 818–822.
95. Altieri, M.A., Funes-Monzote, F.R., and Petersen, P. (2012). Agroecologically efficient agricultural systems for smallholder farmers: contributions to food sovereignty. *Agronomy for Sustainable Development* 32 (1): 1–3.
96. TEEB (2010). *The Economics of Ecosystems and Biodiversity: Ecological and Economic Foundations* (ed. P. Kumar (ed.)). London/Washington: Earthscan.
97. Balmford, A., Rodrigues, A.S.L., Walpole, M. et al. (2008). *The Economics of Ecosystems and Biodiversity; Scoping the Science*. Cambridge, UK: European Commission.
98. EASAC (2009). Ecosystem Services and Biodiversity in Europe. European Academies Science Advisory Council, Policy Report 9. The Royal Society, London.
99. Fischer, J., Mueller, T., Spatz, A.K. et al. (2014). Neonicotinoids interfere with specific components of navigation in honeybees. *PLoS One* 9 (3): e91364.
100. Wolf, S., McMahon, D.P., Lim, K.S. et al. (2014). So near and yet so far: harmonic radar reveals reduced homing ability of *Nosema* infected honeybees. *PLoS One* 9 (8): e103989.
101. Greenleaf, S.S. and Kremen, C. (2006). Wild bees enhance honey bees' pollination of hybrid sunflower. *Proceedings of the National Academy of Sciences* 103 (37): 13890–13895.
102. Arrhenius, S. (1896). On the influence of carbonic acid in the air upon the temperature of the ground. *Philosophical Magazine and Journal of Science* 5 (41): 237–276.
103. Keeling, C.D. (1958). The concentration and isotopic abundances of atmospheric carbon dioxide in rural areas. *Geochimica et Cosmochimica Acta* 13 (4): 322–334.
104. Keeling, C.D., Bacastow, R.B., Bainbridge, A.E. et al. (1976). Atmospheric carbon dioxide variations at Mauna Loa observatory, Hawaii. *Tellus* 28 (6): 538–551.
105. Baes, C.F., Goeller, H.E., Olson, J.S., and Rotty, R.M. (1977). Carbon dioxide and climate: the uncontrolled experiment: possibly severe consequences of growing CO_2 release from fossil fuels require a much better understanding of the carbon cycle, climate change, and the resulting impacts on the atmosphere. *American Scientist* 65 (3): 310–320.

106. Stern, N., Peters, S., Bakhshi, V. et al. (2006). *Stern Review on the Economics of Climate Change*. London: HM Treasury.

107. Rockström, J., Steffen, W., Noone, K. et al. (2009). A safe operating space for humanity. *Nature* 461: 472–475.

108. Steffen, W., Richardson, K., Rockström, J. et al. (2015). Planetary boundaries: guiding human development on a changing planet. *Science* 347 (6223): 1259855.

109. Whitmee, S., Haines, A., Beyrer, C. et al. (2015). Safeguarding human health in the Anthropocene epoch: report of the Rockefeller Foundation – *Lancet* commission on planetary health. *The Lancet* 386 (10007): 1973–2028.

110. Ehrlich, P.R. and Ehrlich, A.H. (2013). Can a collapse of global civilization be avoided? *Proceedings of the Royal Society B* 280 (1754): 20122845.

111. Schramski, J.R., Gattie, D.K., and Brown, J.H. (2015). Human domination of the biosphere: rapid discharge of the earth-space battery foretells the future of humankind. *Proceedings of the National Academy of Sciences* 112 (31): 9511–9517.

112. Godfray, H.C., Beddington, J.R., Crute, I.R. et al. (2010). Food security: the challenge of feeding 9 billion people. *Science* 327 (5967): 812–818.

113. Roggeri, H. (2009). Wetland evaluation in developing countries. In: *The Wetlands Handbook* (ed. E. Maltby and T. Barker), 569–600. Oxford: Wiley-Blackwell.

114. Richter, B.D., Postel, S., Revenga, C. et al. (2010). Lost in development's shadow: the downstream human consequences of dams. *Water Alternatives* 3 (2): 14.

115. Opperman, J.J., Luster, R., McKenney, B.A. et al. (2010). Ecologically functional floodplains: connectivity, flow regime, and scale. *Journal of the American Water Resources Association* 46 (2): 211–226.

116. Gordon, L.J., Finlayson, C.M., and Falkenmark, M. (2010). Managing water in agriculture for food production and other ecosystem services. *Agricultural Water Management* 97 (4): 512–519.

117. Rhodes, C.J. (2017). The imperative for regenerative agriculture. *Science Progress* 100 (1): 80–129.

118. Holt-Giménez, E. and Altieri, M.A. (2013). Agroecology, food sovereignty, and the new green revolution. *Agroecology and Sustainable Food Systems* 37 (1): 90–102.

119. Carolan, M. (2013). *Reclaiming Food Security*. Abindon: Earthscan (Routledge).

120. Mason, P. and Lang, T. (2017). *Sustainable Diets; How Ecological Nutrition Can Transform Consumption and the Food System*. Abingdon: Earthscan (Routledge).

121. Méndez, V.E., Bacon, C.M., and Cohen, R. (2013). Agroecology as a transdisciplinary, participatory, and action-oriented approach. *Agroecology and Sustainable Food Systems* 37 (1): 3–18.

122. O'Rourke, M.E., DeLonge, M.S., and Salvador, R. (2017). Insights from agroecology and a critical next step: integrating human health. *Agroecology and Sustainable Food Systems* 41 (7): 880–884.

123. Koohafkan, P., Altieri, M.A., and Gimenez, E.H. (2012). Green agriculture: foundations for biodiverse, resilient and productive agricultural systems. *International Journal of Agricultural Sustainability* 10 (1): 61–75.

124. (a) ostanza, R., d'Arge, R., de Groot, R., Farber, S., Grasso, M., Hannon, B., Limburg, K., Naeem, S., O'Neill, R.V., Paruello, J., Raskin, R.G., Sutton, P., and van der Belt, M., 1997. The value of the world's ecosystem services and natural capital. Nature 387: 253–260. (b) Costanza, R., de Groot, R., Sutton, P. et al. (2014). Changes in the global value of ecosystem services. *Global Environmental Change* 26: 152–158.

125. Cooke, G.D., Welch, E.B., Peterson, S., and Nichols, S.A. (2016). *Restoration and Management of Lakes and Reservoirs*. Boca Raton: CRC Press.

126. Schumacher, E.F. (1973). *Small Is Beautiful. A Study of Economics as if People Mattered*. London: Blond and Briggs Republished in 1999 by Hartley & Marks Publishers, Vancouver.

127. Weiskel P.K., Barlow L.K. and Smieszek T.W. (2005). Water Resources and the urban Environment, Lower Charles River Watershed, Massachusetts, 1630-2005. US Geological Survey Circular 1280.

128. Gunawardena, M. and Rowan, J.S. (2005). Economic valuation of a mangrove ecosystem threatened by shrimp aquaculture in Sri Lanka. *Environmental Management* 36 (4): 535–550.

129. Barbier, E.B. (2006). Natural barriers to natural disasters: replanting mangroves after the tsunami. *Frontiers in Ecology and the Environment* 4 (3): 124–131.

130. Spalding, M., McIvor, A., Tonneijck, F.H. et al. (2014). *Mangroves for Coastal Defence. Guidelines for Coastal Managers and Policy Makers*, 42. Wetlands International and The Nature Conservancy.

131. Balmford, A., Bruner, A., Cooper, P. et al. (2002). Economic reasons for conserving wild nature. *Science* 297 (5583): 950–953.

132. Myers, S.S., Gaffikin, L., Golden, C.D. et al. (2013). Human health impacts of ecosystem alteration. *Proceedings of the National Academy of Sciences* 110 (47): 18753–18760.

133. IPCC (2014b). *Climate Change 2014 – Impacts, Adaptation and Vulnerability: Regional Aspects. Intergovernmental Panel on Climate Change*. Cambridge University Press 2014, December 29.

134. Brouwer, R., Akter, S., Brander, L., and Haque, E. (2007). Socioeconomic vulnerability and adaptation to environmental risk: a case study of climate change and flooding in Bangladesh. *Risk Analysis* 27 (2): 313–326.

135. Chappells, H., Medd, W., and Shove, E. (2011). Disruption and change: drought and the inconspicuous dynamics of garden lives. *Social and Cultural Geography* 12 (7): 701–715.

136. Downey, L., Bonds, E., and Clark, K. (2010). Natural resource extraction, armed violence, and environmental degradation. *Organization and Environment* 23 (4): 417–445.

137. Franco, J., Mehta, L., and Veldwisch, G.J. (2013). The global politics of water grabbing. *Third World Quarterly* 34 (9): 1651–1675.

138. Gibbs, H.K., Ruesch, A.S., Achard, F. et al. (2010). Tropical forests were the primary sources of new agricultural land in the 1980s and 1990s. *Proceedings of the National Academy of Sciences* 107 (38): 16732–16737.

139. Paul, B.G. and Vogl, C.R. (2011). Impacts of shrimp farming in Bangladesh: challenges and alternatives. *Ocean and Coastal Management* 54 (3): 201–211.

140. Veuthey, S. and Gerber, J.F. (2012). Accumulation by dispossession in coastal Ecuador: shrimp farming, local resistance and the gender structure of mobilizations. *Global Environmental Change* 22 (3): 611–622.
141. Biermann, F. and Boas, I. (2010). Preparing for a warmer world: towards a global governance system to protect climate refugees. *Global Environmental Politics* 10 (1): 60–88.
142. Scheffran, J. and Battaglini, A. (2011). Climate and conflicts: the security risks of global warming. *Regional Environmental Change* 11 (1): 27–39.
143. Pálfai, I. (2003). *The Hydrology and Rehabilitation of Oxbow Lakes of the Hungarian Section of the River Tisza*. Szeged: Ministry of Environmental Control and Water Management.
144. Macklin, M.G., Brewer, P.A., Balteanu, D. et al. (2003). The long term fate and environmental significance of contaminant metals released by the January and March 2000 mining tailings dam failures in Maramureş County, upper Tisa Basin, Romania. *Applied Geochemistry* 18 (2): 241–257.
145. Vittor, A.Y., Pan, W., Gilman, R.H. et al. (2009). Linking deforestation to malaria in the Amazon: characterization of the breeding habitat of the principal malaria vector, *Anopheles darlingi. The American Journal of Tropical Medicine and Hygiene* 81 (1): 5–12.
146. Keiser, J., Singer, B.H., and Utzinger, J. (2005). Reducing the burden of malaria in different eco-epidemiological settings with environmental management: a systematic review. *Lancet Infectious Diseases* 5 (11): 695–708.
147. Epstein, P.R. (2001). Climate change and emerging infectious diseases. *Microbes and Infection* 3 (9): 747–754.
148. Rose, J.R., Epstein, P.R., Lipp, E.K. et al. (2001). Climate variability and change in the United States: potential impacts on water- and food-borne diseases caused by microbiologic agents. *Environmental Health Perspectives* 109 ((Suppl.)): 211–221.
149. Prüss-Ustün, A., Bonjour, S., and Corvalán, C. (2008). The impact of the environment on health by country: a metasynthesis. *Environmental Health* 7 (1): 7.
150. Crump, A., Morel, C.M., and Omura, S. (2012). The onchocerciasis chronicle: from the beginning to the end? *Trends in Parasitology* 28 (7): 280–288. doi: 10.1016/j.pt.2012.04.005.
151. Leach, M., Bett, B., Said, M. et al. (2017). Local disease – ecosystem – livelihood dynamics: reflections from comparative case studies in Africa. *Philosophical Transactions of the Royal Society B* 372: 20160163. doi: 10.1098/rstb.2016.0163.
152. Friess, D.A. (2016). Ecosystem services and disservices of mangrove forests: insights from historical colonial observations. *Forests* 7 (9): 183–198. doi: 10.3390/f7090183.
153. Jentsch, A., Kreyling, J., and Beierkuhnlein, C. (2007). A new generation of climate-change experiments: events, not trends. *Frontiers in Ecology and the Environment* 5: 365–374.
154. McMichael, A. and Githeko, A. (2001). Human health. Climate change 2001: impacts, adaptation and vulnerability. In: *Contributions of the Working Group II to the Third Assessment of the Intergovernmental Panel on Climate Change* (ed. J.J. McCarthy, N.A. Leary, O.F. Canziani, et al.). Cambridge: Cambridge University Press.
155. Caminade, C., Ndione, J.A., Diallo, M. et al. (2014). Rift Valley fever outbreaks in Mauritania and related environmental conditions. *International Journal of Environmental Research and Public Health* 11 (1): 903–918.
156. Murdiyarso, D., Purbopuspito, J., Kauffman, J.B. et al. (2015). The potential of Indonesian mangrove forests for global climate change mitigation. *Nature Climate Change* 5: 1089–1092.
157. Downing, J.A., Cole, J.J., Middelburg, J.J. et al. (2008). Sediment organic carbon burial in agriculturally eutrophic impoundments over the last century. *Global Biogeochemical Cycles* 22 (1): GB1018.
158. Gorham, E. (1998). The biochemistry of northern peatlands and its possible responses to global warming. In: *Biotic Feedbacks in the Global Climatic Systems* (ed. G.M. Woodwell and F.T. (e.) Mackenzie), 169–187. New York: Oxford University Press.
159. Moss, B. (2010). Climate change, nutrient pollution and the bargain of Dr Faustus. *Freshwater Biology* 55 (s1): 175–187.
160. Schneider, S.H. (1989). The changing climate. *Scientific American* 261 (3): 70–79.
161. Lundberg, J. and Moberg, F. (2003). Mobile link organisms and ecosystem functioning: implications for ecosystem resilience and management. *Ecosystems* 6: 87–98. doi: 10.1007/s10021-002-0150-4.
162. Simon, J. (1981). *The Ultimate Resource*. Princeton: Princeton University Press.
163. WMO 2014. *Greenhouse Gas Bulletin, No. 10*. 9 September 2014. World Meteorological Organization. Available from: www.wmo.int.
164. Hendriks, I.E., Duarte, C.M., and Álvarez, M. (2010). Vulnerability of marine biodiversity to ocean acidification: a meta-analysis. *Estuarine, Coastal and Shelf Science* 86 (2): 157–164.
165. Pimm, S.L., Jenkins, C.N., Abell, R. et al. (2014). The biodiversity of species and their rates of extinction, distribution, and protection. *Science* 344 (6187): 1246752.
166. Fleming, D. (2016). *Lean Logic. A Dictionary for the Future and How to Survive It* (ed. S. Chamberlin). Vermont: Chelsea Green See also a linear precis of the dictionary: Fleming D. (2016) *Surviving the Future. Culture, Carnival and Capital in the Aftermath of the Market Economy*, also edited by Shaun Chamberlin and published by Chelsea Green. (www.flemingpolicycentre.org.uk).
167. Monbiot, G. (2013). *Feral; Searching for Enchantment on the Frontiers of Rewilding*. London.: Penguin.
168. Porritt, J. (2007). *Capitalism as if the World Matters*. London: Earthscan.
169. Trainer, T. (1995). *The Conserver Society. Alternatives for Sustainability*. London: Zed Books.
170. Adger, W.N., Eakin, H., and Winkels, A. (2009). Nested and teleconnected vulnerabilities to environmental change. *Frontiers in Ecology and the Environment* 7 (3): 150–157.

171. Arrow, K., Bolin, B., Costanza, R. et al. (1995). Economic growth, carrying capacity, and the environment. *Ecological Economics* 15 (2): 91–95.

172. Childers, D.L., Cadenasso, M.L., Grove, J.M. et al. (2015). An ecology for cities: a transformational nexus of design and ecology to advance climate change resilience and urban sustainability. *Sustainability* 7 (4): 3774–3791.

173. Costanza, R., Graumlich, L., Steffen, W. et al. (2007). Sustainability or collapse: what can we learn from integrating the history of humans and the rest of nature? *Ambio* 36 (7): 522–527.

174. Daly, H.E. and Cobb, J.B. (1990). *For the Common Good: Redirecting the Economy toward Community, the Environment, and a Sustainable Future*, 480. Boston: Beacon Press.

175. McAlpine, C., Seabrook, L., Ryan, J. et al. (2015). Transformational change: creating a safe operating space for humanity. *Ecology and Society* 20 (1): Article 56.

176. Norgaard, R.B. (1988). Sustainable development: a co-evolutionary view. *Futures* 20 (6): 606–620.

177. Oja, A. (1970). Estonia as a pilot for a sustainable society: utopia or opportunity? *WIT Transactions on Ecology and the Environment* 46: 31–40.

178. Raven, P.H. (1990). The politics of preserving biodiversity. *BioScience* 40 (10): 769–775.

179. Luckin, W. (1977). The final catastrophe – cholera in London, 1866. *Medical History* 21 (1): 32.

180. Ostfeld, R.S. and Holt, R.D. (2004). Are predators good for your health? Evaluating evidence for top-down regulation of zoonotic disease reservoirs. *Frontiers in Ecology and the Environment* 2: 13–20. doi: 10.1890/1540-9295(2004)002.

181. Garchitorena, A., Sokolow, S.H., Roche, B. et al. (2017). Disease ecology, health and the environment: a framework to account for ecological and socio-economic drivers in the control of neglected tropical diseases. *Philosophical Transactions of the Royal Society B* 372: 20160128.

182. Morlando, S., Schmidt, S.J., and LoGiudice, K. (2012). Reduction in Lyme disease risk as an economic benefit of habitat restoration. *Restoration Ecology* 20 (4): 498–504.

183. Seaman, W. (2007). Artificial habitats and the restoration of degraded marine ecosystems and fisheries. *Hydrobiologia* 580 (1): 143–155.

184. Edgar, G.J., Russ, G.R., and Babcock, R.C. (2007). Marine protected areas. In: *Marine Ecology* (ed. S.D. Connell and B.M. Gillanders), 534–565. Oxford: Oxford University Press. ISBN: 0195553020.

185. Speldewinde, P., Slaney, D., and Weinstein, P. (2015). Is restoring an ecosystem good for your health? *Science of the Total Environment* 502: 276–279. doi: 10.1016/j.scitotenv.2014.09.028.

186. Clapp, J., Newell, P., and Brent, Z.W. (2017). The global political economy of climate change, agriculture and food systems. *The Journal of Peasant Studies* 45 (1): 80–88. doi: 10.1080/03066150.2017.1381602.

187. Guerry, A.D., Polasky, S., Lubchenco, J. et al. (2015). Natural capital and ecosystem services informing decisions: from promise to practice. *Proceedings of the National Academy of Sciences* 112 (24): 7348–7355.

188. Gupta, J., Pouw, N.R., and Ros-Tonen, M.A. (2015). Towards an elaborated theory of inclusive development. *The European Journal of Development Research* 27 (4): 541–559.

189. Spangenberg, J.H. (2016). The world we see shapes the world we create: how the underlying worldviews lead to different recommendations from environmental and ecological economics-the green economy example. *International Journal of Sustainable Development* 19 (2): 127–146.

190. Allen P., Blake L., Harper P., Hooker-Stroud A., James P. and Kellner T. (2013). *Zero Carbon Britain: Rethinking the Future*. Centre for Alternative Technology, Machynlleth. See also updates and developments from this project: www.zerocarbonbritain.com.

191. Costanza, R., Alperovitz, G., Daly, H. et al. (2017). Building a sustainable and desirable economy-in-society-in-nature. In: *Green Economy Reader. Studies in Ecological Economics*, vol. vol. 6 (ed. S. (e.) Shmelev), 367–454. Cham: Springer.

192. Shove, E. (2010). Beyond the ABC: climate change policy and theories of social change. *Environment and Planning A* 42: 1273–1285.

193. Webb, R., Bai, X., Smith, M.S. et al. (2018). Sustainable urban systems: co-design and framing for transformation. *Ambio* 47 (1): 57–77.

194. Smith, A. (1776). *An Inquiry into the Nature and Causes of the Wealth of Nations*. London: W. Strahan and T. Cadell.

195. Easterlin R.A. and Angelescu L. (2009). *Happiness and Growth the World Over: Time Series Evidence on the Happiness–Income paradox*. IZA Discussion Paper No. 4060.

196. Abdallah, S., Thompson, S., and Marks, N. (2008). Estimating worldwide life satisfaction. *Ecological Economics* 65: 35–47.

197. Bartolini, S. and Sarracino, F. (2014). Happy for how long? How social capital and economic growth relate to happiness over time. *Ecological Economics* 108: 242–256.

198. Kubiszewski, I., Costanza, R., Franco, C. et al. (2013). Beyond GDP: measuring and achieving global genuine progress. *Ecological Economics* 93: 57–68.

199. Spangenberg, J.H. (2010). World civilisations at crossroads: towards an expansionist or a sustainable future – lessons from history. *Futures* 42 (6): 565–573.

200. Sukhdev, P. (2010). *Preface: The Economics of Ecosystems and Biodiversity; Ecological and Economic Foundations* (ed. P. Kumar). London: Earthscan.

201. Folke, C., Biggs, R., Norström, A. et al. (2016). Social-ecological resilience and biosphere-based sustainability science. *Ecology and Society* 21 (3): 41.

202. Weisz, H., Suh, S., and Graedel, T.E. (2015). Industrial ecology: the role of manufactured capital in sustainability. *Proceedings of the National Academy of Sciences* 112 (20): 6260–6264.

203. Wall, D. (2017). *Elinor Ostrom's Rules for Radicals. Cooperative Alternatives Beyond Markets and States*. London: Pluto Press.

204. Reuters (2012). Website: http://www.reuters.com/article/us-maldives/ousted-at-gunpoint-ex-president-of-maldives-takes-to-streets-idUSTRE8170AO20120208. Accessed 22 October 2017.

205. Stern, P.C., Young, O.R., and Druckman, D. (eds.) (1992). *Global Environmental Change: Understanding the Human Dimensions*. Washington, DC: Committee on the Human Dimensions of Global Change NAP.

206. Miller, S. (2001). Public understanding of science at the crossroads. *Public Understanding of Science* 10 (1): 115–120.

20

ADDRESSING THE NEXUS OF WATER, SANITATION, HEALTH, AND CLIMATE CHANGE THROUGH MULTISTAKEHOLDER PARTNERSHIPS

NIKHIL CHANDAVARKAR

Thersus Sustainability, Austin, TX, USA and
Outreach and Communications for Sustainable Development, United Nations, and UN-Water, United Nations, New York, NY, USA

20.1 ADDRESSING THE NEXUS OF WATER, SANITATION, HEALTH, AND CLIMATE CHANGE THROUGH MULTISTAKEHOLDER PARTNERSHIPS

There is growing recognition among governments, civil society, academia, and the private sector that water, sanitation, health, and climate change challenges are closely linked and must be tackled together and treated as a "nexus" or interconnected set of goals. Holistic approaches to development challenges lie behind the new guidance from the global community. The 2030 Agenda [1] with its 17 sustainable development goals approved by the United Nations in 2015 and the 2016 Paris Agreement [2] on climate change both represent landmarks in the global governance of sustainability, including water, sanitation, health, and climate change. A major means of converting this ambitious global agenda into concrete people-centered action on the ground is through voluntary multistakeholder partnerships and commitments, which represent an effective complement to traditional official development aid projects, rather than a substitute for them. Partnerships are voluntary initiatives where actors from different public and private groups come together to achieve specific aims within a set timeframe. The present chapter reviews current global policy, goals, and targets on water, sanitation, health, and climate change and then examines practical efforts to address these interconnected challenges through multistakeholder partnerships that have been registered with the partnerships for the sustainable development goals (SDGs) on-line platform [3] of the United Nations. The chapter concludes with recommendations for future partnerships to cover the full range of this set of goals.

20.2 GLOBAL POLICY GOALS AND TARGETS ON WATER, SANITATION, HEALTH, AND CLIMATE CHANGE

The global governance of and actions for water, sanitation, health, and climate change are guided by sustainable development goal (SDG) 3 on health and well-being for all [4], SDG 6 on water and sanitation, and SDG 13 on climate change and its impacts.

SDG 6, "ensure availability and sustainable management of water and sanitation for all" [5], seeks to complete the task of MDG 7 in water and sanitation by ensuring access for all as well as focusing on the planetary, ecosystems side of water resources, something that the MDGs did not do. The Millennium Development Goals (MDGs) sought to reduce by one-half the proportion of those lacking access to safe water and sanitation between 1990 and 2015. The water target was achieved, but 10% of the world's population still lacked access to improved drinking water sources in 2015. The MDGs also did not address waste management or hygiene. The MDG sanitation target was not met. Some 2.5

Water and Sanitation-Related Diseases and the Changing Environment: Challenges, Interventions, and Preventive Measures, Second Edition. Edited by Janine M. H. Selendy.

Companion website: https://www.wiley.com/legacy/wileychi/selendy/

billion people, roughly 40% of the global population, still lacked access to improved sanitation in 2015.

The six substantive targets of SDG 6 seek to fill the lacunae of the MDGs by addressing drinking water for all, sanitation and hygiene, eliminating open defecation, water quality and pollution, water-use efficiency, integrated water resources management (IWRM), and water ecosystem services. The two means of implementation targets address national capacity building and participation of local communities.

SDG 3, "ensure healthy lives and promote well-being for all at all ages," [6] rounds out and expands considerably the health agenda that was at the heart of the MDGs. The nine substantive targets and four means of implementation targets now include non-communicable diseases and universal health coverage, which were missing from the MDGs. A comprehensive treatment of how SDG 3 goes beyond the MDGs is contained in the 2015 World Health Organization (WHO) report, Health in 2015: from the MDGs to the SDGs [7]. The report identifies universal health coverage as the key to all the health-related SDGs. The research found that progress on the health MDGs fell short on several global targets, but was impressive overall, with significant reductions in child and maternal mortality and progress in combatting diseases such as HIV/AIDS, tuberculosis, and malaria. These results were achieved through, among other means, a 100% increase in global funding for health, new funding methods and partnerships, with a greater involvement of civil society, academia, foundations, and other non-state actors.

The WHO report presents the latest data and in-depth analysis for the key areas outlined in the health SDGs: reproductive, maternal, newborn, child, and adolescent health; infectious diseases including HIV, tuberculosis, malaria, hepatitis, and neglected tropical diseases; non-communicable diseases (NCDs) including heart disease, cancer, and diabetes; mental health and substance use including narcotics and harmful use of alcohol; injuries and violence; and universal health coverage.

SDG 13 on climate change [8] seeks to "take urgent action to combat climate change and its impacts," a goal that was not part of the MDGs. SDG 13 was left deliberately general, modest, and concise in Agenda 2030 in deference to the UN Framework Convention on Climate Change (UNFCCC) and the related 2016 Paris Agreement on Climate Change [9]. SDG 13 consists of three substantive targets covering (i) resilience and adaptive capacity to climate-related hazards and natural disasters, (ii) climate change measures in national policies, strategies, and planning, and (iii) education, awareness-raising, and human and institutional capacity. SDG 13 includes two means of implementation targets, exhorting countries to meet the $100 billion per year funding for climate change action by 2020 agreed under the UNFCCC process and capacity building for climate-related planning and management in developing countries.

A UNFCCC synthesis report [10] analyzing national climate plans and commitments of 119 countries identified water as their top climate change adaptation priority, followed by agriculture, health, and ecosystems. Similarly, business leaders and academics rank water crises as the risk of greatest concern to them over the next decade. The 2016 World Economic Forum's Global Risks Report [11], a survey of some 750 business leaders and specialists, emphasizes the close link between climate change, water, and extreme weather events, as well as the urgency of climate change adaptation measures. The report links water scarcity to other risks including hunger, intercountry conflict, social unrest, and rapid urbanization.

Responsibilities for addressing water, sanitation, health, and climate change challenges often rest in different government ministries and offices. Contact and consultation among these ministries and offices is often minimal or nonexistent. Multistakeholder partnerships addressing one or more of these areas are a practical and effective means of addressing SDG6, SDG3, and SDG 13 by bringing together public and private partners from different sectors.

20.3 MULTISTAKEHOLDER PARTNERSHIPS FOR WATER, SANITATION, HEALTH, AND CLIMATE CHANGE

For all of the 17 sustainable development goals at the time of this writing there was a total of some 2517 partnerships and voluntary commitments registered on the Partnerships for Sustainable Development Goals (SDGs) Platform of the United Nations. The exact number of partnerships on the platform changes in real time as new partnerships are continually registered. Of these partnerships, 117 partnerships are devoted to Goal 6 (water and sanitation for all), 146 to Goal 3 (health and well-being), and 280 to Goal 13 (climate change and its impacts). Of this total of 533 partnerships, some 419 are aimed at two or three of the above goals. In this section, we will examine the existing portfolio of partnerships for these three goals in the light of the foregoing global policy discussion and point out areas that could become the focus of future partnerships or additional commitments under existing partnerships (Table 20.1 and Figure 20.1).

20.3.1 Partnerships for Goal 6 – Water and Sanitation for All

The 146 partnerships that address SDG 6 cover a broader range of targets than the original water and sanitation targets of the earlier millennium development goal (MDG) 7, but address only a part of the rich and balanced agenda contained in the six substantive targets and two means of implementation targets.

TABLE 20.1 Number of partnerships per goal.

Goal 3 (health and well-being)	117
Goal 6 (water and sanitation)	146
Goal 13 (climate change and its impacts)	280
Total	533

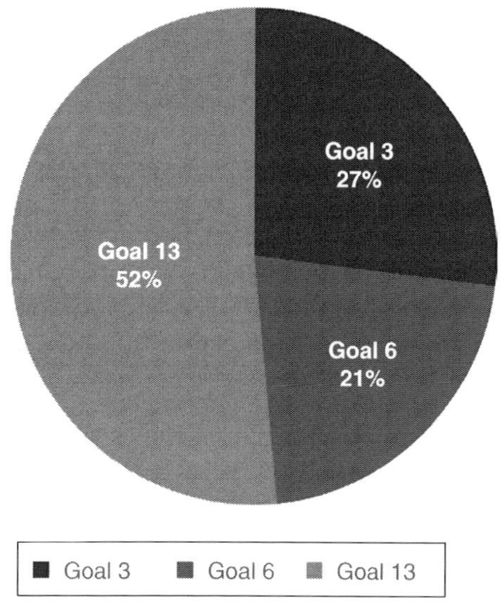

FIGURE 20.1 Partnerships for Water, Sanitation, Health, and Climate Change.

The registered partnerships on the Partnerships for SDGs Platform [12] cover such initiatives as mobile technology for ensuring freshwater availability at grass roots level (e-Water Pay); the Global Waste Water Initiative (GW21); the BKT Tomorrow Water Project; the WAKA water filter for rural communities in Nigeria; the municipal zero waste program for 2015 in Panama; Every Woman, Every Child; citizen-generated data to monitor progress against the SDGs (Everyone Counts); the baby WASH coalition; the toilet board coalition; community school-led sanitation; school mobilization on Lassa-fever prevention and WASH; water for life and against disease; integrated water and waste water management in the small island developing states; a food safety, nutrient, and WASHQ campaign in Nigeria; a peri-urban WASH program in Karonda, Malawi; and green university clubs.

The registered partnerships show a strong concentration in the Water, Sanitation, and Hygiene (WASH) area, building on momentum from earlier projects of the MDG era. Underserved areas lacking sufficient partnerships are sustainable water management and ecosystem services (SDG target 6.6).

20.3.2 Partnerships for Goal 3 – Health and Well-being for All

The 117 partnerships that address SDG 3 cover a broader range of SDG targets than the original maternal and child health areas of MDG 5, but not the entire range of nine substantive targets and four means-of-implementation targets of SDG 3. These broader areas include such initiatives as African–Hungarian health centers, with Hungary as an emerging donor; worldwide hospice palliative care; cancer care in Africa; scholarships for health; Pacific health development teams; coalition; sea and marine ambulance service; a community-based antisexual violence campaign for youth; the international alliance for responsible drinking (IARD); Project Last Mile against tuberculosis and malaria; and Blow Outreach in Nigeria, an awareness-raising campaign for public health and the use of ICTs for awareness-raising in schools about infectious diseases.

The registered partnerships tend to be geographically concentrated in regions such as the Pacific or in countries of interest to partners such as Nigeria, Kenya, or West Africa. More partnerships are needed for other regions of the world and for the full range of targets of Goal 3, and specifically connecting the areas of water, sanitation and health, and climate change.

20.3.3 Partnerships for Goal 13 – Climate Change and its Impacts

Goal 13 climate change was addressed in 280 partnerships, far more than those aimed at Goals 3 and 6, largely because of the efforts related to the preparations and follow-up to the 2016 Paris Climate Agreement.

The 280 partnerships [13] addressing Goal 13 included innovative initiatives that specifically address the nexus of goals 3, 6, and 13, such as the one-planet education network (World Wildlife Fund); promoting environmental activism among youth (USAID); a community of smart neighborhoods in Scandinavia; the carbon zero scheme; Dodder Action monthly river clean-up in Ireland; and Teens' Dream, a program for youth inspired by the legacy of Dr Martin Luther King.

20.4 CONCLUSIONS AND RECOMMENDATIONS FOR FUTURE PARTNERSHIPS

Some 533 or 21% of the total portfolio of 2517 partnerships at the time of writing address one or more goals of the nexus of water, sanitation, health, and climate change. Of these, only some 419 or 16.6% of the total address two or three goals of the nexus. At the same time these partnerships address only a portion of the variety of targets and richness of these goals of Agenda 2030. These partnerships also tend to be concentrated in specific geographic regions of strategic, geopolitical, or other interest to partners and donors. Many of these targets still represent aspirations yet to be incorporated into concrete partnerships or commitments.

The following recommendations identify the principal areas of these three goals where new partnerships are called for to address the broader range of targets included under the goals. Partnerships forums and outreach efforts should draw the attention of stakeholders and potential partners and donors to these underserved areas to encourage the launching of new partnerships in these areas and in geographic regions beyond those of traditional strategic interest to donors and other funding partners.

- Partnerships to promote adequate and equitable sanitation and hygiene for all and end open defecation (target 6.2), paying special attention to the needs of women and girls and those in vulnerable situations, relying on gendered indicators. The currently approved indicator 6.2.1, proportion of population using safely managed sanitation services, including a hand-washing facility with soap and water, is not broken down by gender.
- Partnerships to promote water- and sanitation-related activities and programs (target 6.A), including water harvesting, desalination, water efficiency, wastewater treatment, recycling, and reuse technologies. Partnerships should include public domain and traditional technologies, the costs of which would not burden the balance of payments and indebtedness of developing countries, as would many cutting-edge commercial technologies.
- Partnerships for sustainable management of water resources and for protecting and restoring water-related ecosystems, including mountains, forests, wetlands, rivers, aquifers, and lakes (target 6.6). An example of such a partnership is the economics of ecosystems and biodiversity in Aruba (UNEP).
- Partnerships to promote the universal health coverage (UHC) (target 3.8) and universal reproductive health-care services (target 3.7), which is far from achieved even in industrial countries. Partnerships should serve to share best practices from countries both in the OECD as well as emerging countries that have successfully spread health coverage.
- Partnerships to promote the implementation of the World Health Organization Framework Convention on Tobacco Control across the world, but particularly in developing countries where tobacco companies have made up for lost sales in industrial countries through aggressive marketing (target 3.a).
- Partnerships to promote organic and traditional knowledge-based farming to balance the many existing chemical-based agricultural partnerships sponsored by private companies, which can indirectly contaminate drinking water sources and food grown, compromising health and well-being of people and ecosystems (target 3.9 on hazardous chemicals and 2.4 on preserving ecosystems).
- Partnerships for the prevention and treatment of substance abuse, including narcotic drug abuse and harmful use of alcohol (target 3.5), including partnerships to promote supervised rehabilitation in conjunction with decriminalization of drug abuse as has been done in several OECD countries.
- Partnerships for mental health promotion (target 3.4) in both poor and rich countries, particularly in urban areas where rural–urban migration, expansion of slums, overcrowding, violence, and other challenges contribute to deteriorating mental health as well as to inadequate water and sanitation access (target 6.2).
- Partnerships for education, awareness-raising, and human and institutional capacity on climate change mitigation, adaptation, impact reduction, and early warning (target 13.3) in line with the examples provided of the one-planet education network and promoting environmental activism among youth.

REFERENCES

1. United Nations, Transforming our World, consulted 28 May 2017.
2. United Nations, Framework Convention on Climate Change, COP21, Paris Agreement, consulted 28 May 2017.
3. United Nations, Partnerships for SDGs Online Platform, consulted 28 May 2017.
4. United Nations, Partnerships for the Sustainable Development Goals, consulted 28 May 2017.
5. United Nations, Sustainable Development Knowledge Platform, SDG 6, consulted 28 May 2017.
6. United Nations, Sustainable Development Knowledge Platform, SDG 3, consulted 28 May 2017.
7. WHO (2015). *Health in 2015: From the MDGs to the SDGs*. Geneva: WHO.
8. United Nations, Sustainable Development Knowledge Platform, SDG 13, consulted 28 May 2017.
9. United Nations, UNFCCC, Paris Agreement, 2016.
10. UNFCC, Synthesis Report of National Climate Plans, 2015 and 2016.
11. World Economic Forum's Global Risks Report, 2016.
12. Visit for Further Information on These Partnerships, the SDG Partnerships On-line Platform, consulted 28 May 2017.
13. United Nations, SDG Partnerships On-line Platform, consulted 28 May 2017.

SECTION VI

SUCCESS STORIES

21

EXTENDING THE RIGHT TO HEALTH CARE AND IMPROVING CHILD SURVIVAL IN MEXICO

JULIO FRENK[1] AND OCTAVIO GÓMEZ-DANTÉS[2]
[1] *University of Miami, Coral Gables, FL, USA*
[2] *Center for Health Systems Research, National Institute of Public Health, Cuernavaca, Mexico*

Mexico achieved Millennium Development Goal (MDG) 4, which set the target of reducing the 1990 child mortality rate by two-thirds no later than 2015 [1]. This success established the foundation for the current efforts to meet the health-related Sustainable Development Goals. Progress in Mexico has been the result of the combined implementation, over the past 25 years, of several initiatives to strengthen the health system and various disease-oriented strategies. This chapter describes in detail this achievement and its determinants, with emphasis on the recent health system reform, which created the conditions to guarantee universal coverage of high-quality health services with financial protection for all.

21.1 CHILD SURVIVAL IN MEXICO: THE DIAGONAL APPROACH

In the past two and a half decades, through the explicit adoption of the primary healthcare paradigm, Mexico designed several initiatives to extend access to essential healthcare by strengthening the public supply of clinical services [2]. These efforts were complemented with the implementation of major community-based programs to address the challenges related to common infections, malnutrition, and maternal and child health.

Health system strengthening efforts started in 1979 with the creation of a program to expand the network of healthcare units in rural areas. This program, originally called *IMSS-Coplamar* and then renamed *IMSS-Prospera*, built, in its initial phase, more than 3000 ambulatory units and 64 rural hospitals [3].

In the early 1980s, two other programs were launched, one to expand the health coverage of the urban poor and the other one to care for the health needs of the rural population [4].

In the early 1990s, the Program to Strengthen Health Services for the Uninsured Population (PASSPA), co-financed with federal resources and a loan from the World Bank, was implemented to improve and extend the health service infrastructure for the uninsured population in the four poorest states of the country [5].

Finally, in 1996, a program to further expand the coverage of essential care in the rural population nationwide was created. The Program for Extension of Coverage (PAC) was designed to provide on a regular basis 12 essential interventions: (i) basic household sanitary measures; (ii) family planning services; (iii) pre-natal, peri-natal, and post-natal care; (iv) nutrition and growth surveillance; (v) immunization; (vi) treatment of diarrhea; (vii) treatment of common parasitic diseases; (viii) treatment of acute respiratory infections; (ix) prevention and treatment of tuberculosis; (x) prevention and control of hypertension and diabetes; (xi) initial treatment of injuries; and (xii) community training for health promotion [6].

These efforts to strengthen the provision of personal health services were complemented with the implementation of several disease-oriented initiatives, mostly community-based interventions where the main objective was to reduce the burden of malnutrition and common infections in children, basically diarrheal diseases and respiratory infections.

Water and Sanitation-Related Diseases and the Changing Environment: Challenges, Interventions, and Preventive Measures, Second Edition. Edited by Janine M. H. Selendy.

Companion website: https://www.wiley.com/legacy/wileychi/selendy/

TABLE 21.1 Changes in access to sanitation facilities and drinking water sources, Mexico 1990–2015.

	Percentage of the population		
Access	1990	2000	2015
Access to improved sanitation facilities	58.6	72.8	91.4
Access to improved drinking water sources	78.4	87.8	95.3

Source: MDGs. How is Mexico doing? Available at: http://www.objetivosdedesarrollodelmilenio.org.mx/Default.aspx?Param=CDR,I. Accessed 3 August 2017.

The first major initiative of this period was the introduction of the oral rehydration therapy (ORT) in public hospitals in 1984 as part of the National Program to Control Diarrheal Diseases [7]. Eventually other measures to control these diseases were widely disseminated: the surveillance of the municipal chlorination of drinking water, the promotion of breastfeeding, and the training of mothers in the use of ORT in cases of diarrhea and in the early detection of alarm signs of dehydration.

In this period there were also important improvements in access to water and sanitation in Mexico (Table 21.1). In fact, target 3 of MDG 7 (to halve, by 2015, the proportion of the population without sustainable access to safe and drinking water and basic sanitation) was also met [8].

All these measures combined had an important impact on the incidence of diarrhea and the number of deaths due to diarrheal diseases. Between 1985 and 2000, diarrhea mortality rates in Mexico in children under 5 declined from 11.5 per 1000 live births to only 1.6 [9].

The second major public health measure adopted in this period was the introduction, in 1991, of the Universal Immunization Program. Mexico had established a National Immunization Program in 1973, but it was until two decades later that this program, launched as a vertical initiative, was placed at the center of the national health agenda. It was implemented in response to a major measles epidemic, the results of a national vaccination survey showing that complete vaccination coverage in children under 5 was below 50%, and the commitments made by Mexico in the 1990 World Summit for Children [10]. The program initially offered six vaccines (BCG, TDP, OPV, and antimeasles), which were traditionally applied during the National Health Weeks, a strategy established in 1993. By 2000 the immunization scheme included 10 vaccines (BCG, TDP, OPV, MRM, hepatitis B, and Hib) and had reached complete coverage of more than 90% in children under 1 and more than 95% in children under 5 [11].

The effects of the Universal Immunization Program were dramatic and immediate. The last cases of polio and diphtheria were notified in 1990 and 1991, respectively, while the last case of autochthonous measles was reported in 1996 [12–14]. Important as they are, the effects of this program were not limited to particular diseases; child survival as a whole was positively affected, as will be discussed further ahead in this chapter.

National Health Weeks, now held three times a year, were used not only to offer all the vaccines of the WHO Expanded Program on Immunization but also to provide a package of interventions particularly useful for children: the promotion of the use of oral rehydration salts and the free distribution of mega-doses of oral vitamin A and the antihelminthic drug albendazol.

Finally, in 1997, the federal government launched a program to enhance basic capabilities of families living in extreme poverty, which has had proven impacts in the health and nutritional status of the poor population [15–18]. Initially called *PROGRESA* and later renamed *Oportunidades* and *Prospera*, this program, which is still active and benefiting over six million households, offers incentives for families to invest in their children's human capital through conditional cash transfers. In order to receive these subsidies families are required to send their children to school rather than work; attend a clinic to receive a specified package of health promotion and disease prevention interventions; and provide a specially formulated nutritional supplement to pregnant and lactating women, all children aged 6 to 23 months, and low-weight children aged 2 to 5 years. Recently, this program was expanded to benefit also adults over 65 years of age.

Prospera has contributed to the improvement of malnutrition in children in Mexico. Advances in average height-for-age and prevalence of stunting were already detected in the 1999 National Nutrition Survey but they could hardly be attributed to this intervention. However, additional and more prominent improvements in these two indicators were also recorded for the period 1999–2006, especially among poor and indigenous children [19].

To improve decision-making, several measures were implemented. Salient among them was the creation, in 1987, of the National Institute of Public Health. This teaching and research institution, in collaboration with the main healthcare delivery organizations, implemented regular surveys to monitor progress in health conditions and healthcare, developed broad research initiatives which generated inputs for public health programs, and built the public health manpower that the national health system was demanding.

The impact of the measures implemented between 1980 and 2000 in child survival was enormous. However, additional innovations had to be designed in order to guarantee the sustainability and expansion of these efforts and meet the health-related MDGs.

21.2 THE MEXICAN HEALTH REFORM

A Constitutional amendment establishing the right to the protection of health was passed in Mexico in 1983 [20]. However, in practice not all individuals had been equally able to exercise this right. Those working in the formal sector of the economy and their families, around half of the population, had social health insurance. The other half was left without access to any form of social protection in health. This meant that almost 50 million people in the country were receiving healthcare on an assistential basis, including 2.5 million families from the poorest segments of the population who received only the set of essential interventions included in PAC.

This reality was well documented. Several national health accounts studies developed in the early 1990s had revealed that more than 50% of health expenditure in Mexico was out-of-pocket. This was due to the fact that half of the population lacked health insurance.

The high levels of out-of-pocket expenditure were exposing Mexican families to ruinous financial episodes. In fact, in 2000, nearly 3 million Mexican households suffered catastrophic health expenditures [21]. Not surprisingly, Mexico performed poorly on the international comparative analysis of fair financing developed by the World Health Organization as part of the World Health Report 2000 [22]. These results prompted the development of additional analysis that showed that catastrophic and impoverishing health expenditures were endemic among the poor and uninsured households.

The products of this analysis generated the advocacy tools to promote a legislative reform that established the System for Social Protection in Health (SSPH), approved by a large majority of the Mexican Congress in the spring of 2003. This system is mobilizing public resources by a full percentage point of GDP over a period of seven years to provide health insurance through a new scheme called *Seguro Popular* to all those ineligible for social security: the self-employed, those out of the labor market, and those working in the informal sector of the economy. This public insurance scheme guarantees regular access to two sets of entitlements: (i) a package of more than 250 essential interventions, which include all services offered in ambulatory units and general hospitals, and (ii) a package of 60 costly interventions, which includes, among other interventions, treatment for cancer in children, cervical and breast cancer, and HIV/AIDS. By January of 2015, over 50 million people were enrolled in the new insurance program [23].

Through the SSPH a protected fund for community health services targeting health promotion and disease prevention interventions was also created. This fund now finances on a regular basis all public health interventions, including those that were critical to reduce child mortality in the last two decades of the past century.

In sum, the formal recognition of healthcare as a social right was not enough to guarantee regular access to comprehensive healthcare. As Lynn Hunt has written, "human rights are still easier to endorse than to enforce" [24]. What was lacking was the definition of the entitlements that ensued from such a claim, and the financial and organizational vehicles to translate them into personal and public health services accessible to all. That is exactly what the SSPH did for half of the Mexican population: create the conditions for the effective exercise of the right to healthcare.

21.3 MEETING MDG 4

A structural reform addressing the broad challenge of financing universal coverage of high-quality services was also demanding a clear sense of priorities. This is an imperative not only in terms of resource allocation but also to gather public support by relating abstract financial and managerial notions to concrete deliverables. Every reform must have a limited number of "flagship initiatives" to focus attention on its specific benefits. From the outset, it was decided that those interventions needed to meet the health-related MDGs would be included among those initiatives.

In order to reduce maternal and neonatal mortality and disability, a novel initiative was launched under the name "Fair Start in Life" (APV or *Arranque Parejo en la Vida*). The name of this initiative was meant to underscore the need for equality of opportunity, a core explicit value of the Mexican health reform [25].

The maternal component of this program included the mobilization of resources to strengthen healthcare networks, equipment, and the supply of drugs and other inputs, most notably, safe blood. The supply of skilled human resources was improved through the reintroduction of obstetric nurses and the training of traditional birth attendants. Efforts were made to extend coverage of antenatal care, with emphasis on early identification of high-risk pregnancies, institutional deliveries, and timely diagnosis and treatment of obstetric emergencies.

The neonatal component of APV included the administration of folic acid to pregnant women for the prevention of neural tube defects, preventive neonatal screening of congenital metabolic diseases, and the strict adherence to the vaccination scheme for this age group.

These measures produced an increase in effective coverage of antenatal care; an expansion of effective coverage of births attended by skilled health workers, which reached 93% in 2006, with small variations among states; improvements in immunization coverage in children under 1; and, most importantly, a significant acceleration in the rate of decline of maternal and neonatal mortality [26,27]. Between 2001 and 2006, neonatal mortality dropped over 15%, the highest decline in the last five quinquennia [28].

Immunization efforts were also bolstered through the expansion of the number of vaccines in the basic scheme and the strengthening of immunization campaigns. By 2006 Mexican children had free access to one of the most comprehensive immunization schemes in the world, which includes

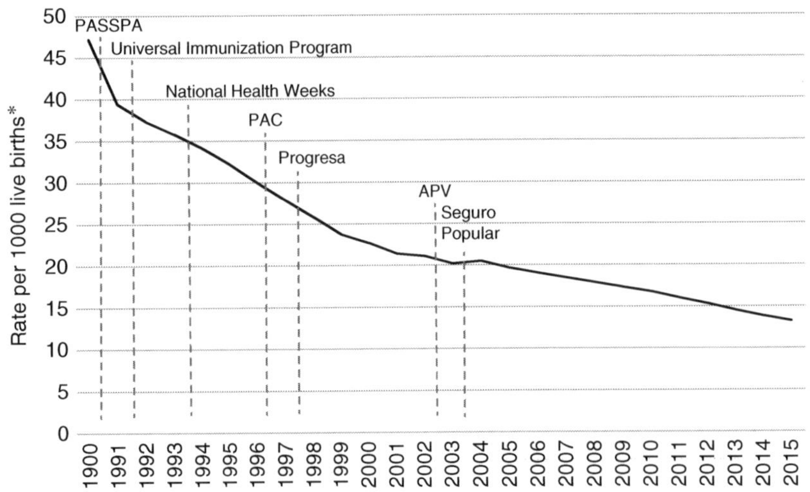

FIGURE 21.1 Evolution of under 5 mortality rate in Mexico, 1990–2015. *Sources*: Ministry of Health, Mexico. Estimates based on projections from the National Population Council (CONAPO). World Bank based on UN Inter-agency Group for Child Mortality Estimation. Tasa de mortalidad, menores de 5 años (por cada 1000) Mexico. Available at: https://datos.bancomundial.org/indicador/SH.DYN.MORT. Accessed 3 August 2017. * Probability of dying by age 5 per 1000 live births.

vaccines against hepatitis B, diphtheria, tetanus, pertussis, *Haemophilus influenzae* type b, polio, measles, mumps, rubella, influenza, TB, and meningitis. That same year, complete immunization coverage reached 95% in children under 1 and 98% in children under 5, one of the highest immunization coverage levels in the Americas [29]. Effective coverage for measles immunization in children aged 18–59 months reached 92.1% [27].

These continuous and increasing efforts explain the positive recent evolution of the under 5 years of age mortality rate in Mexico, which declined from 47.1 per 1000 live deaths in 1990 to 13.2 in 2015 (Figure 21.1), reaching the target of MDG 4.

21.4 CONCLUSIONS

The consistent decline of child mortality in Mexico over a period of 25 years offers several lessons for other countries.

First of all, this achievement was the product of the combined implementation of health system strengthening efforts and disease-oriented interventions, which proves that the dilemma between horizontal and vertical approaches is a false one [30]. In the past decades, Mexico implemented several strategies to upgrade general health services while also focusing on cost-effective interventions for specific priorities, such as child mortality, that reinforced the overall structure and performance of the health system.

Second, improvements in child mortality were also strongly dependent on the development and continuous expansion of a base of evidence for policy design, policy making, and policy evaluation. Most of the initiatives implemented in this period to confront the challenges related to common infections, malnutrition, and maternal and child health were supported by evidence generated by large-scale health surveys. The development of institutions such as the National Institute of Public Health was also crucial. This center of excellence has produced relevant research and policy analysis, trained researchers who occupy key policy-making positions, carried out independent and credible evaluations, and greatly enriched the quality of information.

Finally, stability of leadership in the Ministry of Health also contributed to this achievement [2]. The story of Mexico in this regard may be unique: since the 1940s, most ministers have served an entire term of office; in many cases they were selected for their technical expertise rather than their political affiliation; and, especially in the past 25 years, they have preserved and even expanded policies from preceding administrations when proved effective.

REFERENCES

1. INEGI. Millenium Development Goals, Mexico. 4. Reduce Child Mortality. Available at: http://www.objetivosdedesarrollodelmilenio.org.mx/Default.aspx?Param=INDODM004000100280,75,0,004,False,False,False,False,False,False,False,0,0,I. Accessed 3 August 2017.
2. Frenk, J., Sepúlveda, J., Gómez-Dantés, O., and Knaul, F. (2003). Evidence-based health policy: three generations of reform in Mexico. *Lancet* 362: 1667–1671.
3. Soberón, G. (2006). Desarrollo de las políticas públicas. In: *La Experiencia Mexicana en Salud pública. Oportunidad Y Rumbo Para el Tercer Milenio* (ed. M. Urbina-Fuentes, A. Moguel-Ancheita, M.E. Muñiz-Martelon and J.A. Solís-Urdaibay), 543–566. Mexico City: Fondo de Cultura Económica.

4. Rodríguez-Domínguez, J. (2006). Salud para todos. Atención primaria de la salud. In: *La Experiencia Mexicana en Salud pública. Oportunidad Y Rumbo Para el Tercer Milenio* (ed. M. Urbina-Fuentes, A. Moguel-Ancheita, M.E. Muñiz-Martelon and J.A. Solís-Urdaibay), 577–594. Mexico City: Fondo de Cultura Económica.
5. Gómez-Dantés, O., Garrido-Latorre, F., López-Moreno, S. et al. (1999). Evaluación de un programa de salud para población no asegurada: Assessment of the health program for the non-insured population. *Rev. Saude Publica* 33: 401–412.
6. de Salud, S. and Ampliación de Cobertura, P. (1996). *Lineamientos de operación*. Mexico City: Secretaría de Salud.
7. Fajurdo-Ortiz, G. (2002). 1982 and 2001: Tiempos de reformas y nuevos avances. In: *Perspectiva histórica de atención a la Salud en México 1902–2002* (ed. G. Fajardo-Ortiz, A.M. Carrillo and R. Neri-Vela), 101–123. Mexico City: PAHO, UNAM, Sociedad Mexicana de Historia y Filosofía de la Medicina.
8. MDGS. How is Mexico doing? Available at: http://www.objetivosdedesarrollodelmilenio.org.mx/Default.aspx?Param=CDR,I. Accessed 3 August 2017.
9. Sepúlveda, J., Bustreo, F., Tapia, R. et al. (2006). Improvement of child survival in Mexico: the diagonal approach. *Lancet* 368: 2017–2027.
10. Valdespino-Gómez, J.L. and García-García, M.L. (2004). 30 aniversario del Programa Nacional de Vacunación contra sarampión en México. Los grandes beneficios y los riesgos potenciales. *Gac. Med. Mex.* 140: 639–641.
11. Santos, J.I. (2006). Cambios en los esquemas de vacunación y la vacunación en adultos. In: *La Experiencia Mexicana en Salud pública. Oportunidad Y Rumbo Para el Tercer Milenio* (ed. M. Urbina-Fuentes, A. Moguel-Ancheita, M.E. Muñiz-Martelon and J.A. Solís-Urdaibay), 191–223. Mexico City: Fondo de Cultura Económica.
12. Organización Panamericana de la Salud. Poliomileitis: cronología 1909–2007. Available from: http://www.mex.ops-oms.org/contenido/polio/crono_polio.htm. Accessed 24 March 2010.
13. Santos, J.I. (1999). Nuevo esquema de vacunación en México. *Salud Publica Mex.* 41: 1–2.
14. Santos, J.I. (2002). El Programa Nacional de Vacunación: orgullo de México. *Rev. Fac. Med. UNAM* 45: 142–153.
15. Rivera, J., Sotres-Alvarez, D., Habicht, J.P. et al. (2004). Impact of the Mexican program for education, health, and nutrition (*Progresa*) on rates of growth and anemia in infants and young children: a randomized effectiveness study. *JAMA* 219: 2563–2570.
16. Fernald, L.C.H., Gertler, P., and Neufeld, L. (2008). Role of cash in conditional cash transfer programmes for child health, growth, and development: an analysis of Mexico's *Oportunidades*. *Lancet* 371: 828–837.
17. Leroy, J.L., García Guerra, A., García, R. et al. (2008). *Oportunidades* program increases the linear growth of children enrolled at young ages in urban Mexico. *J. Nutr.* 138: 793–798.
18. Secretaría de Desarrollo Social. Oportunidades. External evaluation site. Available from: http://evaluacion.oportunidades.gob.mx:8010/index2.php?a=800. Accessed 23 March 2010.
19. González de Cossio, T., Rivera, J., González-Castell, D. et al. (2009). Child malnutrition in Mexico in the last two decades: prevalence using the new WHO 2006 growth standards. *Salud Publica Mex.* 4 (Suppl. 51): S494–S506.
20. Soberón, G. (1987). El cambio estructural en la salud. *Salud Publica Mex.* 29: 127–140.
21. Secretaría de Salud (2001). *Programa Nacional de Salud 2001–2006. La Democratización de la Salud en México. Hacia Un Sistema Universal de Salud*, 57. Mexico City: Secretaría de Salud.
22. World Health Organization. World Health Report (2000). *Health Systems: improving Performance*, 2000. Geneva: WHO.
23. Secretaría de Salud (2009). *Sistema de Protección Social en Salud. Informe de Resultados. Primer Semestre de 2009.* Mexico City: Secretaría de Salud.
24. Hunt, L. (2007). *Inventing Human Rights. A History*, 21408. New York and London: WW Norton & Company.
25. Frenk, J. and Gómez-Dantés, O. (2009). Ideas and ideals: ethical basis of health reform in Mexico. *Lancet* 373: 1406–1408.
26. Lozano, R., Soliz, P., Gakidou, E. et al. (2006). Benchmarking of performance of Mexican states with effective coverage. *Lancet* 368: 1729–1741.
27. Gakidou, E., Lozano, R., González-Pier, E. et al. (2006). Assessing the effect of the 2001–2006 Mexican health reform: an interim report card. *Lancet* 368: 1920–1935.
28. Secretaría de Salud (2007). Salud: México 2006. Información Para la rendición de Cuentas, p. 19. Mexico City: Secretaría de Salud.
29. Secretaría de Salud (2007). Salud: México 2006. Información Para la rendición de Cuentas, p. 190. Mexico City: Secretaría de Salud.
30. Sepúlveda, J. (2006). Foreword. In: *Disease Control Priorities in Developing Countries*, 2e (ed. D.T. Jamison, J.G. Breman, A.R. Measham, et al.), xiii–xv. Washington, DC: Oxford University Press.

22

DRACUNCULIASIS (GUINEA WORM DISEASE): CASE STUDY OF THE EFFORT TO ERADICATE GUINEA WORM

Donald R. Hopkins[1] and Ernesto Ruiz-Tiben[2]

[1] *The Carter Center, Atlanta, Georgia, USA*

[2] *Dracunculiasis Eradication Program, The Carter Center, Atlanta, Georgia, USA*

22.1 INTRODUCTION

Dracunculiasis (Guinea worm disease) is a waterborne parasitic infection that is intimately tied to the environment and to human behavior. The campaign to eradicate this disease has lasted over 30 years and provides many lessons that may be adapted for attacking other problems successfully, particularly where a change in human habits is required in order to mitigate or interrupt transmission.

The adult parasite *Dracunculus medinensis* normally occurs only in humans, who become infected after drinking stagnant surface water from sources such as ponds where microscopic water fleas (copepods) have been infected by ingesting immature forms of the parasite. The parasite matures in humans over a period of one year, during which the infection evokes no symptoms or outward signs. After a year, the thin white adult worm, which measures up to 3 ft (1 m) long, emerges slowly and painfully through a ruptured blister that it raises on the skin (Figure 22.1).

Most worms emerge on the leg, ankle, or foot, but they may emerge from anywhere on the body, and a patient may have up to a dozen or more worms emerge around the same time. The emerging adult worms are all females, which spew hundreds of thousands of immature larvae into the water when the infected person enters a surface source such as a pond for any reason. Thus, the life cycle of the worm is perpetuated by humans who drink untreated water from such contaminated sources, as well as by those who enter sources of drinking water while a Guinea worm is emerging or about to emerge from their body.

Since only a few centimeters of a worm can usually be pulled out manually by rolling it on a small stick or rolled gauze in a day, the painful incapacitation associated with this infection normally lasts up to one or two months or more. The site of worm emergence often becomes infected secondarily with bacteria, increasing the local inflammation and pain that are the hallmarks of this disease. Infection with tetanus bacilli at the site of the ulcer, which is the base of the ruptured blister caused by the emerging worm, is the most dangerous secondary complication of Guinea worm disease, which otherwise is not usually fatal. In about one-half of 1% of cases, a worm that emerges in or near a major joint such as the elbow or knee can cause permanent freezing of that joint, with consequences very similar to crippling damage from paralytic poliomyelitis. Past infections do not confer immunity, so people can be reinfected year after year.

Before 2012 no animal or any other environmental reservoir of *D. medinensis* was known outside of human beings, although similar infections occurred in dogs and some other animals occasionally in endemic countries, and dogs and some other mammals were infected experimentally in laboratory studies. Frequent infections of dogs with worms indistinguishable from *D. medinensis* have been found in Chad in recent years, however. The disease has been recognized in ancient sources such as Egyptian mummies and the Old Testament. The immature larvae last only a few weeks

Water and Sanitation-Related Diseases and the Changing Environment: Challenges, Interventions, and Preventive Measures, Second Edition. Edited by Janine M. H. Selendy.

Companion website: https://www.wiley.com/legacy/wileychi/selendy/

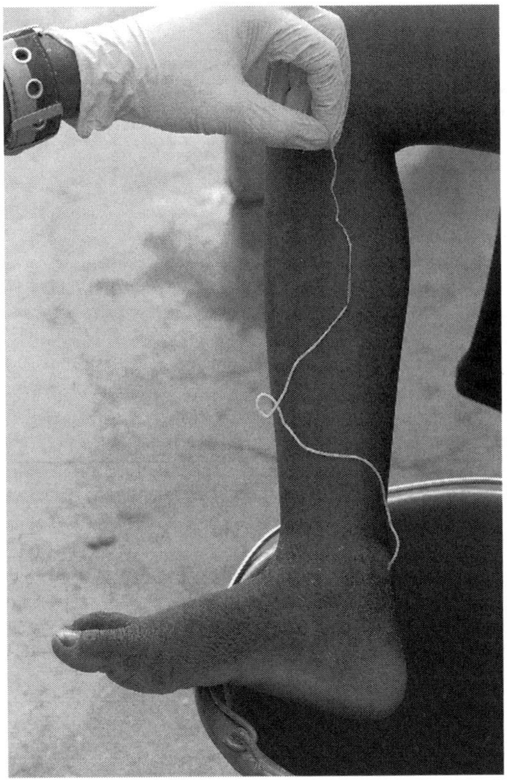

FIGURE 22.1 Emerging Guinea worm. *Source*: Photo credit: The Carter Center/Louise Gubb. (*See insert for color representation of this figure.*)

outside the human body, and only if they are ingested by a copepod. That is why this disease is vulnerable to complete eradication, even though copepods are ubiquitous in stagnant surface sources of freshwater.

The impact of Guinea worm disease on affected communities can be devastating, even though the infection is not usually fatal. This is a disease of remote, neglected rural populations that do not have easy access to clean drinking water. Before the eradication campaign began, up to 50% or more of a village's population could be infected at the same time, for periods averaging one to two months. Working age adults of both sexes are affected most of all because they tend to drink larger volumes of water. The worms emerge seasonally, precisely during the time of year of peak agricultural labor, when villagers need to harvest or plant their crops. Hence, Guinea worm disease negatively impacts not only people's health but also their agricultural productivity and school attendance [1].

In drier areas such as the Sahel just below the Sahara Desert, the "Guinea worm season" coincides with the brief rainy season, because that is the time of year when sources of surface water are abundant and available to receive and transmit new parasites. In better watered climates nearer the Atlantic coast in West Africa, for example, the dry season is the optimal period for the disease because that is when stagnant water sources are shrinking and most contaminated, the abundant rainy season causing rivers to flow (non-stagnant) and diluting any contamination. Areas where the period of peak prevalence of Guinea worm disease coincides with the rainy season are doubly handicapped, since that is also the time of year when travel into remote areas, and hence provision and supervision of control measures, is most difficult.

There is no treatment, cure, or vaccine for Guinea worm disease, except to provide analgesics to relieve the pain and antibiotics to mitigate secondary infections. The ancient practice of carefully winding the emerging worm around a small stick is still the best way to try to shorten the period of disability. If a worm is broken while being coaxed to emerge, however, the remainder of the worm retracts into the body and spills larvae into the tissues, thereby causing severe inflammation and abscess formation, which is very painful and requires incisions and drainage by medically trained personnel, ideally at a clinic facility.

Although Guinea worm disease is not treatable, fortunately transmission of the infection can be prevented in several ways, by protecting drinking water from contamination by infected humans and/or by protecting humans from drinking contaminated water. The most basic intervention is to teach people in endemic areas to avoid entering sources of drinking water when a worm is emerging or about to emerge from their body and to always filter any water they drink from an open pond or other stagnant source through a finely woven cloth. Convincing conservative villagers that this disease is caused by water they drank a year ago, and that it is not because of any number of long and strongly held traditional beliefs, is easier said than done. One of the most effective ways to promote such understanding is to filter water from the usual local pond through a cloth and backwash the filtered material into a clear glass or jar, which is then held up to the light (with or without a magnifying glass) so villagers can see the tiny organisms moving around in the water they are drinking. That revelation is instinctively repulsive and is persuasive to many. Distributing cloth filters free of charge for villagers to use to protect themselves and their families, and demonstrating how to use them properly, also serves as a tangible reminder to persons at risk of what they should do. Special attention is paid to educating women and girls, who are most often the ones who collect water for household use. As of 2008, about 40% of the workforce in the national Guinea Worm Eradication Programs (GWEPs) was female. Specially woven nylon (100 μm mesh) material, which originally was donated by the E.I. DuPont Company and more recently by LifeStraw (Vestergaard), and which lasts longer and filters faster than cotton cloth, is most commonly used in the eradication campaign. Portable "pipe filters," a type of large plastic straw with a 100 μm stainless steel mesh filter inside, can be hung on a string around the neck and are provided to allow persons to filter the water they drink when traveling or farming away from home (Figure 22.2).

(a) (b)

FIGURE 22.2 (a) Cloth filter. Savelugu, Northern Ghana: Technical Assistant Eugene Yeng and his young Community Based Surveillance Volunteer, Abdulhai Idiris. *Source*: Photo credit: The Carter Center/Louise Gubb. (b) Pipe filters. *Source*: Photo credit: The Carter Center/Emily Staub.

Another effective intervention is to treat potentially contaminated ponds with ABATE® Larvicide, which at the recommended concentration is harmless to humans, fish, and plants but lethal to copepods and the parasites they convey. This larvicide, which is donated to The Carter Center by BASF Corporation for use in the eradication campaign, is applied at a concentration of one part per million every four weeks during the transmission season. Providing clean water from underground sources by borehole wells or similar means is the ideal intervention, since such clean water prevents many other infections in addition to Guinea worm disease and is often much more accessible than the traditional pond, which may be a mile or more away from the village. Unfortunately, this is also the slowest and most expensive (circa US$ 15 000 per functional borehole) of the available preventive measures and assuring priority to specific villages for any reason, including Guinea worm disease, is difficult because the geology of the place may preclude extraction of safe drinking water or because of political influences, corruption, and powerlessness of residents in the remote villages concerned.

Once village based interventions are in place in as many affected communities as possible, an additional barrier to transmission can be "case containment," designed to prevent transmission from each patient. This is done by individual health education to urge the patient not to enter any water source, by manual extraction of the worm (slowly wrapping around a stick) and daily bandaging and treatment of the wound, and by caring for the patient in a home, clinic, special "Case Containment Center," or other facility. This provides less disability and early return to normalcy for the patient and discourages further contamination of water sources by ensuring voluntary isolation of infected individuals. Specific criteria, including detection of the patient before or within 24 hours of emergence of the worm, have been developed for classifying cases as contained or uncontained [2].

22.2 IMPLEMENTING THE ERADICATION CAMPAIGN

The global Guinea Worm Eradication Program was conceived and initiated at the Centers for Disease Control and Prevention in Atlanta, Georgia, in October 1980, originally as an adjunct to the International Drinking Water Supply and Sanitation Decade (1981–1990), one of the goals of which was to provide safe drinking water to all who did not yet have safe water. Early in the twentieth century, dracunculiasis was widely prevalent in many Asian and African countries, from southern parts of the Soviet Union to British India, Persia, and Saudi Arabia, and much of North, East, and West Africa. By the time The Carter Center began leading the global campaign in 1986, an estimated 3.5 million persons were being infected annually, in 3 Asian and 17 African countries [3] (Figure 22.3). Because of its remote habitat, marginalized affected population, and lack of available treatment, dracunculiasis was vastly underreported before the campaign began. Although Nigeria and Ghana each officially reported about 3000–5000 cases of the disease to the World Health Organization annually in the 1980s, for example, when the two countries conducted nationwide village-by-village searches for the disease in 1989, they enumerated over 650 000 and almost 180 000 cases, respectively.

Village volunteers (VVs) are the bedrock of the GWEP. Sometimes selected by village chiefs or government health workers, but often chosen by fellow villagers, the VVs

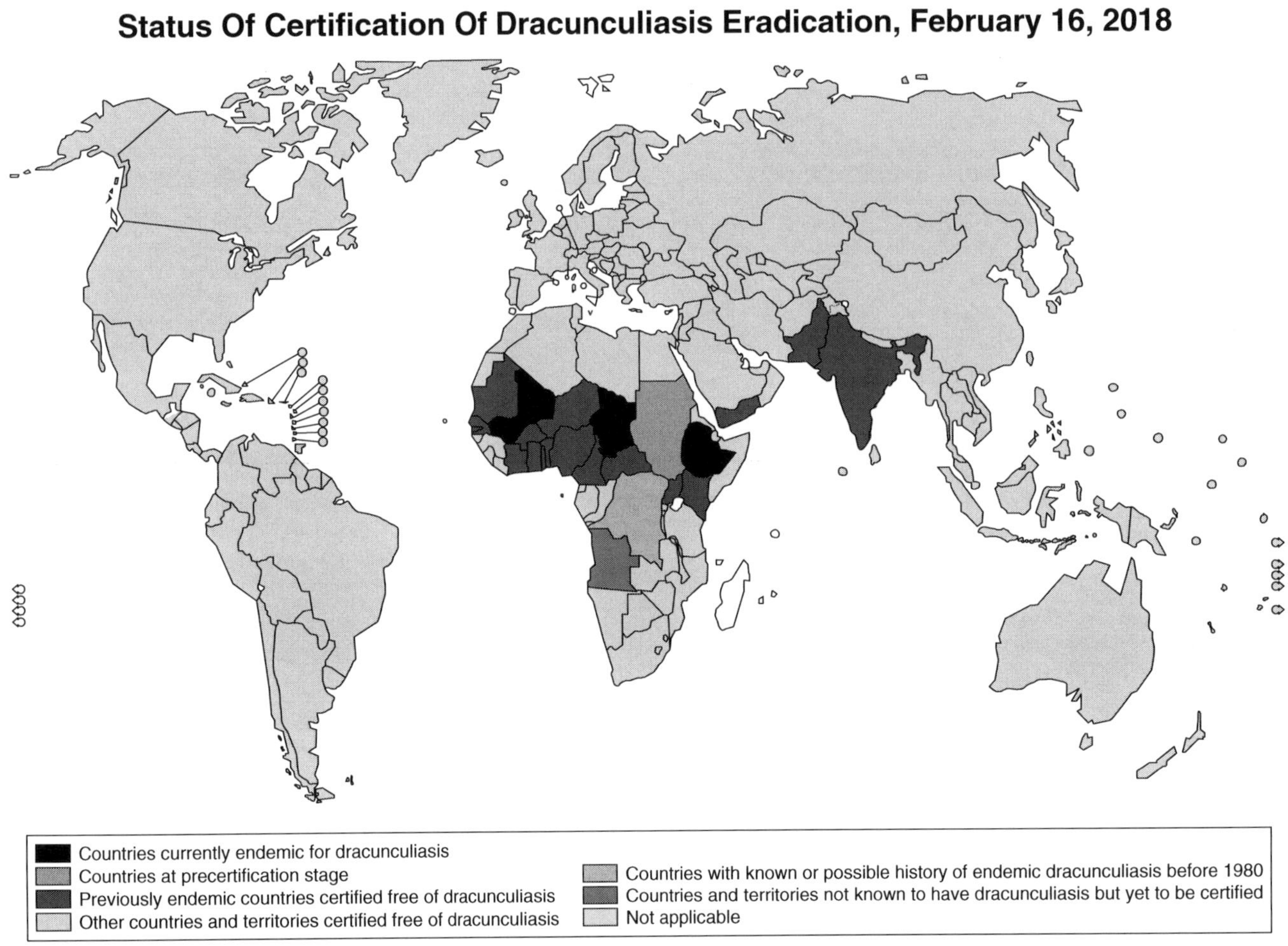

FIGURE 22.3 Distribution of 30 indigenous cases of Dracunculiasis reported during 2017. For the first time since the global Guinea Worm Eradication Program began, only two countries, Chad and Ethiopia, reported cases of Guinea worm disease (dracunculiasis) in humans during 2017 (Figure 22.1). *Source*: Reproduced from The Carter Center and the Centers for Disease Control and Prevention. (9 March 2018), "Guinea Worm Wrap-Up #253". Available from: https://www.cartercenter.org/resources/pdfs/news/health_publications/guinea_worm/wrap-up/253.pdf. (*See insert for color representation of this figure.*)

are trained by the program to do four things: educate their neighbors about how to prevent Guinea worm disease, distribute cloth filters, record and report cases monthly to a supervisor, and provide first aid care to villagers with Guinea worm disease. The immediate supervisors are peripheral governmental health workers or are national and international technical advisors who are specially hired and supported by The Carter Center for the national GWEP. This much-envied system comprised more than 6000 VVs each in Ghana and Nigeria in the early 1990s, for example, pioneering unprecedented monthly reporting and surveillance for cases of Guinea worm disease from every known endemic village, including some of the most remote villages imaginable. A few such villages were previously unknown to government authorities. The key to sustaining this system is and was assiduous attention to training, retraining, supply, constructive supervision, encouragement, and feedback of the VVs and their supervisors. Compilation, display, and regular monitoring of monthly surveillance data and of operational indices are other vital ingredients that are particularly important to motivating workers and mobilizing financial support for the program. This combination of dedicated volunteers, their support, supply, and supervision is what has brought this ancient loathsome disease to the brink of eradication.

Villagers are bombarded with simple messages via as many channels as possible (radio, videos, discussions illustrated with flip charts, drama, songs, posters, etc.), encouraging them to always filter unsafe drinking water, to avoid entering or allowing others to enter sources of drinking water when a Guinea worm is emerging or about to emerge, and to report any cases of the disease to a health worker. By these means, the GWEP seeks to engender a climate whereby undesirable behavior is communally regarded as an unneighborly act, without stigmatizing those infected, and where repeated messaging provides constant and involuntary reminders of what villagers should and should not do when they encounter a source of drinking water. More recently, behavior change communication techniques have been taught to more effectively engage communities in seeking agreement on what to do on their own to prevent Guinea worm disease from affecting their community. Political mobilization is equally as important as community mobilization. Here, one seeks to motivate political, traditional, and religious leaders to actively support eradication efforts, by using data from the campaign to inspire, encourage, and, if necessary, make the right people uncomfortable. The fact that Guinea worm disease is so visible, has important economic impact, and can be prevented, even *eradicated*, provides powerful ammunition for such political advocacy [4].

As mentioned earlier, the ecology and epidemiology of dracunculiasis is intimately bound to local environments where the infection thrives. Copepods are ubiquitous, but vary in their ability to transmit Guinea worms to humans. By definition, all humans must have some source of drinking water, but these also vary in their susceptibility to harbor copepods, which require stagnant, not flowing, fresh water. Streams and small rivers that contain flowing water during the rainy season but become almost dry riverbeds from which humans collect drinking water from stagnant intermittent pools and excavated holes during the dry season are simultaneously transformed into effective potential sources of infection. In northern Ghana, small dams constructed in the 1950s and 1960s for watering cattle and irrigating crops inadvertently provided the ideal habitat for undesirable contact between Guinea worms, receptive copepods, and many thirsty people [5]. Moreover, many such sources of impounded water are too large to reasonably treat with ABATE. This well-intended ecological alteration caused much of northern Ghana to stand out most unfavorably as far as Guinea worm disease was concerned, from otherwise similar neighboring areas of Burkina Faso, Cote d'Ivoire, and Ghana itself. In addition, just as surface sources of drinking water are determined by the environment and rainfall, the availability of safer underground sources, which are not at risk of contamination by people with emerging Guinea worms, is conditioned by geology. Some populations still manage to subsist in areas underlain by rock that prohibits or severely limits the drilling of wells, by collecting rainwater and/or traveling long distances to fetch surface water from far away. A result of these many factors is that Guinea worm disease is distributed sporadically, often being highly endemic or absent altogether, in adjacent villages and districts.

Despite the many environmental determinants of Guinea worm disease's occurrence and intensity, human agency has been a major factor in the eradication campaign in other ways besides health-related behavior, primarily because of insecurity and sporadic violence engendered neither by Guinea worms nor by copepods but by people. Such disruptive insecurity has been most prominent in the GWEP during the civil war in Sudan and later after a *coup d'etat* in Mali, but also has been a significant impediment to control measures in most other affected countries in Africa at one time or another. In Ghana, an outbreak of ethnic violence in the Northern Region during the peak transmission season in 1994 and 1995 disrupted interventions, led to loss of supplies and equipment, and caused villagers and health workers to flee to district capitals, where some fleeing villagers contaminated the drinking water sources at local dams, causing the number of Guinea worm cases to double between January and April 1994 and the same period of 1995. Violence involving nomadic Tuaregs in northern Mali delayed the reporting and deployment of remedial measures to control an outbreak in 2006 and 2007. Between January and August 2009, 27 incidents of sporadic violence or petty banditry disrupted operations of the GWEP in southern Sudan, causing Guinea worm workers to be confined to their dwellings or evacuated for various periods, often in highly endemic areas. In addition to civil war in Cote d'Ivoire, other ethnic, political, or religious conflicts have also impeded Guinea worm eradication activities in Chad, Ethiopia, Mali, Niger, Nigeria, Pakistan, Togo, and Uganda.

Addressing the political issues effectively has been a critical component of the campaign that requires engagement of persons outside of the traditional public health arena. Involvement of former US President Jimmy Carter has been key to mitigating or removing political barriers in the Guinea Worm Eradication Program. In 1995, as the global campaign sought to help Sudan launch its program despite the ongoing civil war between the north and south of that country, by drawing on his prior contacts with both sides, President Carter was able to negotiate a cease-fire. By then, Sudan was known to be among the most highly endemic countries and it seemed wise to try to start working in accessible areas of the country, which included most of the north and some of the more highly affected southern states where most of the fighting was occurring. Eventually known as the "Guinea Worm Cease-Fire," the hiatus in the civil war lasted nearly six months and was even more successful than we expected. Sudanese health workers and their foreign Non-Governmental Organization (NGO) partners on both sides surprised themselves by what they were able to accomplish, visiting almost 2000 endemic villages to inquire whether Guinea worm disease occurred there and beginning to educate local villagers where it did, as well as distributing over 200 000 cloth filters [6]. Despite resumption of the fighting, in 1996 Sudan's GWEP distributed even more cloth filters without a cease-fire than it did the year before.

Another masterstroke was to engage two prominent African leaders in the campaign. In 1992 President Carter persuaded then former Malian head of state General Amadou Toumani Toure to help energize his country's listless eradication effort. General Toure eventually toured all the major endemic areas of Mali and visited all nine other endemic francophone countries to lobby their heads of state and ministers on behalf of the campaign as well. His popularity enhanced, Toure was later elected president of Mali (he had ousted a military dictator in a *coup d'etat* in 1991 and turned the government over to an elected civilian a year later) and was himself ousted in a *coup d'etat* near the end of his second term. Popular former Nigerian head of state General (Dr) Yakubu Gowon was recruited in 1998 to help reverse five years of stagnation due partly to sporadic insecurity and ineffectiveness of the military government in Nigeria, which had begun its program in 1988 with more reported cases than any other country. By the end of 2005, General Gowon had personally visited over a hundred endemic villages in 18 Nigerian states, urging local medical, traditional, and political leaders to intensify control measures, including giving priority to endemic villages for provision of safe drinking water [1]. Nigeria's reported cases began declining again dramatically in 2000 and it reported its last case in 2008.

Marginalization and migration are two other human factors affecting Guinea worm disease and its eradication. Many of the most resistant foci of disease are thus characterized because the poor rural populations in them are disenfranchised and long neglected by others in the same country. Salient examples are the Nyangatom in southern Ethiopia, Konkomba in northern Ghana, and black Tuareg in Mali and Niger. Overcoming decades or centuries of such enmity in order to do the necessary in such circumstances, by paying careful attention to ethnicity of health workers, to the language(s) used for written and verbal messages, and to placement of case containment centers and borehole wells, for example, can be a challenge. The frequency and unpredictability of human travel to attend weddings and funeral ceremonies (often in distant neighboring villages or even other districts), and to farm or trade pose other challenges, in addition to more predictable transhumance and annual movements of some pastoral groups. In Mali, an infected Koranic student triggered an explosion of 85 cases in one village in a previously non-endemic area after having walked over 300 km from his home district. Similar exceptional treks by infected individuals have been reported in persons moving from southern Sudan into Ethiopia.

For all its potential benefits and seemingly irresistible logic, the global GWEP got underway very slowly, because of insufficient funding during the 1980s and early 1990s for assisting national eradication programs. Although it was conceived in 1980 just before the launch of the International Drinking Water and Sanitation Decade (1981–1990), by 1990 only 4 of the 20 endemic countries (India, Pakistan, Ghana, and Nigeria) had actually launched nationwide Guinea Worm Eradication Programs. Twelve countries began their national programs in 1992–1994 and all but one (Central African Republic) got underway by the end of 1995 [1].

22.3 IMPACT AND FINAL CHALLENGES

As already noted, the numbers of reported cases of dracunculiasis increased rapidly as individual countries conducted their first nationwide surveys to count cases of the disease, starting in 1988. As the same countries began implementing control measures successively, their numbers of reported cases began to decline significantly, but not as abruptly as they had risen. The relatively rapid initial reductions in cases reflected impact of control measures in the most receptive populations, gradually revealing the most resistant and difficult clusters of endemic villages, families, and individuals as the campaign progressed at the frustratingly stately pace dictated by the infection's one year long incubation period and aided by human mistakes, neglect, bureaucratic and financial shortcomings, and by the sporadic insecurity described above.

Although the global GWEP missed the original target date of 1995 for interrupting transmission of dracunculiasis, by 2003 the campaign had reduced the number of cases to 32 193 cases reported from 4659 villages in 12 countries, of which 20 299 cases were reported from Sudan, 8285 from Ghana, and 1459 from Nigeria. Heartened by the recent results, representatives of the 12 remaining endemic countries resolved during the World Health Assembly in May 2004 to intensify implementation of control measures so as "to free the world of *dracunculiasis* by the end of 2009" [7]. Another major advance occurred in January 2005 when the Comprehensive Peace Agreement was signed to end the two-decade long civil war in Sudan, thus removing the single most important obstacle to complete eradication. Sustained and generous financial support provided to The Carter Center for the campaign by the Bill & Melinda Gates Foundation and other donors provided the wherewithal to capitalize on the favorable political developments. After a brief upsurge in the number of cases reported from Sudan in 2006 as that program gained access to previously inaccessible endemic areas, the newly constituted South Sudan Guinea Worm Eradication Program (SSGWEP) quickly began building on the momentum and experience gained since the Guinea Worm Cease-Fire in 1995 [8].

By the end of 2016, dracunculiasis had been eliminated from Asia altogether, the number of endemic countries had been reduced from 20 to 3, the number of endemic villages was down from over 23 000 in 1991 to 19, and the number of cases reported was reduced from an estimated 3.5 million in 1986 (and over 900 000 counted in 1989) to 25. Most of the cases reported in 2016 were from Chad (16 cases), with South Sudan [6] and Ethiopia [3] reporting the remainder (Figure 22.4). The discovery of infected domestic dogs in Chad in 2012 with Guinea worms that are genetically indistinguishable from Guinea worms removed from humans followed by rapid increase in the number of infected dogs in subsequent years is a major challenge to the campaign. An unusual life cycle involving a paratenic host is suspected and

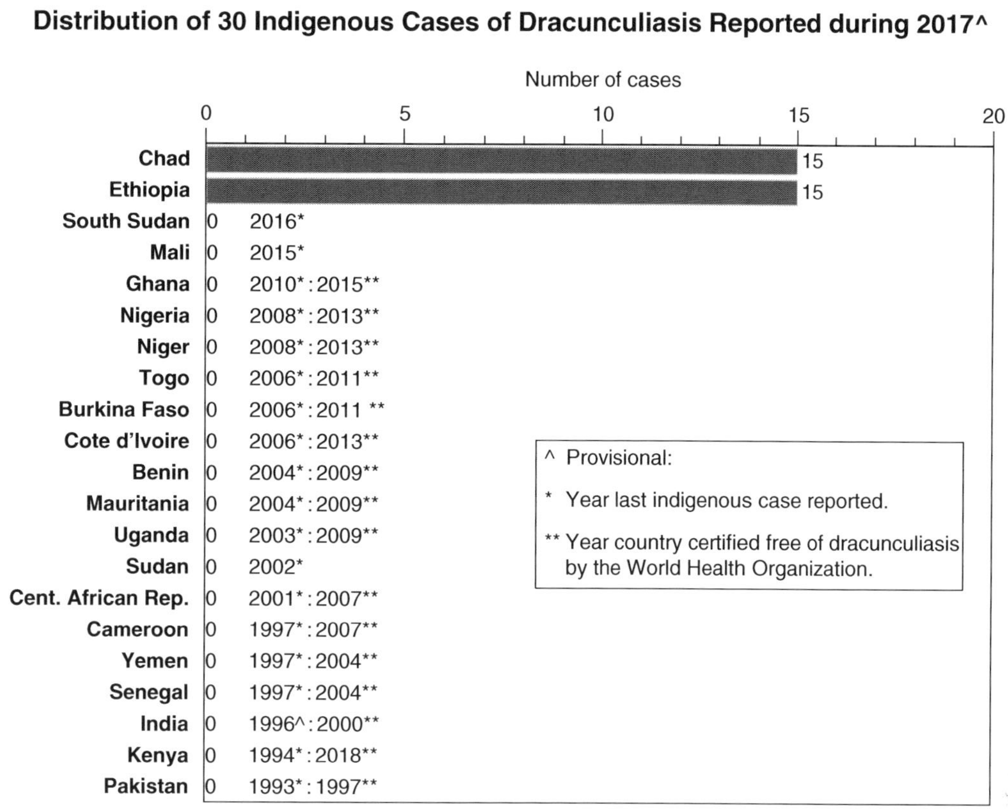

FIGURE 22.4 Status of Certification of Dracunculiasis Eradication, 16 February 2018. The boundaries and names shown and the designations used on maps do not imply the expression of any opinion whatsoever on the part of the World Health Organization concerning the legal status of any country, territory, city or area or of its authorities, or concerning the delimitation of its frontiers or boundaries. Dotted and dashed lines on maps represent approximate border lines for which there may not yet be full agreement. *Data Source*: World Health Organization Map Production: Guinea Worm Eradication, World Health Organization.

additional interventions and studies are underway to meet this new challenge [9]. The number of cases exported from one country to another declined from 154 in 2002 to 6 in 2008 and zero since 2012. Between 2014 and 2015, the number of cases reported globally was reduced by an unprecedented 83%, from 126 to 22 [10]. Between 2015 and 2016, the number of cases increased slightly to 25, but the number of endemic countries was reduced from four to three [11]. At this stage of the campaign, the proportion of cases that were discovered before or within 24 hours of emergence of the Guinea worm and are successfully contained is the single most important indicator of the effectiveness of national programs. In 2016, Chad contained 56% of its cases, South Sudan contained 50%, and Ethiopia 67% [11].

The World Health Organization bears sole responsibility for certifying interruption of transmission of dracunculiasis. To do so, it requires adequate surveillance, having detected no indigenous case of dracunculiasis in a recently endemic country for at least three consecutive years after the last known indigenous case. An International Commission for the Certification of Dracunculiasis Eradication reviews data submitted by countries and the reports of teams sent to verify the status of surveillance in selected countries, and makes recommendations to the Director General of WHO. As of February 2018, WHO had officially certified the absence of dracunculiasis from 199 countries and 7 countries remain to be certified [10].

The global Guinea Worm Eradication Program is poised to achieve its ultimate goal soon, despite many challenges described above, provided sporadic insecurity in South Sudan and Mali is contained.

REFERENCES

1. Ruiz-Tiben, E. and Hopkins, D.R. (2008). Helminthic diseases: dracunculiasis. In: *International Encyclopedia of Public Health*, vol. 3, 294–311. Elsevier Inc.
2. Hopkins, D.R. and Ruiz-Tiben, E. (1991). Strategies for eradication of dracunculiasis. *Bull. World Health Organ.* 69 (5): 533–540.
3. Watts, S.J. (1987). Dracunculiasis in Africa: its geographic extent, incidence, and at-risk population. *Am. J. Trop. Med. Hyg.* 37: 119–125.

4. Edungbola, L.D. et al. (1992). Mobilization strategy for Guinea worm eradication in Nigeria. *Am. J. Trop. Med. Hyg.* 47 (5): 529–538.

5. Hunter, J.M. (1997). Bore holes and the vanishing of Guinea worm disease in Ghana's upper region. *Soc. Sci. Med.* 45 (1): 71–89.

6. Hopkins, D.R. and Withers, P.C. Jr. (2002). Sudan's war and eradication of dracunculiasis. *Lancet* (Suppl. 360): s21–s22.

7. World Health Organization (2004). Dracunculiasis eradication: Geneva declaration on Guinea-worm eradication, Geneva, 2004. *Wkly. Epidemiol. Rec.* 79: 234–235.

8. Hopkins, D.R., Ruiz-Tiben, E., Downs, P. et al. (2008). Dracunculiasis eradication: neglected no longer. *Am. J. Trop. Med. Hyg.* 79 (4): 474–479.

9. Eberhard, M.L., Ruiz-Tiben, E., Hopkins, D.R. et al. (2014). The peculiar epidemiology of Dracunculiasis in Chad. *Am. J. Trop. Med. Hyg.* 90: 61–70.

10. World Health Organization (2018). Dracunculiasis eradication – global surveillance summary, 2017. *Wkly Epidemiol. Rec.* 93:305–320.

11. Centers for Disease Control and Prevention. Guinea Worm Wrap-Up #245. 12 January 2017. Available online from: http://www.cdc.gov/ncidod/dpd/parasites/dracunculiasis/moreinfo_ dracunculiasis.htm#wrap.

23

SANITATION CASE STUDIES

Anoop Jain[1] and Jay Graham[2]

[1] *University of California, Berkeley, CA, USA*

[2] *Public Health Institute, Oakland, CA, USA*

23.1 RURAL SANITATION IN THAILAND: PERSISTENCE AND CREATIVITY PAYS OFF

23.1.1 Summary

In 1960, less than 1% of Thailand's rural residents had access to at least basic sanitation, but by 2015, this figure had increased to 96% [1]. This dramatic increase is the result of a concerted, decades-long campaign by the Thai government, which successfully employed a range of tools – health extension workers, revolving loans, competitions, local artisans, and new policies – to expand sanitation coverage. While Thailand's experiences are unique and challenges still remain, the country's successful sanitation improvement efforts present potentially relevant lessons for other countries still struggling to improve basic sanitation coverage (Figure 23.1).

23.1.2 Background

The Kingdom of Thailand (Thailand) is located in Southeast Asia and has a population of nearly 66 million people – two-thirds of whom reside in rural areas. The country supports a free-market economy and has experienced relatively high economic growth over the past 15 years, with only two years of losses, both of which resulted from the 1997 Asian financial crisis. Thailand has effectively managed population growth, with growth rates dropping from 3.2% in 1960 to a rate of 0.6% today. The portion of the population living in rural areas has fallen relatively slowly – dropping roughly by 13% since 1960 to just over 47% in 2017. Rural development has been a top priority of the Thai Government since the 1960s, when the first five-year National Economic and Social Development Plan began [2]. A number of Ministries have been involved in rural development and have substantially raised educational attainment, improved family planning, expanded access to safe water and adequate sanitation, and delivered basic health services and nutritional assistance to the poor [3,4]. Since 1960, Thailand has experienced one of the fastest declines in its infant mortality rate globally (Figure 23.2).

23.1.3 Sanitation Progress Begins

Thailand's first significant steps towards universal rural sanitation came with the start of the Village Health and Sanitation Project in 1960, when basic sanitation coverage was less than 1% (Graph below) [2]. Directed by the Ministry of Public Health, the project aimed to reduce the incidence of diarrheal diseases associated with poor sanitation and hygiene. In 1961, the project was expanded and integrated into the National Economic and Social Development Plan and subsequently renamed the Rural Environmental Sanitation Program (RESP) [1]. Community health workers from the provincial health departments and village volunteers formed the foundation of the program. They promoted a variety of participatory activities and low-cost technologies converging

Water and Sanitation-Related Diseases and the Changing Environment: Challenges, Interventions, and Preventive Measures, Second Edition. Edited by Janine M. H. Selendy.

Companion website: https://www.wiley.com/legacy/wileychi/selendy/

Kingdom of Thailand Snapshot

- Population: 68 million[1]
- Rural population: 47%[1]
- Life expectancy (M/F): 71/78 years[1]
- Literacy: 92.9%[1]
- Per capita GDP (PPP): $16,900[1]
- Sanitation coverage: 89.9% urban[1]/96.1% rural[1]
- Infant mortality rate: 9.2 per 1,000[1]

1) "Thailand" *The World Factbook* Central Inteligence Agency, 2018. Web. 13 January 2018.

FIGURE 23.1 Key demographic information for Thailand.

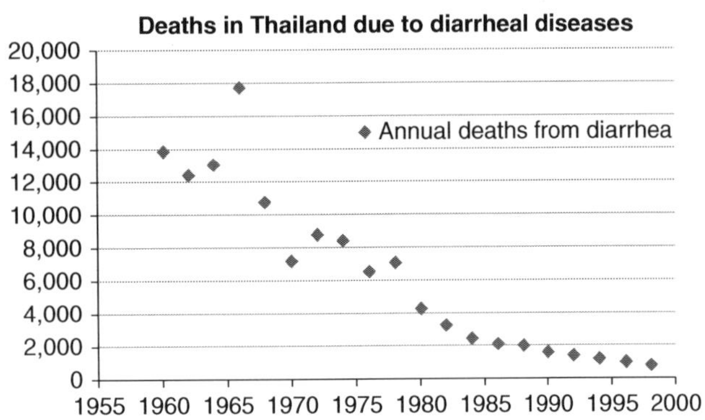

FIGURE 23.2 Deaths in Thailand due to diarrheal disease.

around water, sanitation, hygiene and solid waste disposal. RESP had its setbacks. Hardware for sanitation, for example, was initially given to households through government subsidies, with the assumption that recipient households would serve as models to the rest of the community. It became evident, however, that the free systems were neither used nor maintained, and health officials and village committees consequently developed a more demand-driven process. In place of free systems, a revolving loan was created for each village so that households would finance their own sanitation improvements, helping ensure user commitment and extend limited resources. Exceptions were made for poorest households who often would provide labor in exchange for participation in the program. At the same time, community health workers helped raise awareness and build demand for improved hygiene and sanitation. In many cases, health workers linked sanitation improvements to a household's desire for better access to water. The need to link sanitation to water supply improvements is generally an indicator of low demand for sanitation, and in some cases sanitation systems subsequently failed or were unused. The Ministry of Public Health, however, persisted in their support and follow-up, ensuring continued use and maintenance. The national government committed financial resources to sanitation, dedicating approximately 2–6 million US dollars to sanitation

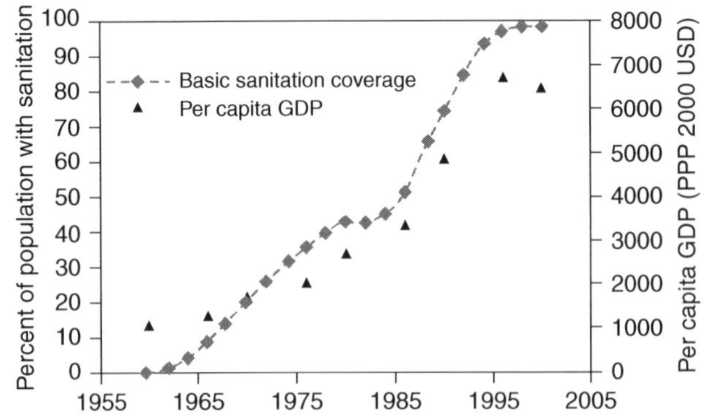

FIGURE 23.3 Sanitation coverage in Thailand over the years.

annually (1990–1998 budget data) and committed health officers to each of the 75 provinces outside of Bangkok to raise awareness, train health workers, and monitor progress [1]. Further, country- and province-level meetings, which involved volunteers, were conducted regularly to gauge progress, identify barriers, and develop solutions. The sanitation technology of choice (selected by the government) was the pour-flush latrine, and although community participation and consumer choices were not part of the selection process, the technology fitted well with Thai preferences. Additionally, the government helped develop the market for sanitation hardware by assisting the local private sector by training artisans to construct and sell products that included an array of low-cost, appropriate technologies (e.g., rainwater catchment systems) to meet the demands of rural consumers.

23.1.4 Recurring Commitments

In 1978, Thailand adopted the World Health Organization's strategy, *Health for All by the Year 2000*, using sanitation and water supply coverage as indicators of success. Nearly a decade later, towards the end of the International Drinking Water Supply and Sanitation Decade (1981–1990), efforts to increase sanitation coverage were accelerated by adoption of the "100% Latrine Coverage Campaign Project." This was a program that offered incentives to local governments to encourage them to achieve 100% sanitation coverage. In addition, a new policy by the Ministry of the Interior (1989) bolstered sanitation efforts by requiring all households to have a sanitary toilet before a house could be registered. Through multiple strategies and projects, Thailand consistently renewed its efforts to increase access to basic sanitation [1] (Figure 23.3).

23.1.5 Sanitation Today

The 2017 Joint Monitoring Programme report highlights that 96% of Thailand's rural population uses at least basic

Events in Thailand's Rural Sanitation History

1961:	The Rural Environmental Sanitation Program begins.
1962–3:	Two successive Prime Ministers chair the National Health and Sanitation Development Conferences, raising awareness at high levels.
1980:	Revolving loan program to help households finance sanitation improvements initiated.
1987:	1) Involvement of the private sector and local artisans began: 2) An award scheme for provincial leaders and competition among provinces implemented.
1988:	100% Latrine Coverage Campaign Project initiated.
1989:	The Ministry of Interior, with support from the Ministry of Public Health, adopts an ordinance requiring all new homes to have a sanitary toilet prior to registration.

FIGURE 23.4 Key events in Thailand's rural sanitation history.

Lessons Learned

- **Persistence** – after 35 years and seven 5-year development plans – consistently including sanitation as part of their development criteria – Thailand has achieved near-universal sanitation
- **High-Level Support** – the importance of basic sanitation was well recognized among government leadership and dedicated resources (financial and human) reflected a broad commitment to achieving universal sanitation.
- **Sanitation Policies** – policies, such as the ordinance requiring households to have a sanitary toilet prior to registering a new house, helped accelerate gains in sanitation
- **Reduced Hardware Subsidies** – the Ministry of Public Health quickly determined that hardware subsidies were undermining sustained use and maintenance of toilets and switched to a more demand-driven tactic that utilized revolving loans to finance hardware
- **Private Producers** – to improve economic sustainability, the Ministry of Public Health provided technical assistance to train community artisans in building and marketing low-cost, appropriate technologies (e.g., pour-flush latrines and rainwater harvesting cisterns)
- **Competition and Awards** – friendly competition among provinces and awards for provincial leaders created incentives to improve sanitation coverage
- **Monitoring** – the government regularly monitored progress to identify barriers and develop solutions
- **Integrated Programs** – sanitation was made part of a larger "basic needs" package of goods aimed at improving the quality of life of rural communities

FIGURE 23.5 Lessons learned.

sanitation, and open defecation in rural areas has been completely eliminated [1]. Diarrheal morbidity, although relatively low, remains a focus of the Ministry of Health – the annual incidence rate is estimated at 1.8 cases per 100. Ongoing efforts by the government – including hygiene awareness campaigns, competitions, new sanitation and hygiene regulations, and a strong cadre of public health inspectors – demonstrate a continued commitment to universal sanitation [5] (Figures 23.4 and 23.5).

23.2 SINGAPORE: FORWARD THINKING AND FORWARD MOVING

At the beginning of the twentieth century, a manual "two-pail" system was the prevailing method for collecting and removing human excreta from homes in Singapore [5]. The two-pails allowed households to have one pail in use while the other was transported to a disposal site where it was emptied and subsequently returned to the household, replacing the pail in use. Around that same time, the annual infant mortality rate was estimated to be 354 per 1000 (Figure 23.6).

Singapore Snapshot

- Population: 5.8 million
- Urban population: 100%
- Life expectancy (M/F): 82/86 years
- Literacy: 97%
- Per capita GDP (PPP): $87,800
- Sanitation coverage: 100%
- Infant mortality rate: 2.4 per 1,000

"Singapore", *The World Factbook*. Central Intelligence Agency, 2018.

FIGURE 23.6 Key demographic information for Singapore.

The development of modern sewage systems in Singapore started in 1910, and by the late 1930s, the population served

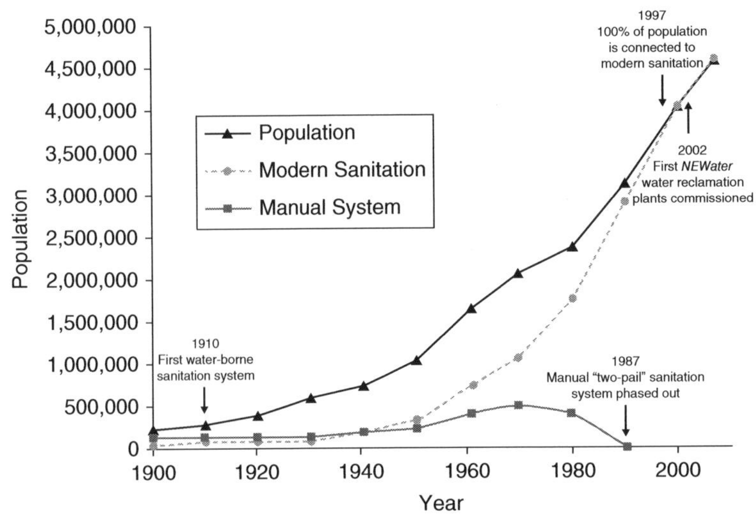

FIGURE 23.7 History of sanitation in Singapore. *Source*: PUB Presentation, 2009.

by modern sanitation surpassed those using the manual disposal system. Singapore gained independence in August 1965 and that same month the Public Utilities Board (PUB), which currently manages the entire water cycle of Singapore, came into being. In 1987, the manual system of excreta collection was completely phased out – at the same time, Singapore was gaining its reputation as a place of orderliness and effective administration. Health-related indicators had much improved – in 1986, the infant mortality rate was 9.1 per 1000 and life expectancy at birth was 71.4 years for males and 76.3 years for females [6]. With many of the health problems associated with a contaminated water supply and lack of basic sanitation solved, environmentally driven efforts in Singapore increased. Efforts to clean up rivers and streams became a primary goal between 1977 and 1987. The Ministry of the Environment made great advances in extending the sewer system – reaching 100% of the population with modern sanitation by 1997. Since then, the population of Singapore has nearly doubled, yet Singapore has maintained 100% sanitation coverage. Furthermore, the authority to manage sewerage and drainage was shifted from the Ministry of the Environment to the PUB in 2001 [7] (Figure 23.7).

The 263 square mile island city-state now has more than 36 square miles of parks, with 16 additional square miles reserved for park space [8]. The whole population is served by modern sewage treatment that includes: 3200 km of sewers, 116 pumping stations and 6 water reclamation plants [7]. Singapore draws on a strategic development framework, the Concept Plan, which is updated every decade. The current plan, established in 2001, sets broad-based development plans for the next half century. It allows for an expected population of 5.5 million within 50 years. The Concept Plan provides details on land-use needs (public, commercial, and industrial), transportation, and recreational facilities. Community participation is a core component of the planning process and neighborhoods are involved in local development planning.

During the National Day Parade on August 9, 2002, Singapore celebrated the start of their NEWater program, which effectively collects wastewater and processes it to a level surpassing international drinking water standards. At the launch of the program, the Prime Minister and the whole Cabinet toasted a crowd of 60 000 parade goers with NEWater, which has grown in demand [9].

Most recently, the Public Utilities Board began implementing the Deep Tunnel Sewerage System (DTSS). The DTSS is intended to help Singapore meet all of its water needs through the collection, treatment, reclamation, and disposal of water from homes, businesses, and industries. The DTSS project was initially spilt into two phases, the first of which was completed in 2008. It cost $2.7 billion and includes over 45 km of sewerage tunnels that collect water from around the island. Phase II will extend this network, ensuring that the entire island is covered, and is scheduled to be completed by 2022. This effort will help eliminate the need for continual upgrading and expansion of the aging sewage system. The new deep tunnel system will allow Singapore to phase out 134 sewage pumping stations and smaller water reclamation plants – freeing up land and resources used for operations and maintenance [9].

REFERENCES

1. Aungkasuvapala, N. (2007). Universal Sanitation – Thailand. East Asia Ministerial Conference on Sanitation and Hygiene, JMP 2017.
2. Luong, T. V., Chanacharnmongkol, O., and Thatsanatheb, T. (2000. Universal Sanitation – Thailand Experiences. 26th WEDC Conference Proceedings, Dhaka, Bangladesh.

3. Anil, B. (2002). Deolalikar. Poverty, Growth, and Inequality in Thailand. ERD Working Paper No. 8, April.
4. World Health Organization (2017). *Progress on Drinking Water, Sanitation and Hygiene: 2017 Update and SDG Baselines*. S.l.: World Health Organization.
5. Personal communication with the Ministry of Public Health, Thailand.
6. Otaki, Y., Otaki, M., and Sakura, O. (2007). Water systems and urban sanitation: a historical comparison of Tokyo and Singapore. *Journal of Water and Health* 5 (2).
7. Tortajada, C. (2006). Water management in Singapore. *Water Resources Development* 22 (2): 227–240.
8. Hinrichsen, D. (2001). Singapore: The Planned City. Population Reports.
9. Soon, T.Y., Jean, L.T., and Tan, K. (2009). *Clean, Green and Blue: Singapore's Journey Towards Environmental and Water Sustainability*. ISEAS Publishing.

24

CATALYZING RURAL SANITATION AT SCALE: LESSONS LEARNED FROM THE GLOBAL SANITATION FUND

PATRICK ENGLAND AND CAROLIEN VAN DER VOORDEN

Technical Support Unit, Water Supply and Sanitation Collaborative Council (WSSCC), Geneva, Switzerland

As part of the Millennium Development Goals (MDGs), between 2000 and 2015 the world aimed to halve the number of people without access to basic water and sanitation. While the target for water was achieved, progress on sanitation lagged: by the end of the MDG period in 2015, 2.4 billion people did not have access to basic sanitation. Disparities between rural and urban areas are stark: 82% of the global urban population versus 51% of the rural population uses basic sanitation facilities, with 9 out of 10 people living in rural areas practicing open defecation [1].

The Sustainable Development Goals (SDGs) aim to raise the threshold for success in two crucial respects.[1] First, SDG target 6.2 is measured by the proportion of people using *safely managed sanitation services.* This means that not only should feces be safely contained within latrines, but that waste is safely treated on- or off-site and that latrines are equipped with hand-washing facilities with soap and water. Second, everyone must be reached, with emphasis placed on women, girls, and those in vulnerable situations. This places new demands on innovating and adapting cost-effective approaches that can deliver on both scale and equity for the SDG era.

This chapter will delve into the experiences and lessons learned on financing large-scale sanitation and hygiene programs through the Global Sanitation Fund (GSF) – a pooled funding mechanism established by the Water Supply and Sanitation Collaborative Council (WSSCC) to support countries improve sanitation and hygiene for the poorest and most vulnerable. Since 2008, the GSF has disbursed over $117 million to 13 supported programs in Africa and Asia and has enabled over 15 million people to live in open-defecation-free (ODF) environments [2].[2] For the GSF, catalyzing the achievement of SDG target 6.2 requires demonstrating success in strategic administrative areas, with a core focus on sustaining collective behavior change, equality and non-discrimination (EQND), and strengthened commitments, capacities, and coordination at all levels.

24.1 TARGETING ENTIRE ADMINISTRATIVE AREAS

Catalytic transformations, rather than small pockets of success, are critical for achieving SDG target 6.2. To achieve this, the GSF uses targeted investments in nationally owned programs to achieve universal usage of safely managed sanitation for the entire population of selected administrative

[1] SDG target 6.2 states: "By 2030, achieve access to adequate and equitable sanitation and hygiene for all and end open defecation, paying special attention to the needs of women and girls and those in vulnerable situations."

[2] Definitions of ODF vary between countries. The GSF's minimum standard for ODF includes (i) no feces in the open; (ii) all households using safe, improved sanitation facilities that separate waste from human contact (as per the JMP definition of "basic service"); and (iii) the presence of a handwashing facility with water and soap (or ash) near each toilet.

Water and Sanitation-Related Diseases and the Changing Environment: Challenges, Interventions, and Preventive Measures, Second Edition. Edited by Janine M. H. Selendy.

Companion website: https://www.wiley.com/legacy/wileychi/selendy/

areas, such as districts, (sub)counties or communes. These areas are predominantly rural, with low sanitation coverage, high rates of open defecation, and a high burden of WASH-related diseases, such as diarrhea, typhoid, and cholera.

Targeting entire administrative areas is key for three reasons. First, health benefits are fully realized when a sufficient number of people in a given area are consistently using safely managed sanitation services [3]. Second, targeting whole populations makes it more likely that positive social norms will develop, and therewith get good sanitation and hygiene behaviors to "stick" [4]. Third, catalytic change has the best chance of success when programming is anchored within defined administrative units that have the mandate for improving sanitation and hygiene at the local level.

Demonstrating success at this level plays a strategic "proof of concept" function that can help persuade governments and external support agencies to replicate successful delivery models to other areas of the country.

24.2 IGNITING AND SUSTAINING COLLECTIVE BEHAVIOR CHANGE

One of the most important ingredients for reaching SDG target 6.2 is a strong focus on igniting and sustaining sanitation and hygiene behavior change. In the past, prescriptive, subsidy-driven, and hardware-focused interventions yielded few gains; while many subsidized latrines were constructed, externally fabricated latrines were frequently unsuited for local conditions, did not reflect end-user's preferences, and did not disrupt deeply engrained habits and behaviors. Consequently, as the number of toilets increased, open defecation continued.

This experience informed GSF's decision to invest in collective behavior change interventions – drawing heavily on the paradigm-shifting Community-Led Total Sanitation (CLTS) approach. Rather than dispensing hardware subsidies or lecturing households on the importance of good sanitation, CLTS involves a series of facilitated community discussions where people make a choice to end, or continue, open defecation [5].[3] CLTS works because it is a participatory approach rooted in collective decision-making, local knowledge, and, in contrast to conventional health education campaigns, draws on evoking feelings of disgust (as a core behavioral motive) to disrupt existing habits and develop new social norms [6].

CLTS (or variants thereof) is now implemented in a growing list of more than 60 countries around the world and is named as the principle approach in several national sanitation and hygiene strategies. However, understanding what it takes to implement effective behavior change programming has been a learning journey over GSF's 10 years of financing sanitation and hygiene programs. Some lessons learned:

- **Facilitation quality is key**. Changing behavior is not easy – especially in areas where open defecation is a culturally engrained practice. Achieving ODF status is highly dependent on the quality engagement with communities. In practice, this means shifting from a prescriptive approach to valuing local decision-making, leaders and activists, and technology solutions.

  There is now a greater understanding that igniting and sustaining behavior change requires more intensive and prolonged interactions with communities. As a result, the GSF-supported Fond d'Appui pour l'Assainissement (FAA)[4] program in Madagascar developed the "Follow-up MANDONA" (FUM) approach – one example of how a greater focus on facilitation quality can enhance both the scale and speed of achieving results. Meaning "to push" in Malagasy, FUM is a CLTS-based approach applied in communities that have already been triggered to undertake action and have started their journey towards ODF status. FUM sessions bring the entire community together for them to assess where and how open defecation (or the use of unsafe latrines) is causing everyone to be exposed unknowingly to each other's "shit".[5] The purpose of exposing and discussing these fecal–oral transmission routes is to provoke feelings of disgust and shock that speak to the community's dignity, and consequently trigger them to undertake small, immediate, and doable corrective actions on-the-spot [7]. FUM has also been an important tool for addressing "slippage" – the return to unhygienic behaviors after ODF status has been declared [8].

- **Dynamic behavior change skill development**. High-quality facilitation is highly dependent on the skills (and mentality) of those engaging with communities to change sanitation and hygiene behavior. This requires moving beyond conventional "training of trainers" and toward flexible, demand-driven, and context-specific capacity building systems guided by a learning-by-doing approach.

  The FAA program in Madagascar initially contracted external CLTS trainers to build the capacity of local facilitators. However, fixed-term trainers were based far from program areas and the services they

[3] For more on the CLTS process, see The Plan International [UK] in reference [5].

[4] For more on the FAA programme see: http://wsscc.org/global-sanitation-fund/madagascar/.

[5] CLTS facilitators are encouraged to use the crudest word for human feaces, such as "shit" or its equivalent in the local language of target communities, to provoke the most emotive response.

A community map being drawn in rural Madagascar as part of the "triggering" meeting. Participants identify key features in their community, where the main open defecation areas are, and how they inadvertently eat each other's feces. *Source*: Credit: WSSCC. (*See insert for color representation of this figure.*)

offered were supply-driven rather than based on the needs of facilitators on-the-ground. After review, the FAA program reoriented its capacity building approach to using successful local facilitators, usually with expertise in specific areas based on their practical experience. These "thematic coaches" are requested to travel to other program areas, based on identified performance issues and skills gaps, to work with other local facilitators in the field. This peer-to-peer coaching system has proven highly successful and has contributed to delivery of accelerated results [9].

- **The importance of community leadership**. Many GSF-supported programs have learned that it is far more effective for local activists and community leaders to implement activities and deliver results themselves. These local actors have more in-depth knowledge of their communities, can more effectively change the behavior of their peers and neighbors, and contribute to the sustainability of sanitation and hygiene improvements as skills and capacities are locally built.

  The Rural Sanitation and Hygiene Promotion in Nigeria (RUSHPIN)[6] program has learned that systematically building a movement of local actors is critical for scale and sustainability. At the village level, emerging "Natural Leaders" (community members who are especially outspoken in their commitment to improve their sanitation and hygiene situation, and willing to push their neighbors to change) work with facilitators from the local government and NGOs to end open defecation in their community. Once communities reach ODF status, communities are invited to formally recognize their Natural Leaders into Water, Sanitation, and Hygiene Committees (WASHCOMs). These government-recognized bodies, which must have equal representation between men and women, are the principle focal point for all WASH interventions in the community, and receive additional training on sanitation and hygiene behavior change, facility maintenance, and basic financial management to ensure that improvements are sustained.

  The RUSHPIN program has learned that the role of Natural Leaders needs to evolve to sustain gains made in their own communities, as well as build a movement that can transform entire administrative areas. Particularly skilled and passionate Natural Leaders are encouraged to become "Community Consultants", who receive additional training and a results-based stipend from the program for supporting their neighboring villages to reach ODF status. The program has learned that Community Consultants often achieve results much faster and are able to visit more remote areas, than facilitators from the local government or NGOs. The role of WASHCOMs has similarly scaled-up. WASHCOMs from different communities are united under a Ward-level WASHCOM, which oversees and organizes sanitation and hygiene improvements for the whole area. These initiatives are tied together by quarterly conferences between Natural Leaders and WASHCOM members across the Local Government Area (LGA). These conferences are a

[6] Read more on the RUSHPIN programme at: http://wsscc.org/global-sanitation-fund/nigeria/.

Benedict Emori (right) with the Chief Eni Idum from Iboni community, Abi LGA. Due to his old age and poor health, Chief Idum was helped by Benedict – one of RUSHPIN's first community consultants – to build his toilet. *Source*: Credit: Jason Florio/United Purpose.

forum for sharing effective behavior-change tactics, exhibiting local technologies, and agreeing on roadmaps for achieving LGA-wide ODF status, and since they are attended by local politicians and traditional authorities, they are also important opportunities to bring key decision-makers on board with the movement. This strategy of scaling-up and organizing Natural Leaders and WASHCOMS intends to create a groundswell of organized, local sanitation champions, and was a key factor in the achievement of Nigeria's first verified and certified ODF LGA [10].

24.3 FOCUSING ON EQUALITY AND NON-DISCRIMINATION

Sanitation and hygiene are intimately linked to equality; women, children, and other groups who may be disadvantaged (for example: the elderly, people with disabilities, the poor, ethnic minorities, sexual and gender minorities, people with mental health conditions) can face stark socioenvironmental barriers to using safely managed sanitation and hygiene services.

Because collective behavior change approaches such as CLTS aim to achieve 100% usage of safe sanitation facilities, it is sometimes assumed that outcomes are automatically equitable. However, unequal power dynamics exist within all communities. Community-led interventions, while purporting to be inclusive, can promote more powerful voices over those who may be marginalized and excluded. To ensure fairness, people who are potentially disadvantaged may require different levels of support and resources and more pro-active inclusion in decision-making to ensure that they are not left behind, or face greater risks to their safety and dignity.

The GSF has undergone a recent scoping and diagnostic exercise to understand the dynamics of inequality affecting access and use of safe sanitation and hygiene, to what extent supported programs are addressing disadvantage, and what improvements can be made [11]. Some of the key recommendations for improving GSF's approach to EQND include:

- **Establish clear EQND principles**. The development of clear principles and a code of conduct are critical for guiding sanitation and hygiene programs to address EQND, and to ensure that the rights of vulnerable people are upheld. Three components are important. First is a recognition that everyone is entitled to their human right to safe adequate sanitation and hygiene without discrimination.[7] Second is establishing the principle of "do no harm". This means that anyone involved in implementing program activities has an obligation to take every precaution necessary to ensure that people are not inadvertently made worse off as a result of the intervention. Third, programs should be guided by the "golden rule:" *nothing about us, without us.* This

[7] In 2010, UN General Assembly resolution 64/292 recognized safe sanitation as a fundamental and distinct human right. See: http://www.un.org/waterforlifedecade/human_right_to_water.shtml.

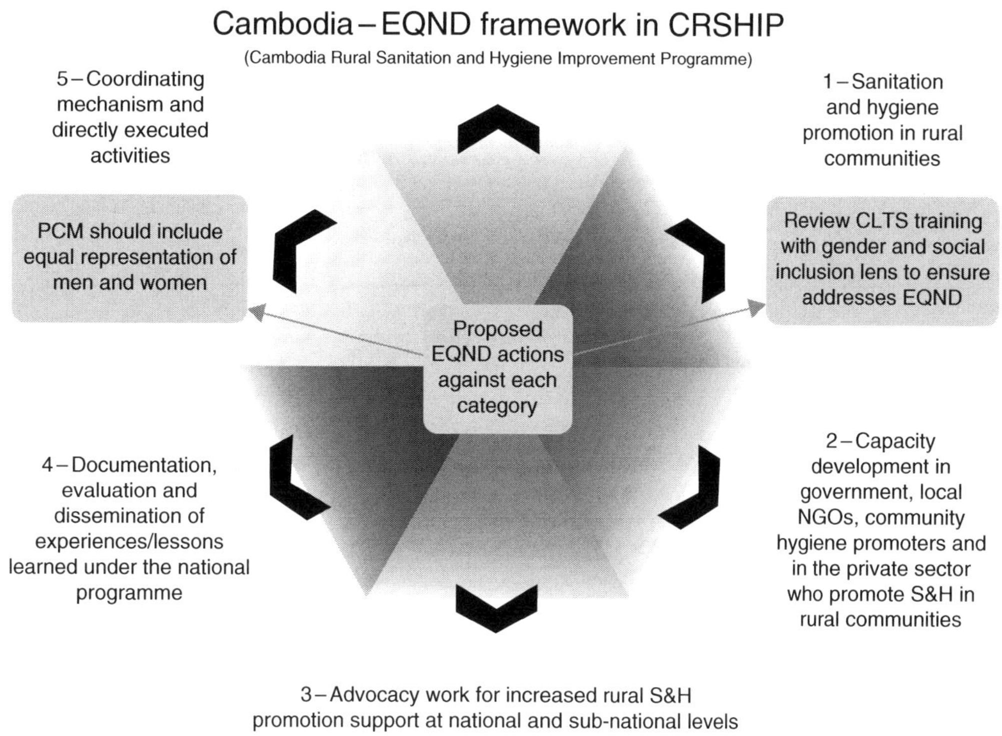

FIGURE 24.1 Components of the Cambodia EQND Framework. PCM refers to the Programme Coordinating Mechanism, a government-led multi-sector body responsible for providing the overall strategic direction for the program.

means that interventions should proactively listen to people who may be disadvantaged, recognize the value their participation can bring, actively work to build their confidence, voice, and agency in the community. Programs should additionally seek out feedback on how interventions are affecting potentially disadvantaged people.

The Cambodia Rural Sanitation and Hygiene Improvement Programme's (CRSHIP)[8] EQND framework is an example of how these principles can be transformed into program guidance. The framework provides an overview of the key principles of the EQND approach in Cambodia, identifies five key programmatic components for mainstreaming EQND principles, and offers practical suggestions and entry points for staff and partners to help integrate EQND into both their work and their organizations (Figure 24.1).

- **Recognize the multiple dimensions of disadvantage**. Multiple, overlapping, and compounding factors impact on a person's ability to realize their human right to safe sanitation and hygiene. Using the "clusters of disadvantage" framework is a way to help categorize the complex web of interlinked factors affecting disadvantage, and to guide program staff to identify people who require special attention.

  Particularly vulnerable are people who fall into more than one category. For example, people with mental health conditions face extreme levels of marginalization, stigma, and are often some of the poorest in communities. Depending on the type and severity of their condition, they may also be unable to change their sanitation and hygiene behavior. Supporting these individuals thus requires special care and attention by facilitators working with communes [12] (Figure 24.2).

- **Leverage appropriate support options**. There are many options available for supporting people who may be disadvantaged, but the nature and degree of support depends on what extent their disadvantage infringes on their ability to build, use, and construct latrines. Wherever possible people should be encouraged to construct their own latrine when they are physically and/or financially able, encouraging their self-efficacy and self-confidence. For those who are unable to manage on their own, support should then be encouraged from their

[8] Read more on the CRSHIP programme at: http://wsscc.org/global-sanitation-fund/cambodia/.

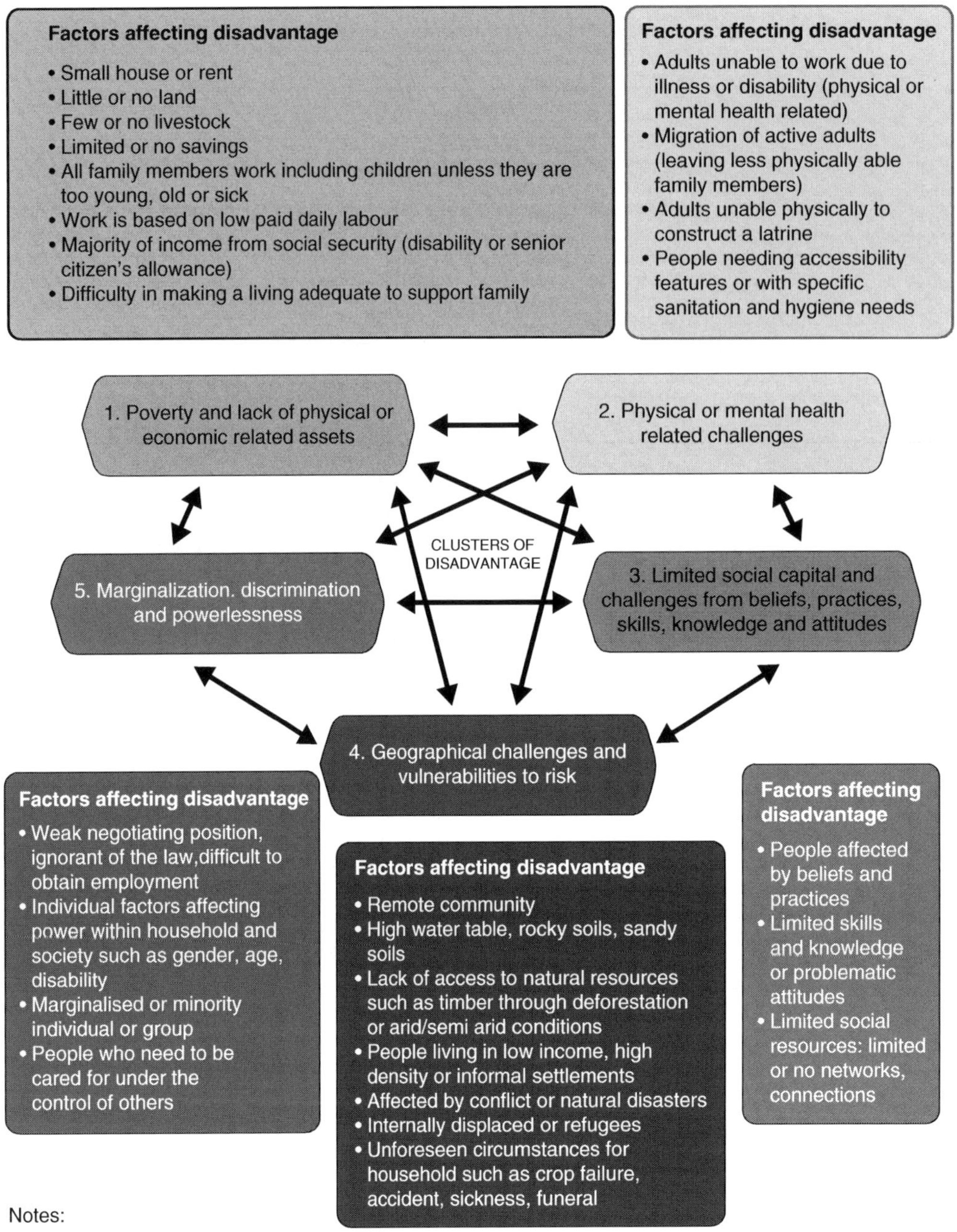

Notes:

1. The arrows indicate the interconnectedness of each factor to the other factors.
2. An individual or group affected by more than one factor is likely to be more disadvantaged than an individual or group affected by just one.
3. This figure has been adapted from Chambers. R (1983) analysis of the deprivation trap related to rural communities.

FIGURE 24.2 Clusters of disadvantage framework [11].

family, and then from the wider community. This is in line with a key CLTS lesson learned: Community problems, community solutions. Even small adjustments to behavior change interventions to proactively integrate EQND can achieve marked gains for people who are considered disadvantaged [13]. In extreme cases, targeted government-sanctioned subsidies (labor, materials, finance) from different sources could be made available for individuals who are unable to support themselves and where support from their family and the wider community is unavailable[9] (Figure 24.3).

[9] Who provides these subsidies, however, is not always clear. Depending on the context, subsidies for extremely vulnerable people could be provided by local or national governments, or, in certain contexts, this could be taken up by external support agencies (for example, in humanitarian emergencies). It is worth noting that the widespread adoption of collective behavior change approaches such as CLTS is partly due to its relative cost-effectiveness vis-à-vis budget limitations faced by most countries.

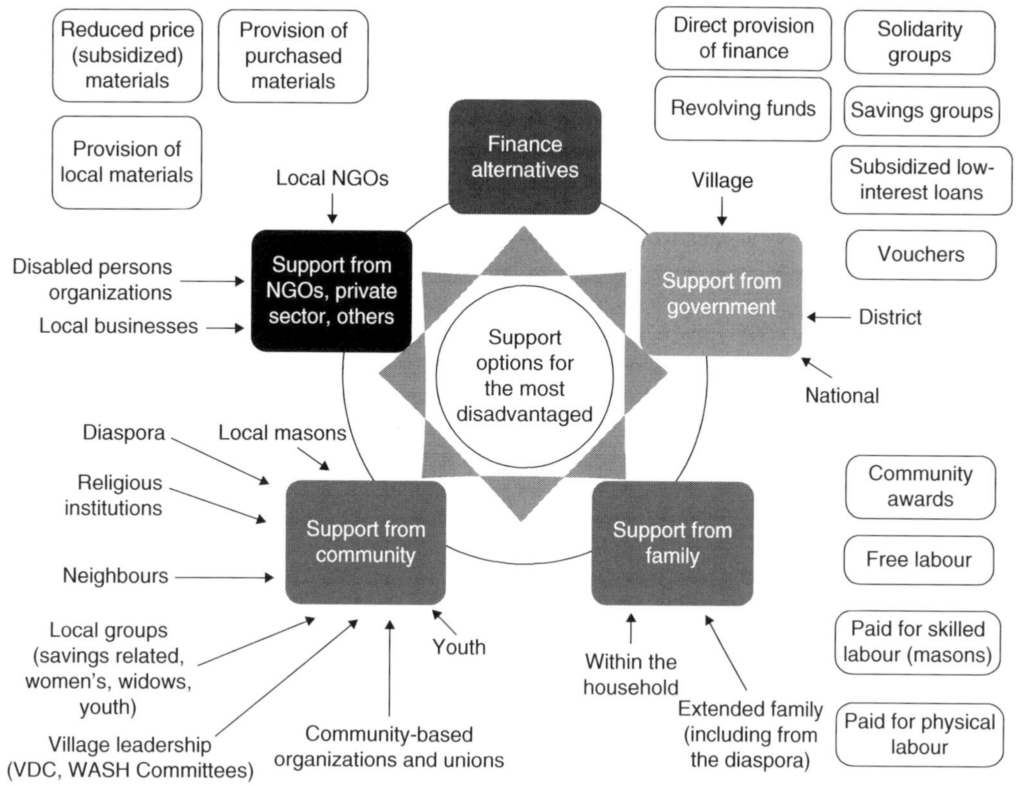

FIGURE 24.3 Options for supporting people who may be disadvantaged [11].

Support for making latrines more accessible often comes from family members. In Malawi, Dorothy's brother decided to build an accessible pit latrine and bathing shelter for her, with a small ramp at the entrance. Source: *Credit*: Sarah House. (*See insert for color representation of this figure.*)

- **Focus on empowerment**. Community-led, collective behavior change approaches such as CLTS should not only aim for full and meaningful participation but also the empowerment of potentially disadvantaged individuals and groups. Empowerment refers to the process of becoming stronger and more confident, especially in controlling one's life and claiming one's rights, and enables them to have a voice in decision-making in their communities. It is transformational in that it aims to alter the structural inequalities that lead to and perpetuate marginalization and exclusion [11]. Because access to safe and dignified sanitation and hygiene is often linked to structural inequalities, contributing to raising the voice, agency, and confidence of people who may be disadvantaged is an important factor for ensuring sustainability.

Nepal is an example where focusing on collective behavior change and community leadership has helped to promote the empowerment of marginalized groups. In Nepal's caste system, Dalits (or "untouchables") face deep-rooted discrimination, exclusion, and extreme poverty. For the GSF-supported program in Nepal, where overcoming inequalities is critical for achieving sanitation and hygiene gains for those usually left behind, the program encourages Dalit communities to challenge stigmas. For example, one Dalit community in Khana VDC, Arghakhanchi District, used the national sanitation campaign to break down their caste stereotype. To show that they would not be considered last, the community organized groups to help each other (four to five households per group) to build latrines and support those who were not able to participate. Working together, the community succeeded in completing their latrines before many people of traditionally "higher" castes. According to a recent sustainability study, Dalit households in areas targeted by the GSF-supported program have comparable levels of sanitation access as other ethnic groups. However, as Dalits are reported to be twice as likely to take out loans from local cooperatives or savings groups to build a latrine, special care needs to be taken to ensure that households are not placed in a more financially precarious situation than before [14].

Sita – a member of a Dalit community in Khana VDC, Nepal – stands outside her newly built latrine. *Source*: Credit: Sue Cavill. (*See insert for color representation of this figure.*)

24.4 ENHANCING THE "ENABLING ENVIRONMENT"

A core component of the scale-up and sustainability of sanitation and hygiene interventions is to strengthen the "enabling environment." This involves working alongside governments, civil society, and the private sector at all levels to secure enhanced political and financial commitments, building technical capacities and systems, and enhancing coordination and learning between different actors.

Ultimately, local ownership and momentum can only be sustained through engagement of mandated local government bodies, supported by a strong national enabling environment with clear lead-ministries, appropriate policies and strategies, budget allocations, technical support, and reporting structures. Delivery mechanisms of the GSF-supported programs still vary, with some programs operating almost entirely via international and local NGOs, and some working through a combination of government and non-government actors. But all programs aim to strengthen the enabling environment at the national and subnational levels.

The GSF currently supports two programs fully run via national and local government systems, including that in Uganda. The Uganda Sanitation Fund (USF) began in 2011 and is currently implemented in and by 38 District Local Governments under the overall management of the Ministry of Health. By December 2017, the USF reported helping over 3.41 million people live in ODF environments, with the government recently committing to scale-up the USF program to eight districts on top of the original 30 [15]. Some of the key program features that successfully enabled the program to reach large numbers of people and leverage government support for scale-up include:

- **Ensure national ownership is at the center**. The USF program was designed to closely align with Uganda's existing policies, sector strategies, and decentralized systems of local governance. This was achieved by ensuring that the USF is nationally owned and sector-led; the program is designed and steered by the National Sanitation Working Group (NSWG), the nationally recognized coordinating body for all key players in the sanitation and hygiene sector through a standing subcommittee. GSF financing is also approved as part of the national budget, with funds being received by the Ministry of Finance, Planning, and Economic Development and disbursed to local government District Health Offices through a dedicated budget line (Figure 24.4).
- **Maximize existing local government structures**. The USF program is situated directly within Uganda's decentralized system of Local Councils, where political leadership and public service delivery at the local level are focused within district, subcounty, and village government structures. District Health Offices take the lead in implementing program activities and monitoring results, with health extension workers, based out of county or subcounty health facilities, rolling out CLTS interventions. At the village level, local health extension staff work closely with Village Health Team (VHT) volunteers. Leveraging these ready-made systems has allowed the USF to extend to a large geographical scale – approximately one-quarter of the country – and reach the grassroots at a significantly reduced cost.
- **Balance local management with central support**. The majority of USF program planning, implementation, and management occurs within District Health Offices, which manage health extension workers based at the subcounty level. District Water and Sanitation

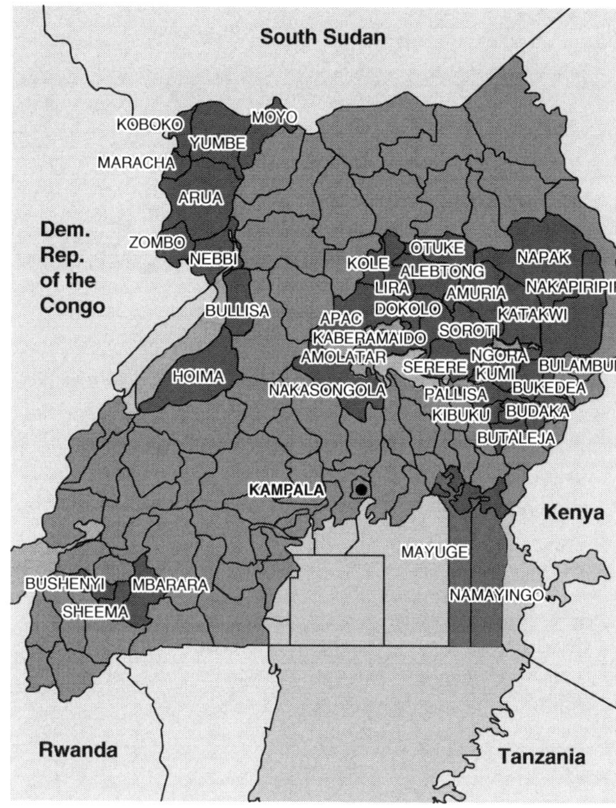

FIGURE 24.4 USF target districts (2017). *Source*: Credit: WSSCC.

Local government health assistants dancing with community members in Arua district, Uganda. After years of using enforcement-based approaches, health assistants have found that building a strong rapport with communities achieves better results. *Source*: Credit: WSSCC.

Local government health assistants (far left and right) with latrine owners in Lira district, Uganda. *Source*: Credit: WSSCC.

Coordinating Committees also play an important role in local-level planning by reducing duplication between organizations, maximizing sector linkages, and reviewing USF work plans from local health departments before they are presented to District Councils and the Ministry of Health.

The Ministry of Health does not exercise direct managerial control over local health departments. Rather, it supports district leadership through regionally-based Field Officers who review work plans, provide technical support, and monitor the use of funds. By providing demand-based mentoring to district health staff, Field Officers help build systems and capacity within local governments.

- **Trigger local political support**. Passionate political leaders acting as champions for the program are strategic resources for enhancing coordination, motivating staff, enlisting the support of other actors, and even leveraging additional human and financial resources to reach an even greater scale. However, improving sanitation and hygiene is not always at the top of the political agenda. This is especially true for non-tangible behavior change approaches with no obvious political appeal often linked to, for example, approaches that provide subsidies for latrines or water supply.

The deployment of several approaches enabled the USF program to harness greater local political support. For example, through using the same CLTS-based methodology that prompts communities to end open defecation, the "institutional triggering" approach aims to ignite shock, shame, and disgust among local leaders and influencers on their home area's sanitation and hygiene situation, and ignite immediate commitments and action to transform the situation. Inviting local politicians and influencers along to community visits is another successful tactic. Not only do they gain an understanding of how open defecation adversely impacts everyone's health and dignity, but while taking part in these energetic events, local leaders sometimes realize that it is in their own interest to be closely associated with the program.

24.5 NEW FRONTIERS FOR SUSTAINABILITY

Collective behavior change, EQND, and a conducive enabling environment are all key ingredients to reaching SDG target 6.2. However, governments and external support agencies must anticipate the next generation of opportunities and challenges for sustainable sanitation and hygiene.

With a growing population using sanitation facilities, and with on-site household-managed sanitation (rather than off-site sewerage systems) the foreseeable future for most of the global population, there is a pressing need to ensure that fecal sludge does not contaminate the wider environment once pits and septic tanks fill up, and that facilities continue to encourage usage [16]. In some contexts, this will involve expanding affordable and locally appropriate emptying, transportation, and treatment services. The challenge is getting the economics right; in remote rural areas where delivery costs are high, or for poor segments of the population where cost-recovery is low, there needs to be sufficient incentive for private sector investment – especially in the absence of public sector involvement.[10] A shifting focus to expanding emptying, transportation, and treatment services will also require closer attention to the rights, safety, and dignity of sanitation workers. Many of these workers operate

[10] For example, in Bangladesh it costs $13 to have a pit latrine emptied – roughly 14% of an average rural family's monthly income. See: https://www.npr.org/sections/goatsandsoda/2017/03/21/520979223/so-what-do-you-do-with-the-poo-in-a-pit-latrine.

Constructing a twin-pit latrine in India. *Source*: Credit: WSSCC. (*See insert for color representation of this figure.*)

A woman in northern Uganda showing her painted dry-pit latrine. *Source*: Credit: WSSCC.

informally outside the protection or regulation of the state and are often already some of the most marginalized members of society.

One possibility is to see the end-use potential of human waste. Encouraging investment in biogas or compost produced from human waste could help incentivize investment (and potentially reduce user costs) in the middle segments of the service chain. This will raise new issues for public safety regulators and enforcement bodies at each point in the supply chain to reduce new public health risks [17].

For many areas, however, practical low-cost solutions for safe treatment will need to be found on-site. For many GSF-supported programs working in remote rural areas, it is anticipated that this will involve continued engagement with communities to ensure that filled pits are correctly capped with soil, with remains contained for approximately one year. One solution, already heavily promoted in South Asia, is the twin-pit system where one pit can be left to compost while the other one fills up, and then emptied safely. However, this is a more expensive option not always in the reach of many rural communities. Achieving community-level safe waste management will therefore involve further work on improving regulations and standards, engaging communities, promoting appropriate technology options, and enhancing local government public health monitoring.

24.6 SLIPPAGE IS ONLY HUMAN

Behavior is never static, and even less so in conditions where people face many challenges to their physical environment, social, economic, and health status. Second, infrastructure will always degrade over time. The poorest people often live in the harshest environments, with sandy or rocky soils, and prolonged periods of drought or annual floods (expected to increase in many areas as a result of climate change) that destroy houses and flood pits. People may be forced to repeatedly reconstruct or improve their latrines, and many factors can influence their ability or desire to do so. The GSF experience has shown that a combination of sustained, community-led behavior change interventions, a strong focus on equality and inclusion, and emphasis on the enabling environment are most likely to lead to sustained behavioral outcomes. Experience from Nepal shows that ODF villages were able to relatively quickly regain that status after the devastating 2015 earthquake [18].

A newly constructed pour-flush latrine with a handwashing facility in Cambodia. *Source*: Credit: WSSCC. (*See insert for color representation of this figure.*)

This sums up perhaps one of the most important lessons learned so far: no village will remain completely ODF at all times, especially as this also includes the continual presence of water and soap in the handwashing station, and that latrines are in good condition indefinitely [8]. What matters is that all people become habituated to using a latrine and washing their hands at critical times, that the latrines improve over time (therefore becoming more resilient and easier to use and maintain), that no-one feels left behind in their ability to construct, maintain, and use public and private facilities, and that there are plans and systems in place to ensure continued service delivery to communities. Not only will this ensure that people are fully able to enjoy their human right to safe sanitation but it will also affect the community health status, empower vulnerable groups, women, and girls, impact on household economies, and much more. In this way sanitation can serve as an entry-point to achieving SDGs far beyond sectoral boundaries.

REFERENCES

1. WHO and UNICEF (2015). *Progress on Sanitation and Drinking Water: 2015 Update and MDG Assessment*. Geneva: WHO and UNICEF Available from: http://www.who.int/water_sanitation_health/publications/jmp-2015-update/en/.
2. WSSCC (2017). *Global Sanitation Fund Progress Report: Championing Sustainable Sanitation and Hygiene for all*. Geneva: WSSCC Available in English and French from: http://wsscc.org/resources-feed/global-sanitation-fund-progress-report-2016/?_sf_s=progress+report.
3. Fuller, J. and Eisenberg, N.S. (2016). Herd protection from drinking water, sanitation, and hygiene interventions. *The American Journal of Tropical Medicine and Hygiene* 95 (5): 1201–1210. Available from: https://www.ncbi.nlm.nih.gov/pmc/articles/PMC5094239/.
4. Bicchieri, C. (2016). *Norms in the Wild: How to Diagnose, Measure, and Change Social Norms*. New York: Oxford University Press Available from: https://books.google.ch/books?id=v214DQAAQBAJ.
5. Kar, K. and Chambers, R. (2008). *Handbook on Community-Led Total Sanitation*. London: Plan International (UK) Available from: http://cltsfoundation.org/upload/publication/pdf/1419852683085170c948.pdf.
6. Aunger, R. and Curtis, V. (2013). The anatomy of motivation: an evolutionary-ecological approach. *Biological Theory* 8 (1): 49–63. Available from: https://link.springer.com/article/10.1007%2Fs13752-013-0101-7#page-1.
7. WSSCC (2016). *Follow-Up MANDONA Handbook*. Geneva: WSSCC Available in English and French from: Global Sanitation Fund Progress Report: Championing Sustainable Sanitation and Hygiene for all.
8. Jerneck, M., Van der Voorden, C., Rudholm, C. et al. (2016). *Sanitation and Hygiene Behaviour Change: Understanding Slippage*. Geneva: WSSCCC Available in English and French from: http://wsscc.org/resources-feed/sanitation-hygiene-behaviour-change-at-scale-understanding-slippage/?_sf_s=slippage.
9. Milward, K., Pradha, S., and Paster, K. (2014). *Promising Pathways: Innovations and Best Practices in CLTS at Scale in Madagascar*. Kolkata: CLTS Foundation Available in English and French from: http://wsscc.org/resources-feed/promising-pathways-innovations-best-practices-clts-scale-madagascar-english/?_sf_s=promising+pathways.
10. Cross River's Obanliku declared first Open Defecation-Free Council in Nigeria. (2016), December 15. In *Premium Times Nigeria*. Retrieved from: http://wsscc.org/2017/04/11/no-one-came-tell-empowered-nigeria-celebrates-first-odf-local-government-area/.
11. House, S., Ferron, S., and Cavill, S. (2017). *Scoping and Diagnosis of the Global Sanitation Fund's Approach to Equality and Non-Discrimination*. Geneva: WSSCC Available in English and French: http://wsscc.org/resources-feed/

scoping-diagnosis-global-sanitation-funds-approach-equality-non-discrimination/?_sf_s=EQND.

12. Cavil, S., England, P., and Ferron, S. (2017). Understanding, respecting, and including people with mental health conditions as part of the CLTS process, 40th WEDC International Conference, Loughborough UK, 2017. Available from: http://wsscc.org/wp-content/uploads/2017/07/Understanding-respecting-and-including-people-with-mental-health_CLTS_Cavill.pdf.
13. Jones, H. E., Singini, W., Holm, R.H., and White, S (2016). CLTS Plus – making CLTS ever more inclusive, 39th WEDC International Conference, Kumasi, Ghana, 2016. Available from: http://www.communityledtotalsanitation.org/sites/communityledtotalsanitation.org/files/Jones_Making_CLTS_ever_more_inclusive.pdf.
14. Kendra, B.S. (2017). *Sustainability of Open Defecation Free Campaign in GSF-Supported Programme Districts, Nepal*. Kathmandu: UN-HABITAT Available from: http://wsscc.org/resources-feed/sustainability-open-defecation-free-campaign-gsf-supported-programme-districts-nepal/.
15. England, P., Adyero, C., Hangi, B. et al. (2017). *Local Governance and Sanitation: Eight Lessons from Community-Led Total Sanitation at Scale through Local Governments in Uganda*. Geneva: WSSCC Available in English and French from: http://wsscc.org/resources-feed/local-governance-and-sanitation-eight-lessons-from-uganda/?_sf_s=uganda.
16. WHO and DfID (2009). *Vision 2030: The Resilience of Water Supply and Sanitation in the Face of Climate Change: Summary and Policy Implications*. Geneva: WHO Available from: http://www.who.int/water_sanitation_health/publications/9789241598422/en/.
17. Andersson, K., Rosemarin, A., Lamizana, B. et al. (2016). *Sanitation, Wastewater Management and Sustainability: From Waste Disposal to Resource Recovery*. Nairobi and Stockholm: United Nations Environment Programme and Stockholm Environment Institute Available from: https://www.sei-international.org/mediamanager/documents/Publications/SEI-UNEP-2016-SanWWM&Sustainability.pdf.
18. UN-HABITAT (2016). *Reviving Sanitation Campaign after the 25 April Earthquake: Global Sanitation Fund Lessons*. Kathmandu: UN-HABITAT Available from: http://wsscc.org/resources-feed/nepal-earthquake-reviving-sanitation-global-sanitation-fund-lessons/?_sf_s=earthquake.

AFTERWORD

ADDRESSING WATER AND SANITATION-RELATED DISEASES IS FUNDAMENTAL TO GLOBAL HEALTH AND SUSTAINABLE DEVELOPMENT

BY WAFAIE FAWZI, AND CHELSEY R CANAVAN

A rural farmer in Tanzania contemplates spending her limited funds on water for the home garden. She aspires to grow plentiful crops of amaranth, okra and eggplant for herself and her family to supplement their diet otherwise based on maize and small amounts of fish. A persistent drought has impeded her efforts. Without clean water and proper nutrition, she worries her three young children will become ill.

We have seen this scenario in our own work and know that it is all too common among vulnerable populations globally. This Tanzanian farmer is one of 2 billion people worldwide affected by water stress and among the 2.4 billion people without access to improved sanitation. The challenge of addressing water and sanitation and its many facets in the context of the changing climate can seem daunting. Throughout the two volumes of this book, the authors have provided insights into the complex challenges related to water and sanitation and offered evidence-based measures to overcome them.

It is an unacceptable fact that diarrheal disease, which is preventable and treatable, remains the second leading cause of death among children under five years of age. It is within our reach to reduce or eliminate diarrheal disease, schistosomiasis, guinea worm, onchocerciasis, soil-transmitted helminths, malaria, and trachoma, among other water and sanitation-related diseases. Many proven strategies already exist, such as installation of improved toilets or latrines and technologies for household water treatment. Breastfeeding is a critical intervention for improving infant and child nutrition and staving off infectious disease. Single interventions, however, are unlikely to bring about large scale change: water and sanitation-related diseases must be tackled from multiple angles. Shifting demographic, social, economic, and environmental contexts require a reformation of our approach to the causes and consequences of inadequate water and sanitation.

These books emphasize the linkages between water and sanitation with a range of infectious diseases, food systems and malnutrition, and water pollutants. Throughout the text, we see how each of these elements interrelates with climate change, extreme weather, and ecosystem changes. Climate change is already impacting food and water systems globally, threatening our ability to provide safe and adequate food and water for the world's growing and urbanizing population. Agricultural production systems are under increased pressure to feed a hungry planet that is also suffering from water pollution and other forms of environmental degradation. While significant progress has been made in reducing the global burden of diseases discussed in these books, risk factors and consequences attributable to climate-related food insecurity, malnutrition, and migration are increasing.

The intertwined nature of human health, water, food, and natural ecosystems is not to be overlooked. The framers and signatories to the Sustainable Development Goals understood this dynamic: "Access to safe water and sanitation and sound management of freshwater ecosystems are essential to human health and to environmental sustainability and economic prosperity." Progress in one area—hunger, sustainable communities, climate action—have implications for others—water and sanitation, health and wellbeing, poverty, and inequality.

The authors of these books have highlighted conditions that must be met to achieve the Sustainable Development Goal to "ensure availability and sustainable management

Water and Sanitation-Related Diseases and the Changing Environment: Challenges, Interventions, and Preventive Measures, Second Edition. Edited by Janine M. H. Selendy.

Companion website: https://www.wiley.com/legacy/wileychi/selendy/

of water and sanitation for all." Today, over 90 percent of the global population has access to an improved drinking water source. Yet, our water sources are at increased risk for contamination and ecosystem change, requiring ever greater effort to maintain and improve upon progress made to date.

The task at hand is urgent but it is not insurmountable. Julio Frenk and Octavio Gómez-Dantés present one example of preventative measures taken in Mexico that successfully improved child health in their chapter on "Extending the Right to Health Care and Improving Child Survival in Mexico." Donald Hopkins and Ernesto Ruiz-Tiben describe the effort to eradicate guinea worm, and how this ambitious task has gone from possible to probable to imminent. The success stories in this book demonstrate that, often, we know what works. And when we fail, the scientific, civic, and activists communities are capable of coming together to overcome obstacles.

Importantly, the global community came together in 2010 and 2015 to formally recognize the human right to water and sanitation. A human rights approach provides a framework for countries to integrate water and sanitation into strategic development plans. It lays the groundwork for reducing the inequalities that contribute to the disease burden and environmental conditions underscored in these pages. Access to adequate water and sanitation increases with income level and urban residence. Thus, targeting rural populations and lower income groups, like our farmer in Tanzania, becomes a priority from a human rights perspective.

The human right to water and sanitation also serves to align various stakeholder groups. Commitment from world leaders must be matched by that from national and local decision-makers, donors and funding agencies, and researchers and program implementers alike. Protection of ecosystems and prevention of pollutants involves collaboration among leaders and participants at global, national and local levels. In several chapters of these books, the authors have reminded us of the necessity of community engagement as a requisite for success. Maintenance of bore holes and provision of optimal and culturally appropriate toilets are much more likely to be effective with meaningful local involvement. This holds true for specific health and nutrition interventions such as cholera case detection, administration of oral rehydration therapy, and micronutrient supplementation.

We know what success in water and sanitation looks like. We've seen a rural farmer in Tanzania plant drought-resistant seed varieties to realize the bounties of her home garden. We've seen new climate science and the development of remote and local sensing technologies for water and sanitation monitoring. While we have much more to learn, we have the tools to apply this knowledge at scale, to develop new implementation strategies that are context-specific, and to work toward a world where everyone can realize their right to sufficient, safe, accessible and affordable water and sanitation.

INDEX

Water and Sanitation-Related Diseases and the Changing Environment: Challenges, Interventions, and Preventive Measures, Second Edition. Edited by Janine M. H. Selendy.

Companion website: https://www.wiley.com/legacy/wileychi/selendy/